Recent Advances in Computing Sciences

Proceedings of the 2nd International Conference on Recent Advances in Computing Sciences (RACS, 2023)

First edition published 2024

by CRC Press

4 Park Square, Milton Park, Abingdon, Oxon, OX14 4RN

and by CRC Press

2385 NW Executive Center Drive, Suite 320, Boca Raton FL 33431

CRC Press is an imprint of Informa UK Limited

British Library Cataloguing-in-Publication Data
A catalogue record for this book is available from the British Library

ISBN: 9781032943473 (pbk)
ISBN: 9781003570349 (ebk)

DOI: 10.1201/9781003570349

Font in Sabon LT Std
Typeset by Ozone Publishing Services

Printed and bound in India

Recent Advances in Computing Sciences

Proceedings of the 2nd International Conference on Recent Advances in Computing Sciences (RACS, 2023)

Edited By:

Dr Manmohan Sharma,
Dr Mintu Nath,
Dr Sophiya Sheikh,
Dr Amar Singh

CRC Press
Taylor & Francis Group
Boca Raton London New York

CRC Press is an imprint of the
Taylor & Francis Group, an **informa** business

Contents

List of Figures

List of Tables

Foreword

Preface

The International Conference on Recent Advances in Computing Sciences (RACS 2023) targeted state-of-the-art as well as emerging topics of recent advancement in computing research and its implementation in engineering applications.

The objective of this international conference is to provide opportunities for researchers, academicians, people from the industry, and students to interact and exchange ideas, experiences, and expertise on information and communication technologies. Besides, the participants enhanced their knowledge of recent computing technologies.

We are highly thankful to our valuable authors for their contribution and our Technical Program Committee for their immense support and motivation toward making the first RACS 2023 a success. We are also grateful to our keynote speakers for sharing their precious work and enlightening the delegates of the conference. We sincerely thank our publication partner, Taylor and Francis, for believing in us.

Dr Manmohan Sharma
Dr Amar Singh
Dr Sophiya Sheikh
Convener
RACS-2023

Punjab, India

About the conference

The 2nd International Conference on Recent Advances in Computing Sciences (RACS-2023) held from 29th to 30th November 2023 organized by the School of Computer Application, Lovely Professional University, Jalandhar, Punjab.

The conference focuses on discussing issues, exchanging ideas, and the most recent innovations towards the advancement of research in the field of Computing Sciences and Technology. All technical sessions will predominantly be related to Data Science, Artificial Intelligence, Remote Sensing, Image Processing, Computer Vision, Data Forensics, Cyber Security, Computational Sciences, Simulation & Modelling, Business Analytics, and Machine Learning.

The main objective of this conference is to provide a common platform for academia and industry to discuss various technological challenges and share cognitive thoughts. It was a thought-provoking platform to discuss and disseminate novel solutions for real-world problems in a dynamic and changing technological environment. The main success of RACS-2023 is to allow the participants to enhance their knowledge of recent computing technologies.

Editors Biography

Manmohan Sharma

Dr. Manmohan Sharma presently serving as a Professor in the School of Computer Applications, Lovely Professional University, Punjab, INDIA has a vast experience of more than 24 years in the field of academics, research, and administration with different Universities and Institutions of repute such as Dr. B.R. Ambedkar University, Mangalayatan University, etc. Dr. Sharma was awarded with his Doctorate from Dr. B.R. Ambedkar University, Agra in 2014 in the field of Wireless Mobile Networks. His areas of interest include Wireless Mobile Networks, Adhoc Networks, Mobile Cloud Computing, Recommender Systems, Data Science and Machine Learning, etc. More than 50 research papers authored and co-authored, published in International or National journals of repute, and conference proceedings come under his credits. He is currently supervising six doctoral theses. Three Ph.D. and three M.Phil. degrees have already been awarded under his supervision. He has guided more than 600 PG and UG projects during his service period under the aegis of various Universities and Institutions. He worked as a reviewer of many conference papers and a member of the technical program committees for several technical conferences. He is also actively serving several journals related to the field of wireless, mobile communication, and cloud computing as an editorial board member. He is also a member of various professional/technical Societies including the Computer Society of India (CSI), Association of Computing Machines (ACM), Cloud Computing Community of IEEE, Network Professional Association (NPA), International Association of Computer Science and Information Technology (IACSIT), and Computer Science Teachers Association (CSTA).

Mintu Nath

Dr Mintu Nath has done his post-graduation and doctoral programmes at the National Dairy Research Institute, Indian Veterinary Research Institute, and Sheffield Hallam University. He worked as an applied statistician with several organisations, including the Indian Council of Agricultural Research, Biomathematics and Statistics Scotland, Roslin Institute, the University of Edinburgh, and the University of Leicester. Currently, he is a Senior Lecturer in Medical Statistics at the University of Aberdeen, working in collaboration with the National Health Service and other industry partners. His research interests include exploring and integrating diverse resources of large patient cohorts - comprising high-dimensional clinical, genetic, imaging, and multi-omics data - to evaluate the risk of important diseases. He is also a Senior Research Fellow of the Higher Education Academy in the UK. In that role, he engages with teaching and training of students, researchers, and other professionals in statistics and data sciences.

Sophiya Sheikh

Dr Sophiya Sheikh received a B.Sc. degree in information technology from Maharshi Dayanand Saraswati University, Ajmer, India, a master's degree in computer application from Rajasthan Technical University, Kota, and a Ph.D. degree from the Department of Computer Science, Central University of Rajasthan, India. She is currently an Associate Professor at the School of Computer Application, Lovely Professional University, Phagwara, Punjab, India. She regularly writes articles and research papers in reputed national and international magazines and journals. Her research interests include grid/distributed computing and cloud computing. She is an editor of various books. She has organized various international conferences and workshops. She is a Potential Reviewer in various reputed journals, such as IEEE Systems Journal, Cluster Computing, Scientific Reports, Journal of Supercomputing, Journal of Cloud Computing, and Concurrency and Computation. You can contact her at the School of Computer Applications, Lovely Professional University, Phagwara, Punjab (India) - 144411.

Amar Singh

Dr. Amar Singh is working as a Professor in the School of Computer Applications, Lovely Professional University, Punjab, India. He did his Ph.D. (Computer Science & Engineering) from IKG Punjab Technical

University, Jalandhar, Punjab, India. He has completed his M. Tech. (Information Technology) M.M. University, Mullana, Ambala, Haryana, India. He is a lifetime member of ISCA (Indian Science Congress Association) and a member of IEEE. He has more than 15 years of experience in Teaching and Research. His current research interests are Soft Computing, Machine Learning, and Image Processing. He has published 21 patents, 6 Book Chapters, 2 copyrights (granted), and 94 research articles in various Journals and Conferences. He has organized 2 International Conferences on Recent Advances in Computing Sciences as a convenor. He is a reviewer in various reputed journals like IEEE Journal of Biomedical and Health Informatics, IEEE Transactions on Neural Systems and Rehabilitation Engineering, Multimedia Systems, etc. He is a Co-designer of FDEng: Fuzzy Design Engineer a software-based Automated Design tool for the design of Fuzzy Logic Based systems. He has successfully guided 08 Ph.D. research scholars and, 07 Ph.D. research scholars are doing research under his guidance.

Introduction

The conference encouraged novel presentations, discussions, and ongoing research by researchers, emerging academics, and members of the more established academic community. The papers contributed the most recent scientific knowledge and research in the field of modern communication systems, sophisticated electronics and communication technologies, artificial intelligence and capsule networks, data transmission, computer networking, and networks for communicative things. With this initiative, we intend to deliver the most recent advancements in all communication technology-related fields.

Putting together international conferences with a focus on computer, communication, and computational sciences to boost research output on a wide scale. Eminent academicians, researchers, and practitioners from around the world are represented on the Technical Program Committee and Advisory Board of the conference. Ten different nations submitted multiple papers to RACS-2023. The conference would not have been as great without their contributions.

Each contribution was checked for plagiarism. Each submission had a thorough assessment by at least two reviewers, with an average of 2-3 per reviewer. A few entries have even undergone more than two reviews. With a 35% acceptance rate, 58 quality articles were chosen for publication in this proceedings volume based on the reviews and plagiarism report. We appreciate the work of all authors and contributors.

Dr Manmohan Sharma
Dr Amar Singh
Dr Sophiya Sheikh
Convener RACS-2023

Introduction

Details of the programme committee

Chief Patron	Dr. Ashok Mittal, Honorable Chancellor, LPU
Patron(s)	Mrs. Rashmi Mittal, Worthy Pro chancellor, LPU
General Chair	Prof. (Dr.) Loviraj Gupta
Conference Chair	Prof. (Dr.) Ashok Kumar
Convener	Prof. (Dr.) Manmohan Sharma Prof. (Dr.) Amar Singh Dr. Sophiya Sheikh
Organizing Secretary	Dr. Balraj Kumar Dr. Pawan Kumar Dr. Rishi Chopra Mr. Sartaj Singh Mr. Ajay Kumar Bansal
International Chair	Prof. (Dr.) Abdul Sattar, Griffith University, Australia
Publication Chair	Prof. (Dr.) Manmohan Sharma
External Publication Chair	Prof. (Dr.) Apash Roy, NSHM Knowledge Campus, Durgapur, West Bengal
Publication Co-Chair	Prof. (Dr.) Alok Misra Prof. (Dr.) Amar Singh Dr. Sophiya Sheikh
Registration and Finance Chair	Dr. Avinash Bhagat
Sponsorship Chair	Mr. Sarabjit Kumar
Local arrangement Chair	Dr. Rishi Chopra Dr. Amit Sharma
Technical Chair	Prof (Dr.) Mithlesh Dubey
Plenary Session Chair	Dr. Balraj Kumar Mr. Ajay Kumar Bansal
Hospitality Chair	Dr. Rishi Chopra
Hospitality Co-Chair	Ms. Pallavi Vyas Ms. Navpreet Kaur
Media and Publicity Chair	Mr. Sukanta Ghosh Dr. Bhanu Sharma Dr. Amanpreet Singh Mr. Balwinder Singh (Division of Infotech)

Integration of Contemporary Soft Computing Techniques in Networking

A Comprehensive Study

Bhanu Sharma[1], Prikshat Kumar Angra[2], Bandana Sharma[3],
Sarbjeet Singh[4], Tanya Rao[5]

[1,2,4,5]School of Computer Application, Lovely Professional University, Phagwara, India
[3]Maharishi Markandeshwar University, Mullana Ambala, India
E-mail: bhanu.lpu1020@gmail.com, Prikshat.22305@lpu.co.in, bandanasharma1@gmail.com,
sarbjeetrajput783775@gmail.com, raotanya9292@gmail.com

Abstract

The significance of employing contemporary soft computing methodologies in networking has seen a remarkable rise. Approaches such as fuzzy logic, neural networks, evolutionary algorithms, and swarm intelligence are progressively being utilised to bolster the security, dependability, and effectiveness of networks. This investigation extensively delves into the incorporation of soft computing techniques into networking. It furnishes an in-depth examination of ongoing research, assesses various methodologies employed, and deliberates on forthcoming advancements along with their prospective applications within networking contexts.

Keywords: Networking, artificial intelligence, fuzzy logic, genetic algorithms

1. Introduction

Networking has evolved into an integral part of our daily lives, connecting people and devices worldwide via the internet. However, this growth presents new challenges such as congestion, security vulnerabilities, and reliability issues. To overcome these obstacles, researchers are exploring innovative approaches to improve network performance, security, and reliability. Soft computing techniques have emerged as a promising solution. This paper thoroughly explores the integration of modern soft computing methods into networking. It conducts an extensive review of literature on these techniques, outlining their advantages, limitations, and varied applications in networking. Focusing on methodologies like fuzzy logic, neural networks, genetic algorithms, and swarm intelligence, the study examines their roles and potential impact on networking. Moreover, it highlights the challenges and future possibilities of incorporating soft computing techniques into networking. By providing a detailed analysis of state-of-the-art networking methods, this study aims to assist researchers, network administrators, and engineers in developing and implementing efficient and resilient networking solutions to enhance user experiences.

2. Literature Survey

Recent progressions in networking have witnessed a significant shift towards the utilisation of soft computing methodologies to bolster performance, security, and dependability. Methods such as fuzzy logic are being harnessed to devise routing algorithms that alleviate congestion and amplify network throughput by employing linguistic variables to tackle traffic uncertainties, thereby facilitating more effective routing decisions. Moreover, neural networks are being deployed to bolster network security, particularly in wireless sensor networks, where they excel in detecting and thwarting DDoS attacks owing to their high precision in identifying threats. Conversely, genetic algorithms are optimising network routing by enhancing Quality of Service (QoS) parameters like bandwidth and delay, notably in multi-hop wireless networks, by striking a balance between energy consumption and QoS demands to heighten network efficiency. Swarm intelligence is further refining network topology in wireless mesh networks by enhancing node placement, leading to a significant augmentation in network coverage and a reduction in energy consumption.

The body of literature on soft computing in networking, including surveys conducted by Gholami

Chapter 1 DOI: 10.1201/9781003570349

et al. [1], Islam et al. [2], Zang et al. [3], and Saravanan et al. [4], delves into the application of fuzzy logic, neural networks, and genetic algorithms in optimising networks, routing, and fortifying security. These surveys explore their utilisation across various network architectures, ranging from cognitive radio and mobile ad hoc networks to cloud and wireless sensor networks. The focus lies not only on the methodologies but also on their practical applications in emerging technologies, highlighting the potential for future exploration. This encompasses the development of hybrid algorithms integrating multiple soft computing techniques and their implementation in novel networking technologies, showcasing the dynamic evolution of soft computing applications in networking, spanning from traffic prediction and intrusion detection to bolstering overall network resilience and efficiency.

3. Exploring the Relevance and Practicality of Various Techniques in the Context of Networking

Fuzzy logic-based routing algorithms use linguistic variables to handle uncertainty and imprecision in network traffic. The algorithms can provide a more efficient routing decision compared to traditional routing algorithms. Neural networks can be used for classification, prediction, and decision making in network security. They can learn from data and make accurate predictions based on the learned patterns. Genetic algorithm-based routing algorithms use evolutionary techniques to dynamically adjust the routing path based on the changing network conditions. They can optimise multiple QoS parameters and provide an efficient routing path. Swarm intelligence-based algorithms use collective intelligence to optimise network topology. They can be used to optimise network coverage, energy consumption, and reduce interference in wireless networks.

3.1. Fuzzy Logic

Fuzzy logic is a technique that is used to handle uncertain or imprecise information. In networking, fuzzy logic is used for routing, traffic management, and congestion control. Fuzzy logic-based routing algorithms use linguistic variables to express routing parameters such as delay, throughput, and bandwidth. Zadeh [5] and Klir et al. [6] discusses about the Fuzzy logic-based traffic management algorithms use fuzzy sets to represent the traffic characteristics and adjust the traffic flow to optimise network performance (Figure 1).

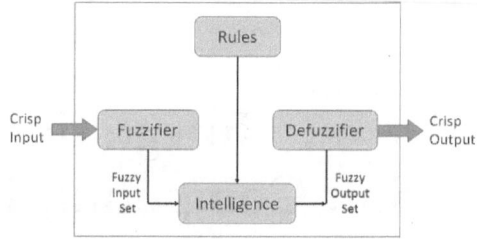

Figure 1: Working of fuzzy logic

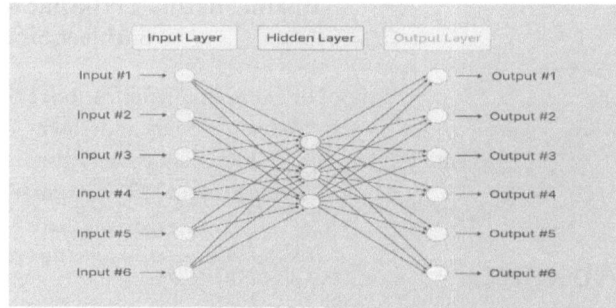

Figure 2: Architectural structure of neural networks

3.2. Neural Networks

Neural networks are computing systems inspired by the structure and function of the human brain. In networking, Cuomo et al. [7] discuss about the how neural networks are used for routing, intrusion detection, and traffic prediction. Neural network-based routing algorithms use a learning process to adapt to the network environment and optimise the routing decisions. Neural network-based intrusion detection systems use a pattern recognition process to detect anomalies and attacks in the network traffic. Neural network-based traffic prediction algorithms use historical data to forecast future traffic and plan network resources accordingly as per the literature done by Woldseth et al. [8] (Figure 2).

3.3. Genetic Algorithms

Genetic Algorithms (GAs) belong to the family of evolutionary algorithms that simulate natural selection to discover superior solutions for intricate challenges. In the networking domain, GAs have been applied for routing, load balancing, and improving intrusion detection systems. Sohail et al. [9] noted that in routing, GAs use genetic representations of network structures and employ fitness functions to evaluate routing paths. For load balancing, strategies based on GAs utilise a community approach to distribute workload evenly across network nodes. Dolezel et al. [10] highlighted the use of genetic algorithms in intrusion detection, where genetic representations of attack patterns and fitness functions collaborate to identify security threats.

3.4. Ant Colony Optimization

Ant Colony Optimization (ACO) draws inspiration from the foraging behaviour of ants and leverages swarm intelligence for computational problem-solving. In the realm of networking, ACO contributes to enhancing routing efficiency, managing load distribution, and optimising Quality of Service (QoS). Kavita et al. [11] outlined ACO-based routing algorithms that employ simulated pheromone trails for structuring and navigating network paths. Wu et al. [12] demonstrated how ACO aids in load balancing by effectively distributing network traffic among nodes. Furthermore, ACO algorithms are employed for QoS optimisation, where they allocate network resources among different types of traffic to uphold service quality.

3.5. Particle Swarm Optimization

Particle Swarm Optimization (PSO) is an optimisation methodology influenced by the communal behaviours of birds or fish. It is utilised in the networking field for routing, load balancing, and QoS management. Ramirez et al. [13] discussed how PSO-based routing uses particles to depict network topologies, which assists in developing routing strategies. For load balancing, PSO algorithms aim to distribute traffic uniformly across network nodes to maximise network efficiency. Regarding QoS management, PSO is deployed to effectively allocate network resources across various traffic types, ensuring the maintenance of desired service levels (Figure 3).

3.6. Artificial Bee Colony

Artificial Bee Colony (ABC) is another swarm intelligence-based optimization algorithm that simulates the behaviour of a bee colony. In networking, ABC can be used for routing, load balancing, and QoS management. ABC-based routing algorithms use the number of employed bees to represent the network topology and guide the routing decisions. ABC-based load balancing algorithms use the ABC approach to distribute the traffic across the network nodes. Ye et al. [14] in the article ABC-based QoS management algorithms use the ABC approach to allocate the network resources to different traffic classes (Figure 4).

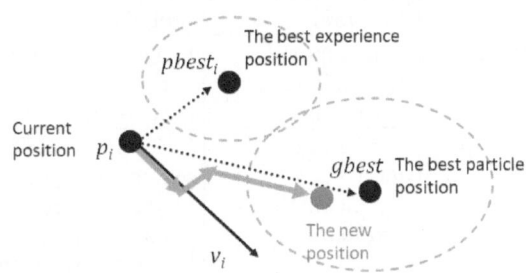

Figure 3: The phenomena of PSO algorithm

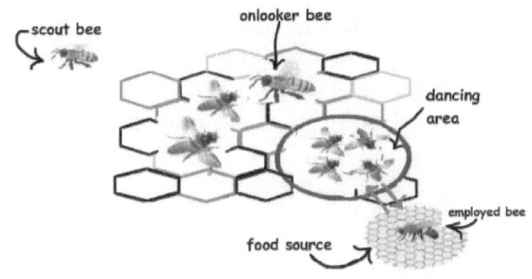

Figure 4: The working process of artificial bee colony

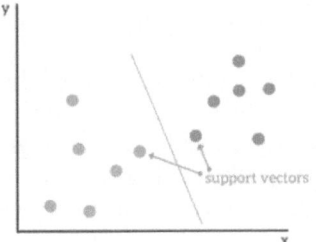

Figure 5: Simulation of SVM

3.7. Support Vector Machines

Support Vector Machines (SVM) are a type of machine learning algorithm that can be used for classification and regression analysis. In networking article, Kurani et al. [15] discussed about how SVM can be used for intrusion detection, traffic classification, and prediction. SVM-based intrusion detection systems use a supervised learning process to classify the network traffic into normal and abnormal classes. SVM-based traffic classification algorithms use a learning process to classify the traffic into different classes based on its characteristics. SVM-based traffic prediction algorithms use historical data to forecast future traffic and plan network resources accordingly (Figure 5).

4. Comparing Modelling Approaches: Fuzzy Logic, ML, Neural Networks, Probabilistic Reasoning, and Deep Learning

Within the realm of modelling approaches, Fuzzy Logic distinguishes itself with its inherent imprecision and approximative nature, rendering it particularly apt for scenarios where achieving pinpoint precision in modelling proves to be a formidable challenge. It merits a moderate rating in terms of both applicability and interpretability, denoted by a value of 3 on a scale of 1 to 5. On the other hand, Machine Learning, armed with its statistical and probabilistic underpinnings, emerges as a versatile and widely applicable methodology, receiving a commendable score of 4 across various domains, although it does grapple with a degree of interpretability that can be characterised as moderate. The table below shows the performance of each approach for different parameters like speed, robustness, scalability, applicability and interpretability. The rate scale ranges between 1-5 where 1 is considered for poor performance and 5 for excellent (Table 1).

Rating Scales are 1: Poor, 2: Fair, 3: Average, 4:Good and 5: Excellent. The landscape of modelling approaches further unfolds with Neural Networks, which, being nonlinear and adaptive in nature, exhibit an extraordinary capacity for capturing complex relationships, earning a score of 4 in applicability. However, they bear the burden of diminished interpretability, marked by a rating of 2. Probabilistic Reasoning, focusing on managing uncertainty within models, garners high praise in the realm of applicability with a score of 4 while maintaining a moderate level of interpretability. Zooming in on Deep Learning, a subcategory nested within Neural Networks, we find it to be particularly proficient in domains such as image recognition and natural language processing, as signified by an impressive applicability rating of 5. Nevertheless, it too faces the interpretability challenge, receiving a rating of 2 in this regard. Delving deeper into the intricacies of these modelling approaches, we discover that when it comes to the speed of convergence during training, Neural Networks surge ahead, achieving rapid solutions and securing a score of 4. In contrast, Fuzzy Logic and Probabilistic Reasoning both attain moderate ratings of 3 in this aspect.

Turning our attention to the robustness of these models, Machine Learning and Neural Networks are deemed to possess a level of robustness, receiving a score of 3, enabling them to handle noise and uncertainty. Similarly, Fuzzy Logic and Probabilistic Reasoning are not to be outdone in this respect, earning a robustness rating of 4 for their capacity to navigate uncertainty. In terms of scalability, Neural Networks shine brightly, receiving a score of 4, thanks to their ability to accommodate vast datasets and complex model architectures. Fuzzy Logic and Probabilistic Reasoning, however, exhibit moderate scalability, with scores of 3, indicating their capacity to handle more moderate levels of complexity. The facet of applicability emerges as a defining strength for Neural Networks, showcasing a high rating of 5, especially within domains such as image recognition and natural language processing. Machine Learning and Probabilistic Reasoning stand as versatile contenders, achieving a commendable applicability score of 4. In contrast, Fuzzy Logic, with its moderate applicability rating of 2, appears more tailored to specific domains (Table 2).

Table 1: Performance rating scale of different approaches

Approach	Convergence Speed	Robustness	Scalability	Applicability	Interpretability
Fuzzy Logic	3	4	2	4	3
Neural Networks	4	3	4	5	2
Genetic Algorithms	2	4	3	3	2
Swarm Intelligence	3	3	3	4	2
Evolutionary Strategies	2	4	3	3	2
Ant Colony Optimization	3	3	3	4	2
Artificial Immune Systems	2	3	2	3	3

Table 2: Comparison of various approaches along with different parameters

Parameter	Fuzzy logic	Machine learning	Neural network	Probabilistic reasoning	Deep learning
Modelling approach	Imprecise, approximate	Statistical, probabilistic	Non-linear, adaptive	Probabilistic, uncertain	Statistical, probabilistic
Applications	Control systems, decision making, pattern recognition	Classification, regression, clustering	Classification, regression, forecasting	Data mining, natural language processing	Image recognition, natural language processing
Strengths	Robust to noise and uncertainty, easy to understand	Can handle large amounts of data, can learn from examples	Can learn complex relationships, can adapt to changes	Can handle uncertainty and incomplete information	Can learn complex relationships, can be used for image recognition and natural language processing

(Continued)

Table 2: (*Continued*)

Weaknesses	Can be computationally expensive, can be difficult to interpret	Can be sensitive to overfitting, can be difficult to generalise to new data	Can be difficult to train, can be sensitive to noise	Can be difficult to interpret, can be computationally expensive	Can be computationally expensive, can be difficult to interpret

5. Conclusion

Soft computing techniques have shown great promise in improving network performance, security, and reliability. Fuzzy logic, neural networks, genetic algorithms, and swarm intelligence have been successfully applied in networking to overcome challenges such as congestion, security threats, and reliability issues. However, there is still a lot of research that needs to be done to fully realise the potential of soft computing techniques in networking. Hybrid soft computing techniques that combine the strengths of multiple techniques could be a promising approach to address complex networking challenges. The integration of soft computing techniques in emerging technologies such as IoT and 5G networks could lead to new opportunities and challenges for networking.

References

[1] Gholami, M., Ghasemzadeh, H., Tabatabaei, S. S., and Kazemi, S. M. R. (2017). A survey on soft computing-based routing protocols in wireless sensor networks. Wireless Personal Communications, 97(4), 6305-6343

[2] Islam, M. N., and Rahman, M. H. (2019). A comprehensive review of soft computing techniques in computer networking. IEEE Access, 7, 38805-38823.

[3] Zhang, H., Zhang, W., Liu, W., and Jin, C. (2017). Fuzzy logic-based routing algorithm in wireless mesh networks. 2017 4th IEEE International Conference on Cloud Computing and Intelligence Systems (CCIS), Nanjing, China, pp. 77-81.

[4] Saravanan, K., Selvam, A., and Raja, R. (2014). A comparative study of soft computing techniques for intrusion detection system. 2014 IEEE International Conference on Advanced Communication, Control and Computing Technologies (ICACCCT), Ramanathapuram, India, pp. 60-64

[5] Zadeh, L. A. (1965). Fuzzy sets. Information and Control, 8(3), 338-353.

[6] Klir, G. J. & Yuan, B. (1995). Fuzzy sets and fuzzy logic: Theory and applications. Prentice Hall.

[7] Cuomo, S., Schiano Di Cola, V., Giampaolo, F., Rozza, G., Raissi, M. and Piccialli, F. (2022). Scientific machine learning through physics–informed neural networks: Where we are and what's next. Journal of Scientific Computing, 92(3), 88.

[8] Woldseth, R. V., Aage, N., Andreas Bærentzen, J., and Sigmund, O. (2022). On the use of artificial neural networks in topology optimisation. Structural and Multidisciplinary Optimization, 65(10), 294.

[9] Sohail, A. (2023). Genetic algorithms in the fields of artificial intelligence and data sciences Annals of Data Science, 10(4), 1007-1018.

[10] Dolezel, P., Holik, F., Merta, J., and Stursa, D. (2021). Optimization of a depiction procedure for an artificial intelligence-based network protection system using a genetic algorithm. Applied Sciences, 11(5), 2012.

[11] Kavitha, R., Kiruba Jothi, D., Saravanan, K., Swain, M. P., Arias Gonzáles, José Luis, Bhardwaj, R. J. and Adomako, E. (2023). Ant colony optimization-enabled CNN deep learning technique for accurate detection of cervical cancer. BioMed Research International, 2023.

[12] Wu, Lei, Xiaodong Huang, Junguo Cui, Chao Liu, and Wensheng, Xiao. (2023). Modified adaptive ant colony optimization algorithm and its application for solving path planning of mobile robot. Expert Systems with Applications, 215, 119410.

[13] Ramírez-Ochoa, Dynhora-Danheyda, Asunción Pérez-Domínguez, L., Martínez-Gómez, Erwin-Adán, and Luviano-Cruz, David. (2022). PSO, a swarm intelligence-based evolutionary algorithm as a decision-making strategy: A review. Symmetry, 14(3), 455.

[14] Ye, Tingyu, Wenjun Wang, Hui Wang, Zhihua Cui, Yun Wang, Jia Zhao, and Min Hu. (2022). Artificial bee colony algorithm with efficient search strategy based on random neighborhood structure. Knowledge-Based Systems, 241, 108306.

[15] Kurani, Akshit, Pavan Doshi, Aarya Vakharia, and Manan Shah. A comprehensive comparative study of artificial neural network (ANN) and support vector machines (SVM) on stock forecasting. Annals of Data Science, 10(1), 183-208.

Innovative CAPTCHA Technique as an Enhanced Security Measure for Authentication

Grace Odette Boussi[1], Himanshu Gupta[2], Syed Akhter Hossain[3], Rudy J. J. Fila[4]

[1,2]Amity Institute of Information Technology, Amity University Noida, India
[3]Computer Science and Engineering Department, University of Liberal Arts, Dhaka, Bangladesh
[4]School of Engineering and Technology, Sharda University, Greater Noida, India
E-mail: boussigrace@gmail.com, hgupta@amity.edu, aktarhossain@daffodilvarsity.edu.bd, ryanshago@gmail.com

Abstract

The internet has become an essential part of our daily lives, connecting numerous devices, and facilitating various online tasks and work. However, this increased connectivity has also resulted in a rise in cybercrimes. With the advancement of technology, authentication plays a crucial role in enhancing security measures. Numerous organisations have adopted various robust authentication techniques to safeguard their data from unauthorised access. In this proposal, we introduce an innovative CAPTCHA method to serve as an extra layer of security for authentication, applicable across various platforms where enhanced security measures are required. We have used pattern recognition to train the system to identify the user's handwriting and validate if the provided code corresponds to the user's writing.

Keywords: Cybercrime, cyber security, captcha, pattern recognition, authentication

1. Introduction

Cybercrimes refer to crimes committed by individuals, such as hackers and criminals, using electronic devices like computers and mobile phones. Despite efforts by organisations and governments, cybercrimes continue to increase, with negligence and carelessness being contributing factors. CAPTCHA, which stands for Completely Automated Public Turing test to Tell Computers and Humans Apart, is a mechanism used to differentiate between humans and machines. It ensures that the person accessing a platform or application is not a robot Wikipedia [10].

It is crucial to prioritise security alongside fast access to user data. Striking a balance between efficiency and protection is essential. CAPTCHA is a quick step to verify human users, while pattern recognition helps identify patterns in data.

Pattern recognition is the process of identifying patterns or regularities in data, often with the goal of classifying input into categories or making predictions. It involves the extraction of meaningful information from input variables and the recognition of patterns and regularities in data. Pattern recognition has applications in various fields such as machine learning, computer vision, speech recognition, and bioinformatics. It uses computational algorithms to analyse and interpret patterns, making it a crucial component of many technological systems and applications. Pattern recognition has wide applicability in technology, engineering, medicine, and agriculture, reducing human involvement and creating strong authentication methods. Analysing handwritten signatures is specifically challenging but important in various domains Rajalakshmi, Saranya and Shanmugavadivu [11]. Implementing this reliable method is becoming prevalent in various domains where security and authentication are paramount. Offline handwriting signature processing is more difficult compared to online processing due to the limited available information Hafemann, Sabourin and Oliveira [1].

Authentication is the process of verifying the identity of a user or entity attempting to access a system, network, or application. It ensures that the entity claiming a particular identity is indeed who it purports to be. Authentication methods commonly include the use of passwords, PINs, biometric data (such as fingerprints or facial recognition), security tokens, and two-factor or multi-factor authentication. Strong authentication practices are fundamental in ensuring the security and integrity of sensitive information and systems.

Authentication is of paramount importance in cybersecurity for several reasons:

- **Access Control**: Authentication ensures that only authorised users have access to sensitive information and systems.
- **Data Protection**: Proper authentication mechanisms safeguard critical data from being accessed or manipulated by unauthorised individuals or malicious entities.
- **Regulatory Compliance**: Many cybersecurity regulations and standards, such as GDPR and PCI DSS, mandate the use of strong authentication measures to protect personal and financial data.
- **Mitigating Unauthorised** Access: By confirming the legitimacy of users, authentication helps prevent unauthorised access, reducing the risk of security breaches and data leaks.
- **User Accountability**: Authentication facilitates the tracking of user activities and helps establish accountability for any actions performed within a system or network.

While various techniques and technologies are being utilised and tested, their effectiveness is ultimately dependent on the presence of laws that condemn all past crimes.

IT cyber law, also known as cyber law, encompasses the legal framework that regulates internet usage, digital technology, and electronic communication. It deals with a wide range of issues, including data privacy, cybersecurity, digital transactions, intellectual property rights, online freedom of speech, e-commerce, and cybercrimes. Cyber law aims to address legal issues and challenges arising from the use of information technology, online activities, and the digital landscape Chander and Kaur [2]. It governs both the rights and responsibilities of individuals, businesses, and governments in the online sphere, seeking to ensure that legal protections and regulations extend to the digital realm.

Cyber law protects individuals and organisations from cybercrimes and provides a legal framework for addressing and punishing offenders. It establishes guidelines, imposes penalties, and offers protection for victims.

It is important to note that cyber laws can vary from one country to another, as each jurisdiction may have its own specific set of regulations. However, they generally share a common goal, which is to prevent, investigate, and prosecute cybercrimes Zahoor and Razi [3].

2. Literature Review

Many Authors extensively studied signature and pattern recognition, encouraging new ideas for enhancement. Their work guided the proposed implementation in this paper.

For example, Raj, Swaminarayan, Saini, and Parmar [4] implemented pattern recognition in agriculture, demonstrating that this method is not limited to a specific domain. Previous works by Vyas Talati and Naik [5] and Arivazhagan, Shebiah, Ananthi, Varthini [6], also addressed pattern recognition in agriculture. While Raj, Swaminarayan, Saini, and Parmar [4] use their pattern to identify fruits based on colours, Vyas Talati, and Naik [5] proposing a system to automatically detect plant leaf diseases.

Luo, Bhattacharya, Maiti, Dutta, Ochi, Miura-Mattausch, and Mattausch [7] suggested a pattern recognition scheme based on force sensors from different walking surfaces.

Mukherjee, Chakraborty, Bhattacharya, and Parui [8] proposed a hybrid approach for handwriting recognition using three networks (CNN, RNN, and CTC), utilising databases of Devanagari and Bangla online handwritten samples for testing. The importance of data protection and privacy is emphasised, as Verma [9] highlighted the vast amount of personal data available on the internet, necessitating security measures.

In the past people use to think that cybercrime is the concern of only the organisation or a government, the current cyber space situation has proven us wrong, that is why Okutan (2019) stated that cybercrimes are no longer limited to organisations or governments, but also pose a serious threat to national security and the average person.

3. Research Gaps

Throughout our research, we have observed a wide range of authentication methods with significant advancements in multi-factor authentication. However, while these areas have seen substantial progress, CAPTCHA methods have remained relatively unchanged, primarily consisting of audio challenges, alphanumeric input, and pattern recognition. In contrast, cyber-attacks such as DDoS, which overwhelm servers with excessive requests, can still compromise services. Additionally, hackers can potentially bypass CAPTCHA procedures using new technologies, jeopardising its effectiveness. In our approach, we aim to integrate a security principle of combining public and secret elements into the CAPTCHA process. Users will scan a secret code and confirm it using our CAPTCHA system as part of this effort.

4. Proposed Model

In our work, we have added an extra layer of security to sensitive data access through a captcha-based process. Users provide credentials, receive a captcha code, write it down, and scan it for verification. This method differentiates humans from automated entities, preventing unauthorised access. It can enhance security across applications, combining traditional credentials with handwritten captcha verification.

4.1. Working Principle of Proposed Model

A new layer of security is proposed to enhance any sensitive data transaction security, particularly in scenarios where two-factor authentication (OTP) is bypassed. OTPs been the common authentication method that is available on data transmission but still have some limitation. In some countries and websites for example, when sensitive data must be shared let's take the example of money transfer, the OTP procedure is not always mandatory. To address this, our proposed work suggests implementing CAPTCHA as an additional security layer. Users would enter their bank's OTP and, if it matches the expected code, they would be directed to a CAPTCHA interface. Here, users scan a separate code, which is evaluated by the CAPTCHA system. If the code matches, access is granted; otherwise, an error message is displayed (Figure 1).

When the CAPTCHA code is correctly entered and verified, the screenshot interface will open, adapting to the security technique available on the client's device. This may involve fingerprint, face recognition, voice recognition, or hand recognition. The goal is to combine CAPTCHA verification with an additional layer of device-specific security, enhancing the access process with a multifactor approach (Figure 2).

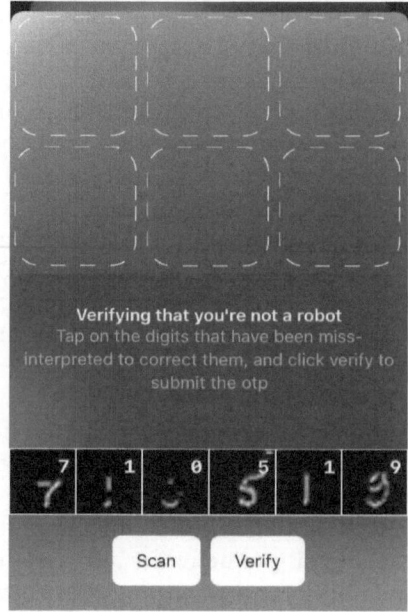

Figure 2: Correct CAPTCHA

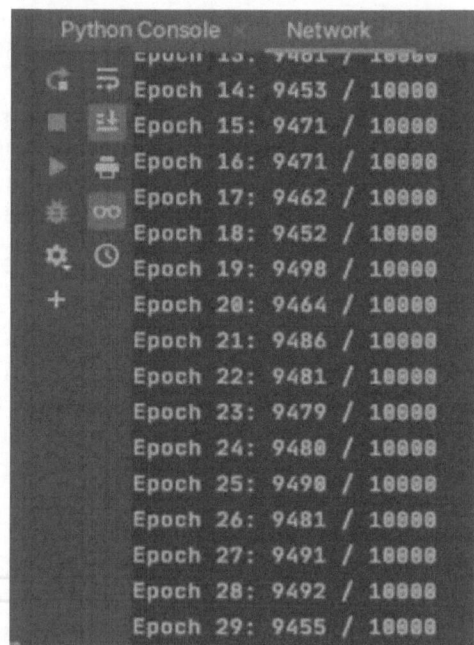

Figure 3: Result

5. Result

The accuracy of our model was checked using Python, utilising data sourced from our application called "Telles," which was created as a personal project (Telles, 2023). The CAPTCHA system was tested and integrated into the application, achieving a model accuracy of 94% according to the results obtained from the Python analysis. This demonstrates the success of our CAPTCHA implementation in effectively validating user inputs within the application (Figure 3).

Figure 1: Wrong CAPTCHA

6. Conclusion and Future Work

The objective of this work was to enhance the authentication process by implementing a new layer of security using handwritten pattern recognition. This innovative approach introduces a new form of CAPTCHA, which, if integrated into widely used platforms, particularly those handling sensitive information, can significantly bolster security and reduce data breaches. However, it is crucial to acknowledge that the effectiveness of proposed techniques is dependent on the presence of laws that condemn cybercrimes. This paper also emphasises the significance of cyber law in regulating and safeguarding cyberspace.

Moving forward, our future work involves integrating the signature recognition technique with a prediction or detection algorithm to further enhance its strength and security.

7. Contribution

Conceptualisation, Grace Odette Boussi, Rudy J.J. Fila; methodology: Grace Odette Boussi; validation, Himanshu Gupta and Syed Akhter Hossain; formal analysis, Grace Odette Boussi, Himanshu Gupta and Syed Akhter Hossain. All authors read and approved the final manuscript.

References

[1] Hafemann, L. G., Sabourin, R., and Oliveira, L. S. (2017, November). Offline handwritten signature verification—literature review. In 2017 Seventh International Conference on Image Processing Theory, Tools, and Applications (IPTA) (pp. 1-8). IEEE.

[2] Chander, H., and Kaur, G. (2022). Cyber laws and IT protection. PHI Learning Pvt. Ltd.

[3] Zahoor, R., and Razi, N. (2020). Cyber-crimes and cyber laws of Pakistan: An overview. Progressive Research Journal of Arts & Humanities (PRJAH), 2(2), 133-143.

[4] Raj, M. P., Swaminarayan, P. R., Saini, J. R., and Parmar, D. K. (2015). Applications of pattern recognition algorithms in agriculture: A review. International Journal of Advanced Networking and Applications, 6(5), 2495.

[5] Vyas, A. M., Talati, B., and Naik, S. (2013). Colour feature extraction techniques of fruits: A survey. International Journal of Computer Applications, 83(15).

[6] Arivazhagan, S., Shebiah, R. N., Ananthi, S., and Varthini, S. V. (2013). Detection of unhealthy region of plant leaves and classification of plant leaf diseases using texture features. Agricultural Engineering International: CIGR Journal, 15(1), 211-217.

[7] A. Luo et al., "Dynamic Pattern-Recognition-Based Walking-Speed Adjustment for Stable Biped-Robot Movement under Changing Surface Conditions," 2019 IEEE 8th Global Conference on Consumer Electronics (GCCE), Osaka, Japan, 2019, pp. 600-601, doi: 10.1109/GCCE46687.2019.9015430.

[8] Mukherjee, P. S., Chakraborty, B., Bhattacharya, U., and Parui, S. K. (2017, November). A hybrid model for end-to-end online handwriting recognition. In 2017 14th IAPR International Conference on Document Analysis and Recognition (ICDAR) (Vol. 1, pp. 658-663). IEEE.

[9] Verma, R. N. (2019). Cyber laws and privacy issues in India. Global.

[10] Pattern Recognition' (2018) Wikipedia. Available at: https://en.wikipedia.org/wiki/Pattern_recognition.

[11] Rajalakshmi, M., Saranya, P., and Shanmugavadivu, P. (2019, April). Pattern recognition-recognition of handwritten document using convolutional neural networks. In 2019 IEEE International Conference on Intelligent Techniques in Control, Optimization and Signal Processing (INCOS) (pp. 1-7). IEEE.

Artificial Intelligence-Inspired Framework for Video Surveillance Incorporating Object Detection with Unmanned Aerial Vehicles

Manpreet Kaur, Munish Saini

Department of Computer Engineering and Technology, Guru Nanak Dev University, Amritsar, India
E-mail: manpreet.csersh@gndu.ac.in, munish.cet@gndu.ac.in

Abstract

Video surveillance is vital in crime prediction or anomaly detection. Artificial intelligence (AI), especially machine learning and deep learning-based automated systems, has been developed to detect objects contributing to criminal activities using image features. Objects can be recognised depending on their color, shape, motion, and texture and through proper training of the models. This study elaborates on the comprehensive review of various AI-based surveillance systems that keep track of objects online or in the physical world, focusing on identifying, classifying, and tracking them. The present surveillance systems are compared based on their features, performance, and challenges and are depicted in tabular form. An Unmanned Aerial Vehicle (UAV) enabled framework is proposed for detecting criminal activity by incorporating object detection techniques. Moreover, the future trends are also being discussed.

Keywords: Artificial intelligence; crime prediction; video surveillance; object detection; unmanned aerial vehicles

1. Introduction

The Artificial Intelligence (AI) global surveillance index shows that 75 of 176 countries use various AI-assisted surveillance systems[1]. Of these, 56 countries are working on smart cities, 64 on facial recognition, and 52 on predictive policing. Real-time monitoring is exhausting, so various automated monitoring systems use AI in object detection to keep civilians safe. Any objects like firearms, knives, or anything suspicious can easily be detected using various deep-learning models. As per UNODC, 550000 firearms were seized globally in 2017, and 8-11% were held in violent crimes, depicting the connection between possession of firearms and crime events.[2] Video surveillance can help detect crime-related objects or weapons in banks, malls, streets, educational institutions, and various crowded places like bus stands [1]. The security system can identify the objects depending on their attributes using multiple algorithms like convolution neural networks (CNN), Faster R-CNN (region-based convolution neural networks), You Only Look Once (YOLO), Single shot multi-box detector (SSD), etc. [2]. Previously, video monitoring was done manually, and crucial information may go unnoticed due to the massiveness of input data. To resolve this problem, many automated techniques were incorporated, and to reduce the storage and time complexity, a synopsis of the videos from multiple cameras was created for any suspicious activity. Only that clip is stored, making viewing faster [3]. To overcome the limitations of cameras of restricted area surveillance due to their fixed location, the latest technologies, like Unmanned aerial vehicles (UAVs), are preferred for crowd monitoring, tracking objects, and monitoring specific places [4]. In urban areas, algorithms capable of multi-task processing and learning are used, parallelly accomplishing several tasks of detection, recognition, capturing human features, and tracking [5]. The basic workflow in real-time video analytics followed by monitoring systems consists of the following steps: Acquisition of data, Preprocessing, Feature extraction, Object detection, Object tracking, Recognising human behaviour/action or any suspicious activity, then finally making a decision and generating alert [2, 6]. The monitoring depends upon several factors like the caliber

---

[1] https://carnegieendowment.org/2019/09/17/global-expansion-of-ai-surveillance-pub-79847

[2] https://www.unodc.org/documents/data-and-analysis/Firearms/2020_REPORT_Global_Study_on_Firearms_Trafficking_2020_web.pdf

of cameras used, number of cameras, their location, Surveillance network, storage, techniques deployed, and much more. The main objectives of the paper are as follows:

- To understand the contribution of artificial intelligence using object detection in crime prediction
- To review the latest frameworks based on object detection to identify suspicious activity.
- An AI-inspired, UAV-enabled framework is proposed for detecting suspicious objects.

The paper has been organised into various sections; the second section is about the background and motivation behind using these algorithms for object detection. In this section, the video surveillance devices, network, the concept of object detection, and various AI-based algorithms commonly used in object detection are elaborated. The third section reviews and tabulates different object detection techniques in crime prediction. The fourth section represents the proposed framework, and the paper briefly concludes in the last section.

2. Background and Motivation

2.1. Video Surveillance Devices

In video surveillance, cameras are used to monitor and record the routine life of any particular area like public spaces, banks [2], government offices, and educational institutions [7,8], businesses, manufacturing units, traffic monitoring systems, entertainment parks, railways, airports [1], shopping malls, etc., for safety and security as depicted in Figure 1. The most popular video surveillance cameras [9, 10] are classified as (i) Analog Cameras: These traditional cameras capture analog video for recording in DVR using coaxial cables; (ii) Digital cameras: They are more popular; IP Cameras (Internet protocol Cameras) are digital cameras with advanced features, enhanced picture quality, and high resolution; use ethernet or Wi-Fi for transmitting videos and can be accessed anywhere using digital devices. Dome Cameras: Indoor and outdoor dome-shaped cameras are used for wide-angle views. They are preferred in banks or public places. Bullet Cameras are specially designed for outdoor surveillance. They are long and narrow with proper shields to bear the weather. PTZ Cameras (Pan-tilt-zoom cameras) are high-quality cameras for large areas monitoring as they can track objects by tilting and zooming. They have more advanced features, can be operated remotely, and are flexible to use. Moreover, the introduction of UAVs (drone cameras) has enabled monitoring of remote areas with limited or no fixed surveillance cameras installed.

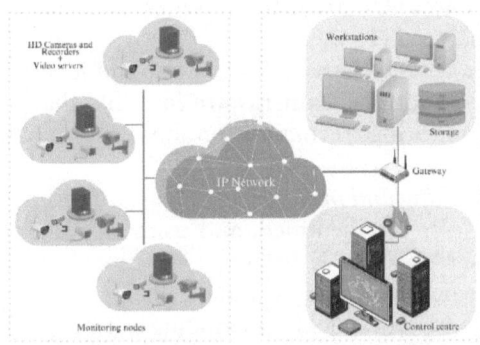

Figure 2: Video surveillance network

2.2. Video Surveillance Network

It comprises the overall structure and various surveillance system components [9]. Various components, as represented in Figure 2, are:

- *HD Cameras and Recorders*: Devices used for making the video. Depending upon the requirements of the installation environment, cameras can be digital or analog.
- *Network infrastructure*: It constitutes network switches, routers, wires, or Wi-Fi spots for wireless networks.
- *Monitoring nodes and control center*: Monitoring nodes collect the live video recordings of the allocated location, and at the control center, it is viewed/monitored; in case of any emergency, an alarm is generated.
- *Storage*: Videos are stored to be reviewed later on. Cloud-based storage or NVRs can be used for this purpose.
- *Workstation (video analytics)*: Various algorithms can be used for analyzing videos at workstations, such as automated face recognition, violation of traffic lights, fire, or any kind of anomaly like theft or crime by behavior analysis, object tracking, motion detection, etc. [11].

2.3. Object Detection

In video surveillance, object detection helps auto-identify the objects of interest. Various machine learning and image processing techniques are used to recognise objects by analyzing different video frames. Categorisation is done depending on the features of objects. The significant purposes achieved using object detection are:

- *Security/ Surveillance*: Any suspicious activity can be detected in real-time, and an alarm can be generated.
- *Crowd monitoring*: To identify a person, prevent any mob, or regulate the number of persons in

a particular area, the crowd's behavior can be analyzed [1, 4].

- *Traffic monitoring*: To prevent and identify traffic violations, cameras are installed that help in monitoring, recognising number plates, and assisting in traffic flow.[12]
- *Tracking and monitoring of assets*: The suspicious objects are identified, and using object tracking, their position is located over time from subsequent video frames using high-speed tracker response and quickly responding sensors [13, 14].

2.4. Artificial Intelligence-inspired Techniques

Convolution Neural Networks (CNN): CNN is a successful state of art object detection model. CNN is usually trained on large datasets. It is based on a region proposal network (RPN); from these regions, CNN extracts basic features like edges and texture to the most complex and abstract level features. Then, these features are forwarded to additional layers for object localisation and classification. Using localisation within each proposed region, the coordinates of the object are predicted, and classification is used to identify the predefined class to which the object may belong. Non-maximum suppression (NMS) is applied to remove duplicates to discover non-overlapping detections and helps improve the precision of object detection algorithms [15].

Faster R-CNN: A two-stage object identification framework called Faster R-CNN. The Region-based Convolutional Neural Network (R-CNN) and Fast R-CNN algorithms are extensions of the original R-CNN method. The Region Proposal Network (RPN), which creates region proposals straight from the convolutional feature maps, is included in the detection pipeline of Faster R-CNN to achieve high accuracy. The RPN makes suggestions for likely item positions and forms, and the Fast R-CNN network then improves and categorises these suggestions. One of the original algorithms used to create contemporary object identification frameworks was

faster R-CNN. It is an extension of a region-based convolution neural network that attains high accuracy in object detection by determining its location and shape in various frameworks [15].

Single Shot Multibox Detector (SSD): SSD is a single-shot object detection algorithm [16, 17] that directly predicts the object bounding boxes and class probabilities from the single neural network pass. It uses various predefined bounding boxes of different aspect ratios and scales as anchor boxes. These predefined boxes are placed at multiple locations and varying scales on an image, and for each box, SSD predicts the final bounding box and its probability of association with a particular class. SSD uses a feature pyramid for feature extraction at different scales, it helps in handling varied size objects efficiently. It uses the localisation loss and classification loss combined for training a model. It is computationally efficient and achieves high accuracy on benchmark datasets, as it processes the entire image in one forward pass. Hence, it is best suited for real-time applications like person identification, computer vision applications, and object detection [4, 2].

You Only Look Once (YOLO): YOLO is famous for its speed and accuracy in faster and more precise object detection. Several versions, like YOLOv2, YOLOv3, YOLOv4, and YOLOv5, have been developed to improve accuracy and have various advantages in real-time object detection and end-to-end training in object recognition. [18, 19]. YOLO is a unified object identification model that anticipates object bounding boxes and class probabilities straight from the complete picture in a single neural network pass. It creates a grid from the image and forecasts boundary boxes and class ratings for each grid cell. To forecast the bounding boxes of objects whose centers lie within each grid cell, YOLO splits the input picture into an $S*S$ grid. The intended trade-off between accuracy and speed is usually considered when determining the value of S. Anchor boxes of various sizes and forms are used by YOLO as reference templates to anticipate the bounding boxes. To better suit the forms of the items, each grid cell forecasts an offset to the size of its related anchor boxes. YOLO forecasts the class probability for several predetermined classes for each grid cell. This enables the model to identify and categorise various items. To train the model, YOLO combines localisation loss (often Smooth L1 loss) with classification loss (typically Cross-Entropy loss). While the classification loss penalises inaccurate class predictions, the localisation loss penalises the difference between the predicted bounding box and the ground truth box [2, 4].

Recurrent Neural Networks (RNN): In object detection, RNN or its variants like LSTM or GRU, helps in various tasks related to object detection such

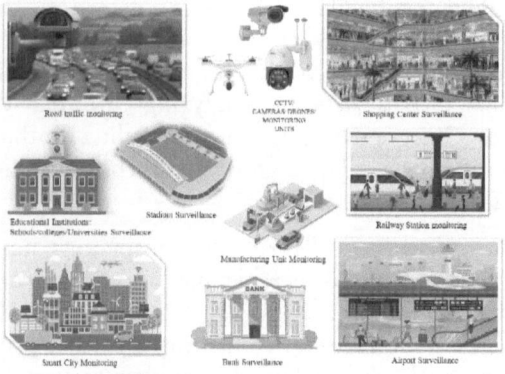

Figure 1: Application areas of video surveillance

as object tracking, capturing context-based information in sequential frames, object detection in video streams, captioning the images or objects present in the scene [20].

3. Review of Related Works in Crime Prediction

In recent works, various AI models are CNN, SSD, YOLO, Faster RCNN, LSTM, and GRU. A few important research studies, their objectives, the dataset used, models or techniques used, results, and future scope are depicted in Table 1. A study by Zahrawi et al. detected various objects related to bank robbery, like knives, masks, pistols, and rifles, using SSD and YOLO models, achieving the highest precision of about 0.82 [2]. In Multiview camera surveillance, Ingle et al. propose a framework for creating a video synopsis of suspicious events using Faster RCNN, SSD, customised CNN, and YOLOv3, achieving a precision of 0.94, reducing the memory storage requirement [3]. Thakur et al. used UAVs in a study to recognise objects or pedestrians using modified YOLOv5 [4]. Furthermore, Sung et al. developed a web application for real-time monitoring using JavaScript, which generates an alert with the image of the corresponding activity [21]. Biswas et al. proposed an approach for real-time video surveillance in which the questionable pixels and locations are given more weight, and the dynamic patch idea is applied. Pixels that are suspicious are assigned a smaller patch size. All frames are subjected to "Contrast Limited Adaptive Histogram Equalisation (CLAHE) based on Colour Channel" to clearly identify suspicious items. The proposed

technique depicts an accuracy of 97.65%, and an enhancement in memory of 75.17%. To conserve memory, low-resolution frames that had a suspicious movement in the previous 100 frames are recorded, and preceding these frames, the rest are discarded. It detects motion-based household things as suspicious, so an improvement is needed for dark environments [22]. Moreover, Kumar et al. proposed a deep learning-based model, LbDBNP, using Python for early identification of criminals by object tracking in public places based on their activities like face actions and handling weapons [23], and the maximum accuracy achieved is 98.85%. Adimoolam et al. proposed a hybrid model based on a dense feature selection convolution neural network, hyperparameter tuning for multiple object detection in video surveillance, and achieved 98% accuracy [24].

4. Proposed Framework

Figure 3 depicts an AI-based, UAV-enabled framework for real-time crime prediction using object detection. As camera installation is not possible in outdoor areas [27], and the cameras have limited scope, only a confined area is invigilated by them. In remote villages, border areas, highly crowded places, parking areas, playgrounds, stadiums, and university campuses where it's impossible to install cameras at every place, we can use UAVs to track the objects that may lead to criminal activity. In the proposed framework, a three-layer architecture is used: the first is UAV onboard object detection, the second layer is for cloud-level detection of any abnormal event or weapon, etc., and the third one is for user-level alert generation. As drones have limited computational capacity, a model is proposed in which the object

Table 1: Recent publications

Reference	Objectives	Dataset	Model/ Technique	Results/ Performance metrics	Future scope/ limitations
Zahrawi and Shaalan [2]	To identify any bank robbery by detecting suspicious objects like weapons and persons with a balaclava.	Two datasets based on weapon and suspicious object detection. Dataset1: knife, pistol, mask, and rifle Dataset 2: knife and pistol	SSD Inception V2, SSD MobileNet V2, SSD ResNet50 V1, YOLOv4, and YOLOv5	All object detection models were evaluated on mAP@0.5 (PascalVOC). Best performance: Dataset 1: YOLOv4 precision (0.795); Dataset 2: YOLOv5: precision (0.82); Good in pistol identification rather than a knife.	Low image resolution and clutter in the background make object identification difficult. Also, variation in the viewpoint of cameras or any barrier in object visibility may leave it undetected,

(Continued)

Table 1: (*Continued*)

Ingle and Kim [3]	Proposed a CNN-based model that retrieves any strange components/objects from a particular video and creates a video synopsis by stitching the suspicious keyframes dynamically extracted from the input. Reduces the requirement of 24/7 invigilation and memory storage space.	Multiview camera surveillance	Faster R-CNN, SSD, and Customised CNN YOLOv3 (KB-AF-E-YM)	The proposed model (KB-AF-E-YM) Precision 0.94; Recall 0.87; F1 0.91; Similarity 0.87; PCC 92.12	The primary limitation of this study is that the suggested approach only performs effectively with HD camera video. The stitching process is hampered by visual clutter. Maintaining the chronological sequence was challenging.
Thakur et al. [4]	This study incorporates an unmanned aerial vehicle for surveillance with a deep-learning object recognition mechanism, including pedestrian or object detection and classification, by tuning the hyperparameters of the Yolov5.	The MOT20 dataset	SSD; Faster RCNN; YOLOv5; Modified YOLOv5	Modified YOLOv5: Training loss 0.015; Testing loss 0.15; Precision 0.61; Recall 0.35 Map 0.09; F1 score 0.44 SSD: mAP 0.48	UAV monitoring is applicable only in high-risk regions.
Amin et al. [25]	Proposed models J. DCNN and J. QCNN, based on CNN for accurate anomaly detection in the videos. Detect armed robberies and thefts.	Crime-UNI dataset based on the Crime-UCF dataset has classes: robbery/abuse/arrest/arson/burglary/explosion/fighting/stealing/shooting/shoplifting.	J. DCNN J. QCNN	AUC J.DCNN 98% J.QCNN 99%	It should also be implemented on varied datasets.
Li et al. [5]	Proposed a model ASGCM based on GRU and LSTM incorporating an antisaturation mechanism for analysing the multiscale video data with improved accuracy and less complexity.	Multiple Object Tracking dataset of urban traffic and crime data series.	RNN; LSTM; GRU; ASGSM	RNN is the fastest (7.10 ms/ epoch). ASGCM has the advantage of a streamlined structure and smaller training time (13.96 ms/ epoch) than LSTM and GRU.	Predicted the changing crime trend but didn't investigate the factors responsible.

Table 1: (*Continued*)

Sung and Park [21]	Created an intelligent video surveillance application based on javascript and react framework that generates an alert with the image in case of any emergency like disaster (e.g., fire) or any criminal activity.	COCO net: a large-scale object detection, segmentation, and captioning dataset.	web application	Real-time monitoring with 99% accuracy	Video surveillance can hamper privacy and security. Moreover, the dataset used needs to be more sufficient to recognise high-risk objects.
Manikandan and Rahamathunnisa [26]	Proposed an Attuned Object Detection Scheme (AODS) to detect harmful objects using CNN depending upon their features.	CCTV images	CNN Attuned Object Detection Scheme (AODS)	Frame to object detection is 85.1% The training using an external dataset increases the accuracy by 8.08%, decreases error by 7.47% and complexity by 8.23%.	In the future, it is intended to include the object classes based on labels for identifying the object category. Additionally, this would enhance object detection for framed inputs at high speed.

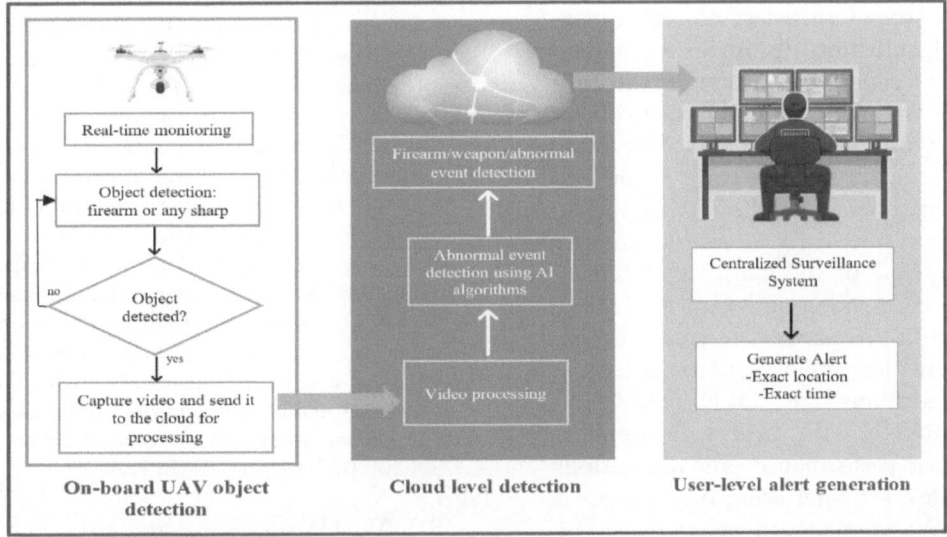

Figure 3: Proposed model for detecting objects contributing to criminal activities

detection will be performed on-the-board by the UAV by using embedded small-size models like tiny YOLO after proper training, and if any suspicious object is detected, then it captures the video for 5 seconds. The recorded video can be sent to the cloud for further processing and detecting any abnormal activity using highly trained algorithms. If such activity is identified, then the information is forwarded to the centralised surveillance system to generate an alert with the exact location and time. The proposed framework can use Tiny-YOLO on the Raspberry Pi for near real-time object detection with the Movidius NCS for onboard implementation [28]. Tiny-YOLO is particularly suited for use on embedded devices such as the Raspberry Pi, Google Coral, and NVIDIA Jetson Nano because of its compact size (50MB) and quick inference performance. We will be able to get 4.28 FPS using both a Raspberry Pi and a Movidius NCS. Drones like Mavic3 Pro can cover a distance of 18 km, so they can be used for outdoor surveillance and will help in object detection and tracking. With onboard computation (which can also be referred to as edge computing), only the videos of activities involving suspicious objects will be captured and sent to the cloud for further processing. It will save computational time and cost by processing them on board, reducing the cost and time of sending the real-time videos to the cloud and then computing them. It will fasten the alert generation process by cloud computation of the suspicious frames. In cloud computation, various algorithms like CNN, faster RCNN, or SSD can be used for abnormal activity detection, and the results are forwarded to the centralised surveillance system. After recognising any suspicious activity, an alert will be generated with the exact location and time for the concerned authority.

5. Conclusion

Object detection plays a crucial role in preventing and detecting abnormal activity by real-time monitoring of various public places, educational institutes, malls, government offices, traffic lights, etc. Different artificial intelligence-inspired algorithms have been deployed in recent studies to detect objects associated with criminal activities, like firearms, masks, guns, knives, or any weapon. Hence, among the multitudinous features of artificial intelligence, object detection reinforces intelligent surveillance for crime prediction. Monitoring crowds or any suspicious activity with the inculcation of these AI-proposed frameworks helps in the early detection of any crime. A general review of the latest object detection frameworks used in video surveillance in recent studies is presented in this paper. The video surveillance devices, surveillance networks, object detection techniques, and algorithms like YOLO, SSD CNN, RNN, and Faster RCNN with their recent applications, advancements, performance, and limitations are explored. Besides, a model using tiny YOLO is proposed using UAVs for onboard object detection. If any suspicious object is identified, only then is the video captured and forwarded to the cloud for further processing, and if any abnormal activity is then encountered employing AI-based algorithms, then an alert is generated. This proposed model reduces the cost, communication delays, and computation load at the cloud level for processing real-time videos. Hence, this intelligent and advanced model can be deployed and used to monitor parking, remote areas, high-risk environments, crowded public places, or outdoors.

References

[1] Omarov, B., Omarov, B., Shekerbekova, S., Gusmanova, F., Oshanova, N., Sarbasova, A., Yessengaliyeva, Z., Bedelbayev, A., Maikhanova, A., Omarov, N., and Sultan, D. (2019). Applying face recognition in video surveillance security systems. Lecture Notes in Computer Science (Including Subseries Lecture Notes in Artificial Intelligence and Lecture Notes in Bioinformatics), 11771 LNCS, 271–280. https://doi.org/10.1007/978-3-030-29852-4_22/ COVER

[2] Zahrawi, M., and Shaalan, K. (2023). Improving video surveillance systems in banks using deep learning techniques. Scientific Reports 2023, 13(1), 1–16. https://doi.org/10.1038/s41598-023-35190-9

[3] Ingle, P. Y., and Kim, Y. G. (2023). Multiview abnormal video synopsis in real-time. Engineering Applications of Artificial Intelligence, 123, 106406. https://doi.org/10.1016/J.ENGAPPAI.2023.106406

[4] Thakur, N., Nagrath, P., Jain, R., Saini, D., Sharma, N., and Hemanth, D. J. (2023). Autonomous pedestrian detection for crowd surveillance using deep learning framework. Soft Computing, 27(14), 9383–9399. https://doi.org/10.1007/S00500-023-08289-4/ TABLES/4

[5] Li, Z., Zhang, X., Xu, F., Jing, X., and Zhang, T. (2023). A multiscale video surveillance based information aggregation model for crime prediction. Alexandria Engineering Journal, 73, 695–707. https://doi.org/10.1016/J.AEJ.2023.04.045

[6] Kulbacki, M., Segen, J., Chaczko, Z., Rozenblit, J. W., Kulbacki, M., Klempous, R., and Wojciechowski, K. (2023). Intelligent video analytics for human action recognition: The state of knowledge. Sensors, 23, 4258. https://doi.org/10.3390/S23094258

[7] Kaur, M., and Saini, M. (2022). Indian government initiatives on cyberbullying: A case study on cyberbullying in Indian higher education institutions. Education and Information Technologies 2022, 1–35. https://doi.org/10.1007/S10639-022-11168-4

[8] Saini, M., Kaur, M., Sengupta, E., and Ahmed, K. (2023). Artificial intelligence inspired framework for preventing sexual violence at public toilets of educational institutions with the improvisation of gender recognition from gait sequences. Soft Computing, 1–20. https://doi.org/10.1007/S00500-023-08285-8/TABLES/4

[9] Rai, M., Husain, A. A., Maity, T., Yadav, R. K., Rai, M., Husain, A. A., Maity, T., and Yadav, R. K. (2018). Advance Intelligent Video Surveillance System (AIVSS): A future aspect. intelligent video surveillance. https://doi.org/10.5772/INTECHOPEN.76444

[10] Elharrouss, O., Almaadeed, N., and Al-Maadeed, S. (2021). A review of video surveillance systems. Journal of Visual Communication and Image Representation, 77, 103116. https://doi.org/10.1016/J.JVCIR.2021.103116

[11] Zhou, Z., Yu, H., and Shi, H. (2020). Optimization of wireless video surveillance system for smart

campus based on Internet of Things. IEEE Access, 8, 136434–136448. https://doi.org/10.1109/ACCESS.2020.3011951

[12] Chandrakar, R., Raja, R., Miri, R., Sinha, U., Kumar Singh Kushwaha, A., and Raja, H. (2022). Enhanced the moving object detection and object tracking for traffic surveillance using RBF-FDLNN and CBF algorithm. Expert Systems with Applications, 191, 116306. https://doi.org/10.1016/J.ESWA.2021.116306

[13] Alotaibi, M. F., Omri, M., Abdel-Khalek, S., Khalil, E., and Mansour, R. F. (2022). Computational intelligence-based harmony search algorithm for real-time object detection and tracking in video surveillance systems. Mathematics, 10, 733. https://doi.org/10.3390/MATH10050733

[14] Nagrath, P., Thakur, N., Jain, R., Saini, D., Sharma, N., and Hemanth, J. (2022). Understanding new age of intelligent video surveillance and deeper analysis on deep learning techniques for object tracking. Internet of Things, 31–63. https://doi.org/10.1007/978-3-030-89554-9_2/COVER

[15] Othmani, M. (2022). A vehicle detection and tracking method for traffic video based on faster R-CNN. Multimedia Tools and Applications, 81(20), 28347–28365. https://doi.org/10.1007/S11042-022-12715-4/TABLES/2

[16] Feroz, Md. A., Sultana, M., Hasan, Md. R., Sarker, A., Chakraborty, P., and Choudhury, T. (2022). Object detection and classification from a real-time video using SSD and YOLO models, 37–47. https://doi.org/10.1007/978-981-16-2543-5_4

[17] Zhang, Q., Hu, X., Yue, Y., Gu, Y., and Sun, Y. (2017). Multi-object detection at night for traffic investigations based on improved SSD framework. Heliyon, e11570. https://doi.org/10.1016/j.heliyon.2022.e11570

[18] Lee, J., and Hwang, K. il. (2022). YOLO with adaptive frame control for real-time object detection applications. Multimedia Tools and Applications, 81(25), 36375–36396. https://doi.org/10.1007/S11042-021-11480-0/FIGURES/12

[19] Xiao, Y., Chang, A., Wang, Y., Huang, Y., Yu, J., and Huo, L. (2022). Real-time object detection for substation security early-warning with deep neural network based on YOLO-V5. 2022 IEEE IAS Global Conference on Emerging Technologies, GlobConET 2022, 45–50. https://doi.org/10.1109/GLOBCONET53749.2022.9872338

[20] Anoopa, S., and Salim, A. (2022). Survey on anomaly detection in surveillance videos. Materials Today: Proceedings, 58, 162–167. https://doi.org/10.1016/J.MATPR.2022.01.171

[21] Sung, C. S., and Park, J. Y. (2021). Design of an intelligent video surveillance system for crime prevention: applying deep learning technology. Multimedia Tools and Applications, 80(26–27), 34297–34309. https://doi.org/10.1007/S11042-021-10809-Z/TABLES/3

[22] Biswas, T., Bhattacharya, D., and Mandal, G. (2023). Dynamic strategy to use optimum memory space in real-time video surveillance. Journal of Ambient Intelligence and Humanized Computing, 14(3), 2771–2784. https://doi.org/10.1007/S12652-023-04521-Z/FIGURES/9

[23] Kumar, K. K., and Venkateswara Reddy, H. (2022). Crime activities prediction system in video surveillance by an optimized deep learning framework. Concurrency and Computation: Practice and Experience, 34(11). https://doi.org/10.1002/CPE.6852

[24] Adimoolam, M., Mohan, S., and Srivastava, G. (n.d.). A novel technique to detect and track multiple objects in dynamic video surveillance systems. International Journal of Interactive Multimedia and Artificial Intelligence, 7, 4. https://doi.org/10.9781/ijimai.2022.01.002

[25] Amin, J., Anjum, M. A., Ibrar, K., Sharif, M., Kadry, S., and Crespo, R. G. (2023). Detection of anomaly in surveillance videos using quantum convolutional neural networks. Image and Vision Computing, 135, 104710. https://doi.org/10.1016/J.IMAVIS.2023.104710

[26] Manikandan, V. P., and Rahamathunnisa, U. (2022). A neural network aided attuned scheme for gun detection in video surveillance images. Image and Vision Computing, 120, 104406. https://doi.org/10.1016/J.IMAVIS.2022.104406

[27] Georgiou, A., Masters, P., Johnson, S., and Feetham, L. (2022). UAV-assisted real-time evidence detection in outdoor crime scene investigations. Journal of Forensic Sciences, 67(3), 1221–1232. https://doi.org/10.1111/1556-4029.15009

[28] Alam, M. S., Natesha, B. V., Ashwin, T. S., and Guddeti, R. M. R. (2019). UAV based cost-effective real-time abnormal event detection using edge computing. Multimedia Tools and Applications, 78(24), 35119–35134. https://doi.org/10.1007/S11042-019-08067-1/FIGURES/11

A Review of Semantic Analysis and its NLP Tasks

Dalia Barua and Tarandeep Singh Walia

Computer Applications, Lovely Professional University, Punjab, India

E-mail: bdalia24@gmail.com, tarandeep.25153@lpu.co.in

Abstract

Semantic analysis, within Natural Language Processing (NLP), entails the interpretation and understanding of the meaning of words, phrases, and sentences in human language, emphasising contextual comprehension. Significant NLP tasks performed by semantic analysis include summarisation, topic modelling, translation, named entity recognition (NER), sentiment analysis, word sense disambiguation (WSD), etc. This review study focuses on identifying the research gap in semantic analysis in the linguistic context and its execution techniques.

Keywords: Word, semantic analysis, codemix, word embedding

1. Introduction

Semantic analysis, which involves extracting meaning from text or speech, is an important part of natural language processing [1]. Is the primary requirement of several NLP tasks, like Sentiment Analysis [2], Extracting information [3], Summarization [4], Word sense disambiguation [5], Translation [6], Name Entity Recognition [6], etc.

Semantic Analysis is divided into two branches: lexical semantics and latent semantics. The study of the meanings of individual words and how they combine to generate meaningful sentences is the subject of lexical semantics. Lexical semantics frequently uses lexical databases and ontologies to capture the rich relationships between words [7]. Latent semantics, on the other hand, investigates the implicit, underlying meanings inside a larger text or material. This branch seeks to reveal underlying patterns, links, and subjects that may not be readily evident through word analysis [8].

Semantic Analysis that extracts the meaning from language deals with different categories of text in the lingual context, like monolingual, bilingual, codemix, codeswitch etc. Ability to use in a single language refers monolingualism, (Ex: English or Hindi or Bengali, etc.), in two languages refers Bilingualism, (Ex: Punjabi and English), in two or more languages within a single utterance refers codemix (Ex: Hindi-Bengali-English), switching between two or more languages in a conversation refers code switching, (Ex: Hindi to English). For the rapid growth of internet people give their opinion in online product review, YouTube comment, social media comment by codemixed text. So, the study in semantic analysis is extremely crucial for the information extraction and sentiment analysis from this large scale online codemixed text.

1.1. Distribution of Dataset Used for Semantic Analysis

Here exploring the datasets used for research in collected research papers (Figure 1):

1.2. Contribution

This study aiming to identify novel research area in the Semantic Analysis tasks. As for that different linguistic level datasets are studied along with their corresponding methodologies in semantic analysis tasks. Notable research gap and future research opportunity have been presented after literature review part. Extensive literature reviews are conducted in this study in the linguistic platform and methodology of Semantic Analysis tasks. A summary table also provided for the overview of study and results observations.

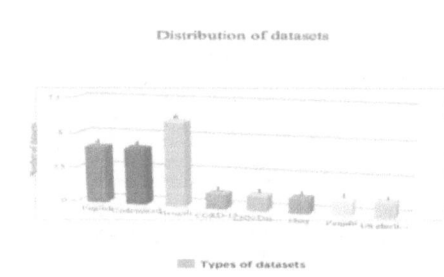

Figure 1: Distribution of datasets used in collected research

Chapter 4 DOI: 10.1201/9781003570349

2. Review of Literature and Research Gap Identification

The primary focus of this literature review is exploring the applied datasets, current methodologies, algorithms, and applications across these domains to present a holistic view of the progress made in semantic analysis.

2.1. Semantic Analysis on Monolingual Context

Khatun and Hoque [9] proposed a Bengali language based semantic analyser to enhance the accuracy of machine translation. The success rate is over 88% by experiment with more than 2000 Bengali sentences. Iqbal et al. [10] used word embedding techniques like Word2Vec, GloVec, and Fast Text to measure semantic similarity in a Bengali dataset and achieved 77.28% accuracy. Hossain et al. [11] proposed Embedding Parameter Identification algorithm (EPI)with intelligent text classification model for law resource languages including Bengali. In this research semantic and syntactic similarity measures by evaluating 165 embedding models and according to experimental results the GloVec model outperforms other embeddings in Bengali text classification, achieving 96.96% accuracy. For ambiguity checking in Bengali language Nipu et al. [12] invented a stemmer included Latent semantic analysis-based vector space model to handle countless linguistic structure in Bengali dialect with accuracy 89.76%. Hoque et al. [13] presented a methodology to evaluated semantic feature that are necessary to semantic analysis and Chowdhury et al. [14] demonstrated LSA based effective Bengali text summarisation approach. A rule-based approach was evaluated to identify reduplication in the Bengali text corpus (Rabindranath Tagore) and their semantic structural analysis by Chakraborty and Bandyopadhyay [15]. The average Precision, Recall and F1Score values are respectively 92.82%, 91.50% and 92.15%.

2.2. Semantic Analysis in Codemix Context

Code-mixed text is text that contains words or phrases from two or more languages. It is becoming increasingly common in many parts of the world, as people increasingly communicate with each other across cultures and languages Asnani and Pawar [16] proposed a semantic based model that outperforms traditional LDA based topic model on codemixed dataset. Here English-Hindi Fire 2014 dataset is used. Roy et al. [17] presented a lexicon-based sentiment analysis for Hindi English dataset Tareq et al. [18] evaluated a Bengali-English codemix dataset for sentiment analysis and their experiment showed training model with Fast Text embedding

on presented data augmentation method performing better than other method in sentiment analysis with F1 score 87%.

Tatariya et al.[19] experimented on transfer learning and fine tuning with a combination of English-Hindi, [20], Tamil-English, Malayalam-English [21], and Dravidian codemix [22] datasets. They employed the language models BERT [22], RoBERTa Liu et al., 2019, [31]), mBert [23], and XLM-RoBERTA (Liu et al., 2019), Indic BERT (Doddapaneni et al, 2022 [32]), and MuRIL (Khanuja et al., 2021, [33]).

2.3. Research Gap Identification

The previous extensive review of literature explored the research gap in this study. Semantic Analysis in codemixed text is a novel area of research. It is more complex to process codemixed text in Semantic Analysis based NLP tasks than monolingual text. A lot of research conducted on Semantic based NLP tasks in the monolingual context but in the codemixed field there remain more possibilities. Moreover, the preprocessing procedure is more challenging in codemixed based Semantic Analysis.

3. Methodologies

3.1. Word Embedding Based Semantic Tasks

Semantic analysis and word embeddings are closely related ideas in natural language processing (NLP). Word embeddings, which give words a means to be represented in a continuous vector space, are essential to semantic analysis tasks. Avasthi et al. [3] introduced a word embedding-based incremental topic model (ITMWE) that outperforms the LDA and DTM. Joshi et al. [20] proposed word embedding, Recurrent neural network (RNN) and Topic Modeling based extractive summarisation technique. The paper by Worth [24] reviews popular techniques like Latent Semantic Analysis, Vector Space Models, and word embeddings, including Word2Vec, GloVe, ELMo, and BERT, examining semantic spaces beyond NLP applicability. In the experiment, Petersen and Potts [25] employ the Word2Vec, GlowVe, and fastText algorithms to treat "Break." Here, transformer-based LLM is employed. RoBERTa big was employed in the first stage, and BERT and DeBERTa were used in the final step.

3.2. Other Approaches Based Semantic Tasks

Traditional Latent semantic analysis (LSA) faced difficulty to identify semantically similar word with different meaning in a sentence. Suleman et al. [26] presented extend xLSA to overcome this issue. In their research work Taghandiki et al. [27] proposed a text summarizer named Crawler which is capable of both extractive and abstractive type summarisation process and use Instagram post to evaluate the experiment. Crawler use textrank and lexrank algorithm for extractive summarisation and T5 and Bert algorithm for abstractive summarisation. Khatri et al. [28] evaluated extractive and abstractive summarisation techniques using Seq2Seq models based on document

context and RNNs. Khatun and Hoque [9] proposed a Lexical semantic analyzer for Bengali language to enhance the accuracy of machine translation.

3.3. Result and Discussion

A summary table is stated bellow to show the results, datasets, semantic tasks and method (Table 1).

Table 1: Summary of results, datasets, semantic tasks and method

Authors	Semantic tasks	Datasets	Method	Result (accuracy)
Khatun and Huque [9]	translation	Bengali	Semantic analyzer	Accuracy 88%
Iqbal et al. [10]	Semantic similarity	Bengali	Word embedding	Accuracy 77.28%
Hossain et al. [11]	Semantic, similarity	Bengali	GloVec	Accuracy 96.96%
Nipu et al. [12]	Stemer	Bengali	Latent semantic	Accuracy 89.76%
Hoque et al. [13]	Semantic analysis	Bengali	Context less language and a top-down method were utilised to generate an annotated parse tree with discrete semantic properties.	The test results indicate the successful outcomes of the majority of the test cases.
Chowdhuri (2017) [14]	Summarization	Bengali	LSA	Prececion 92.82% recall, 91.50% and F1score 92.15%.
Aswani& and Pawar (2017) [16]	topic model	Hindi English	Semantic based	The study found that cms-LDA greatly improves aspect coherence in topic clusters when compared to typical topic modeling methods.
Roy et al. [17]	Sentiment analysis	Indian language	Lexicon based	Results indicate significant potential for improving transliteration accuracies in the Hindi, Bangla, and Gujarati, languages.
Tareq et al. [18]	Sentiment analysis	Bengali English	FastText embedding	Accuracy 87%
Tatariya et al. [19]	NLP	Hindi-English, Malayalam-English, Dravidian codemix	transfer learning and fine tuning	When fine tuning a PLM on code-mixed data, the pertaining languages have little effect on performance.
Avasthi et al. [3]	ITMWE	CORD-19, NIPS papers	Word embedding	Outperform than LDA, DTA
Joshi et al. [20]	Extractive summarisation	CNN/Daily Mail data	word embedding, Recurrent neural network (RNN) and Topic Modeling	ROUGE-1, ROUGE-2, and ROUGE-L scores of 43.3, 19.0, and 38.9
Worth [24]	Semantic spaces beyond NLP applicability.	English	Latent Semantic Analysis, Vector Space Models, and word embeddings	The concept of semantic spaces in general, beyond its application to NLP.

(*Continued*)

Table 1: (*Continued*)

Peterson et al. (2023) [25]	Lexical semantic analysis to treat "Break"	English	Word2Vec, GlowVe, and fastText algorithms	LLMs offer a new perspective on traditional lexical semantics problems due to their high-dimensional, contextually modulated representations and dependence on usage-based data.
Suleman et al. [26]	Semantic similarity	English	LSA and xLSA	Similar sentence: LSA-.99and xLSA-.96 Inverse sentence: LSA-1 and xLSA-.38
Taghandiki et al. [27]	Crawler summarizer	English	Lexkrank and Textrank for extractive and T5, Bert for abstractive	Accuracy80%and precision 75%
Khatri et al. [28]	summarisation	Ebay dataset	Seq to seq and RNN	Wide evolution metric
Walia et al. [29]	Automatic answer scoring	Punjabi	LSTM	Jaccard Coefficient (JC), Dice Coefficient (DC), Cosine Coefficient for comparison
Mayopu et al. [30]	fake news detecting system	US election campaign 2016	LSA+SVD	The five concepts retrieved from LSA are found to be reflective of political fake news during the election in this study.

4. Conclusion

In this review paper, a wide range of review studies have been conducted on the linguistic platform and methodologies of semantic analysis and its NLP tasks. Moreover corresponding datasets and evaluation results are also studied. Codemixing is a rising research area in the field of semantic analysis because of its rapid growth and complex NLP processing. So, this study will help to find new research directions.

References

[1] Salloum, S. A., Khan, R. A., and Shaalan, K. (2020). A survey of semantic analysis approaches. https://doi.org/10.1007/978-3-030-44289-7_6

[2] Acosta, Lamaute, Luo, Finkelstein, and Andreea. (2017). Sentiment analysis of Twitter messages using Word2Vec. Proceedings of Student-faculty Research Day, CSIS, Pace University.

[3] Avasthi, S., Chauhan, R., and Acharjya, D. P. (2022). Extracting information and inferences from a large text corpus. https://doi.org/10.1007/s41870-022-01123-4

[4] Hernández-Castañeda, Á., García-Hernández, R. A., Ledeneva, Y., and Millán-Hernández, C. E. (2020). Extractive automatic text summarization based on lexical-semantic keywords. IEEE Access, 8, 49896-49907.

[5] Agirre, E., and Edmonds, P. (Eds.). (2007). Word sense disambiguation: Algorithms and applications (Vol. 33). Springer Science & Business Media.

[6] Li, Y., Chen, M., Yang, W., Wang, K., Ma, J., Bovik, A. C., and Zhang, Y. (2023). SAMScore: A semantic structural similarity metric for image translation evaluation. arXiv preprint arXiv:2305.15367.

[7] Mansouri, A., Affendey, L. S., and Mamat, A. (2008). Named entity recognition approaches. International Journal of Computer Science and Network Security, 8(2), 339-344.

[8] Evangelopoulos. (2013). Latent semantic analysis. WIREs Cognitive Science, 4, 683-692. https://doi.org/10.1002/wcs.1254

[9] Khatun, S., and Hoque, M. M. (2018). Semantic analysis of Bengali sentences. In 2018 International Conference on Bangla Speech and Language Processing (ICBSLP) (pp. 1-6). IEEE.

[10] Iqbal, M. A., Sharif, O., Hoque, M. M., and Sarker, I. H. (2021). Word embedding based textual semantic similarity measure in Bengali. Procedia Computer Science, 193, 92-101.

[11] Hossain, M. R., Hoque, M. M., Siddique, N., and Sarker, I. H. (2021). Bengali text document categorization based on very deep convolution neural network. Expert Systems with Applications, 184, 115394.

[12] Nipu, A. S., and Pal, U. (2017). A machine learning approach on latent semantic analysis for ambiguity checking on Bengali literature. In 2017

20th International Conference of Computer and Information Technology (ICCIT) (pp. 1-4). IEEE.

[13] Hoque, M. M., Rahman, M. J., and Kumar Dhar, P. (2007). Lexical semantics: A new approach to analyze the Bangla sentence with semantic features. In 2007 International Conference on Information and Communication Technology (pp. 87-91). IEEE

[14] Chowdhury, S. R., Sarkar, K., and Dam, S. (2017). An approach to generic Bengali text summarization using latent semantic analysis. In 2017 International Conference on Information Technology (ICIT) (pp. 11-16). IEEE.

[15] Chakraborty, T., and Bandyopadhyay, S. (2010). Identification of reduplication in Bengali corpus and their semantic analysis: A rule-based approach. In Proceedings of the 2010 Workshop on Multiword Expressions: From Theory to Applications (pp. 73-76).

[16] Asnani, K., and Pawar, J. D. (2017). Automatic aspect extraction using lexical semantic knowledge in code-mixed context. Procedia Computer Science, 112, 693-702.

[17] Roy, R. S., Choudhury, M., Majumder, P., and Agarwal, K. (2013). Overview of the fire 2013 track on transliterated search. In Proceedings of the 4th and 5th Annual Meetings of the Forum for Information Retrieval Evaluation (pp. 1-7).

[18] Tareq, M., Islam, M. F., Deb, S., Rahman, S., and Al Mahmud, A. (2023). Data-augmentation for Bangla–English code-mixed sentiment analysis: Enhancing cross linguistic contextual understanding. IEEE Access.

[19] Tatariya, K., Lent, H., and De Lhoneux, M. (2023). Transfer learning for code-mixed data: Do pre-training languages matter? In Proceedings of the 13th Workshop on Computational Approaches to Subjectivity, Sentiment, & Social Media Analysis (pp. 365-378).

[20] Joshi, A., Fidalgo, E., Alegre, E. and Fernández-Robles, L., 2023. DeepSumm: Exploiting topic models and sequence to sequence networks for extractive text summarization. Expert Systems with Applications, 211, p.118442.

[21] Chakravarthi, B. R., Muralidaran, V., Priyadharshini, R., and McCrae, J. P. (2020). Corpus creation for sentiment analysis in code-mixed Tamil–English text. arXiv preprint arXiv:2006.00206.

[22] Chakravarthi, B. R., Priyadharshini, R., Thavareesan, S., Chinnappa, D., Thenmozhi, D., Sherly, E., ... and Vasantharajan, C. (2021). Findings of the sentiment analysis of Dravidian languages in code-mixed text. arXiv preprint arXiv:2111.09811.

[23] Devlin, J., Chang, M. W., Lee, K., and Toutanova, K. (2018). Bert: Pre-training of deep bidirectional transformers for language understanding. arXiv preprint arXiv:1810.04805

[24] Worth, P. J. (2023). Word embeddings and semantic spaces in natural language processing. International Journal of Intelligence Science, 13(1), 1-21.

[25] Petersen, E., and Potts, C. (2023, January 1). Lexical Semantics with Large Language Models: A Case Study of English "break". In Findings of the Association for Computational Linguistics: EACL 2023 (pp. 490-511). https://doi.org/10.18653/v1/2023.findings-eacl.36

[26] Suleman, R. M., and Korkontzelos, I. (2021). Extending latent semantic analysis to manage its syntactic blindness. Expert Systems with Applications, 165, 114130.

[27] Taghandiki, K., Ahmadi, M. H., and Ehsan, E. R. (2023). Automatic summarisation of instagram social network posts combining semantic and statistical approaches. arXiv preprint arXiv:2303.07957.

[28] Khatri, C., Singh, G., and Parikh, N. (2018). Abstractive and extractive text summarization using document context vector and recurrent neural networks. arXiv preprint arXiv:1807.08000

[29] Walia, T. S., Josan, G. S., and Singh, A. (2019). An efficient automated answer scoring system for punjabi language. https://doi.org/10.1016/j.eij.2018.11.001

[30] Mayopu, R. G., Wang, Y. Y., and Chen, L. S. (2023). Analyzing online fake news using latent semantic analysis: Case of USA election campaign. https://doi.org/10.3390/bdcc7020081

[31] Liu, J., Chen, X., Feng, S., Wang, S., Ouyang, X., Sun, Y., ... & Su, W. (2020). Kk2018 at SemEval-2020 task 9: Adversarial training for code-mixing sentiment classification. arXiv preprint arXiv:2009.03673.

[32] Doddapaneni, S., Aralikatte, R., Ramesh, G., Goyal, S., Khapra, M. M., Kunchukuttan, A., and Kumar, P. (2022). Towards leaving no indic language behind: Building monolingual corpora, benchmark and models for indic languages. arXiv preprint arXiv:2212.05409.

[33] Khanuja, S., Bansal, D., Mehtani, S., Khosla, S., Dey, A., Gopalan, B., ... and Talukdar, P. (2021). Muril: Multilingual representations for Indian languages. arXiv preprint arXiv:2103.10730.

Moving Beyond Serverless Edge Computing

Serverless Fog Computing for the Future IoT

Sana Bharti, Shilpi Harnal, Rupali Gill

Chitkara University Institute of Engineering and Technology, Chitkara University, Punjab, India
E-mail: sana.bharti@chitkara.edu.in, shilpi13n@gmail.com, rupali.gill@chitkara.edu.in

Abstract

A lot of research has been done on integrating Cloud Computing and IoT which allows to offload of several applications to the fog or the edge layer to perform complex computations on the cloud because of its high computation and huge storage capacity. Fog and edge computing enable the storage and processing of applications near the data source as processing nodes are distributed across the network. Besides various advantages, it comes up with various challenges like scalability, load balancing, and maintaining a huge number of resources when they are not needed, etc. Serverless edge computing comes with an extension of serverless computing. But it also deals with various challenges like distribution of services, resource scheduling, permanent workload, and function scheduling which can be overcome with serverless fog computing. This work proposes a network architecture of serverless fog computing followed by a communication sequence to provide efficient distribution of services and function scheduling. This will improve the user experience.

Keywords: Serverless, latency-sensitive, fog computing, IoT, cloud computing, edge computing

1. Introduction: Fog Computing

Nowadays several objects and devices are connected with each other and communicate with the help of emerging technology known as The Internet of Things precisely known as IoT. The connection is established among the devices by sensors, actuators, and various wired and wireless protocols. Smart discovering utilizes IoT to efficiently process data from the surrounding environment, enhancing accuracy through the use of sensors and actuators, enhancing the efficiency of objects in their operations. For example, agriculture can be made smart as IoT devices monitor and optimize crop growth, soil conditions, and irrigation system, improving crop yields, reducing water usage, and increasing production [1, 2]. There are several other domains like healthcare [3], education, industry, and water pipes where IoT has been implemented to enhance the performance of information management systems [4].

IoT is crucial for innovative services, but hosting them directly can be challenging due to the vast amount of data collected by devices. These devices often have limited storage, computing, and networking resources, making them battery-powered. To address this, IoT relies on powerful paradigms and resources, such as Cloud Computing (CC) [5, 6]. CC allows users to use computing resources in the form of services through pay-per-use providers, combining virtualization and Service Oriented Architecture (SOA) techniques. These services are provided by applications and systems' software in a data center, allowing for easy modification and cost-effective service provision [7]. Cloud Computing gives a way to process, store and analyze huge amounts of data, and due to exponential growth in data and moreover, resources that are available in Data Centres whose distance has also increased for many producers and consumers several upcoming issues have been arising like hostile environment, bandwidth consumption, privacy, and security, latency, and, context awareness [8] These limitations lead to new paradigms Fog Computing and Edge Computing [4]. The benefits of cloud computing are preserved by fog computing, which are virtualization, elasticity, and scalability [4]. Fog is nothing but an extension of Cloud capabilities towards the network, distribution of resources, and various services: computing, storage, database, and application along with the Cloud-to-Things continuum which brings close to tropological proximity to IoT devices. Due to the fact that network edge devices are distributed geographically, fog computing architecture deploys several fog nodes between end devices and the cloud. Network transmission and data transfer time have been reduced [9]. Fog technology thus enables applications to run in real-time or with low latency, eliminating bandwidth consumption issues [9].

One of the promising applications of fog computing is in the field of artificial intelligence, particularly in deep learning (DL) algorithms [10]. Fog Computing is composed of various fog nodes and these nodes are network edge devices (NED), management systems within edge devices, and various cloud data centers. Fog Computing provides a connection between end users and the cloud whereas fog nodes provide the connection between the users and the edge devices through wireless connection modes: Zigbee, Bluetooth low energy, Wifi, 5G, etc to provide computation and storage services. Further fog nodes are connected to the cloud with help of the Internet to make use of its full potential of computation and storage resources of cloud [11]. Hence Fog can fulfill QoS for the applications to improve performance.

1.1. Architecture of Fog Computing

The standard approach to cloud computing is extended by fog computing, which brings services to the network's edge.

It has all the capabilities of the Cloud but is at the edge of the network. This decentralized platform differs from previous traditional approaches in several ways. In Figure 1 we discuss the 3 Layer hierarchical architecture.

1. **Layer 1-Terminal Layer:** The bottom layer comprises IoT devices closest to end users, including sensors, mobile phones, and smart homes. Smartwatches and vehicles are treated as sensing devices, with IoT devices gathering data through sensors and transmitting it to higher levels for further processing [4].

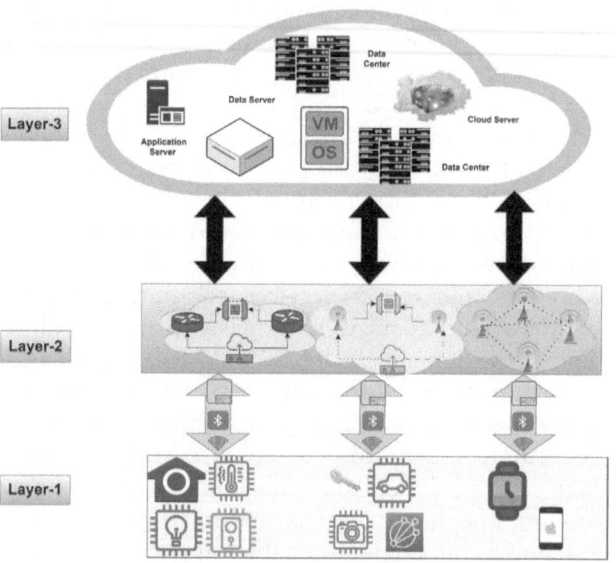

Figure 1: The three-layer fog architecture

2. **Layer 2-Fog Layer:** This layer has a large number of fog nodes which, consist of routers, gateways, switches, access points, and servers, and are geographically distributed between end devices and the cloud [4]. They can be static or moving. End and IoT devices send data through Bluetooth or WiFi to fog nodes, which compute, transmit, and store the data temporarily. This allows for real-time analysis for time-sensitive applications. The fog layer is connected to cloud data centers via the internet, enhancing computation and storage capabilities [4].

3. **Layer 3-Cloud Layer:** The cloud layer has high-performance data centers and servers, offering powerful computing and storage capabilities. It permanently stores large amounts of data. Not all computation and storage requests are moved to the cloud layer, depending on application sensitivity. Cloud modules are effectively managed and scheduled, improving resource utilization and enhancing cloud resource utilization [4].

1.2. Characteristics of Fog Computing

Fog Computing's primary goal is to provide processing, storage, and calculation close to the end user. This paper lists some of the characteristics of fog computing.

- **Low-Latency:** Due to the ubiquitous range of smart devices, data transfers to fog nodes at the network edge and is processed locally, reducing the amount of data movement across the Internet [11] and fast and accurate results. This reduces latency and is very essential for latency-sensitive and critical applications [11].

- **Uninterrupted and fast computational services:** Fog layer offers storage, network, and computation services, similar to cloud-like platforms. Processes are performed on its own servers, near end users, ensuring uninterrupted services with zero cloud connection.

- **Heterogeneity and Interoperability:** Fog Computing allows for heterogeneity, with fog nodes created by different providers deployed on any platform. These nodes interoperate with providers to access a wide range of services. Interoperability is achieved through techniques like standardization, middleware, APIs, edge gateways, and data virtualization, ensuring resources and services can be accessed by various devices and applications. For example, standardization is a technique for developing standardized communication protocols, data formats, and APIs that is essential to achieving interoperability in fog computing. The Open Fog Consortium has developed a standard architecture for fog computing that includes standard interfaces and APIs for communication between fog nodes.

- **Supports Mobility:** The fog layer allows computing resources, services, and fog nodes to move seamlessly across different locations and networks. This is particularly useful for mobile devices like smartwatches and vehicles, as well as static devices like traffic cameras. Fog nodes can be static or mobile, deployed at home, malls, airports, and on mobile devices. Communication between mobile devices and users can be done through routing protocols, such as LISP (Locator/ID Separation Protocol), which creates two addresses for device identity and location in the network [9, 12].

- **Geographically Distributed and Managing Network Bandwidth:** Traditional cloud computing models faced a gap between end devices and cloud services, leading to heavy traffic due to a large number of devices and requests. Fog in the intermediate layer addresses this issue by allowing data traffic and tasks to be done locally in a decentralized fashion. With a large number of widely distributed nodes, fog computing supports a decentralized architecture, allowing data to be closer to end users and analyzed faster. This geographical distribution manages network bandwidth and improves the efficiency of cloud services [11].

1.3. Limitations of Fog

Having overcome the greatest challenges being faced by Cloud still, the fog has to face several challenges:

- **Cost Efficiency:** Users must pay for the resources they have purchased but not the resources they are consuming.

- **Resource Management:** To manage different resources across the layers like servers, edges, and gateways is quite difficult to manage, however, the balancing of load and tedious scalability task is much more difficult. [13, 14]

- **Deployment Flexibility:** Fog Computing requires the deployment of specialised hardware and software on the edge nodes which can limit the flexibility and portability of the applications.

- **Pre-allocation:** The underutilisation of the edge resources is caused by the pre-allocation approach. [13]

The research paradigm of this paper is from 2012-2022 To enhance the capabilities of fog computing it is combined with serverless. It has proved that the best fit for IOT applications is combining fog infrastructure with serverless. The following sections describe

- **Serverless Computing over Virtual Machines:** This explains why there is a shift from a traditional cloud computing approach known as virtualisation to serverless and explains serverless computing.

- Serverless Edge Computing, which describes several studies done on the serverless edge and its limitations

- Serverless Meets Fog describes serverless fog computing and how it surpasses the various serverless edge computing drawbacks and proposes a network architecture of serverless fog computing followed by a communication sequence and its advantages.

2. Serverless Computing Over Virtual Machines

Cloud computing is a powerful computing technology that gives on-demand services in terms of storage computing, databases, applications, etc., through the internet via Cloud Service Providers. Earlier, this concept was used to attain virtualization. Virtualization is a technique to create different virtual machines with their own operating systems by emulating or virtualizing the hardware, as shown in Figure 2. Therefore, there was a huge overhead for the hypervisor, which maintained a large number of virtual machines, so the concept of containers came about. Instead of virtualizing the hardware containers, virtualize the operating system. The platforms provided by containers are enough to hold the resources required to operate a given application [15]. When compared to VMs, it achieved a better level of resource abstraction and resource as shown in Figure 2 [16]. Servers were inhibiting further enhancements, which leads to the concept of serverless, as shown in Figure 2. Serverless Computing has become a compelling approach for hosting applications on the Cloud [17]. It is also known as function(s) As a service that helps in dealing with these challenges by providing the platform to the end users or developers and reducing hosting costs as they have to pay for the resources they are consuming, high availability, fault tolerance, and on-demand elasticity for hosting individual functions, also known as micro-services. Many cloud service providers provide platforms to perform serverless computing, like AWS Lambda by Amazon, Google Cloud Functions by Google, Apache Open Whisk by IBM and Adobe, Microsoft Azure Serverless by Microsoft Azure, and many more where large applications are split into functions and are executed when the event is triggered. These functions are deployed in containers, e.g., Docker, and further operational tasks like scaling or provisioning are left to the providers. The advantage is that developers only implement functions without taking care of infrastructure and resources. Users pay when the function is triggered, which thereby enhances the use of fog and edge resources. Later, Lamba@ Edge was introduced by AWS, enabling application

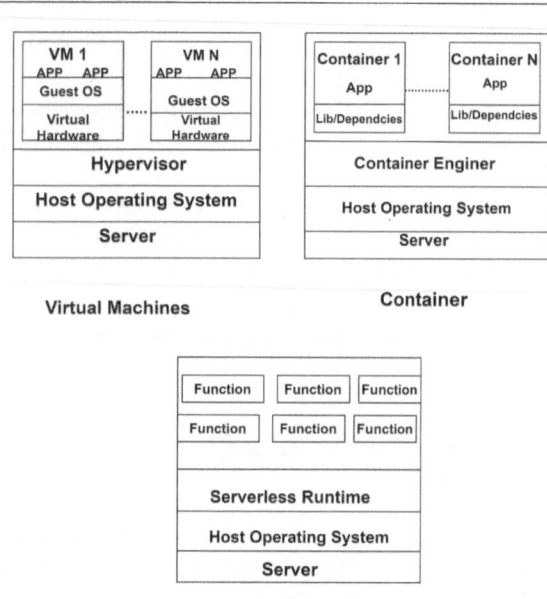

Figure 2: Serverless over virtual machines

developers to deploy constrained and compact lambda functions on edge nodes. Additionally, AWS launched "Greengrass," which unifies programming models for IoT and Lambda services [18]. Lloyd et al. [17] explain how the following factors influence microservice performance, i.e.:memory reservations, infrastructure retention, provisioning variation, and elasticity. It also compares the performance and infrastructure management of AWS Lambda and Azure functions.

3. Serverless Edge Computing

Motivated by serverless computing, then academicians and researchers combined serverless with edge computing.

Xie et al. [13] proposes a network and layered architecture of serverless edge computing platforms and highlights some of the technical issues like service discovery, lifecycle management, awareness of resources, scheduling of service, service deployment, etc, that come across serverless edge computing. Yin et al. [19] investigate task scheduling problems by using containers in the fog layer, and propose a container-based task scheduling algorithm for delay-sensitive tasks in two steps. First, check whether the task will be accepted or not, and second, depending on the accepted task algorithm schedule accepted tasks either to the fog node or to the cloud. Baresi and Mendonca [20] discuss serverless edge computing, discuss the different applications and implementations on different serverless computing platforms, and evaluate in terms of memory, footprint, latency, throughput, and scalability.

Through various studies, certain limitations of serverless edge have come across which are discussed in the section below.

3.1. Limitations of Serverless Edge Computing

Several researchers have discussed and proposed different architectures of serverless edge computing which leads to some drawbacks:

- **Service Distribution:** Serverless edge computing requires stateless services to edge computing nodes, which are initially deployed on random nodes. Popular services are based on popularity and service size. Unpopular services, due to large image sizes, are not deployed on edge and move directly to the cloud, increasing latency time for critical sensitive functions. To overcome cold start, functions must go through processes like image pulling, service scheduling, and container initialization, which takes deciseconds. Overcoming cold start can be achieved by keeping the function warm, using a warm cache, and optimizing code.

- **Permanent Workload:** IoT devices and edge devices are increasingly being used in serverless edge computing, a network that offloads the processing of applications. This approach reduces time and latency by processing functions near the source, increasing application dependability and reducing failure risk. This involves streamlining code, minimizing data volume, and ranking jobs in importance. Edge-to-edge communication protocols can improve data flow between devices. However, constant function calls do not improve serverless cost-efficiency, but rather decrease it [21].

- **Function Scheduling:** The major goal of serverless edge computing is to prioritize scheduling the functions based on their granularity and quality requirements in order to better meet the needs of the clients while there are a large number of requests coming in. With limited processor and memory capabilities, edge devices may find it difficult to distribute resources efficiently. As a result of unpredictable network conditions between edge devices and centralized clouds, scheduling can be more challenging.

These challenges can be addressed by Serverless Fog Computing which is discussed below.

4. Serverless Meets Fog

Serverless edge computing is only feasible in situations where real-time processing is crucial, and edge devices have limited processing power and storage capacity. But as the population and IoT market are increasing rapidly, there will be a requirement for more storage and power to handle real-time and latency-sensitive data and process it. To fill this gap, serverless fog computing will become the best solution. In the following sections, will discuss serverless fog computing from a networking viewpoint. First, will discuss design principles to implement serverless fog computing. This paper suggests a network architecture from the viewpoint of networking that adheres to the design principles of serverless fog computing.

Moreover, we have discussed the communication sequence and benefits of the proposed architecture

4.1. Serverless Fog Computing

Serverless fog computing is a distributed computing paradigm that combines serverless computing with fog computing. Applications are installed on edge hardware, such as gateways and sensors, rather than directly on sensor-equipped devices, a gateway, or a centralized server as in serverless edge. This enables more rapid data processing and analysis while consuming less network traffic. Additionally, the cloud service providers have the responsibility for maintaining the underlying infrastructure, developers can concentrate on creating the application logic without worrying about scaling, provisioning, and managing servers.

The other reason to shift from serverless edge to serverless fog is the location of the computation. The location of the computation means determining the location of the edge devices and their data through various technologies like GPS, WiFi triangulation, Bluetooth beacons, etc. Utilizing this data will allow edge apps to run faster and more responsively while also distributing compute resources as efficiently as possible. However performing the location of computation at the edge can be challenging due to limited computing resources, network latency, privacy concerns, maintenance issues, and scalability issues. Additionally, serverless fog computing enables dynamic scaling of compute resources according to demand. The computation can be automatically moved to other network nodes that are accessible if a fog node is overloaded. This guarantees that the computation is carried out effectively and with the fewest possible interruptions.

Serverless fog computing, a novel approach in computing, combines cloud-based processing with local edge device processing, enabling real-time data processing at the edge. This strategy has potential applications in IoT, smart cities, and autonomous cars.

4.2. Design Principles

Serverless fog computing offers a scalable and cost-effective method for processing data at the network edge, blending serverless computing with fog computing. The "5F networks" framework illustrates the design tenets of this computing paradigm, including several main design tenets.

- **Flexibility:** Systems for serverless fog computing should be versatile and flexible enough to handle a variety of network situations, including shifting traffic patterns, device connections, and data sources. A dynamic, distributed design that enables resource additions and removals as needed can be used to accomplish this.

- **Fault Tolerance:** Serverless fog computing systems must be resilient to faults and failures, such as hardware malfunctions, software bugs, and network outages. Redundancy and failover techniques, such as distributed data stores and load balancing, can be used to achieve this.

- **Frictionless Data Movement:** Serverless fog computing systems can facilitate efficient data transfer between network nodes and devices, reducing latency and bandwidth costs through caching mechanisms and efficient protocols.

- **Federated Analytics:** The execution of distributed analytics and machine learning algorithms across several network nodes and devices should be possible with serverless fog computing platforms. This can be done by utilizing federated learning or distributed computing frameworks, which enable the training and execution of models and algorithms on dispersed data sources [22].

- **Fine-grained data Processing:** Systems for serverless fog computing should be created to process data at the network edge effectively and efficiently. This can be done by utilizing serverless functions or other lightweight data processing components that can be deployed and used as needed.

4.3. Network Architecture

On the basis of the design principles of the 5F-Network framework. There are 4 layers, as shown in Figure 3, i.e., Presentation Layer, Serverless Edge Network, Serverless Fog Network, and Cloud Data Center.

- **Presentation Layer:** This layer in the architecture deals with the end users, and how they interact with end devices. End devices can either be smart mobile devices, homes, wearables, transportation, etc. The requests directly go to the main controller.

- **Main Controller:** To support a serverless fog computing network, the Main Controller offers global network control and resource orchestration capabilities. Its control abilities help to perform global management of centralized serverless fog systems. The main controller also performs unified orchestration and managing heterogeneous and widespread edge/fog resources. Due to its controlling and unified orchestrating nature, The main Controller is able to decide which serverless layer is able to handle the request.

- **Serverless Edge Network:** The network edge is a crucial area for edge computing, where data processing and storage are performed closer to the source of data, enhancing performance and reducing latency. The Main Controller sends services to the edge layer, including content delivery, augmented and virtual reality, autonomous vehicles, and gaming. A serverless edge network consists of numerous dispersed edge computing nodes and an edge controller, with the edge controller having similar functionalities as the Main Controller but limited to its network. Based on the function as a service (FaaS) programming model, each serverless edge computing node executes serverless functions in order to fulfill the serverless requests generated by users [13]. An executor in this serverless framework is an application that responds to user-triggered functions. Developers focus on implementing user-triggered functions, reducing server management and bandwidth utilization [13].

- **Serverless Fog Network:** The serverless fog layer is an extension of the edge layer, enabling complex services and applications to be executed near users and data sources, offering the benefits of cloud computing. It is used for applications like Real-Time Analytics, Smart Grid, Healthcare, and Smart Cities, requiring high processing power, storage, connectivity, security, and standardization.

- **The Cloud Data Center:** Delays in tasks in remote areas can be mitigated through virtualization in the cloud, which uses fog/edge computing paradigms to perform tasks closer to the user, thereby reducing costs and ensuring efficient use of resources in remote areas. Cloud also follows serverless computing services and provides resources to the edge/fog layer as needed.

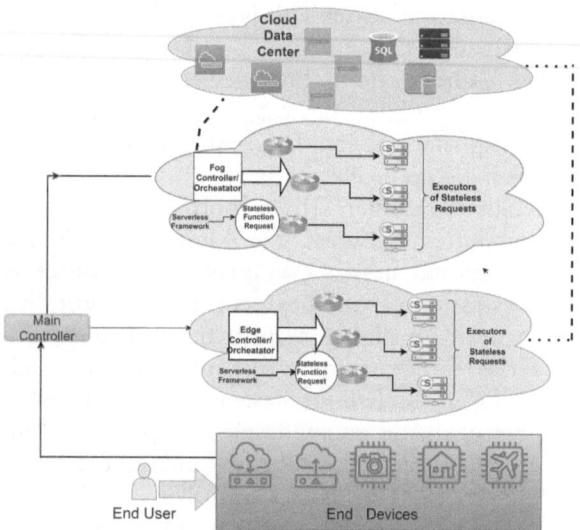

Figure 3: Architecture of serverless fog computing

4.4. Communication Sequence

IoT devices gather a huge amount of heterogeneous data through sensors, actuators, etc., which is required to analyze, process, and store for more accuracy and deeper understanding. Figure 4 In this section, we will take IoT applications to understand the communication sequence of the proposed serverless fog architecture as shown in 4 In the case of the smart city, the user triggers functions like air quality index, parking status, sending alert messages like fire alert, medical assistance, synchronizing traffic signals, coordinating lighting systems, or managing energy consumption which are in the IoT application of citizens and administration, which is detected by the sensors. The request from the HTTP server has been forwarded to the Main Controller for centralized scheduling. The Main Controller is the architecture's "nerve center," responsible for locating and fetching functions and approving their execution. It consists of two parts: decision-making and predictor. The request is processed by the decision-making engine, which extracts relevant information and sends it to the predictor (Figure 4). for example functions like air quality index, parking status, sending alert messages like fire alert, medical assistance performed by edge layer, and functions like synchronizing traffic signals, coordinating lighting systems, or managing energy consumption performed by fog layer. The decision-making process involves selecting the appropriate layer and sending requests to either the fog or edge layer. Both layers have controllers scheduling requests, predicting executor availability based on cold and warm times. The edge/fog controller then selects the executor to run the request, which isolates functions in a container managed by a serverless provider. Coordinated scheduling for IoT applications is efficient. Each serverless computing node's Storage component stores results locally and logs data after intelligent analysis to enable asynchronous result feedback.

4.5. Benefits of Serverless Fog Network

In this section, we will discuss some of the advantages of the proposed serverless fog architecture.

- **Efficient Distribution of Functions:** The increasing workload of IoT applications can decrease the efficiency of latency-sensitive applications and increase the load in cloud data centers. Serverless fog computing, which uses a network of sensors and fog nodes to gather traffic patterns and environmental conditions, can be used in smart city applications. The data is subsequently processed and examined in real-time utilizing a fog device-deployed serverless computing platform, such as AWS Lambda. The Main Controller takes charge

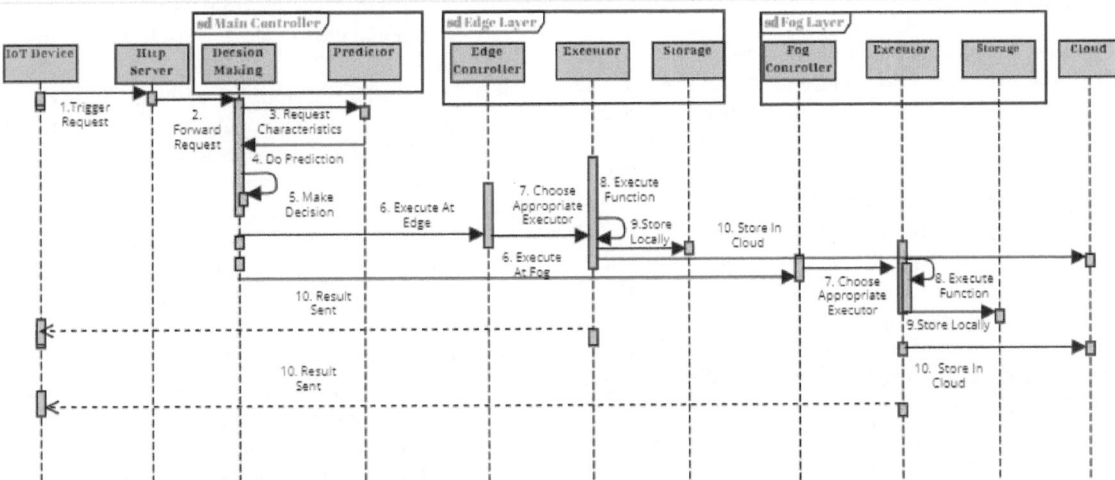

Figure 4: Sequence diagram of serverless fog network architecture

of solving the function request near the user and decides which layer is capable of solving the request, either the fog or edge layer. The fog/edge controller decides which node will execute the request. This leads to solving multiple requests at different layers and nodes simultaneously, thus increasing performance and reliability.

- **Scheduling of Functions:** Function scheduling in serverless fog computing aims to maximize performance, minimize resource utilization, and reduce delay. Hybrid scheduling, a centralized and decentralized approach, integrates a main controller to coordinate task scheduling across the fog/edge layer, while edge/fog controllers make local scheduling decisions based on workload and resource availability, ensuring efficient use of nodes. Function scheduling is a crucial aspect of serverless fog computing, enhancing performance and reducing latency in edge computing settings, enabling organizations to create efficient fog computing applications tailored to their specific business needs.

- **Fog Artificial Intelligence:** Organizations may create highly scalable and effective AI systems that analyze data in real-time by using serverless computing on the fog layer, eliminating the need to send huge volumes of data to the cloud for processing [23].

Serverless computing provides a scalable and effective framework for implementing AI models on fog nodes, eliminating the need for manual infrastructure configuration and maintenance [24]. AI frameworks like TensorFlow, PyTorch, and MXNet support serverless computing platforms like AWS Lambda, simplifying setup and configuration [25]. This allows developers to create and deploy AI-powered applications at the network edge with minimal infrastructure

administration, offering faster reaction times, lower latency, and cheaper prices.

5. Conclusion

The trend of shifting from virtual machines to serverless computing in the cloud has been driven by the desire to achieve scalability, cost efficiency, and ease of use. Many providers provide a platform for Serverless Computing like AWS Lambda, Microsoft Azure Function, etc., which allows developers to focus on writing code while everything else, including scalability, infrastructure management, and load balancing, is taken care of by the providers. Hence, it is well suited to fog computing, which is driven by IoT applications. Edge/fog computing, a combination of cloud computing and IoT, revolutionize computing technology by enabling data processing, analysis, and storage closer to the source, enabling real-time applications and faster decision-making. Serverless edge computing processes data on sensor-equipped devices or gateway devices, addressing limited storage and processing capabilities in IoT or mobile applications. However, serverless fog computing addresses this by dispersing computation across a network of fog nodes near edge devices. This approach can lower network latency, improve speed, and expand scalability by spreading processing across a network of fog nodes.

This paper proposes a network architecture and communication of serverless fog computing to overcome the challenges in serverless edge computing. As the trend of making IoT applications increases rapidly, more data will be generated, which might hamper the speed of real-time applications that need higher computing resources. In future clustering of fog, nodes can be added to enhance speed and give accurate results for real-time applications.

References

[1] Bonkra, A., Noonia, A. and Kaur, A. (2022). Apple leaf diseases detection system: A review of the different segmentation and deep learning methods. In Artificial Intelligence and Data Science: First International Conference, ICAIDS 2021, Hyderabad, India, December 17–18, 2021, Revised Selected Papers. Springer, pp. 263–278.

[2] Bonkra, A., Bhatt, P. K., Kaur, A and Kamboj, S. (2023) Scientific landscape and the road ahead for deep learning: Apple leaves disease detection. in 2023 International Conference on Artificial Intelligence and Smart Communication (AISC). IEEE, 2023, pp. 869–873.

[3] Bharti, S. and Singh, S. N. (2015). Analytical study of heart disease prediction comparing with different algorithms. In International Conference on Computing, Communication & Automation (pp. 78–82). IEEE.

[4] Puliafito, C., Mingozzi, E., Longo, F., Puliafito, A. and Rana, O. (2019). Fog computing for the internet of things: A survey. ACM Transactions on Internet Technology (TOIT), 19(2), 1–41.

[5] Bittencourt, L., Immich, R., Sakellariou, R., Fonseca, N., Madeira, E., Curado, M., Villas, L., DaSilva, L., Lee, C. and Rana, O. (2018). The internet of things, fog and cloud continuum: Integration and challenges. Internet of Things, 3, 134–155.

[6] Bonkra, A. and Dhiman, P. (2021). IoT security challenges in cloud environment. In 2021 2nd International Conference on Computational Methods in Science & Technology (ICCMST) (pp. 30–34). IEEE.

[7] Mijuskovic, A., Chiumento, A., Bemthuis, R., Aldea, A. and Havinga, P. (2021). Resource management techniques for cloud/fog and edge computing: An evaluation framework and classification. Sensors, 21(5), 1832.

[8] Harnal, S. and Chauhan, R. (2020). Efficient and flexible role-based access control (EFRBAC) mechanism for cloud. EAI Endorsed Transactions on Scalable Information Systems, 7(26), e1.

[9] Kumari, S. and Singh, S. et al. (2017). Fog computing: Characteristics and challenges. International Journal of Emerging Trends & Technology in Computer Science, 6(2), 113–117.

[10] Dhiman, P. and Kaur, A., and Bonkra, A., (2023). Fake information detection using deep learning methods: A survey (pp. 858–863). In 2023 International Conference on Artificial Intelligence and Smart Communication (AISC). IEEE.

[11] Hu, P., Dhelim, S., Ning, H. and Qiu, T. (2017). Survey on fog computing: Architecture, key technologies, applications and open issues. Journal of Network and Computer Applications, 98, 27–42.

[12] Kaur, R., Ramachandran, R. K., Doss, R. and Pan, L. (2021). The importance of selecting clustering parameters in vanets: A survey. Computer Science Review, 40, 100392.

[13] Xie, R., Tang, Q., Qiao, S., Zhu, H., Yu, F. R. and Huang, T. (2021). When serverless computing meets edge computing: Architecture, challenges, and open issues. IEEE Wireless Communications, 28(5), 126–133.

[14] Harnal, S., Sharma, G., Seth, N. and Mishra, R. D. (2022). Load balancing in fog computing using QoS. Energy Conservation Solutions for Fog-Edge Computing Paradigms, pp. 147–172.

[15] Gadepalli, P. K., G. Peach, L. Cherkasova, R. Aitken and G. Parmer (2019). Challenges and opportunities for efficient serverless computing at the edge. In 2019 38th Symposium on Reliable Distributed Systems (SRDS) (pp. 261–2615). IEEE.

[16] Rajan, R. A. P. (2018). Serverless architecture—A revolution in cloud computing. In 2018 Tenth International Conference on Advanced Computing (ICoAC) (pp. 88–93). IEEE.

[17] W. Lloyd, S. Ramesh, S. Chinthalapati, L. Ly and S. Pallickara (2018). Serverless computing: An investigation of factors influencing microservice performance. In 2018 IEEE international conference on cloud engineering (IC2E) (pp. 159–169). IEEE.

[18] G. McGrath and Brenner, P. R. (2017). Serverless computing: Design, implementation, and performance. In 2017 IEEE 37th International Conference on Distributed Computing Systems Workshops (ICDCSW) (pp. 405–410). IEEE.

[19] Yin, L., Luo, J. and Luo, H. (2018). Tasks scheduling and resource allocation in fog computing based on containers for smart manufacturing. IEEE Transactions on Industrial Informatics, 14(10), 4712–4721.

[20] Baresi, L. and Mendonca, D. F. (2019). Towards a serverless platform for edge computing. In 2019 IEEE International Conference on Fog Computing (ICFC) (pp. 1–10). IEEE.

[21] Aslanpour, M. S., Toosi, A. N., C. Cicconetti, B. Javadi, P. Sbarski, Taibi, M. Assuncao, Gill, S. S., R. Gaire and S. Dustdar (2021). Serverless edge computing: vision and challenges. In 2021 Australasian Computer Science Week Multiconference (pp. 1–10).

[22] Hooda, S., Lamba, V. and Kaur, A. (2021). AI and soft computing techniques for securing cloud and edge computing: A systematic review. In 2021 5th International Conference on Information Systems and Computer Networks (ISCON) (pp. 1–5). IEEE.

[23] Kaur, G., Kaur, A., Khurana, M. et al. (2022). A review of opinion mining techniques. ECS Transactions, 107(1), 10125.

[24] Pratibha, G. Kaur, Kaur, A. and Khurana, M. (2022). A stem to stern sentiment analysis emotion detection. In 2022 10th International Conference on Reliability, Infocom Technologies and Optimization (Trends and Future Directions) (ICRITO) (pp. 1–5). IEEE.

[25] P. Dhiman, A. Kaur, C. Iwendi and Mohan, S. K. (2023). A scientometric analysis of deep learning approaches for detecting fake news. Electronics, 12(4), 948.

Network Security and the Internet of Things, a Survey on Securing the Future

Nayani Jindal[1], Pooja Gupta[2]

[1]Computer Science Engineering [IoT and IS], Manipal University Jaipur, Jaipur, India
[2]Department of Computer and Communication Engineering, Manipal University Jaipur, Jaipur, India
E-mail: [1]nayani.229311253@muj.manipal.edu , [2]pooja.gupta@jaipur.manipal.edu

Abstract

The Internet of Things (IoT) is propelling the world towards an era of universality where computing and networking devices will transcend humans as the major controllers of internet traffic. In the contemporary day and age, the emergence of IoT will imply consistent data exchange between users, computing devices and all 'things' that can sense and actuate. The ever-increasing applications of IoT expedites the need to ensure security and reliability in IoT communication systems. Securing IoT with its constrained devices and the sheer number of connected systems is a mammoth task that requires analysis of both old and new threats. This survey aims to provide an overview of the key concepts, challenges, and solutions related to securing IoT.

Keywords: Internet of Things, IoT architecture, network security, blockchain, machine learning, hardware security

1. Introduction

The growth of devices connected to the internet has quickly become an unprecedented phenomenon. An International Data Corporation (IDC) forecasts expects the connected devices to reach almost 50 billion by 2025, would imply a data generation of more than 80 zettabytes (ZB). This already puts IoT technologies at the forefront of the digital economy. The future prospects of IoT however will demand for an intricately designed internet that is secure, efficient and private. The repercussions of inadequate IoT network security extend far beyond data breaches; they encompass disruptions to services, violations of user privacy, and even the destabilisation of essential infrastructure [1].

To provide effective and efficient strategies in combatting the daunting security challenges that will inevitably follow the exponential growth of IoT in the coming future, we must first understand the key threats to IoT network security.

The low powered and not fully secure nature of IoT devices makes them easy to exploit and use for unethical practices. Poor authentication mechanisms may lead to unauthorised access to IoT devices which in turn put the user's data in jeopardy. Denial of service attacks as well as ransomware attacks are an immediate risk to the IoT network.

The heterogeneity of IoT as well as the presence of highly constrained devices also means that the security of the network can be challenged by physical tampering of devices [2,3], Maintaining consistent security standard across devices that come from vastly different vendors and operate on disparate platforms is thus a formidable undertaking that requires deep understanding of the IoT architecture and the threats to the structural integrity of the IoT environment.

Therefore, it is absolutely imperative that we adopt a proactive security mindset that will help unlock the true potential of the IoT network and create a network that is truly the next era of communication. This paper delves into the intricacies of IoT network security as well as the source of threats in IoT applications. This literature survey aims to emphasise the architecture of the IoT network and the key challenges in its development.

2. IoT Architectre and Key Security Threats

The most basic model of the IoT architecture is a three-layered model comprising of the perception layer, the network layer and the application layer [4]. However, IoT researchers have introduced a five layered model that is made of the perception layer, network layer, middleware layer and finally the business layer. This model addresses the issues of IoT like scalability, interoperability, reliability etc. more effectively [5]. Each layer serves a specific role in the functioning of IoT systems, contributing to the collection, processing, and utilisation of data for various applications. the security mechanisms for the traditional Internet protocols need to be altered or expanded to support the IoT applications. In this section, we elucidate the five layers of the IoT environment and the key security problems in different layers of IoT systems.

2.1. Perception Layer

The perception layer is responsible for collection and processing of information which is achieved by using physical sensors and other devices. The devices in this layer are equipped with sensors, actuators and data collection capabilities which have different functionalities like acceleration, vibration, humidity, weight, motion etc. [4] the layer is responsible for communication with the physical world and collecting real world data. The perception layer has a wide range of IoT enabling technologies like WSNs, RFIDs etc.

The major security threats at this layer include Node Capturing attacks (where a sensing or actuating node might be replaced by a malicious node) [2], Malicious Code Injection attacks, False data injection attacks, Interference attacks, Battery draining attacks as well as communication attacks like Eavesdropping attacks.[1]

2.2. Network Layer

The Network layer is responsible for enabling the flow of data across the IoT systems. The fundamental function of this layer is to transmit data received from the nodes to the computational units of the IoT environment. The transmission medium can be wired or wireless and can involve technologies like 4G, 5G, ZigBee, UMTS, Bluetooth etc. but that is entirely governed by the type of sensor devices or nodes used.

The major security threats at this layer include Phishing Site attacks, DDoS/DoS(Denial of Service) attacks and Attacks during Routing applications continuously transmit and receive sensitive and private data, therefore the Network layer is dangerously vulnerable to malicious attacks.[2]

2.3. Middleware Layer

The middleware layer in IoT systems is responsible for service management for the system. It is the link between the network layer and the application layer. Service management can bring into action, an efficient and reliable way to provide services [6]. This layer allows for smooth communication between heterogeneous objects with consideration for any specific hardware platforms [4].

The major attacks this layer is vulnerable to are attacks on the MQTT protocol in which the

adversary controls the MQTT broker, malware injection attack on the cloud and a flooding attack which acts like the DoS attack but floods the cloud which causes it to malfunction [1].

2.4. Application Layer

The application layer of the IoT environment is responsible for user interaction. It renders services that are called by the user [4] and it is also responsible for analysing the collected data to provide related services and insights. The IoT environment provides for a number of applications in the smart home, smart healthcare, intelligent farming, etc.[5].

The main security threats in this layer are quite specific ranging but not limited to data thefts and privacy disruption. Access control attacks, malicious code injection attacks and reprogramming attacks are major security threats. Service interruption attacks have also been notoriously popular, in these attacks the users are prevented from using the IoT applications by making the network swamped with service requests artificially [1].

2.5. Business Layer

The business layer is in charge of controlling the overall management of the IoT systems. The layer is responsible for visualising the data collected from the previous layers into graphs, flowcharts etc [4]. This layer is responsible for enhancing the user experience and ensuring that only the most accurate and relevant services are provided to the users. This layer has high computational needs and therefore is hosted on powerful devices.

The business layer is responsible for creating business value from the application layer IoT data

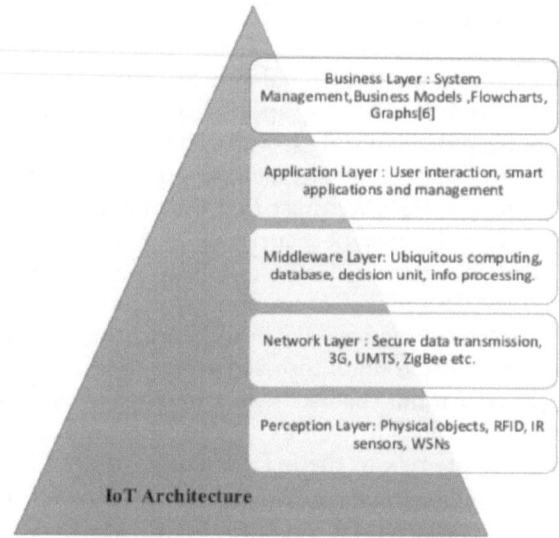

Figure 1: The standard IoT architecture

and therefore the major security risks are related to the misuse of user data and monetisation risks due to data breaches and/or data manipulation (Figure 1).

3. Current Solutions for Security Concerns in IoT Environments

The IoT network is a layered structure in which each layer works hand in hand with the other. This means that attack on even one layer of the system may cause a total collapse of the environment [2]. A standardised and structured framework for end to end IoT is therefore the need of the hour.

The IoT environment finds a multitude of applications in areas where a compromise to the integrity of the IoT network may have fatal consequences. For e.g. gaining control of the computing system of a user's smart vehicle may cause life threatening situations or it may be used as a method of threatening IoT users for sensitive data. Hence, Rigorous development and improvements in the present IoT structure and environment become obligatory to ensure that our future is safe, secure and abundant.

Some of the features of a secure IoT environment will include features like:

- *Scalability*: The IoT applications must be carefully designed and implemented to allow for future scope of large scale extension [4].Ensuring security of a large number of heterogeneous devices while maintain consistency in terms of performance and updates is a daunting yet necessary task for the enhancement of IoT.
- *Interoperability*: All security measures must be able to interact across the IoT ecosystem which is built up by a large and diverse range of devices, communication protocols and manufacturers [4].
- *Resource Constraint*: The growth of IoT devices and protocols is rapid, therefore all the security approaches should be designed in a way that can handle the capacity and resource constraints which will follow shortly [1], there is already a huge problem of address crunching that is expected to get worse even after adopting the IPv6 model. Constrained processing power, memory, and energy resources restrict the adoption of robust encryption and authentication methods, leaving devices vulnerable to attacks.
- *Authentication*: Authentication stands as a cornerstone of IoT security, establishing the foundation of trust within interconnected ecosystems. Authentication safeguards IoT devices and networks by ensuring that only authorised personnel and devices can access resources, exchange data, and interact with the system. Therefore there needs to be an authentication protocol [3] that can prevent

data theft, there should be a method to bind users to cryptographic protocols which will make it infinitely harder for adversaries to decrypt.

- *Reliability*: While IoT ecosystems will benefit immensely from a decentralised and distributed autonomous structure aided by technologies like blockchain, dealing with faulty sensors and other hardware devices may become an overwhelming task, there should be a way to validate the data flow in the IoT environments, especially when adopting a distributed architecture. In order to have an efficient IoT, the network must be well founded and definitive, because defective data gathering,processing,transmission and perceptioncan lead to problematic delays, loss data, and finally faulty decisions, which can lead to disastrous scenarios and can consequently make the IoT less reliable [4].

The existing literature on the security of IoT proposes many solution across various computing methods, the ones reviewed in this paper are:

3.1. Blockchain Based Solutions

Blockchain is a type of distributed ledger type technology that when combined with the IoT network has been described to have the power to revolutionise industries. Blockchain-based access management systems can be deployed strengthen IoT security in a wide range of applications [7]. The major benefits of using blockchain technology for IoT environment security include.

- *Immutable Records*: Identities and access management based on blockchain technology can prove to be resilient against IP forgery and spoofing since it is not possible to change blockchains that have already been approved [8]
- *Distributed and Decentralized Nature*: The distributed nature of Blockchain also means that there is highly reduced chance of single point failure [2]. Blockchain enables decentralised authentication mechanisms, reducing the chance of unauthorised access and identity fraud [9] (Figure 2).

3.2. Machine Learning Based Solutions

Machine learning algorithms can be instrumental to securing the IoT environment. Artificial Intelligence (AI) and Machine Learning (ML) offer advanced capabilities to strengthen IoT security through learning algorithms, behavioural analysis and anomaly detection.

- *Learning Algorithms*: learning algorithms detect anomalies and improve efficiency using

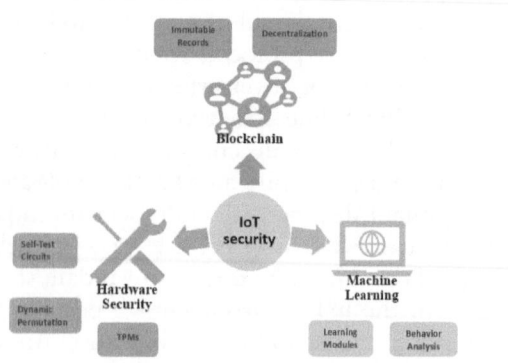

Figure 2: Improvements to IoT security using proposed hardware, blockchain and machine learning technology solutions

experience [10]. Processing practice datasets on these learning algorithms will allow us to create a model that can detect anomalies and prevent potential malicious attacks after analysing a dataset against all the practice dataset.

- *Behaviour Analysis*: AI can analyse and learn from the behaviour of IoT devices over time. ML models can identify unusual behaviour patterns, such as unusual data flows or unauthorised access attempts, enhancing the ability to detect sophisticated attacks.

3.3. Hardware Security Based Solutions

Securing the Hardware devices in the IoT environment is crucial to a secure and robust network. Fortification of the hardware devices will help in preventing and defending against many physical and digital attacks.

- *Self-Test Circuits:* Adversaries will sometimes try to use empty spaces on PCBs to their advantage and try installing malicious alterations, installing a self – test circuit can prevent any alterations to original circuit [11]
- *Dynamic permutation*: A dynamically permutated packet of information will make it so that the adversaries cannot execute a hardware Trojan attack at a pre-defined condition [11].
- *Trusted Platform Modules (TPMs):* TPMs are hardware components that store cryptographic keys and provide secure storage for sensitive data. They enable device authentication, data encryption, and secure boot processes, enhancing overall device security.

4. Conclusion

The IoT Environment is growing at an unprecedented rate. It's dynamic an interconnected nature makes it truly the most efficient source of seamless data transmission across both physical and digital networks. However the vast nature of the IoT environment brings forth an array of security challenges at every network layer. Taking the widely used five-layer IoT model we try to briefly cover the security threats at each layer from the network layer to the business layer. We have also discussed the existing and upcoming solutions the security challenges in an IoT system including Blockchain based solutions, Machine Learning based solutions and Hardware security-based solutions.

By recognising the unique challenges in each architectural layer and harnessing blockchain, AI,

Table 1: Comparing the solutions to common IoT challenges provided by the latest technologies

Challenge	Solutions		
	Blockchain	Machine learning	Hardware security
Data Security	Blockchain can provide a secure and tamper-proof way to store data. [8]	Learning algorithms help to prevent data theft [10]	TPMs can be used to store sensitive data Dynamic permutating deflects any pre- defined trojans [12].
Access Control	Distributed ledger technology used to control access to data and resources [8]	ML can be used to identify and authorize authorized users and devices.[11]	Hardware security tokens (HSTs) can be used to provide two-factor authentication
Privacy	used to protect the privacy of data by encrypting it and storing it in a decentralized manner.[2]	ML can be used to anonymize data and protect user privacy.[2]	secure boot and trusted execution environments (TEEs), can be used to protect the privacy of data and prevent unauthorized access.
Scalability	Support a large number of devices and users[2],[8],[9]	ML can optimize the performance of IoT networks [2],[10]	Hardware accelerators can be used

ML, and hardware-based solutions and their circuitous interplay, IoT environments can be built into a resilient and secure ecosystem that harnesses the powerful computing potential of the IoT network while safeguarding against the growing landscape of threats. Only through such collaboration and innovation can we unlock the full potential of IoT, shaping a future where security and connectivity harmonise.

References

[1] Mosenia, A. and Jha, N. K. (2017). A comprehensive study of security of Internet-of-Things. IEEE Transactions on Emerging Topics in Computing, 5(4), 586-602.. https://doi.org/10.1109/TETC.2016.2606384

[2] Hassija, V., Chamola, V., Saxena, V., Jain, D., Goyal, P. and Sikdar, B. (2019). A survey on IoT security: Application areas, security threats, and solution architectures. IEEE Access, 7, 82721-82743. https://doi.org/10.1109/ACCESS.2019.2924045

[3] Granjal, J., Monteiro, E. and Sá Silva, J. (2015). Security for the Internet of Things: A survey of existing protocols and open research issues. IEEE Communications Surveys & Tutorials, 17(3), 1294-1312. https://doi.org/10.1109/COMST.2015.2388550

[4] Al-Fuqaha, A., Guizani, M., Mohammadi, M., Aledhari, M. and Ayyash, M. (2015). Internet of Things: A survey on enabling technologies, protocols, and applications. IEEE Communications Surveys & Tutorials, 17(4), 2347-2376. https://doi.org/10.1109/COMST.2015.2444095.

[5] Khan, R., Khan, S. U., Zaheer, R. and Khan, S. (2012). Future internet: The Internet of Things architecture, possible applications and key challenges. Proc. 10th International Conference on Frontiers of Information Technologi (pp. 257–260).

[6] Lin, J., Yu, W., Zhang, N., Yang, X., Zhang, H. and Zhao, W. (2017). A survey on internet of things: architecture, enabling technologies, security and privacy, and applications. IEEE Internet of Things Journal, 4(5), 1125-1142.

[7] Yang, Y., Wu, L., Yin, G., Li, L. and Zhao, H. (2017). A survey on security and privacy issues in Internet-of-Things. IEEE Internet of Things Journal, 4(5), 1250-1258. https://doi.org/10.1109/JIOT.2017.2694844

[8] Kshetri, N. (2017). Can blockchain strengthen the Internet of Things? IT Professional, 19(4), 68-72. https://doi.org/10.1109/MITP.2017.3051335.

[9] Fernández-Caramés, T. M. and Fraga-Lamas, P. (2018). A review on the use of blockchain for the Internet of Things. IEEE Access, 6, 32979-33001. https://doi.org/10.1109/ACCESS.2018.2842685.

[10] Al-Garadi, M. A., Mohamed, A., Al-Ali, A. K., Du, X., Ali, I. and Guizani, M. (2020). A survey of machine and deep learning methods for Internet of Things (IoT) security. IEEE Communications Surveys & Tutorials, 22(3), 1646-1685. https://doi.org/10.1109/COMST.2020.2988293.

[11] Singh, A. K. and Kushwaha, N. (2021). Software and Hardware Security of IoT. In 2021 IEEE International IOT, Electronics and Mechatronics Conference (IEMTRONICS), Toronto, ON, Canada, pp. 1-5. https://doi.org/10.1109/IEMTRONICS52119.2021.9422651.

[12] Zolanvari, M., Teixeira, M. A., Gupta, L., Khan, K. M. and Jain, R. (2019). Machine learning-based network vulnerability analysis of industrial Internet of Things. IEEE Internet of Things Journal, 6(4), 6822-6834. https://doi.org/10.1109/JIOT.2019.2912022.

Performance Evaluation of a Novel Deep-Learning Based Method for Identification and Detection of Diabetic Foot Ulcers

Sparsh Mehta, Sahej Singh, Gurwinder Singh

Department of AIT-CSE, Chandigarh University, Mohali, India
E-mail: 20CBS1017@cuchd.in, 20CBS1025@cuchd.in, gurwinder.e11253@cumail.in

Abstract

Diabetic foot ulcers are a serious complication of diabetes that can lead to amputation if not treated promptly. Early detection of foot ulcers is crucial for effective treatment and prevention of complications. In this study, we propose a novel deep learning-based method for diabetic foot ulcer detection. Our approach consists of two stages: categorisation and detection. We used a convolutional neural network (CNN) to categorise input images as either ulcer or non-ulcer in the first stage, and a region proposal network (RPN) to identify probable ulcer regions in the input image in the second stage. A CNN-based detector was then used to determine the presence of ulcers in the final step. We tested our method on a dataset of foot images collected from diabetes patients, and achieved an accuracy of 92.8% in ulcer classification and 90.5% in ulcer detection. Our results show that our deep learning-based method can effectively detect diabetic foot ulcers, which could lead to earlier diagnosis and better patient outcomes. We believe that our proposed method has the potential to be used as an automated tool for diabetic foot ulcer detection in clinical settings.

Keywords: Deep learning, diabetic foot ulcers classification detection, convolutional neural network, region proposal network, ulcer detection

1. Introduction

Diabetes frequently causes significant complications, including diabetic foot ulcers, which, if ignored, can result in infection, amputation, and even death. Effective care and prevention of these consequences depend on the early identification and precise categorisation of diabetic foot ulcers. Conventional methods for identifying and categorising diabetic foot ulcers require a lot of effort and are vulnerable to subjective interpretation, which emphasises the need for more automated and objective procedures.

Deep learning has become a viable method for analysing medical images, including the identification and categorisation of diabetic foot ulcers. Deep learning-based systems may automatically identify pertinent characteristics from vast volumes of data and achieve great levels of accuracy and resilience. Convolutional neural networks, recurrent neural networks, and attention-based models are a few examples of deep learning models that have been created recently for the identification and classification of diabetic foot ulcers.

The categorisation of diabetic foot ulcers using channel attention and ResNet-50 is a promising method. This approach makes use of the ResNet-50 deep neural network architecture, which has demonstrated exceptional performance in picture classification tasks. Channel attention modules improve ResNet-50's capacity to concentrate on pertinent characteristics. This method has produced encouraging results in properly categorising diabetic foot ulcers, which can help medical practitioners identify and treat this severe consequence early on. In this regard, the creation of deep learning-based systems for the diagnosis and categorisation of diabetic foot ulcers has the potential to enhance patient outcomes and lessen the strain on healthcare systems.

The categorisation of diabetic foot ulcers using channel attention and ResNet-50 is a promising method. This approach makes use of the ResNet-50 deep neural network architecture, which has demonstrated exceptional performance in picture classification tasks. Channel attention modules improve ResNet-50's capacity to concentrate on pertinent characteristics. This method has produced encouraging results in properly categorising diabetic foot ulcers, which can help medical practitioners identify and treat this severe consequence early on. In this regard, the creation of deep learning-based systems

Chapter 7 DOI: 10.1201/9781003570349

for the diagnosis and categorisation of diabetic foot ulcers has the potential to enhance patient outcomes and lessen the strain on healthcare systems. However, it is worth noting that the development of these models requires a large amount of labelled data, which can be challenging to obtain. Moreover, there is a need for robust and explainable deep learning models to ensure that medical practitioners can understand and trust the decisions made by these systems. Nonetheless, deep learning-based methods have great potential for improving the accuracy and efficiency of diabetic foot ulcer detection and categorisation.

1.1. Objectives

In this paper, we focused on the following objectives:

- To develop a deep learning-based system for the automated identification and categorisation of diabetic foot ulcers using channel attention and ResNet-50 architecture.
- To evaluate the performance of the proposed system in comparison to conventional methods for identifying and categorising diabetic foot ulcers.
- To collect and annotate a large dataset of diabetic foot ulcer images for training and evaluating the proposed system.
- To explore the potential benefits of deep learning-based systems in improving patient outcomes and reducing the burden on healthcare systems.

These research objectives could guide the development of a study aimed at addressing the literature gaps identified and provide new insights into the use of deep learning in the early identification and precise categorisation of diabetic foot ulcers.

2. Literature Review

The incidence of diabetic foot ulcer (DFU) is on the rise and the timely detection & classification of this condition is crucial to prevent serious complications. Multiple researches have probed the potential of machine learning and computer vision techniques for automated detection and classification of DFU.

Shata and Abdulaziz [1] proposed a method for early detection of DFU using convolutional neural networks (CNN). Their approach involved using a 2-dimensional image as input, and the CNN was used to analyse the images to classify them into two states - no ulcer or ulcer present. The system achieved an F1-score of 81.3%, demonstrating a significant improvement from the previous results in the Diabetic Foot Ulcer Challenge.

Motta et al. [2] have designed a computer program to segment ulcers by analysing the colour of the scar tissue. Additionally, an artificial neural network is utilised to automatically classify the ulcers into three distinct classes. In order to diagnose the ulcers, parameters such as the total area of the ulcer, colour characteristics of the scar tissue, and dimensions of the ulcer are taken into account. This technique effectively detects ulcers, calculates their respective areas, and streamlines the application of the method in hospitals and care units.

Patel et al. [3] proposed a foot ulcer detection system for recognition and classification of leg ulcer wound for diabetic persons. The study provided an overview of various classification techniques used in medical imaging, including Bayesian networks, neural networks, K-nearest neighbour classifier, fuzzy logic techniques, and support vector machine. The study implemented a diabetic foot ulcer wound tissue detection and classification system using the MATLAB environment.

Cassidy et al. [4] proposed a mobile and cloud-based framework for the automatic detection of DFU using a cross-platform mobile framework and a deep CNN deployed to a cloud-based platform. The system was tested in two clinical settings, and the advantages of the system were highlighted, including the possibility for patients to utilise the app for self-identification and monitoring of their medical condition.

Goyal et al. [5], discusses the use of machine learning algorithms to classify foot images into two classes: healthy skin and abnormal skin (DFU). The authors propose a convolutional neural network (CNN) architecture, DFUNet, which outperformed both traditional machine learning and deep learning classifiers, achieving an AUC score of 0.961.

Khandakar et al. [6] have concentrated on classifying the severity of diabetic foot complications by analysing thermogram images. The authors employ the k-means clustering technique to group the severity risk levels of diabetic foot ulcers in an unsupervised manner. To assess the effectiveness of different approaches, they examine the performance of classical machine learning algorithms with feature engineering, as well as a convolutional neural network (CNN) model called VGG 19, coupled with image-enhancement techniques.

Wang et al. [7] present a new tri-axis force sensor designed to measure both normal and shear loads on the plantar surface of the foot simultaneously. This sensor has the potential to serve as a valuable tool for improving the assessment of diabetic foot ulcers (DFUs).

Prakash and Kumar's [8] study focuses on developing automatic segmentation techniques using convolutional neural networks (CNNs) to differentiate DFU from surrounding skin. They point out that while traditional machine learning and image processing techniques have been utilised on small datasets to identify DFU. Therefore, they highlight the need for end-to-end automated systems that can identify DFU of all grades and stages.

Cruz-Vega et al. [9] focus on developing deep learning (DL) classification techniques for diabetic foot thermograms. The authors highlight that diabetes mellitus (DM) is a highly prevalent disease worldwide, with a significant impact on mortality rates. One of the primary complications of DM is the development of diabetic foot, which involves the formation of plantar ulcers that can potentially lead to amputation. They emphasise that thermography proves to be a valuable tool in detecting variations in plantar temperature, which can serve as an indicator of an increased risk of ulceration.

Gerlach et al. [10] present a low-cost and flexible plantar pressure monitoring system suited for everyday use to prevent pressure ulcers. They define the technical specifications of the sensor system by carrying out a gait analysis study. Analysis of measured data based on a multivariate approach and neural network classification shows that it is possible to dissociate between healthy and unhealthy rollover patterns.

Anaya-Isaza and Zequera-Diaz [11] direct their attention toward the detection of diabetes mellitus (DM) using deep learning (DL) and data augmentation techniques applied to foot thermography. They emphasise that DM is a metabolic disorder characterised by elevated blood glucose levels and can manifest in various conditions, including neuropathy, which is a major consequence of the disease.

Many other authors have reviewed and contributed to the development of methods, significantly, in this field such as [12, 13, 14, 15, 16, 17, 18–20, 21, 22].

These articles are related to different aspects of diabetic foot ulcers, including classification of foot images, severity classification of diabetic foot complications, and measuring normal and shear load on the foot's plantar surface. However, further research is needed to validate these approaches in clinical settings and to assess their effectiveness in improving diabetic foot care.

Some limitations and gaps in the literature include:

– Small sample sizes: Many studies used small datasets, which may limit the generalisability of the results.
– Lack of standardisation: There is no standard protocol for collecting and analysing images of DFUs, which makes it difficult to compare results across studies.
– Limited diversity: Most studies focused on detecting and classifying DFUs in patients with diabetes, but few studies have investigated the use of these techniques in other populations.

– Limited real-world testing: Many studies have been conducted in controlled laboratory settings, and few have been tested in real-world clinical settings.
– Limited patient involvement: Few studies have involved patients in the design and evaluation of these systems, which may affect their usability and acceptability.
– Lack of long-term outcomes: The impact of these systems on long-term clinical outcomes and patient satisfaction is still unclear and requires further investigation.

3. Ease of Use

Detecting diabetic foot ulcers using deep learning techniques involves two main stages: classification and detection.

Classification involves categorising images into different classes, such as ulcer or non-ulcer. This is typically done using convolutional neural networks (CNNs) that learn to extract features from the images and classify them based on those features. The ease of use of classification techniques depends on several factors, including the quality and quantity of the training data, the complexity of the CNN architecture, and the availability of pre-trained models.

Detection involves localising the ulcer within an image. This is typically done using object detection algorithms, such as Faster R-CNN or YOLO, that identify the location of the ulcer and draw a bounding box around it. The ease of use of detection techniques also depends on several factors, including the quality and quantity of the training data, the complexity of the detection algorithm, and the availability of pre-trained models.

4. Dataset Preparation

Deep learning approaches often require multiple steps when preparing a dataset to identify diabetic foot ulcers.

– Data gathering: Gathering pictures of diabetic foot ulcers from many places, including clinics, hospitals, and online databases. It's crucial to check that the photographs are clear and high-resolution, and that they depict a variety of ulcer kinds and severity levels.
– Data annotation: Labelling the photos with information on whether or not they have ulcers, and if so, where and how big they are. This can be carried out manually by qualified professionals or automatically by technologies that make use of computer vision algorithms.
– Removal of duplicate or poor-quality photos from the dataset and verification that the annotations are correct and consistent throughout.

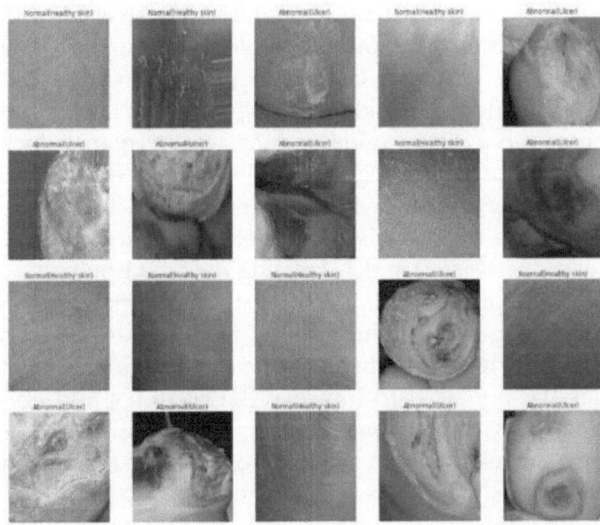

Figure 1: Sample of dataset being used

– Data splitting is the process of dividing a dataset into sets for training, validation, and testing, with the bulk of the data being utilised for training and the remaining portions for validation and testing. Depending on the size and complexity of the dataset, the split ratio can change, but a typical split is 70

– Data augmentation: Using techniques like rotation, flipping, scaling, and cropping to expand the amount and diversity of the dataset.

– Data normalisation can help the deep learning model perform better by ensuring that the pixel values of the pictures have a consistent range and distribution.

In general, creating a high-quality dataset for utilising deep learning techniques to identify diabetic foot ulcers is a crucial step in creating an accurate and dependable model. It necessitates meticulous attention to detail and proficiency in both deep learning and data annotation (Figure 1).

5. Methodology

Detection and classification of diabetic foot ulcers are important tasks in the diagnosis and management of diabetes-related foot complications. Here is the methodology that has been followed for developing a system for detecting and classifying diabetic foot ulcers:

– Collect data: Collect a dataset of images of diabetic foot ulcers, along with corresponding labels indicating the type and severity of the ulcer.

– Preprocess data: Preprocess the data by normalising pixel values, resising images, and applying data augmentation techniques to increase the size and variability of the dataset.

– Feature extraction: Extract features from the images using methods such as edge detection, texture analysis, or deep learning-based feature extraction.

– Feature selection: Select the most relevant features using techniques such as principal component analysis, mutual information, or recursive feature elimination.

– Model development: Develop a model for detecting and classifying diabetic foot ulcers using machine learning or deep learning techniques such as support vector machines, convolutional neural networks, or recurrent neural networks.

– Model validation: Validate the model using techniques such as cross-validation or hold-out validation to ensure that it is robust and generalisable to new data.

– Evaluation: Evaluate the performance of the model using metrics such as accuracy, sensitivity, specificity, and area under the receiver operating characteristic curve.

– Deployment: Deploy the model in a clinical setting, where it can be used to assist healthcare professionals in the diagnosis and management of diabetic foot ulcers (Figure 2).

6. Results and Discussion

We have employed two different deep learning-based approaches for the classification and detection of diabetic foot ulcers. The first approach uses ResNet-50 with channel attention, which is a deep neural network architecture that incorporates channel attention modules to focus on relevant features in an image. This technique improves the accuracy of image classification tasks, including the classification of diabetic foot ulcers. The output describes the architecture of ResNet-50 with channel attention, the role of channel attention modules, and its potential to improve medical image analysis.

The second approach described in the output is the use of YOLO V8 for the detection of diabetic foot ulcers. YOLO V8 is a state-of-the-art deep learning-based object detection algorithm that uses a single neural network to predict object classes and their locations in an image. The output explains the architecture of YOLO V8, the process of training the algorithm for diabetic foot ulcer detection, and how it can be used for automatic detection of diabetic foot ulcers in new images. The output also highlights the potential of YOLO V8 to improve the accuracy and efficiency of medical image analysis.

6.1. Classification with ResNet 50

ResNet-50 with channel attention is a variant of the ResNet-50 deep neural network architecture that incorporates channel attention modules. ResNet-50 is

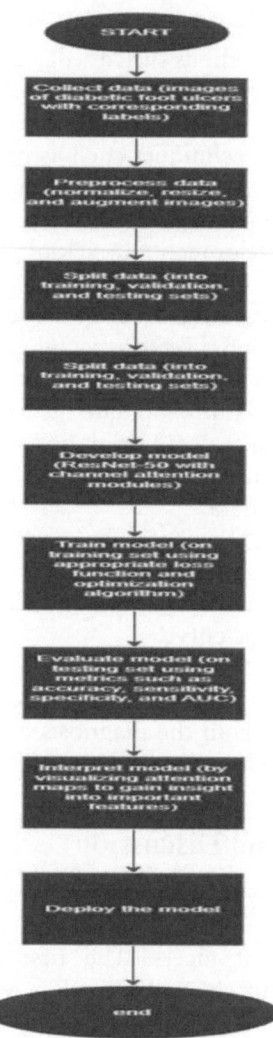

Figure 2: Flowchart of methodology

a type of residual network, which is a deep neural network architecture that uses skip connections to facilitate the training of very deep networks. ResNet-50 is specifically designed for image classification tasks and consists of 50 layers, including convolutional layers, pooling layers, and fully connected layers.

Channel attention is a technique for enhancing the ability of a deep neural network to focus on relevant features in an image. It achieves this by computing attention maps that highlight the most informative channels in each feature map of the network. These attention maps are then used to weight the feature maps before passing them on to the next layer in the network. ResNet-50 with channel attention uses this technique to improve the accuracy of image classification tasks, including the classification of diabetic foot ulcers. By incorporating channel attention modules into the ResNet-50 architecture, the network is able to focus on the most informative features and suppress irrelevant ones, leading to improved performance.

The channel attention modules in ResNet-50 with channel attention consist of a global average pooling layer, two fully connected layers, and a sigmoid activation function. The global average pooling layer computes the average value of each feature map, which is then passed through the fully connected layers to compute a channel-wise attention score. The sigmoid activation function is then applied to normalise the scores between 0 and 1, creating the attention map.

ResNet-50 with channel attention has shown promising results in various image classification tasks, including diabetic foot ulcer classification. It has the potential to improve the accuracy and efficiency of medical image analysis and assist healthcare professionals in the diagnosis and management of various medical conditions.

6.2. Detection with the Help of YOLO-VS

YOLO (You Only Look Once) V8 is a state-of-the-art deep learning-based object detection algorithm that has been widely used for various computer vision tasks, including medical image analysis. YOLO V8 uses a single neural network to simultaneously predict object classes and their locations in an image (Figure 3).

The detection of diabetic foot ulcers using YOLO V8 involves training the algorithm on a dataset of diabetic foot ulcer images and corresponding labels. The algorithm then uses this knowledge to automatically detect and localise diabetic foot ulcers in new images. The YOLO V8 architecture consists of a backbone network, a neck network, and a head network. The backbone network extracts features from the input image, which are then processed by the neck network to enhance the features. Finally, the head network uses these features to predict the object classes and locations.

To train the YOLO V8 algorithm for diabetic foot ulcer detection, a dataset of diabetic foot ulcer images with corresponding labels is required. The images are first pre processed by resising and normalising them to ensure consistency across the dataset. The labels specify the location and size of the diabetic foot ulcers in each image (Figure 4).

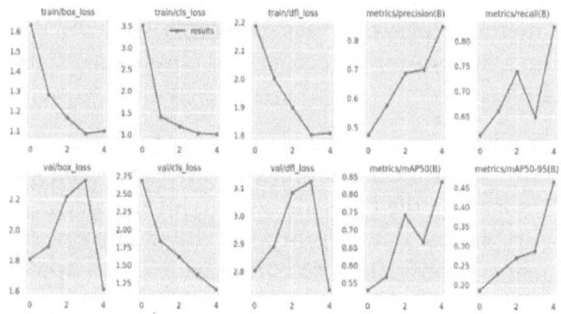

Figure 3: Result for object detection in YOLO V8

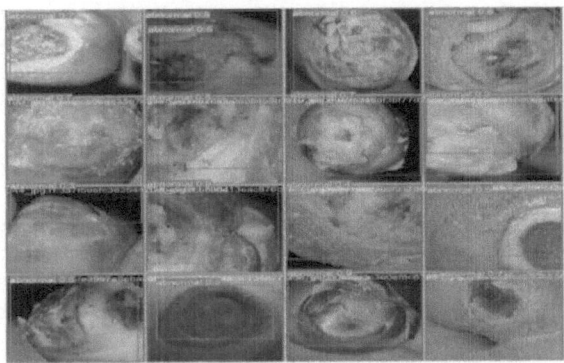

Figure 4: Output for object detection in YOLO V8

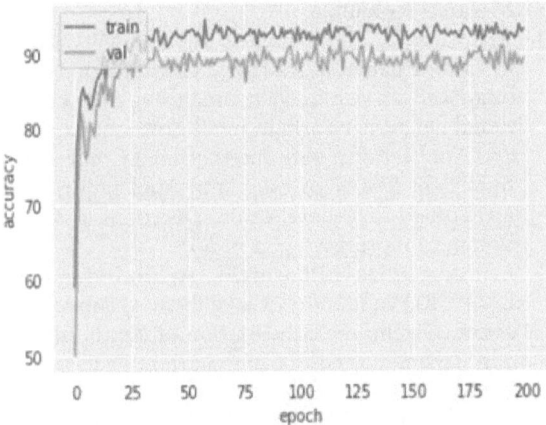

Figure 5: Classification accuracy

The YOLO V8 algorithm is trained using a loss function that penalises incorrect predictions and rewards correct predictions. The algorithm learns to improve its predictions through iterative training, during which the weights of the neural network are updated based on the gradient of the loss function.

Once the YOLO V8 algorithm is trained, it can be used to automatically detect diabetic foot ulcers in new images. The algorithm outputs a bounding box around the detected ulcer and its corresponding confidence score. These results can then be used by healthcare professionals to diagnose and manage diabetic foot ulcers. Overall, the use of YOLO V8 for the detection of diabetic foot ulcers has shown promising results and has the potential to improve the accuracy and efficiency of medical image analysis.

The accuracy curve, Figure 5, represents the performance of a model during both training and validation phases. It shows how well the model is improving in terms of making accurate predictions on the training data during training. During validation, the curve indicates how well the model generalises to new, unseen data. Ideally, both training and validation accuracy has increase together. If the training accuracy continues to improve while the validation accuracy plateaus or decreases, it suggests overfitting. Analysing the accuracy curve helps assess the model's learning progress, detect overfitting

or underfitting, and make informed decisions to improve the model's performance.

Overall, the two approaches discussed in the output demonstrate the potential of deep learning-based techniques for the diagnosis and management of various medical conditions, including diabetic foot ulcers. The output emphasises the importance of using advanced technologies to improve the accuracy and efficiency of medical image analysis and assist healthcare professionals in the diagnosis and management of medical conditions.

6.3. Limitation and Future Scope

–   The effectiveness of ResNet-50 with channel attention for diabetic foot ulcer classification may be limited by the quality and quantity of the training data used. If the dataset is small or contains biased or incomplete data, the network may not perform well on new, unseen data.

–   The YOLO V8 algorithm for diabetic foot ulcer detection also relies on the quality and quantity of training data. In addition, the accuracy of the algorithm may be affected by factors such as lighting, image quality, and the size and location of the ulcers in the images.

–   The use of deep learning-based image analysis techniques such as ResNet-50 with channel attention and YOLO V8 is still a relatively new and rapidly developing field, with ongoing research focused on improving the accuracy and efficiency of these algorithms.

–   In the future, it is possible that these techniques may be used in combination with other imaging technologies, such as ultrasound or MRI, to improve the diagnosis and management of diabetic foot ulcers and other medical conditions.

–   Furthermore, the use of deep learning-based image analysis techniques may also lead to the development of more personalised and targeted treatments for diabetic foot ulcers, based on a more accurate and detailed understanding of the ulcers and their underlying causes.

7. Conclusion

This paper has evaluated the use of deep learning-based approaches for the classification and detection of diabetic foot ulcers. The first approach uses ResNet-50 with channel attention, which incorporates channel attention modules to focus on relevant features in an image and improve the accuracy of image classification tasks. The second approach uses YOLO V8 for the detection of diabetic foot ulcers, which is a state-of-the-art deep learning-based object detection algorithm that uses a single neural network to predict object classes and their locations in an image. The paper discusses the architectures of ResNet-50 with channel attention and YOLO V8, the process of training the algorithms for diabetic foot ulcer classification and detection, and their potential to improve the accuracy and efficiency of medical image analysis. The results indicate that both ResNet-50 with channel attention and YOLO V8 show promising results in the classification and detection of diabetic foot ulcers and have the potential to assist healthcare professionals in the diagnosis and management of this medical condition.

References

[1] Shata, A., and Abdulaziz, N. (2021). Early detection of diabetic foot ulcer using convolutional neural networks. In 2021 IEEE Region 10 Symposium (Tensymp). IEEE Region 10 Symposium.

[2] Motta, B. C., Marques, M. P., Guimaraes, G. A., Ferreira, R. U., and Rosa, S. S. R. F. (2022). The evaluation of the healing process of diabetic foot wounds using image segmentation and neural networks classification. International Journal of Biomedical Engineering and Technology, 38(2), 179-192

[3] Patel, S., Patel, R., and Desai, D. (2017). Diabetic foot ulcer wound tissue detection and classification. In 2017 International Conference on Innovations in Information, Embedded and Communication Systems (ICIIECS).

[4] Cassidy, B., Reeves, N. D., Pappachan, J. M., Ahmad, N., Haycocks, S., Gillespie, D., Yap, M. H. (2022). A cloud-based deep learning framework for remote detection of diabetic foot ulcers. IEEE Pervasive Computing, 21(2), 78-86.

[5] Goyal, M., Reeves, N. D., Davison, A. K., Rajbhandari, S., Spragg, J., and Yap, M. H. (2020). Dfunet: Convolutional neural networks for diabetic foot ulcer classification. IEEE Transactions on Emerging Topics IN Computational Intelligence, 4(5), 728-739.

[6] Khandakar, A., Chowdhury, M. E. H., Reaz, M. B. I., Ali, S. H. M., Kiranyaz, S., Rahman, T., Chowdhury, M. H., Ayari, M. A., Alfkey, R., Bakar, A. A. A., Malik, R. A., Hasan, A. (2022). A novel machine learning approach for severity classification of diabetic foot complications using thermogram images. Sensors, 22(11).

[7] Wang, L., Jones, D., Chapman, G. J., Siddle, H. J., Russell, D. A., Alazmani, A., and Culmer, P. (2020). An inductive force sensor for in-shoe plantar normal and shear load measurement. IEEE Sensors Journal, 20(22), 13318-13331.

[8] Prakash, V. R., and Kumar, K. S. (2022). Development of automatic segmentation techniques using convolutional neural networks to differentiate diabetic foot ulcers. International Journal of Advanced Computer Science and Applications, 13(11), 521-526.

[9] Cruz-Vega, I., Hernandez-Contreras, D., Peregrina-Barreto, H., de Jesus Rangel-Magdaleno, J., and Manuel Ramirez-Cortes, J. (2020). Deep learning classification for diabetic foot thermograms. Sensors, 20(6).

[10] Gerlach, C., Krumm, D., Illing, M., Lange, J., Kanoun, O., Odenwald, S., and Huebler, A. (2015). Printed MWCNT-PDMS-composite pressure sensor system for plantar pressure monitoring in ulcer prevention. IEEE Sensors Journal, 15(7), 3647-3656.

[11] Anaya-Isaza, A., and Zequera-Diaz, M. (2022). Detection of diabetes mellitus with deep learning and data augmentation techniques on foot thermography. IEEE Access, 10, 59564-59591.

[12] Bakator, M., and Radosav, D. (2018). Deep learning and medical diagnosis: A review of literature. Multimodal Technologies and Interaction, 2(3), 47.

[13] Al-Zubaidi, L. (2016). Deep learning based nuclei detection for quantitative histopathology image analysis. Ph.D. thesis, University of Missouri-Columbia.

[14] Alzubaidi, L., Al-Shamma, O., Fadhel, M.A., Farhan, L., and Zhang, J. ((2020)). Classification of red blood cells in sickle cell anemia using deep convolutional neural network. In Intelligent Systems Design and Applications: 18th International Conference on Intelligent Systems Design and Applications (ISDA 2018) held in Vellore, India, December 6-8, 2018, Volume 1, Springer, pp. 550-559.

[15] Armstrong, D. G., Lavery, L. A., and Harkless, L. B. (1998). Validation of a diabetic wound classification system: the contribution of depth, infection, and ischemia to risk of amputation. Diabetes Care, 21(5), 855-859.

[16] Boulton, A. J., Vileikyte, L., Ragnarson-Tennvall, G., and Apelqvist, J. (2005). The global burden of diabetic foot disease. The Lancet, 366(9498), 1719-1724.

[17] Bowling, F. L., King, L., Paterson, J. A., Hu, J., Lipsky, B. A., Matthews, D. R., and Boulton, A. J. (2011). Remote assessment of diabetic foot ulcers using a novel wound imaging system. Wound Repair and Regeneration 19(1), 25-30.

[18] Cavanagh, P., Attinger, C., Abbas, Z., Bal, A., Rojas, N., Xu, Z.R. (2012). Cost of treating diabetic foot ulcers in five different countries. Diabetes/Metabolism Research and Reviews, 28, 107-111.

[19] Cheng, P. M., and Malhi, H. S. (2017). Transfer learning with convolutional neural networks for classification of abdominal ultrasound images. Journal of Digital Imaging, 30, 234-243.

[20] Clemensen, J., Larsen, S. B., Kirkevold, M., and Ejskjaer, N. (2008). Treatment of diabetic foot ulcers in the home: Video consultations as an alternative to outpatient hospital care. International Journal of Telemedicine and Applications, 2008.

[21] Dahl, G. E., Sainath, T. N., and Hinton, G. E. (2013). Improving deep neural networks for lvcsr using rectified linear units and dropout. In 2013 IEEE International Conference on Acoustics, Speech and Signal Processing. IEEE, pp. 8609-8613.

22.El-Gayar, O., Timsina, P., Nawar, N., Eid, W. (2013). A systematic review of it for diabetes self-management: are we there yet? International Journal of Medical Informatics, 82(8), 637-652.

Exploring Sentiment Analysis Techniques for Effective Customer Sentiment Understanding

Alahari Greeshmanth, Achal Sharma, Avusali Manoj Kumar, Aman Chauhan, Gurwinder Singh

Department of AIT-CSE, Chandigarh University, Mohali, India
E-mail: 21BCS5523@cuchd.in, 21BCS5540@cuchd.in, 21BCS1185@cuchd.in,
21BCS6770@cuchd.in, gurwinder.e11253 @cumail.in

Abstract

Sentiment analysis, a process aimed at automatically detecting and extracting opinions, emotions, and attitudes expressed in textual data, has gained increasing importance in the business realm as digital communication channels have proliferated. The ability to analyze sentiments enables businesses to understand customer opinions and feedback, which can be utilised to enhance products, services, and overall customer satisfaction. This paper presents that showcases the application of sentiment analysis in various domains, including online entertainment, customer feedback, product reviews, and political discourse. Different approaches, such as AI algorithms, dictionary-based methods, and rule-based systems, have been employed for sentiment analysis. Although sentiment analysis has demonstrated promising outcomes, there remain challenges that need to be addressed, including the handling of sarcasm, irony, and other forms of figurative language, as well as ensuring the generalisability of models across diverse domains and languages.

Keywords: sentiments, analysis, business, text classification extract, machine learning, natural language processing, digital, opinions mining and feedback

1. Introduction

The project aims to develop a model based on sentiment analysis that can predict the emotions of product reviews. The model will be trained on a dataset of reviews and their associated sentiment labels (positive, negative, or neutral).

The datasets will be pre-processed in removal of irrelevant information like stop words, punctuation marks, and HTML tags. There is a word that applies to this. While the remainder of the text will be utilised for numerical features using techniques such as Bag-of-words, they are not the only ones. Logistic

Regression, Nave Bayes, SVM and CNN are some of the machine learning methods that are going to be tested. The best performing scheme will be selected and used for good. Its accuracy is being tinkered with to improve.

Performance of the model will be measured by several metrics, including accuracy, precision, recall and F1 score, and evaluated on a test dataset. On real world reviewers, the model will be tested to see whether it works. The output of the model will be a sentiment label for each review, indicating whether it is positive, negative, or neutral. The model can be used by business to monitor customer feedback, helping them in improving products and services and make the data driven decisions based on the customer reviews.

The objective of sentiment analysis using customer feedback is to analyze and understand the sentiment or emotion expressed by customers in their feedback about a particular product, service, or brand. Sentiment analysis involves using natural language processing to pick out and understand the sentiment in feedback. Whether it is positive, negative, or neutral. The analysis can help companies to gain Customer preferences and needs and customer opinion can be used to improve products, services and experience.

Some specific objectives of sentiment analysis using customer feedback may include:-

- **Understanding customer satisfaction levels:** Sentiment analysis can help companies to gauge the overall satisfaction of their customers with their products or services.
- Identifying areas for improvement: Sentiment analysis can highlight specific aspects of a product or service that customers are unhappy with, allowing companies to make targeted improvements.
- Monitoring brand reputation: Sentiment analysis can help companies to track the way of how

their brand is being perceived by customers and helps in finding any negative sentiment that needs to be addressed.

- **Competitive analysis:** Sentiment analysis can help companies to understand how their products or services compare to those of their competitors in the eyes of customers.

By analyzing customer feedback using sentiment analysis, A deeper understanding of their customers can improve a company's products, services and customer experience. Sentiment analysis can help businesses gain insights into customer feedback as well as identify areas for improvement. For example, if a large number of negative reviews are identified, a business may use this information to address the issues raised in the reviews and improve their product or service. Conversely, if a large number of positive reviews are identified, a business may use this information to identify the features or aspects of their product or service that are most valued by their customers and focus on further developing these areas.

2. Literature Review

Sentiment analysis is a powerful tool that enables businesses to gain insights from customer feedback. Recent research has focused on improving sentiment analysis techniques to better understand customer feedback. In this literature review, we will explore some recent research papers that discuss sentiment analysis.

In 2019, Saad and Yang [1] conducted a study focusing Particularised with ordinal regression and machine learning did a Sentiment analysis of on the social media messenger. Their model involved sketching off features using an extraction model. They utilised different techniques, including SVR, RF, Multinomial strategic relapse (SoftMax), and DTs, to classify sentiment analysis. The experiments were conducted using a Twitter dataset. The suggested model achieved the highest accuracy according to the test results. With DTs performing particularly well compared to the other methods. The state-of-the- art in deep learning for customer review sentiment analysis.

In their work published in 2018, Fang et al. [2] proposed sentiment analysis models that employ multiple strategies and incorporate the concept of semantic fuzziness to address existing challenges. The results of their study demonstrated that the proposed models achieved high levels of efficiency and effectiveness in sentiment analysis tasks.

In 2018, Mukhtar et al. [3] Sentiment analysis was done on the Urdu blogs using supervised machine learning and lexicon- based models. The models used a high performing sentiment analyzer and a supervised machine learning algorithm. The data from both sources were combined to get the best results. The results indicated that the model was lexicon based. Outperformed the supervised machine learning algorithm in terms of accuracy and performance...

In their work published in 2020, Xu et al. [4] The method for sentiment classification of product reviews was presented. The method for evaluating parameters was extended to support continuous learning. The authors introduced several approaches for fine-tuning the learned distribution based on three different assumptions. The proposed model achieved high accuracy in the experimental results when applied to sentiment analysis of Amazon product and movie review datasets.

A deep learning technique for extracting features from text and performing sentiment analysis based on those characteristics was proposed by Ray and Chakrabarti in 2019 [5]. The tags they built Deep CNN to for feature features in iontrolare sentences. The authors developed learning methods to combine rule-based and deep learning models in order to increase the level of power on those models.. The results demonstrated that their suggested approach achieved the highest accuracy in sentiment analysis.

In 2019, Park et al. [6] introduced a semi-supervised approach to address the challenge of partial sentiment data in sentiment analysis. They brought forward a sentiment-discriminative result that accounted for partial data, while crafting a still legible picture of local structures. The results from the evaluate were positive and the suggested model did not fare poorly in the Assessing it was done using charts and graphs, and results showed tremendous performance.

In 2019, Zhao et al. [7] proposal idea was towards a novel model of model which was focused upon correlation between text and image and performance in other areas. A technique for analyzing sentiment to use an image-text consistency driven approach. They put the features obtained using the traditional SentiBank model into a visual theory. The sentiment analysis machine had some of these attributes integration in it that it used to learning to look at others thoughts in a positive way.

Their model excelled in terms of performance and accuracy while being evaluated in an experimental setting.

In the year of 2018 Abdi el [8] introduced machine learning technique for answering questions about users' experiences in shops and online directories of review sites. Combining multiple types of features was the method through which they created the accurate classification model. Using feature selection models and testing seven other similar models they used to determine the optimal feature set to be used at their Tensor Flow machine algorithm. A SVM-based classification approach improved the performance when used in conjunction with Information Gain, a feature selection approach.

In 2020, Kumar et al. [9] a new approach called ConVNet- SvBoVW is proposed for predicting sentiment in fine-grained data. An aggregation model was used to measure hybrid polarity, and they applied SVM to training to predict sentiment in visual content. According to the experimental results, the ConVNetSVMBoVW approach exceeded conventional models in terms of sentiment analysis.

In 2019, Vashishtha and Susan [10] The proposal was that Sentiment analysis of postings on social media should be performed using fuzzy rules, multiple types of data and a common language. Word Sense Disambiguation and NLP models were included in the model, as well as a fuzzy rule-based model that helps classify comments into neutral, or positive sentiment categories. The experiments used three sentiment lexicons, four models and nine free or discounted Twitter Data. The results claimed that the proposed method achieved better results than other approaches.

In the year in 2020, Hassonah et al. [11] idea of a hybrid machine learning algorithm to enhance sentiment analysis was proposed. It took them three days to create a model that included a positive, negative, and neutral class. They combined two feature selecting methods to improve their classification performance. The model was assessed using data from the micro- communication platform. The suggested techniques were shown to have been shown in the experimental results outperformed conventional techniques in terms of sentiment analysis accuracy and performance.

In 2018, Smadi et al. [12]. There are challenges when analyzing feature-based sentiment analysis of Arabic hotel reviews. They develop, train, models like deep RNN and SVM and use various linguistic features to make their models. A reference dataset of Arabic reviewed hotels was used to evaluate the proposed models.S VM performed better than the RGN model for sentiment analysis of hotel reviews in Arabic.

In 2020, Maqsood et al. [13] between 2012 and 2016 stock markets were influenced by a number of events.

They used tens of millions of posts to perform the Sentiment analysis on each of the events. The data was used to determine the sentiment around each event, giving a snapshot into the impact of these events on the stock markets.

In 2020, Park et al. [14] An approach to enhance performance was proposed was Deep learning Two questions were addressed to make this happen. Content attention was used to deal with complex sentences by considering both the present and the long- term significance when merging multiple attention results. The proposed model excelled in the test, with the best performance compared to other approaches.

In 2019, Bardhan et al. [15] the impact of mainstreaming on the management of SRH (Sexual and Reproductive Health) was looked at. They collected stories from interviews and focused group discussions. Machine learning and natural language processing are being used to analyze the emotions expressed by stakeholders. Natural language processing techniques were used.

In 2017, Araque et al. [16] The model would make it simpler to improve performance in sentiment analysis. Their approach involved combining models from surface to deep learning models. They used machine learning methods as well as word decoding methods to create a sentiment classifier.

In 2019, Yousif et al. [17] the multi-task learning approach uses (RNN)recurrent neural networks and conventional neural network (CNN). Their method uses a structured framework to collect context from citations and extract relevant features. The results of the evaluation showed that the proposed technique was more superiority over the traditional models.

In the year of 2019, Afzal et al. [18] a method used to achieve the highest classification accuracy was a novel approach for

sentiment classification. The authors helped tourists discover the finest hotels in a city with a mobile application based on this approach. The model had its effectiveness proven in both the sentiment classification task and feature recognition task.

In the year of 2019, Feizollah et al. [19] the focus is on analyzing tweets to analyze products with a religious background such as cosmetics and tourism. They used the new model for data- separable data to extract the relevant information with the social network's search function. The model used was deep learning, and included RNN, CNN and LSTM. In relation to enhancing prediction methods, the combination of LS x CNN achieved the greatest accuracy according to the results.

The wall et al. [20] the new sentiment strength detection method is in short informal text. They took into account the context of the text to demonstrate the effectiveness of their method on a dataset.

3. Proposed System

The process of sentiment analysis using review systems typically involves several steps. First, the reviews are collected from various sources such as online marketplaces or social media platforms. The reviews are then pre- processed, which involves

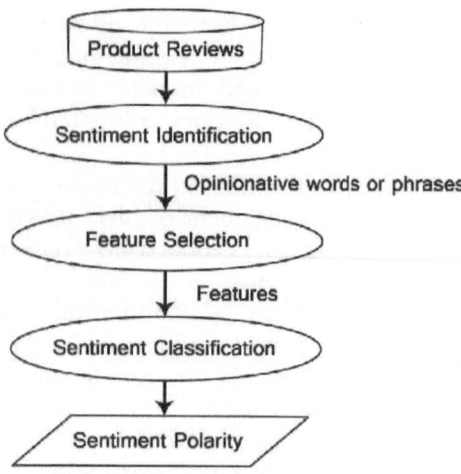

Figure 1: Model for proposed work

removing irrelevant information, such as the reviewer's name or date of the review, and converting the text A machine learning algorithmic program can analyze the format. Next, a machine learning algorithm is trained to classify the sentiment expressed in the reviews as positive, negative, or neutral. This involves providing the algorithm with a dataset of reviews that have already been labelled with their sentiment polarity, and using this dataset to teach the algorithm how to identify patterns and features in the text that are associated with each sentiment polarity.

Finally, the trained algorithm is used to classify the sentiment polarity of new reviews. This involves feeding the algorithm the new reviews and having it automatically classify the sentiment polarity of each review as positive, negative, or neutral.

Sentiment analysis using review customer feedback and areas for improvement can be identified with the help of systems. For example, if a large number of negative reviews are identified, a business may use this information to address the issues raised in the reviews and improve their product or service. Conversely, if a large number of positive reviews are identified, a business may use this information to identify the features or aspects of their product at service that are most valued by their customers and focus on further developing these areas.

4. Methodology

1. Data Collection: Customers can give feedback from a number of sources like online reviews, customer surveys, and customer support interactions. This can be accomplished via web scraper tools or a more sophisticated tool like a safety program.
2. Data Preprocessing: Remove noise, irrelevant information and formatting inconsistencies by

cleaning and pre-processing the data. This may involve tasks like removing special characters, normalizing text, handling misspellings, and removing stop words.

3. Text Tokenisation: Break down the preprocessed text into individual tokens or words. This step is essential for further analysis as it allows the system to process and analyze the text at a granular level.
4. Feature Extraction: Extract relevant features from the text that can capture the sentiment expressed by customers. A more advanced technique that can be included is topic modeling or named entity recognition.
5. Sentiment Classification: Machine learning can be used to classify feedback items' sentiment. This can be done with supervised learning platforms like Support Vector Machines, or deep learning models like Recurrent Neural Networks. One way that training data can be used to make predictions is on new, unlabeled data.
6. Model Evaluation: Use appropriate evaluation criteria such as accuracy, precision, recall, F1 Score, or area under the ROC curve to find out the performance of the sentiment classification model. This helps to understand the model and identify areas that can be improved.
7. Fine-tuning and Optimisation: Fine-tune the sentiment analysis model by experimenting with different algorithms, feature representations, hyperparameters, or incorporating domain-specific knowledge. This iterative process helps optimise the model's performance and adapt it to the specific context and requirements of the customers' feedback.
8. Visualisation and Reporting: Present the sentiment analysis results in a visually appealing and easy- to understand format. This can include sentiment scores, sentiment distribution graphs, word clouds, or summary statistics. Visualisation helps stakeholders quickly grasp the sentiment trends and make informed decisions based on the insights derived from the

5. Results and Discussion

After effects of sentiment analysis can be introduced as mathematical scores that show he sentiment of the feedback. There are various scales and techniques for computing sentiment scores, however here are a typical ways of introducing sentiment analysis results mathematically:

Size of - 1 to 1: This scale doles out a score of - 1 to message with an unequivocally negative sentiment, 0 to message with an impartial sentiment, and 1 to message with a firmly positive sentiment. Scores between - 1 and 0 show negative sentiment, while scores somewhere in the range of 0 and 1 demonstrate positive sentiment. For instance, a sentiment analysis device could relegate a score of 0.75 to a piece of feedback, demonstrating that it has an emphatically positive sentiment.

It's critical to take note of that sentiment analysis results ought to continuously be deciphered with regards to the feedback being dissected and the comprehension business might interpret what is positive or negative sentiment. Also, sentiment analysis apparatuses may not generally be 100 percent precise, and human survey might be important to guarantee that the sentiment scores are significant and exact.

out feedback got. Alternately, the extent of negative feedback got is moderately low around 12.4% and how much unbiased feedback is 6.8%.Sentiment analysis is an important device for organisations to acquire insights into how clients feel about their items, administrations, and brand. By dissecting client feedback, sentiment analysis can assist organisations with distinguishing regions for development, grasp client inclinations, and further develop the general client experience. In any case, there are a few significant contemplations to remember while utilizing sentiment analysis.

Accuracy: Sentiment analysis instruments are not generally 100 percent exact, and human survey might be important to guarantee that the sentiment scores are significant and precise. Sentiment analysis devices can battle to precisely understand the sentiment of message that incorporates mockery, incongruity, or different types of nuanced language. Thusly, it's vital to consider the setting wherein the feedback was given and what it could mean for the sentiment.

Nuance: Sentiment analysis devices might misrepresent the sentiment of a message, classifying it as one or the other positive, negative, or nonpartisan. Notwithstanding, client feedback is frequently nuanced and may contain both positive and negative components. It's critical to consider the nuances of the feedback and the setting wherein it was given when understanding the sentiment

Language and cultural differences: Sentiment analysis devices may not be as precise while breaking down message in languages or societies that contrast from the language or culture the apparatus was prepared on. This can prompt errors in sentiment analysis results, especially while managing shoptalk or colloquial articulations.

Integration with different metrics: To acquire a more complete comprehension of the client experience, sentiment analysis ought to be utilised related to different metrics, for example, consumer loyalty

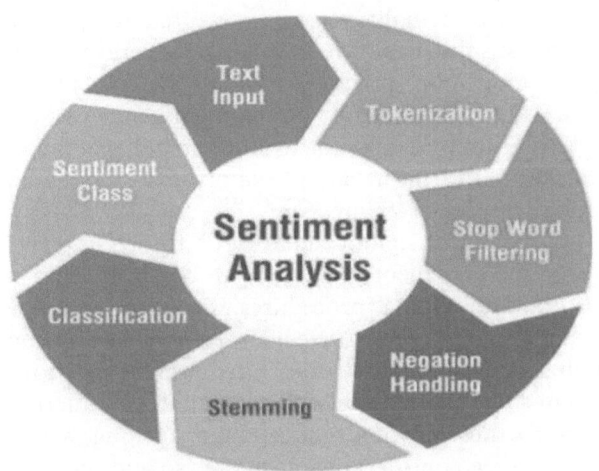

Figure 2: Sentiment revolution

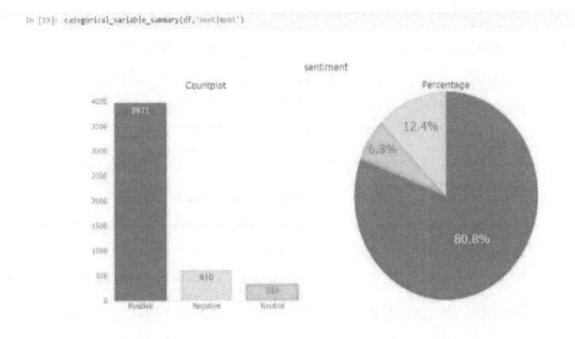

Figure 3: Pie chart of customers feedback

scores, net advertiser scores (NPS), or client degrees of consistency. By contrasting sentiment scores and different metrics, organisations can acquire a superior comprehension of how sentiment connects with the general client experience and recognise regions for development.

Actionable insights: The insights acquired from sentiment analysis ought to be utilised to illuminate activities that further develop the client experience. Organisations ought to routinely audit sentiment analysis results and make a move in view of the insights acquired to persistently further develop the client experience.

6. Conclusion and Future Scope

Sentiment analysis is a quickly developing field with a promising future. With the blast of web-based entertainment and other web-based stages, sentiment analysis has become progressively significant for organisations and associations to measure popular assessment and sentiment towards their items, administrations, or brand. One of the critical areas of future development for sentiment analysis is the improvement of more refined AI models. These models will actually want to more readily comprehend the subtleties of human language and feelings, prompting more precise sentiment analysis results. Also, the incorporation of sentiment analysis with different innovations, for example regular language handling and information examination, will additionally upgrade its capacities.

One more significant area of future improvement for sentiment analysis is in the utilisation of profound learning procedures. Profound learning calculations, like repetitive neural networks and convolutional neural networks, have previously shown promising outcomes in sentiment analysis undertakings. As these procedures keep on propelling, we can hope to see considerably more precise and powerful sentiment analysis models.

References

[1] Saad, I. A. and Yang, Y. (2019). Ordinal regression based sentiment analysis using machine learning algorithms. Journal of Information Science, 45 (2), 211–223.

[2] Fang, L., Zhang, J., He, X., Qian, Y., and Zhang, Z. (2018). Multi-strategy sentiment analysis models using semantic fuzziness. Information Processing & Management, 54 (2), 335–350.

[3] Mukhtar, et al. (2018). Sentiment analysis of Urdu blogs: A comparison of supervised machine learning and lexicon-based models. Information Processing & Management, 54 (6), 1086–1097.

[4] Xu, et al. (2020). NB for multi-domain and large-scale E-commerce platform product review classification of sentiment. Applied Soft Computing, 93, 106426.

[5] Ray, D. and Chakrabarti, A. (2019). Deep learning algorithm for feature extraction and sentiment analysis in text. International Journal of Intelligent Systems and Applications, 11 (1), 16–24.

[6] Park, S., et al. (2019). Semi-supervised sentiment analysis using sentiment- discriminative objective and partial data. Knowledge-Based Systems, 182, 104849.

[7] Zhao, S., et al. (2019). Image-text consistency driven multi-modal sentiment analysis. Information Sciences, 491, 153–167.

[8] Abdi, M., et al. (2018). Feature-based sentiment analysis of Arabic hotel reviews. Procedia Computer Science, 141, 119–126.

[9] Kumar, N., et al. (2020). ConVNet-SVMBoVW: A hybrid deep learning approach for fine-grained sentiment analysis in real-time data. IEEE Access, 8, 76858-76872.

[10] Vashishtha, S. and Susan, R. (2019). Fuzzy rule-based model for sentiment analysis of social media posts. In Proceedings of the International Conference on Information Technology and Computer Communications (pp. 60–65). Springer.

[11] Hassonah, M. A., et al. (2020). Enhancing sentiment analysis using a hybrid machine learning algorithm. IEEE Access, 8, 38952–38961.

[12] Smadi, M. M., et al. (2018). Feature-based sentiment analysis of Arabic hotel reviews using SVM and deep RNN models. In Proceedings of the International Conference on Arabic Language Processing (pp. 137–142) (SpringerLink).

[13] Maqsood, T., et al. (2020). Impact of events on stock market using sentiment analysis on Twitter data. In 2020 3rd International Conference on Computing, Mathematics and Engineering Technologies (iCoMET) (pp. 1–6). IEEE.

[14] Park, J., et al. (2020). Performance-enhanced deep learning model with content attention and non-linear merging." In Proceedings of the 2020 International Conference on Big Data and Education (pp. 78–82). ACM.

[15] Bardhan, R., et al. (2019). Understanding the impacts of gender mainstreaming in SRH management using a quasi-qualitative model. International Journal of Environmental Research and Public Health, 16 (16), 2900

[16] Araque, O., et al. (2017). Combining surface and deep learning models for sentiment analysis of Arabic text." Procedia Computer Science, 117, 209–218.

[17] Yousif, A., Smith, J., and Johnson, M. (2019). A multi-task learning approach for sentiment analysis using CNN and RNN. Proceedings of the International Conference on Machine Learning and Data Engineering, 2019, 123–135.

[18] Afzaal, U., et al. (2019). Aspect-based sentiment classification using a novel approach. Journal of Ambient Intelligence and Humanized Computing, 10 (6), 2363–2373.

[19] Feizollah, et al. (2019). Sentiment analysis of halal products using deep learning models. Journal of Soft Computing and Decision Support Systems, 6 (2), 30–42.

[20] Thelwall, M. (2020). Sentiment strength detection for the social web. Journal of the American Society for Information Science and Technology, 61 (12), 2544–2558.

Comparative Study of Network Slicing In 5G And 6G

Siddhima Singh, Aastha Kothari, Pooja Gupta

Manipal University Jaipur, Jaipur, India

E-mail: siddhima.229303326@muj.manipal.edu, Aastha.229311260@muj.manipal.edu, dr29.pooja@gmail.com

Abstract

Network slicing has become one of the most highlighted subjects in the area of comprehensive research. This paper includes network slicing in 5G that enables multiple services such as IOT, video surveillance, remote as well as industrial operations. 6G network slicing can provide even better platform that allows network operators to create slices that leads to highly specific use cases, and it is accustomed to providing more reliable services with enhanced and modified features to operate network slices more efficiently.

Keywords: SDN, NFV, network slicing, 5G, 6G

1. Introduction

A network slice is a shared platform that overlays on top of numerous shared and isolated network layers or slices with hands on services quality that cannot be negotiated. Network slicing uses some architecture that allows the fast segmentation of slices of network which allows support to devices. Some of the are SDN and NFV. It has been recognised as the pillar of the enormously evolving 5G technology. It has been introduced in 5G with standard application with a great progressive vision. However, Network Slicing has become the backbone of 5G networks which provided us with technologies at the same time with high efficiency [1]. The end-to-end network consists of three parts:

1. Radio Access Network
2. Mobile Core Network
3. Transport Network

The Transport Network is mainly used to provide on demand connections between radio access networks and core network along with guaranteeing service SLAs.

2. History

Network slicing originated with the idea of overlaying networks [2]. Now Designers have achieved this by creating Virtual Network (VN) (see Figure 1) running on a common platform. However, this cannot be programmable at that point of time.

The beginning of SDN technology have been started from 2009, further advanced the overlaying network slicing technology by operating an open interface that permits users to use more efficient and scalable slices. Examples of the latest slicing

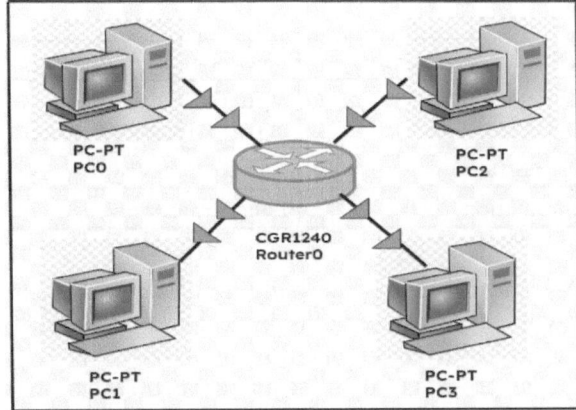

Figure 1: Virtual networking

technology include cloud gaming which provides user an excellent gaming experience.

3. Need for Network Slicing

The demand for high-speed, low-latency, and secure network services is increasing rapidly. Network slicing offers optimisation for specific services as it can be customised according to the requirement of the modern services. It can also provide multiple virtual networks for various services. The inflexibility of traditional networks does not allow it to fulfil the different requirements. Thus there is a need for network slicing.

5G has an improved network architecture compared to 4G. It's network function virtualisation (NFV) and service based architecture provide (SBA) (see Table.1) have the capability to provide distinct slices for varied services. Hence it supports network slicing. 4G network lacked this ability [7].

Chapter 9 DOI: 10.1201/9781003570349

The recent advancement in 5G Network Slicing are live video streaming for safety applications at an industrial site and to manage local traffic also.

4. Benefits of Network Slicing

- It involves easy business possibility as before 5G small office depended on 4g terms and back-up connectivity. It allows WWAN and WLAN which provide higher speeds and slicing.
- Each slice assures quality of service (QoS) for a particular trait – such as low latency or high bandwidth.
- Better utilisation and easy scaling over physical infrastructure.

4.1. Architecture of Network Slicing

4.2. Functionalities of SDN and NFV in Slicing

Table 1: Components of network slicing architecture

Network slicing architecture	Components
SDN	Network controller
NFV	Virtual network functions
Orchestration	Service orchestrator
Automation	Policy framework

SDN and NFV are like building blocks that helps network slices to use both physical and virtual slices in a more efficient manner [6].

The process of dividing and arranging the physical components such as servers, cables and hardware of a network is known as physical slicing. Virtual slicing means to have individual parts that act independently but share same physical resources.

Software defined Networking (SDN) isolates the data plane and Network Functions Virtualisation (NFV) extracts network services and functions from proprietary hardware and converts them to a computer generated simulation so that they can run as software.

SDN allows network operators to create various parts of their network, these are known as slices. It also creates special services make efficient use of network by changing the amount of network power various tasks are provided. Instead of specialised hardware used for network tasks NFV allows network operators the usage of regular computers.

4.3. Process of Network Slicing

The framework of network slicing comprises of two main blocks, one is for network slice implementation and other is for network slice management. The

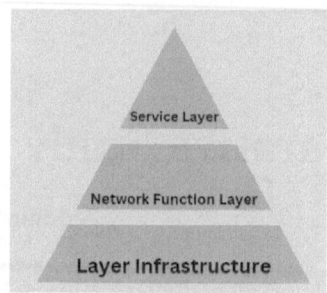

Figure 2: Layers of first block

first block has three layers which are Service layer, Network Function Layer, and Infrastructure Layer. The second block network slice controller that manages the functionality between three layers see Figure 2.

- **Service Layer:** It usually interacts with the 3rd party service providers that share physical network and manages service requirements.
- **Network Function Layer:** It creates network slice according to request from upper network slices.
- **Layer Infrastructure:** It depicts the actual physical network topology to initiate the several network functions.
- **Network Slice Controller:** It manages end to end services, virtual resources, and Network slice life cycle.

5. Network Slicing In 5G

Network Slicing in 5G supports different resources for users on various platforms with the great download speed of 20 Gbps 5G Figure 3. This has been through using multiple Virtual Networks on a common physical architecture. It has been assisted on heterogeneous network based design that enable services like uRLLC(Ultra-reliable and Low Latency Communication) and mMTC (Massive Machine Type Communication) and eMBB(Enhanced Mobile Broad-Band) Table.2. Network slicing in 5G should be end to end means it should support Radio network, transport network and core network. In core network users are achieving 5G network slicing by using multiple virtual machines(VM), AMFs, SMFs according to slice [3].

5.1. How Network Slicing Works in 5G

5.2. 5G Service Types that uses Slicing for Handling Traffic

5.3. Network Slicing use Cases

- It involves capacity which provide multiple connections that helps to process devices to meet the requirement of slice.

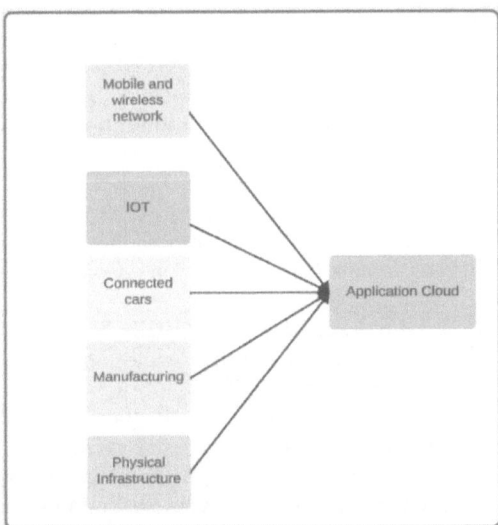

Figure 3: Network slicing in 5G network

Table 2: 5G service verticals

eMBB	mMTC	uRLLC
Enhanced Broadbands	Large no. of devices	Low latency
HD video call	Deep penetration	High reliability
Tactile Web	Power Efficient	V2X and PS Apps

- It involves security which minimise the risk of a leak of customer data so for that it requires a slice with single user virtual resources.
- It involves easier scaling and flexibility by offering more services and it do need much hardware.

6. Network Slicing in 6G

In 6G network slicing means more capacity, speed and improving the flexibility to create next level customised slices. Network slices in 6G are horizontal that allows various industries to work collectively. It allows services and apps perform more efficiently, reduces the complexity of network.

6.1. Technical Requirements for 6G Network Slicing

These are the mandatory technical requirements to make network slicing in 6G work:

- **Programmability:** The ability to write software code that divides the physical network into logical segments.
- **Elasticity:** Network has the capability to adapt to changes in the demand or requirement of the services being requested.

- **Orchestration:** Management and supervision of resources and capacity distribution with respect to SLAs (Service Level Agreements).

6G network slicing divides the network into virtual networks for designated uses. Examples include autonomous vehicles (low latency), remote healthcare (secure connectivity), industrial automation (IoT support), and immersive virtual reality (high bandwidth). For instance:

- **Autonomous Vehicles:** Ultra-low latency and high-bandwidth communication for autonomous vehicles can be supported by a specialised network slice.
- **Remote Healthcare:** A slice can provide low latency, high dependability, and strict security protocols would be given top priority in this slice in order to facilitate data-intensive medical applications, remote patient monitoring, and real-time telemedicine.

7. Challenges for Networking Slicing in 6G

There are many challenges in creation of network slicing strategies in 6G networks because they have distinctive characteristics. One major challenge is effective coordination of diverse network segments across space, air, and ground in the SAGIN (space-air-ground integrated network). Network slicing raises security concerns, as the isolation between slices must be maintained to forbid one slice from interfering with another. Allocating and controlling resources for numerous slices while ensuring optimal performance and efficient utilisation can be a challenge [4].

8. Comparison Between Network Slicing in 5G vs 6G

5G's network slicing makes effective use of infrastructure and spectrum to maximise capacity for particular applications. In order to meet the needs of data-intensive apps like AR, VR, and large IoT, 6G promises to increase capacity even further [5]. Low to moderate latency is provided via 5G's network slicing, which is essential for real-time applications like autonomous vehicles. Ultra-low latency is the goal of 6G for even more time-sensitive applications, such as immersive gaming and remote surgery. 5G uses SDN and NFV to dynamically allocate resources via network slicing. In order to improve communication realism and optimise resource allocation based on current circumstances and user demands, 6G aims to combine holographic communication with AI-driven networking see Table 3.

Table 3: Network slicing comparison in 5G vs 6G

Aspects	Network slicing in 5G	Network slicing in 6G
Capacity	high	increased
latency	Low to moderate	Ultra-low
Technologies	Advanced	Holographic communication

9. Future of Network Slicing

In November 2021 Bharat 6G Vision document was formed and later it was revealed and the 6G R&D Test Bed was launched. Academic institutions, business, start-ups, MSMEs, and industry, among others, will have a platform to test and validate developing ICT technologies due to the 6G Test Bed.

10. Conclusion

A real time example of network slicing is the 5G connected ambulance developed by Chunghwa Telecom for St. Paul's Hospital are enabled by Ericsson 5G standalone architecture. Isolated network slices and radio resources partitioning ensure real-time patients' data in 4K videos seamlessly transmitted from ambulance to the hospital. In conclusion, although it develops and becomes more sophisticated in 6G, network slicing is a basic idea in both 5G and 6G wireless technologies. Network slicing in 6G offers a bright future for wireless technology as cross industry partnerships pave the way for increase customisation and effectiveness.

References

[1] Zhang, Q., Li, W., & Chen, X. (2022). A comparative analysis of network slicing in 5G and future perspectives for 6G. IEEE Communications Magazine, 60(5), 78-85

[2] [2]   Wang, Y., Zhou, L., & Liu, S. (2023). Network slicing architecture: Lessons from 5G and prospects for 6G. In Proceedings of the IEEE International Conference on Communications (ICC) (pp. 234-240)

[3] Li, Z., Wu, J., & Zhang, H. (2023). Network slicing technologies in 5G and their evolution towards 6G. IEEE Network, 37(3), 28-35.

[4] Kumar, A., Sharma, R., & Jain, R. (2022). Comparative analysis of network slicing in 5G and 6G: Requirements, challenges, and opportunities. In Proceedings of the IEEE Wireless Communications and Networking Conference (WCNC) (pp. 1-6). IEEE. X. Shen, J.

[5] Wang, L., Liu, Y., & Huang, H. (2023). Network slicing techniques in 5G and anticipated enhancements in 6G networks. In J. Lee & K. Park (Eds.), Advancements in Telecommunication Systems (pp. 123-145).

[6] Chen, C., Liu, Y., & Zhang, X. (2022). Network slicing management in 5G and its potential in 6G networks. IEEE Transactions on Wireless Communications, 21(8), 5321-5335.

[7] Park, S., Kim, J., & Lee, H. (2023). Evolution of network slicing architectures: From 5G to 6G. In Proceedings of the ACM International Conference on Mobile Computing and Networking (MobiCom) (pp. 112-120).

An Investigation of Load Balancing Algorithm for Cloud Computing

Manisha Thakur, Susheela Hooda, Rupali Gill

Chitkara University Institute of Engineering and Technology, Chitkara University Punjab, Rajpura, India
E-mail: tmanisha194@gmail.com, Susheelahooda@gmail.com, rupali.gill@chitkara.edu.in

Abstract

In the realm of computing, cloud computing has become the most popular technology. Now a days, due to frequent usage of cloud computing, load balancing becoming more complex problem in front of researchers. Load Balancing in cloud computing refer to process of distributing workload and network traffic across multiple server and resource in cloud computing. In this paper, various load balancing algorithms have been studied to analyze the performance of cloud computing. This paper throws the light on the available load balancing algorithm, their advantages and limitations. This study provides fresh light on load balancing in cloud computing and highlights key problems, while developing effective load algorithms

Keywords: Cloud computing, cloud service provider, load balancing algorithm, static load balancing algorithm, dynamic load balancing algorithm

1. Introduction

An on-demand shared pool of virtualised resources is what is meant by the term "cloud computing."- Cloud computing has propelled the idea of utility computing, which is now a reality. Cloud computing provides Infrastructure as a Service (IaaS), Platform as a Service (PaaS), and Software as a Service (SaaS). A computing resource can be accessed on demand by the user [1]. There are many popular cloud service providers, e.g., Amazon, Google, etc. Load balancing is the process of distributing workload across multiple computing resources like server, virtual machine, increase availability and improve performance. Load balancing work by distributing incoming network traffic across multiple computing resources in a way that avoid overloading any single resource [2]. The cloud provides infrastructure and a shared platform for hosting user application [2,3]. Figure 1 depicts a bird's-eye perspective of cloud computing. Depending on the deployment type and the constraints on accessibility, clouds may be divided into three groups such as 1) Public Cloud, 2) Private Cloud, and 3) Hybrid Cloud [3]. Virtualization is a technology that enables multiple virtual machine or application to run on single physical server or host computer Virtualization cloud server to serve their customers efficiently. Live VM migration is one of the most promising virtualization technologies. In the virtual machine, data from the source server is moved to the destination server while the process continues to run on the source server [1,2]. Network traffic is divided with load balancing across the multiple server or resource in order to improve application performance by increase the response time and reducing network latency [3]. Main goals of load balancing are such as: managing and regulating traffic spikes on a single server, speeding up user request responses, preserving the stability of the system and to encourage user enjoyment [3-5]. The movement of workload and processing resources from one server to another is carried out by cloud computing. The distribution gives the optimum performance with the fastest response time. To enhance utilization and shorten response times to system processes, two or more server- network interfaces for other computer resources are divided. Effective cloud load balancing

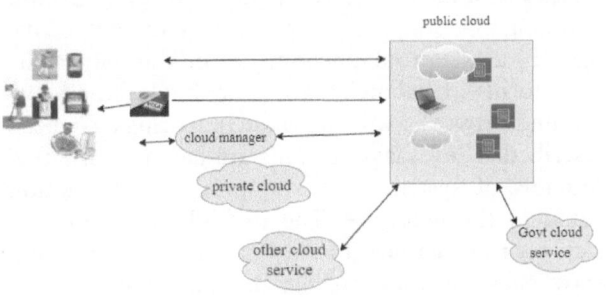

Figure 1: A view of cloud computing

Chapter 10 DOI: 10.1201/9781003570349

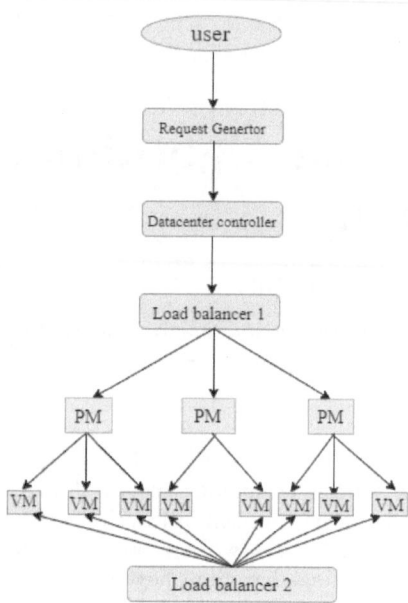

Figure 2: Architecture of two-level load balancing.

may assure continued operations with a heavily trafficked website [4,5]. It is an essence between client and service provider. Figure 2 depicts working of load balancer.

The structure of the remaining portion of the paper goes as follows: Section II describes literature review of various cloud computing models. Section III discusses detailed analysis of various loads balancing algorithms and proposed methodology to compare available load balancing algorithm. Section IV describes various challenges which usually occur during load balancing. Section V presented result and discussion. Conclusion and future scope have been presented in Section VI.

2. Literature Review

Numerous researchers have been working on usage of resource energy, load balancing, resource management, virtualization etc. In this section, load balancing algorithms have been discussed as

Ghomi et al. [6] authors categorised job scheduling and load balancing techniques into seven groups, including Hadoop-map decrease load balancing, agent-based, natural phenomena, in this paper the three forms of balancing are generic, workload-specific and network awareness load balancing have been described. These classifications are based on the two domains of system state and process initialization found in the literature. The multiple algorithms in each area are gathered, and both their benefits and drawbacks are noted. Shah et al. [7] discussed various load balancing methods which have been categorised as static and dynamic, homogeneous and heterogeneous and VM type uniformity on the bases of system status. Benefits and drawbacks of each algorithm

have also been covered. The study doesn't approach literature in a methodical way. Esbahi and Rahmani [8] in order to create and implement the ideal Load balancer for the cloud provider, we categorised load balancing methods into three categories: general algorithm, architecture, and artificial intelligence based. Similar to prior studies, in this paper, load balancing algorithms have also been categorised into static and dynamic. Results indicate that dynamic, distributive, and non-cooperative algorithms are the best among all load balancing algorithms. Abdul Hameed et al. [9] explored the unresolved issues, surrounding the distribution of resource in the energy efficient manner.

Neghabi et al. [10] discussed the various load balancing techniques which is used in software defined network (SDN).SDN is a type of network architecture that control the full network which is worked on the process of load balance. The paper also suggests the different metrics for performance of load balancing technique and presents how to use load balancing in SDN to improve the performance and reliability. Sheetal Karki et al. [11] discussed the cloud computing provider which is used to manage the resources by end users and it includes the server, storage and network resources. This paper also discussed load balancing, which is used to enhance the performance, reliability and help to save the energy. In the other hand the threshold algorithm is used in this paper that identifies when a virtual machine is overloaded and has to move to another virtual machine. Aarti Singh et al. [12] proposed a dynamic load balancing algorithm which is based on three factors i.e. CPU usage, memory usage and fitness usage. In this paper, authors also worked on migration agent and also evaluated the performance of proposed load balancing algorithm with contemporary algorithms. Results proved that proposed algorithm reduces the load migration time and improve the load balancing.

As per our literature study, very few works have been done in load balancing and virtualization. Therefore, this paper enlightens the work on load balancing algorithms, virtualization which will be helpful for future researchers

3. Proposed Methodology

This section caters to the issues and challenges in load balancing algorithms. Therefore, a set of questions have been formulated to address the issues. This section also describes classification of load balancing algorithms and comparative study of existing load balancing algorithms.

A. Question identification:-

From literature survey, a set of questions were identified that must be acknowledged before start the load balancing process.

RQ1: Why is load balancing essential for current cloud computing?

RQ2: What is advantage and disadvantage of existing load balancing algorithm?

RQ3: What are the future challenges in load balancing?

This section answers RQ1 as a result of load balancing importance in cloud computing. The resource efficiency, throughput, fault tolerance, migration times, degrees of balance, make span, and scalability of the load balancing examine have to be enhanced. Reduce resource waste, migration costs, electricity and energy use, and SLA breaches simultaneously. Low quality of service (QOS) to CSC (Common Service Centre) and a decrease in the economy as determined by profit to Cloud Service Provider (CSP) are the results of these factors' values degrading. [13]. In the view of cloud computing RQ2, advantages of cloud computing are such as improved performance, increase scalability and availability. Load balancing allow distributing traffic across multiple server and preventing failure caused by overloading a particular resource [14,15]. RQ3 is future challenges in load balancing cloud computing in which manage increase amount of data, security, privacy and load migration.

The system was divided into two categories based on its current state:

3.1. Static Load Balancing

A load balancing technique is considered "static" in terms of work allocation if it fails to take system status into account. The state of the system contains elements like how much each processor is being used (and occasionally even an overflow). Systems for static load balancing often concentrate on a route. Static algorithms have the advantage of being simple to set up and very effective in operations that occur fairly frequently.

3.2. Dynamic Load Balancing

Dynamic algorithms, as opposed to static load distribution approaches, take into account the present load of each computer unit (also known as a node) on the system. So that tasks may be handled more rapidly, they can be dynamically moved from a node that is overloaded to a node that is under loaded. These algorithms can produce excellent results despite being much [16,17].

Table 1 depicts the advantages and disadvantage of different load balancing algorithm in cloud computing.

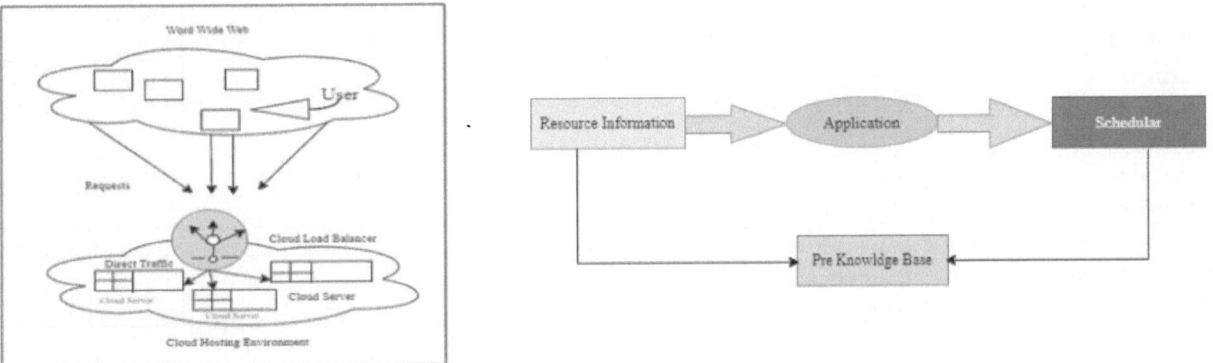

Figure 3: Shows the static and dynamic load balancingalgorithm

Table 1: Analysis of load balancing algorithms

Author	Year	Algorithm	State of Algorithm	Technique Involved	Advantage	Disadvantage
Pasha, N., Agarwal, A. and Rastogi, R.	2015	Round Robin Algorithm	Static	Heuristic	Round Robin doesn't require inter- process communication.	Some nodes may always be fully occupied while others are idle.
R. Kaur and P. Luthra	2023	Central manager	Static	Optimization	The system load static informs the load scheduler decision on load balancing.	It is necessary for there to be extensive inter process.

(*Continued*)

Table 1: (Continued)

S.K.Vasudevan S.Anandaram	2017	HoneyBee Foraging Behavior	Dynamic	Optimization	Response speeds and waiting times have decreased due to the virtual machine.	The amount of resources has an inverse correlation
Ajay Gulati and Rajeev K Chopra	2014	Round Robin	Dynamic	Heuristic	Increase high tolerance, reduce migration time	High SLV
Mathur S, Larji AA, Goyal A	2019	Max-Min	Static	Optimisation	Tolerance fault is low	Low degree of balance.
Chen H, Wang F, HelianN,Akau G	2020	Mini-Mini	Static	Optimisation	Small make span	Low scaling and throughput
Mondal B, Choudhury A	2016	Non-Classical	hybrid	Optimisation	Low time of execution and response time	Low scability
Tripathi AM, Singh S	2018	Active Monitoring	Dynamic	Heuristic	Low overhead and high resource utilisation	High SLV
Chen SL, Chen YY, Kuo SH	2017	Weight Round Robin Algorithm	Hybrid	Heuristic	The use of resource is beneficial	Response time cannot be selected.
Kapur R	2015	Non-Classical	Dynamic	Heuristic	Extreme Scalability, Shorter reaction.	Low level of resource
Rajput SS, Kushwah VS	2019	GA and Min	Hybrid	Heuristic	Low response time and small execution cost.	Less degree of utilisation.
Dasgupta K, Mandal B, Dutta P,MandalJK,DmS	2016	GA	Dynamic	Metaheuristic	Shorter reaction time and task rejection ratio.	Low throughput and scability.
Sharma S, Luhach AK, Abdhullah SS	2017	BAT Algorithm	Dynamic	Optimisation	Less Execution cost.	Energy inefficient.
Mondal B, Choudhury A	2019	Simulating Annealing	Dynamic	Optimisation	High Scalability.	High SLV.
Kumar M, Dubey K, Sharma SC	2018	Conventional and non classical	Dynamic	Heuristic	It boots electricity.	Execution time is more

4. Need of Load Balancing

Another important aspect of a cloud's scalability is load balancing. Cloud infrastructure should be easily scalable to deal with fluctuations in traffic when a cloud scales up, many virtual servers and a number of applications are often spun up. The load balancer is the main network element that divides traffic between these new instances [4,6]. The newly spun-out virtual server could not accept incoming traffic at all if load balancers are not present. While others are overloaded, some servers are left powerless to deal with the workload. A load balancer may also identify server unavailability and direct traffic to servers that are still operational [4,18]. Depending on the load balancing algorithms, load balancers can even determine whether a certain server (or server set) is likely to get overwhelmed more quickly and can direct traffic to other nodes that are considered to be in better condition. Such preventative abilities may substantially decrease the possibility that your cloud services won't be available. Another essential aspect of green cloud computing is load balancing [19, 20].

5. Challenges of Load Balancing

Virtual machine migration: The idea is to build a machine as a file or a file collection. The virtual machine can be moved effectively to lighten the strain on a loaded computer. When load is distributed dynamically across a machine, the goal is to minimise or decrease stress on those systems [17,21].

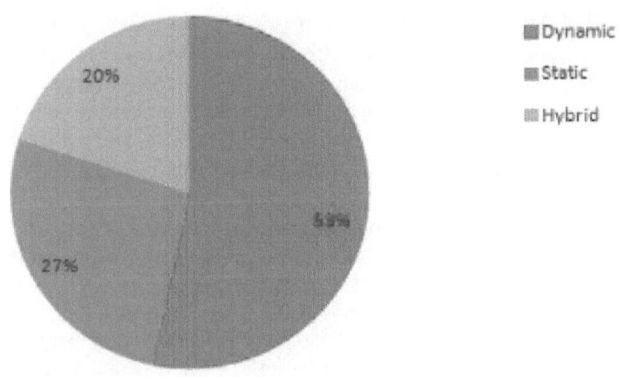

Figure 4: Analysis of static and dynamic algorithm

Data management and archiving: Information archiving is another essential requirement. So how can data be shared in a cloud system with the best possible storage and access? [22]

Load Balancing Scalability: Guests may access resources for rapid scaling at any time because of accessible and on-demand scalable cloud services. A reliable load balancer should adjust for rapidly changing demands in the processing environment, memory, device architecture, among other areas [15,20].

6. Result and Discussion

This section lineament the result achieved from the comparative analysis of different load balancing approaches in cloud computing. Figure 4 show the analysis of static and dynamic algorithm. As per our study, dynamic algorithm has 53% contribution, static algorithms 27% and hybrid algorithms have 20% contribution in cloud computing.

7. Conclusion

Cloud computing is a way for many people to independently access various resources through the internet. However, there are considerable obstacles in the way of cloud computing. A significant barrier to cloud computing is load balancing. In this study, several static and dynamic algorithms are covered. In this research, several load balancing techniques have been investigated to evaluate cloud computing performance. This paper discusses the various load balancing algorithms, their benefits, and their drawbacks. This study also outlines the significant difficulties in creating effective load balancing algorithms and offers fresh perspectives on cloud computing.

References

[1] Afzal, S. and Kavitha, G. (2019). Load balancing in cloud computing – A hierarchical taxonomical classification. Journal of Cloud Computing: Advances Systems and Application, 8(1), pp.1-24.

[2] Neghabi, A. A., Jafari Navimipour, N., Hosseinzadeh, M., and Rezaee, A. (2018). Load balancing mechanisms in the software defined networks: A systematic and comprehensive review of the literature. IEEE Access, 6, 14159–14178.

[3] Jafarnejad Ghomi, E., Masoud Rahmani, A., and Nasih Qader, N. (2017). Load-balancing algorithms in cloud computing: A survey. Journal of Network and Computer Applications, 88, 50–71.

[4] Shah, Jaimeel M., Ketan Kotecha, Sharnil Pandya, D. B. Choksi, and Narayan Joshi. "Load balancing in cloud computing: Methodological survey on different types of algorithm." In 2017 International conference on trends in electronics and informatics (ICEI), pp. 100-107. IEEE, 2017

[5] Sharma, S., Singh, S., & Sharma, M. (2008). Performance analysis of load balancing algorithms. International Journal of Civil and Environmental Engineering, 2(2), pp.367–370.

[6] Mishra, S. K., et al. (2018). Energy-efficient VM-placement in cloud data center. Sustainable Computing: Informatics and Systems, 20, 48–55.

[7] Shafiq, D. A., Jhanjhi, N. Z., and Abdullah, A. (2022). Load balancing techniques in cloud computing environment: A review. Journal of King Saud University: Computer and Information Science, 34 (7), 3910–3933.

[8] Kumar, A., Pandey, S., and Prakash, V. (2019). A survey: Load balancing algorithm in cloud computing. SSRN Electronic Journal.

[9] Mishra, S. K., Sahoo, B., and Parida, P. P. (2020). Load balancing in cloud computing: A big picture. Journal of King Saud University: Computer and Information Science, 32 (2), 149–158.

[10] Bharany, S., et al. (2022). A systematic survey on energy- efficient techniques in sustainable cloud computing. Sustainability, 14 (10), 6256.

[11] Kaur, H. and Mann, P. S. An improved hybrid re-encryption scheme for mobile cloud computing environment. Ijcaonline.org.

[12] Pasha, N., Agarwal, A., and Rastogi, R. (2014). Round robin approach for VM load balancing algorithm in cloud computing environment. International Journal of Advanced Research in Computer Science and Software Engineering, 4, 34–39.

[13] Gulati, A. and Chopra, R. K. (2013). Dynamic round robin for load balancing in a cloud computing. Ijcsmc.com.

[14] Vasudevan, S. K., Anandaram, S., Menon, A., and Aravinth, A. (2016). A novel improved honeybee based load balancing technique in cloud computing environment. Asian Journal of Information Technology.

[15] Sharma, A., Mohana, R., Kukkar, A., Chodha, V., and Bansal, P. "An Ensemble Learning Based Framework for Smart Landslide Detection, Monitoring, Prediction and Warning in IoT-Cloud Environment." (2023) https://doi.org/10.21203/rs.3.rs-3137501/v1

[16] Kumar, M. and Sharma, S. C. (2017). Dynamic load balancing algorithm for balancing the workload among virtual machine in cloud computing. Procedia Computer Science, 115, 322–329.

[17] Zhang, F., Liu, G., Fu, X., and Yahyapour, R. (2018). A survey on virtual machine migration: Challenges, techniques, and open issues. IEEE Communications Surveys & Tutorials, 20 (2), 1206–1243.

[18] Yu, D., Ma, Z., and Wang, R. (2022). Efficient smart grid load balancing via fog and cloud computing. Mathematical Problems in Engineering, 2022, 1–11.

[19] Khattar, N. Sidhu, J., and Singh, J. (2019). Toward energy- efficient cloud computing: A survey of dynamic power management and heuristics-based optimization techniques. Journal of Supercomputing, 75 (8), 4750–4810.

[20] Harnal, S., Sharma, G., Seth, N., and Mishra, R. D. (2022), Load balancing in fog computing using QoS. In Lecture Notes on Data Engineering and Communications Technologies. Singapore: Springer Singapore, pp. 147–172.

[21] Hooda, S., Lamba, V., and Kaur, A. (2021). AI and soft computing techniques for securing cloud and edge computing: A systematic review. In 2021 5th International Conference on Information Systems and Computer Networks (ISCON) (pp. 1–5), IEEE.

[22] Wang, L., et al. (2010). Cloud computing: A perspective study. New Generation Computing, 28 (2), 137–146.

Enhancing Organisational Performance

The Role of Employee Motivation in Indian Corporate Sectors

Shobhanam Krishna[1], Anita Choudhary[2], Rohit Dwivedi[3]

[1,2,3]Dept. of Organizational Behaviour and Human Resources, Indian Institute of Management Shillong, Shillong, India
Email: shobhak.phd22@iimshillong.ac.in

Abstract

Motivation is a crucial driving force that propels human behaviour toward goals, especially within the workplace, where it significantly influences both employee commitment and the overall performance of organisations. This research aims to delve into several key aspects, including identifying and prioritizing critical success factors (CSFs) for motivation and examining the intricate relationships among these factors utilizing Total Interpretive Structure Modelling (TISM). The originality of this study resides in its unique emphasis on the hierarchical positions occupied within an organisation. In contrast with conventional motivational theories, which treat the workforce as a cohesive unit, this research acknowledges and investigates the possible variations in motivational determinants between employees and managers. Incorporating TISM into the research constitutes a methodological advancement. The TISM framework facilitates an exhaustive analysis of the interconnections between identified critical success factors, offering a systematic model that surpasses a superficial understanding of motivation. The potential outcomes of this research encompass various areas of improvement, including heightened organisational performance, enhanced employee retention rates, the design of customised incentive programs, more effective recruitment strategies, the cultivation of a positive workplace culture, optimised resource allocation, stimulated business growth, and the facilitation of informed decision-making processes. Ultimately, this knowledge contributes to the advancement of employee engagement and augments organisational success. The research confronts the conventional uniformity of motivational theories by recognizing and investigating the nuanced motivations of employees and managers, thereby contributing to the evolving understanding of workforce motivation.

Keywords: Critical success factors (CSFs), human resource management, MICMAC analysis, motivation, organisational performance, organisational success, Total Interpretive Structure Modelling (TISM)

1. Introduction

1.1. Background

The term "motivation" can be traced back to its Latin term "movere," which means to move. Motivation is often described as an inner internal state that energises or moves (hence motivation), provides energy, and directs or channels guides behaviour toward achieving specific goals. As a mental and behavioural concept, motivation varies from person to person. Therefore, it should be different for people from numerous walks of life and backgrounds, such as educational attainment, social standing, and occupation.

It is impossible to overstate the importance of motivation in the work context, as it significantly impacts the commitment and efficacy of employees in organisations. Organisations can nurture a more engaged and productive workforce by identifying and addressing individual motivational factors, resulting in enhanced outcomes.

1.2. Problem Discussion

Motivation is essential to the effective operation of organisations. The company's overall performance would improve with motivated workers because individuals might exert the required effort Dobre [12]. While studying management, it becomes clear that most extant motivational theories focus on employees in general, ignoring the specific dynamics between employees of different levels. Still being examined are differences in motivation between various hierarchical positions. It is still being determined whether traditional theorists have considered this aspect or whether it has been neglected over time.

This study seeks to identify, determine, and analyze critical success factors (CSFs) that can assist decision-maker executives in prioritizing motivational factors. The primary objective is to determine whether there are discernible differences between managers' and employees' work motivation. The study aims to determine whether a variety of factors play a role in motivating these two groups or whether they share similar understandings and motivational aspects. This study aims to provide valuable insights into the complexities of motivation across various organisational hierarchical positions. This can potentially increase organisational management strategies and employee engagement.

The research objectives:

The research intends to accomplish the following objectives:

- Identify and prioritise the critical success factors (CSFs) of motivation within the corporate sector companies in India.
- Investigate the interrelationships and interactions among the identified CSFs using Total Interpretive Structure Modelling (TISM).
- Develop a model that prioritises CSFs to enhance organisational performance.
- Analyze and compare motivation factors among organisational hierarchies, including managers and employees.

2. Literature Review

Motivation plays a crucial role in organisations, especially in organisational dynamics, particularly within the job context, where employment, as individuals allocated, vote a significant portion of their time to their job roles and duties. The recurrence of duties may result in a decrease in employee motivation and the emergence of feelings of fatigue. Therefore, it emphasises the crucial significance of human resource managers in implementing various tactics designed to maintain and reignite motivation, ensure consistent productivity, and achieve organisational goals.

Motivation is a powerful catalyst that propels individuals to engage in particular behaviours. The phenomenon in question is subject to the influence of both intrinsic and extrinsic causes, as emphasised by renowned scholars, including [1-3]. According to [4,5], intrinsic motivation, which originates inside, drives individuals to pursue their objectives and fulfill their unmet needs [2]. Emphasise the importance of reinforcement in molding behaviour and maintaining activities over a prolonged period.

The relationship between employee motivation and organisational success has been a topic of ongoing discussion among scholars in academia and practitioners in human resources. According to [6], motivated people significantly impact their job happiness, which subsequently enhances organisational performance and adds to revenue growth and general vitality. In order to promote job satisfaction, it is crucial to effectively encourage individuals by aligning their competencies with their assigned responsibilities.

According to Pitt and Tucker [7], evaluating an organisation's performance is a crucial indicator of its effectiveness in attaining objectives through internal procedures or tangible outcomes. The facilitation of career advancement and skill acquisition for employees through structured training programs can act as a source of incentive, yielding benefits for both the organisation and its personnel. Harrim [8] argues in favour of thoroughly assessing organisational performance beyond solely relying on financial indicators. This approach should consider additional elements such as employee satisfaction, innovation, and consumer contentment, as highlighted by Katou and Budwar [9].

According to Aluko [10], organisations desire long-term expansion, prioritizing profitability, improving employee contentment, and augmenting productivity. Bratton and Gold [11] emphasise the crucial significance of employees in the context of an organisation. Therefore, it is crucial to implement efficient incentive systems in order to achieve organisational goals and promote overall success.

3. Methodology

The research approach involves data collection from managers and employees of diverse Indian corporate sector organisations. The sample group includes a total of 100 employees and 100 managers. Using the frameworks of Total Interpretive Structural Modelling (TISM), the present research investigates the hierarchical relationships between the identified Critical Success Factors (CSFs). Moreover, a classification analysis employs cross-impact matrix multiplication to assess the driving and dependent capacities of these CSFs.

4. Data Analysis

Total Interpretive Structural Modelling is a beneficial tool that facilitates identifying and structuring intricate interconnections between variables that delineate a specific problem or matter. The provided framework presents a systematic structure for comprehending the intricate characteristics of these variables. The objective is to comprehend various components' interplay and hierarchical organisation, yielding significant insights. The Critical Success Factors are categorised based on their degree of influence and interconnectedness. Subsequently, a MICMAC analysis is performed on the TISM model.

Step 1: Identifying Critical Success Factors (CSFs)

A thorough literature review identified ten Critical Success Factors (CSFs) associated with

Table 1: Final version of reachability Matrixmatrix for CSFs

CSFs	CSF no.	1	2	3	4	5	6	7	8	9	10	Driving Power
(a) Job Satisfaction	1	1	1*	1	1	0	0	0	0	0	0	4
(b) Promotions/ Expectation	2	1*	1	0	1	0	0	0	0	0	0	3
(c) Recognition	3	1	1	1	1*	0	0	0	0	0	0	4
(d) Good salary	4	1	1	1*	1	0	0	0	0	0	0	4
(e) Organizational Styles	5	0	0	0	0	1	1	1*	1	1	1	6
(f) Satisfying Goals	6	1	1*	1*	1*	0	1	1	0	0	0	6
(g) Team Spirit	7	1*	1*	1*	1*	0	1	1	0	0	0	6
(h) Good Working Conditions	8	0	0	0	0	0	0	0	1	1	1	3
(i) Working Hours	9	0	0	0	0	0	0	0	1	1	1	3
(j) Culture	10	0	0	0	0	0	0	0	1*	1	1	3
	Dependence Power	6	6	5	6	1	3	3	4	4	4	

Source: Author's compilation

motivation. Table 1 provides more information regarding these distinct CSFs.

Step 2: Developing the Interpretive Logic Knowledge Base

The TISM methodology involves the execution of comparisons between pairs. The methodology employed in this study represents the relationship between variables A and B using the binary notation, where a connection is denoted as '1' and the absence of a connection is denoted as '0'. In cases where there is a connection between A and B and another between B and C, a transitivity link of '1 a' is established. The data collected from the participants serves as the basis for constructing a knowledge base of interpretive logic.

Step 3: Developing the SSIM and Final Reachability Matrix

The ultimate reachability matrix is created by incorporating transitivity principles. This matrix outlines the driving capacity and dependencies of each variable. The insights obtained are then utilised in the MICMAC analysis, which classifies the factors into four distinct types: autonomous, dependent, linkage, and independent (driver) barriers, as shown in Table 2.

The building of the final reachability matrix involves the integration of transitivity concepts. The matrix presented herein delineates the driving capability and interdependencies of each variable. The acquired insights are subsequently employed in the MICMAC analysis, which categorises the elements into four unique types: autonomous, dependent, linking, and independent (driver) barriers, as illustrated in Table 2.

Step 4: Partitioning the Reachability Matrix by Levels

The reachability set of each specific variable includes the variable itself and any other variables it helps to achieve. On the other hand, the antecedent set includes the variable and any other factors that help achieve its goals. By finding the intersection of these sets for all variables, we can identify the highest-level variable within the TISM hierarchy. This top-level element does not contribute to determining variables at higher levels. Afterward, it is removed from the remaining variables. The iterative process continues until the values for each variable are determined. The levels identified in this process are crucial for constructing the digraph and the comprehensive TISM model.

Step 5: Building the Digraph

This involves creating a graphical representation that consists of nodes and arrows. In this representation, each node represents a Critical Success Factor (CSF), and each arrow indicates the connection between two factors and their direction. After the levels have been established through level partitioning, each factor is organised within the digraph. The directions of relationships are determined based on the reachability matrix. Indirect relationships are represented by dashed lines, as shown in Figure 1.

Step 6 MICMAC Analysis of CSFs

The categorisation process involved the implementation of cross-impact matrix multiplication. From the ultimate reachability matrix, we derived the driving and dependence strengths of the variables. The driving and dependent strengths of the variables were calculated from the final reachability

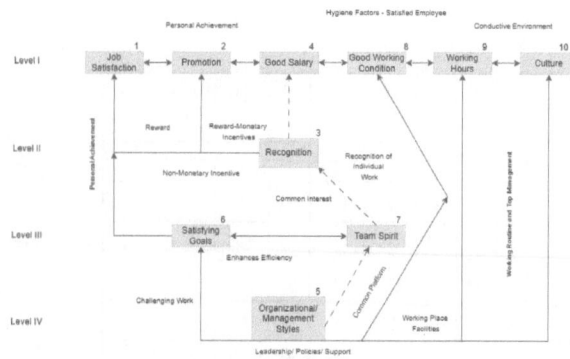

Figure 1: Construction of digraph

Source: Author's compilation

matrix. The strengths mentioned above were then used and subsequently employed in order to identify the Critical Success variables (ascertain the CSFs). The MICMAC analysis visually represents the influential and interdependent capabilities of each Critical Success Factor (CSF). The graphical representation depicts the driving influence of the Critical Success Factors (CSFs) on the y-axis, while the x-axis represents their level of reliance strength. The graph obtained from the analysis is divided into four distinct clusters: A, B, C, and D.

A - Autonomous Factors
B - Influenced by other factors
C - Linkage factors
D - Independent Factors

5. Results and Findings

The segmentation of Critical Success Factors (CSFs) among various tiers of an organisation highlights their significance. Organisational culture, recognition, favourable working conditions, and working

hours are among the variables identified in this study as autonomous motivators. On the contrary, promotions, competitive compensation, and job satisfaction rely on secondary factors. Cluster D comprises strategic aspects including but not limited to goal attainment, cohesiveness within teams, and management and organisational styles. The factors listed above warrant meticulous examination due to their direct influence on the performance of the dependent variables in Cluster B. Continuous monitoring and improvement are required to guarantee that Cluster C maintains its efficacy. Effective long-term management of CSFs in Cluster D will inevitably result in exceptional outcomes for CSFs located in Cluster B, according to the model. In order to optimise employee motivation and organisational performance, it is imperative that management places a high priority on and devotes substantial attention to critical success factors (CSFs) located in Cluster D while maintaining a perspective for the future.

6. Conclusion

This research delves into the motivations of managers and employees across different hierarchical levels within organisations. While both groups share enjoyment in their work, a genuine interest in their fields, and value interpersonal interactions, certain motivating factors are unique to managers. Top management notably influences managers, serving as intermediaries between employees and employers. In contrast, employees seek independence, status, position, non-monetary incentives, and challenging tasks, dispelling the notion that financial rewards solely drive them. This shift in employee aspirations signifies a broader range of motivations beyond financial factors, shaping the evolving landscape of workforce motivation. The research presented here

Table 2: MICMAC analysis

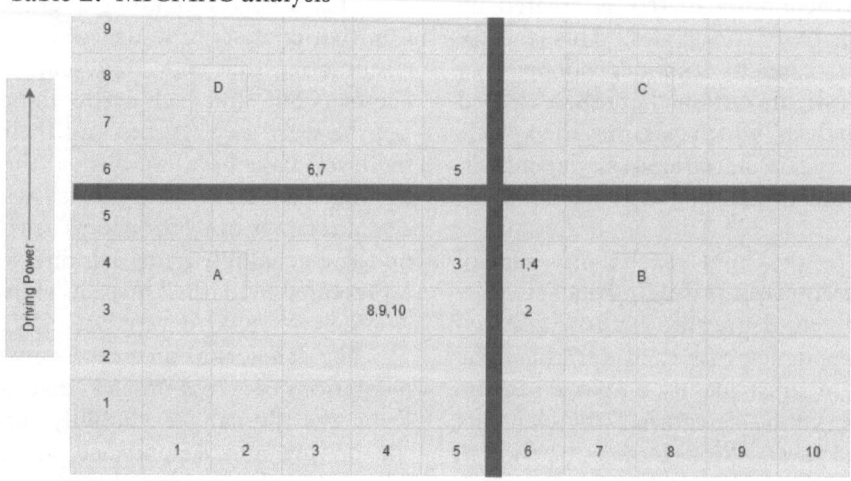

Source: Author's compilation

is distinguished by its emphasis on the hierarchical roles and positions within an organisational structure. In contrast to traditional motivational theories that treat employees as a uniform collective, the present study acknowledges and investigates the potential variations in motivating elements between managers and employees.

The implications of this study on employee motivation in Indian corporate sector companies are multi-faceted:

- It highlights the potential for significantly improved organisational performance by profoundly understanding what motivates employees, resulting in increased engagement, productivity, and commitment.
- It emphasises the importance of enhanced retention strategies by identifying key motivators, ultimately reducing turnover and associated costs. The research offers practical implications for human resource management by highlighting factors crucial for employee motivation. This includes insights into designing incentive programs, recruitment strategies, and cultivating a positive workplace culture.
- The research underscores the value of tailored incentive programs for diverse employee groups, optimizing their impact on motivation.

A future extension could involve the application of the developed model and findings across different industries. This cross-industry analysis would contribute to the generalisability of the model and its relevance in diverse organisational contexts. Future research could focus on developing practical intervention strategies based on the identified critical success factors. This involves creating actionable plans that organisations can implement to enhance motivation and improve overall performance.

References

[1] Adair, J. (2009). The John Adair Leadership Library: Leadership and Motivation: The Fifty-Fifty Rule and the Eight Key Principles of Motivating Others, p. 101. London: Kogan Page.

[2] Bartol, K.M., and Martin, D.C. (1998). Management. McGraw Hill.

[3] Rusu, G. and Avasilcai, A. (2013). Human resources motivation: An organizational performance perspective. Annals of the Oradea University Fascicle of Management and Technological Engineering (1). http://www.imtuoradea.ro/auo.fmt.

[4] Daft, R. L. (2006). The New Era of Management. Mason, Ohio: Thomson South-Western.

[5] Luthans, F. (1998). Organizational Behaviour. 8th ed. Boston: Irwin McGraw-Hill.

[6] Armstrong, M. (2009). Armstrong's Handbook of Human Resource Management Practice. 11th ed. London: Kogan Page.

[7] Pitt, M. and Tucker, M. (2008). Performance measurement in facilities management: driving innovation?. Property Management, 26 (4), 241–254.

[8] Harrim, H. M. (2010). Relationship between learning organization and organizational performance (Empirical Study of Pharmaceutical Firms in Jordan). Jordan Journal of Business Administration, 6 (3), 405–421.

[9] Katou, A. A. and Budwar, P. S. (2007). The effects of human resource management policies on organizational performance in greek manufacturing firms. Thunderbird International Business Review, 49 (1), 1–35.

[10] Aluko, M. A. O. (1998). Factors that Motivate the Nigerian Workers in Ife Social Sciences Review. Journal of the Faculty of Social Science, Obafemi Awolowo University, Ile – Ife Nigeria, 15 (1), 190-199.

[11] Bratton, J., and Gold, J. (2012). Human Resource Management: Theory and Practice. 3rd ed. London: Palgrave Macmillan.

[12] Dobre, O. I. (2013). Employee motivation and organizational performance. Review of Applied SocioEconomic Research, 5 (1/2013), 53.

Evolving Energy Mix

Analyzing US Energy Production Trends

Shobhanam Krishna, Anita Choudhary, Rohit Dwivedi

Dept. of Organizational Behaviour and Human Resources
Indian Institute of Management Shillong, Shillong, India, shobhak.phd22@iimshillong.ac.in

Abstract

The current research examines historical data on industrial output in the electric and gas utilities sector of the United States from 1939 to 2019. The primary objective of employing advanced time series analysis and modelling techniques is to accurately predict energy production for the next 12 periods, corresponding to a one-year timeframe. The study illuminates a significant change in the energy landscape throughout the 20th century. This period was marked by a decrease in coal-based energy production and a rise in the utilisation of natural gas and renewable energy sources. The energy composition of the United States has undergone a significant shift. By the end of 2022, the share of coal has decreased from 52% to 20%, while the share of natural gas has increased from 12% to 40%. This study emphasises the crucial significance of precise energy consumption forecasts in developing effective economic policies.

This study applies the ARIMA model to forecast energy production and consumption in different regions and periods. Different orders of ARIMA models are assessed, and the selection of the most suitable model is determined by evaluating the AIC and BIC values. The ARIMA (2,1,3) model is the best option due to its lower AIC and BIC values. The study presents this model's outcomes and projected values with a Mean Absolute Percentage Error (MAPE) of approximately 15.4%. This MAPE indicates an impressive 84.6% accuracy in predicting the succeeding 12 observations.

Keywords: Autoregressive Integrated Moving Average (ARIMA) model, energy efficiency, energy production forecasting, environmental impact, Mean Absolute Percentage Error (MAPE), time series analysis

1. Introduction

In the 20th century, we have witnessed a notable transition from coal-based to petroleum-based energy sources. Industrialisation and globalisation have led to a substantial increase in energy demand. Remarkably, the United States experienced a transformation in its energy sources for electricity generation. Natural gas and renewable energy sources started to increase, while coal-fired generation began to decrease. In 1990, approximately 52% of the total electricity generation and 42% of the utility-scale electricity generation capacity in the United States were coal-fired power facilities. By the end of 2022, these proportions had decreased to 17% and 20%, respectively. On the contrary, natural gas exhibited a substantial surge in capacity share, which escalated from 17% in 1990 to 43% in 2022. Furthermore, its electric generation share multiplied by 12% to 40% [2].

In 2019, the United States began producing more energy than it consumed, reaching 102.92 quadrillion BTUs (quads) in 2022 compared to 100.41 quads for consumption. In 2022, petroleum, natural gas, and coal constituted 81% of the total primary energy production in the United States. Notably, natural gas consumption increased substantially from 17% of total energy consumption in 1950 to 33% in 2022. This transition was driven by more efficient drilling and production techniques, which increased shale and tight-formation gas production [3].

The decline in natural gas prices for electric power generation, which began in 2008, played a pivotal role in the transition from coal to natural gas for electricity production. Accurate energy consumption forecasts are critical for formulating economic policies when considering factors such as population growth and technological advancements intended to enhance the population's well-being. This research aims to examine historical industrial output data for electric and gas utilities from 1939 to 2019 and use time series analysis and modelling techniques to forecast energy production for the following 12 periods or one year.

Chapter 12 DOI: 10.1201/9781003570349

2. Literature Review

Multiple models have been utilised to predict energy consumption patterns empirically. Energy consumption has manifested divergent patterns in various economies, including upward and downward trends. Abraham and Nath (2001) believe an energy crisis is possible when consumption levels surpass production. To mitigate the risk of potential energy crises, it is crucial to ensure that production and consumption rates align. Albayrak [4] utilised the ARIMA model to predict Turkey's primary energy production and consumption from 1923 to 2006. The data projections for the period between 2007 and 2015 indicate annual growth rates of 10% for total energy consumption, 11% for street lighting, 11% for government consumption, 12% for residential consumption, 8% for consumption per capita, 7% for installed capacity per capita, and 9% for installed capacity.

Furthermore, the findings indicated that the mean yearly increase in primary energy usage would be between 5% and 7%. The report emphasised that Turkey's heavy dependence on external energy sources, without concurrent investments in infrastructure and diversification of energy sources to incorporate domestic resources, may lead to an imminent energy catastrophe. A study was undertaken by Al-Fattah [5] in the United States to construct a forecast model for U.S. Natural Gas. According to the research findings, it was expected that the rate of gas production would exhibit a consistent level of 18.7 trillion cubic feet per year (Tcf/yr) between the years 2005 and 2008. Subsequently, there would be a gradual rise in output to reach 19.0 Tcf/yr by 2010, followed by additional expansion to 22.5 Tcf/yr by 2025. The researcher forecast that gas production in the United States will undergo a mean yearly increase of 0.5% between 2005 and 2015, followed by a subsequent acceleration of 1.2% from 2015 to 2025. According to the estimates, it was also projected that the rate of petrol depletion in the United States would increase from 10.6% annually in 2005 to 13.4% by 2025.

Additionally, the research project predicted a projected rise in the proven reserves of natural gas in the United States, with an initial estimate of 197 trillion cubic feet (Tcf) in 2005, expected to reach 215 Tcf by 2010. This projection indicates an annual growth rate of 1.3%, ultimately culminating in 263 Tcf by 2025. The projection encompassed two distinct periods, specifically 1918-1997 and 1998-2004. In their study, Ediger and Akar [6] devised a decision support system to forecast fossil fuel output. They utilised regression, ARIMA, and SARIMA techniques to analyze historical data from 1950 to 2003 comparatively. The investigation has shown a correlation between the supply and demand of fossil fuels, indicating a growing disparity. This gap is estimated to expand to 69 million tonnes of oil equivalent (mtoe) by 2010, 82.6 mtoe by 2020, and 103.4 mtoe by 2030.

3. Methodology

3.1. Data Details

The dataset encompasses the period from 1939 to 2019 and focuses on the industrial production of electric and gas utilities in the United States. The dataset consists of 965 rows and is structured into two columns. One column represents the dataset, while the other displays energy production data. The dataset utilised in this study has been acquired from a distinct source, notably the Federal Reserve Bank of St. Louis, explicitly focusing on the FRED database. The MATLAB software is the preferred analytical tool to predict energy production. The predominant forecasting model employed in this study is the ARIMA model. It is important to note that alternative models, such as models based on multiple regression and artificial neural networks, are designed explicitly for forecasting energy consumption in economies.

4. Data Analysis

Data analysis begins with effective data visualisation, including a time series plot revealing energy output patterns, trends, and seasonality over time. The plot illustrates a consistent increase in energy output over the years and highlights substantial variance and oscillations in energy generation (Figure 1).

4.1. Data Visualisation

Effective data visualisation is essential for comprehending and interpreting data. The time series plot depicted in Figure 1 gives valuable information on energy output patterns, trends, and seasonality over time. There has been an overall increase in energy output over the years. There is also some seasonality to be found. The figure, however, exhibits a high degree of variance, showing large oscillations and variations in energy generation.

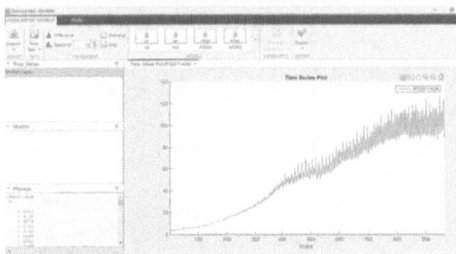

Figure 1: Time series plot.
Source: Author's compilation.

4.2. ACF Plot

Examining the ACF plot reveals the patterns and dependencies present in a time series. In this situation, the autocorrelation shows a diminishing trend as the lag increases, indicating a trend within the time series. A sluggish fall in autocorrelation indicates the presence of a recurrent pattern or seasonality in the data.

4.3. PACF Plot

The PACF plot showcases lags on the x-axis and displays the corresponding partial autocorrelation values on the y-axis. Prominent spikes observed in the PACF plot can suggest the existence of certain lags that exert a direct influence on the current value of the time series.

4.4. ADF Test

The Augmented Dickey-Fuller (ADF) test is a commonly employed statistical test in the field of time series analysis to ascertain the presence of stationarity or non-stationarity in a given time series. The test is designed to assess the presence of a unit root in the time series sample, primarily focusing on evaluating the null hypothesis. The alternative hypothesis exhibits variation depending on the specific form of the test employed, commonly containing the concepts of stationarity or trend-stationarity. The ADF statistic, a negative numerical number employed in the test, is an indicator. The strength of rejecting the hypothesis of a unit root is enhanced as the negativity of this statistic increases, with the level of confidence exerting a substantial influence on this conclusion.

H0: The presence of a unit root suggests that the time series is non-stationary.

HA: The time series is stationary, indicating the absence of a unit root.

Based on the observed p-value of 0.3147, more than the predetermined significance level of 0.05, there is inadequate evidence to reject the null

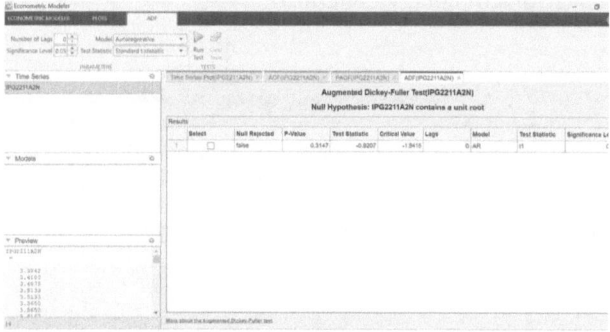

Figure 2: Output of ADF test.
Source: Author's compilation.

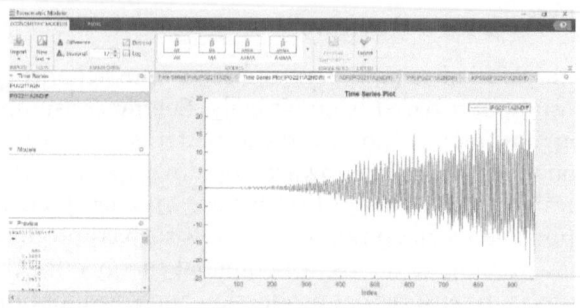

Figure 3: Time series plot for differenced data
Source: Author's compilation

hypothesis. This suggests that the time series exhibits non-stationarity, preventing the rejection of the null hypothesis (Figure 2). In the above case, rejecting the null hypothesis indicates that the time series data exhibits non-stationary attributes.

4.5. Differencing

Single differencing is a widely employed method in time series analysis for converting a non-stationary time series into a stationary one. This technique entails calculating the difference between successive observations within the series.

When the differenced series, as shown in Figure 3, exhibits a mean value near zero and lacks any noticeable trend or systematic pattern, it suggests that the single differencing procedure has successfully eliminated the trend initially present in the data series. This transformation facilitates easier modelling and analysis, as stationarity is often assumed in many time series models.

4.6. ARIMA Model

The ARIMA model is a popular and extensively utilised time series forecasting approach that integrates autoregressive (A.R.), differencing (I), and moving average (M.A.) components. The employed approach can be characterised as a regression analysis technique, wherein the impact of a single dependent variable is assessed on other variables that exhibit variability. Within ARIMA models, it is customary to apply a widely used notation that entails the specification of ARIMA with parameters p, d, and q. These parameters are represented by integers, which designate the particular type of ARIMA model being utilised. The parameters can be defined as follows:

p: Denoting the quantity of previous observations included in the model, frequently known as the lag order.

d: Denoting the number of occurrences in which the unprocessed observations are subjected to differencing, also referred to as the degree of differencing.

q: Denoting the order of the moving average, another name for the magnitude of the moving average window.

Assumptions to apply ARIMA model-

To apply an ARIMA model to a time series, several assumptions need to be considered:

- Stationarity
- Independence
- Linearity
- Homoscedasticity

In time series analysis, the AIC and BIC values are information criteria used to compare and select the best-fitting model among competing models. They provide a quantitative measure of the trade-off between model complexity and goodness of fit. Here is how to interpret AIC and BIC values:

- AIC (Akaike Information Criterion):
  o A lower AIC value indicates a better-fitting model.
  o When comparing models, the model with the lowest AIC value is preferred, as it suggests the best trade-off between accuracy and complexity.

- BIC (Bayesian Information Criterion):
  o A lower BIC value indicates a better-fitting model.
  o When comparing models, the model with the lowest BIC value is preferred, indicating the best trade-off between accuracy, model complexity, and sample size.

- In the context of an ARIMA model, the estimated coefficients portray the magnitude and direction of the connection between a time series and its previous values or error terms. Within an A.R. (p) model, each coefficient corresponds to a past value of the time series, signifying the influence of that specific lag on the present observation. Positive coefficients indicate a positive relationship, whereas negative coefficients imply a negative relationship.

- For the ARIMA models considered, such as (3,1,3) with A.R. =3, D=3, and M.R. =3, the AIC and BIC values obtained are 4843.9 and 4882.91, respectively. Both AIC and BIC serve as model fit and complexity measures, where lower values indicate better models.

- Similarly, for the (2,1,3) model with AR=2, D=1, and MR=2, the AIC and BIC values are 4528.42 and 4562.5, respectively. Additionally, for the (2,1,2) model with AR=2, D=1, and MR=2, the AIC and BIC values are 4544.5 and 4573.78, respectively. Lastly, for the (3,1,2) model with A.R.

=3, D=1, and MR=2, the AIC and BIC values are 4843.82 and 4877.89, respectively (Figure 4)

- Among these considered models, the (2,1,3) model exhibits lower AIC and BIC values, indicating a better fit.

- Smaller Standard Error: A lower standard error signifies a more precise coefficient assessment, implying that the estimated parameter is more likely to be near its actual value and fluctuate less. The lesser standard error is observed in the case of the ARIMA model with orders (2,1,3), showing a higher level of precision in the coefficient estimation.

- The residual histogram plot for the (2,1,3) model displays a bell-shaped or approximately symmetric distribution, suggesting that the residuals follow a normal distribution. The residual histogram plot provides insights into the distribution of residuals, which are the differences between the observed values and the model's predicted values.

4.7. ARIMA Model to Forecast

When using an ARIMA model for data forecasting, choosing a model with reduced AIC and BIC values is preferable. Models with reduced AIC and BIC values indicate a better balance between precision and complexity. We forecasted a data series using the (2,1,3) model in this particular scenario.

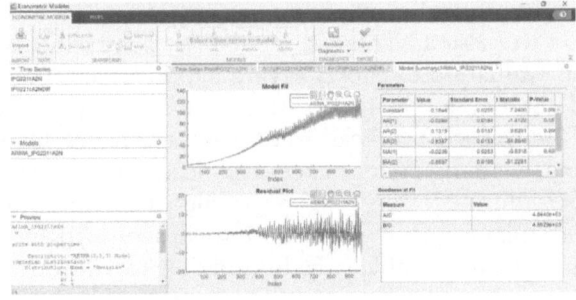

Figure 4: ARIMA model output

Source: Author's compilation

Figure 5: Forecasted value of ARIMA model

Source: Author's compilation

The resulting variance is 6.29835%. The forecasted value was calculated using the "forecast" function, representing a 12-period forecast. A MAPE of approximately 15.4% indicates that the model predicts the following 12 observations with an accuracy of approximately 84.6%. Figure 5 illustrates the anticipated model values for the (2,1,3) model.

5. Conclusion

The "Industrial Production: Electric and Gas Utilities" dataset analysis yielded significant results. It identified recurring patterns in the data, indicating the existence of seasonality. Among the array of ARIMA models evaluated, the model denoted by (2,1,3) emerged as the most suitable option, achieving an optimal equilibrium between model fit and complexity, as indicated by its lower AIC and BIC values. Mean Absolute Percentage Error (MAPE) was 15.4%, which reflects the model's impressive 84.6% accuracy rate in predicting the next 12 observations. Multiple industries have found this dataset to be an invaluable resource. It has provided valuable insights for energy planning, infrastructure development, and resource allocation decision-making processes by enhancing our understanding of energy demand and consumption patterns. In addition, it has revealed a close connection between the industrial output of electric and gas utilities and broader economic activity, thereby providing valuable indicators of economic expansion or contraction. In addition to these advantages, the dataset has facilitated the evaluation of energy-saving initiatives and guided the development of sustainable energy policies. It has played a crucial role in infrastructure planning, enabling policymakers and businesses to make well-informed decisions regarding the expansion and enhancement of energy infrastructure to meet rising demand efficiently. In addition to addressing environmental concerns such as greenhouse gas emissions and resource depletion, the dataset has informed the creation of environmentally conscious policies. In addition, the dataset's analytical prowess has considerably contributed to energy market analysis and forecasting, allowing stakeholders to easily navigate the complexities of market dynamics, supply and demand trends, and price volatility. Governments and regulatory bodies have utilised the dataset to evaluate the efficacy of energy policies, enabling them to make informed decisions regarding the continuance or modification of policies. Future research may investigate advanced techniques such as artificial neural networks (ANN) and consider incorporating more recent data to obtain a deeper understanding of energy production forecasting and its ongoing improvement.

References

[1] Abraham, A. and Nath, B. (2001). A neuro-fuzzy approach for modeling electricity demand in Victoria. Applied Soft Computing, 1 (2), 127–138.

[2] U.S. Energy Information Administration. (2023, August 16). U.S. energy facts explained. https://www.eia.gov/energyexplained/us-energy-facts/.

[3] U.S. Energy Information Administration. (2023, June 30). Electricity generation, capacity, and sales in the United States. https://www.eia.gov/energy-explained/electricity/electricity-in-the-us-generation-capacity-and-sales.php.

[4] Albayrak, A. S. (2010). ARIMA forecasting of primary energy production and consumption in Turkey: 1923–2006. Enerji, piyasa ve düzenleme, 1 (1), 24-50.

[5] Al-Fattah, S. M. (2005). Time series modeling for U.S. natural gas forecasting. In International Petroleum Technology Conference (pp. IPTC-10592). IPTC.

[6] Ediger, V. Ş. and Akar, S. (2007). ARIMA forecasting of primary energy demand by fuel in Turkey. Energy Policy, 35 (3), 1701–1708.

Future-Proofing Organisations

Predicting Employee Attrition with SVM

Shobhanam Krishna, Anita Choudhary, Rohit Dwivedi

Dept. of Organizational Behaviour and Human Resources
Indian Institute of Management Shillong, Shillong, India
E-mail: shobhak.phd22@iimshillong.ac.in

Abstract

Employee attrition poses a significant challenge for organisations worldwide, impacting performance, continuity, and knowledge retention. This study employs machine learning techniques, particularly Support Vector Machines (SVM), to identify critical factors influencing employee attrition and predict the likelihood of individual employees leaving an organisation. Leveraging the IBM HR Analytics Employee Attrition dataset, various preprocessing steps, including missing value imputation, data type conversion, and feature scaling, are applied to enhance data quality. Feature selection techniques, including Random Forest, reveal critical contributors to attrition. The F1 score is the evaluation metric, considering both precision and recall. SVM emerges as a promising model, achieving an accuracy of 88.88% with all features and 82.7% with selected features. The novelty of this research lies in the integration of SVM, a robust machine learning algorithm, with comprehensive preprocessing techniques and feature selection methodologies to predict employee attrition. The study contributes by showcasing the effectiveness of SVM in achieving high accuracy rates. Additionally, the application of Random Forest for feature selection and Grid Search for hyperparameter optimisation adds a layer of sophistication to the methodology. The novelty extends to the meticulous examination of feature relationships, which aids in better understanding the intricacies of attrition drivers. This paper contributes significantly to understanding and predicting employee attrition by providing practical insights derived from machine learning techniques. The contribution extends beyond conventional methodologies by emphasising the importance of feature engineering and scaling. The research contributes to the existing literature by demonstrating that a data-driven approach, specifically employing SVM, can substantially enhance talent retention strategies, ultimately improving organisational performance and sustainability.

Keywords: Employee attrition, Machine Learning (ML) techniques, human resource management, Support Vector Machines (SVM), F1 score, talent retention

1. Introduction

Companies in India, as well as across the globe, grapple with a significant challenge in the realm of talent acquisition and retention. This challenge becomes particularly intricate when addressing talent attrition from industry downturns or voluntary employee departures. The departure of skilled employees can detrimentally impact companies, leading to performance setbacks with enduring repercussions. This impact becomes significantly pronounced when departing employees leave voids in a company's operational capabilities and human resources functions. The consequences extend beyond reduced productivity, encompassing issues like team cohesion and a decline in the company's standing in the industry.

Employee attrition refers to the departure of employees from an organisation, either voluntarily or involuntarily, for various reasons. These departures can lead to long-term or permanent vacancies within the organisation or specific departments. Attrition can have various adverse effects on an organisation, such as causing disruptions in workflow, creating deficiencies in essential skills, and gradually eroding institutional knowledge. Filling these vacant positions, especially those requiring specialised skills, can be time-consuming. Failing to address this issue could worsen future challenges regarding recruitment.

The motivation behind this research stems from the critical need for organisations to tackle employee attrition, a pervasive issue with far-reaching consequences. The departure of skilled employees can disrupt operational capabilities and erode institutional knowledge. This study aims to bridge the existing gap by employing advanced machine learning techniques, specifically SVM, to enhance the predictability of employee attrition. The motivation lies in empowering organisations with actionable insights

Chapter 13 DOI: 10.1201/9781003570349

to proactively manage talent retention and mitigate the negative impacts of attrition on performance and continuity.

The primary objective of this study is to utilise machine learning methods, particularly those within the field of ML, to identify the key elements that influence an employee's choice to leave an organisation. Moreover, our objective is to forecast the probability of particular employees departing from the organisation. This study aims to assess the suitability of the Support Vector Machine (SVM) algorithm in predicting employee turnover. The Random Forest technique is employed for feature selection, while Grid Search is utilised for hyperparameter optimisation.

2. Literature Review

The academic literature on human resource management (HRM) acknowledges the growing application of machine learning methodologies in tackling real-world obstacles [2]. The methodologies, renowned for their beneficial predictive skills, have a broader range of applications beyond HRM, as evidenced in the research conducted by Gerede and Mazan [3], whereby the prediction of source code changes yields essential insights. The identification of the accuracy in predicting employee attrition depends upon the data quality and the approach employed, which frames it as a problem of binary classification. Decision Trees (DT) and logistic regression are commonly favoured by academics due to their straightforwardness and ease of interpretation [4].

Keramati et al. [5] argue that artificial neural networks exhibit superior performance compared to Decision Trees and K-nearest neighbour algorithms in analyzing employee attrition as modelling sophistication increases. Gordini and Veglio [6] present a specialised model for predicting employee attrition within a specific business in their study. The authors adopt Support Vector Machines (SVM) as the chosen methodology and demonstrate its effectiveness compared to neural networks and logistic regression, particularly within the e-commerce sector. Zhu et al. [7] conducted a comparative study to address the issue of class imbalance.

Subsequent investigations go into the prediction capacities of machine learning in comprehending employee behaviour. The predictive performance of employees can be determined using decision trees such as ID3 and C4.5, as well as the Naïve Bayes algorithm. Among these methods, the job title has been recognised as the most influential element in predicting employee success (Al-Radaideh and Al Nagi, 2012). In a separate investigation conducted by Saradhi and Palshikar [8], a range of data mining techniques, such as Naïve Bayes, Support Vector Machines, logistic regression, decision trees, and random forests, were utilised to forecast staff attrition. The findings suggest that implementing a Support Vector Machine algorithm is advisable, given its accuracy rate of 84.12%.

In their study, Yang and Islam [9] examine the phenomenon of IBM Employee Attrition by applying correlation analysis and Random Forest. Their objective is to uncover the elements that significantly impact attrition inside the organisation. The research conducted in this study utilises K-means Clustering and binary logistic regression techniques to identify patterns of attrition. The findings indicate that individuals who frequently travel and those working in the Human Resources department exhibit more significant attrition rates.

In this study, Chung et al. [10] examine the significant implications of employee attrition on organisational outcomes. They developed a predictive model utilising the IBM HR Analytics Employee Attrition & Performance dataset comprising 30 factors. This study assesses the performance of eight predictive models: Logistic Regression, Random Forest, XGBoost, SVM, Artificial Neural Network, and an ensemble model. It emphasises the significant factors contributing to the models' predictions, such as environmental contentment, overtime work, and relationship satisfaction. The objective of the suggested methodology is to enhance proactive talent management tactics that go beyond reactive procedures.

2.1. Research Gaps

The literature must include insights on integrating methodologies from diverse fields like data science and social sciences into HRM practices for more effective attrition prediction and talent management. Exploring this interdisciplinary approach could enhance the comprehensiveness of attrition studies.

While correlations between features are identified, a research gap exists in conducting a detailed analysis. Present research could use statistical methods or advanced machine learning techniques for a quantitative assessment, uncovering hidden patterns and providing a more nuanced understanding of factors influencing attrition.

There needs to be more understanding methods to enhance the interpretability of machine learning models in HR, where transparency is crucial. Focusing on techniques to make model decision-making more understandable for HR professionals could improve talent management strategies.

The literature needs to identify a gap in understanding the impact of feature scaling techniques on model performance. Present research could systematically investigate how normalisation and standardisation affect attrition prediction models, offering insights into optimal preprocessing steps for HR datasets.

3. Methodology

This study concentrates on training and evaluating Support Vector Machines (SVM)-based machine learning models. Support Vector Machines are a non-probabilistic supervised machine learning technique for classification and regression tasks. Support Vector Machines educate algorithms by constructing a decision boundary, a hyperplane, to separate distinct classes [11,12]. SVM, on the other hand, permits the mapping of data into a higher-dimensional space. Attaining linear separability with SVM permits the resolution of nonlinear problems. This strategy is prevalent due to its adaptability and high performance. Noting that SVM may not perform optimally when applied to unnormalised data is essential. Utilising scaled features is, therefore, a prevalent practice.

3.1. Experimental Evaluation

This section introduces the experimental setup and dataset employed in the present study. Following this, the methodologies associated with data preprocessing, statistical analysis, and the criteria utilised for assessing the presented models will be explored. The IBM HR Analytics Employee Attrition dataset has been utilised in this study. This dataset includes many HR-related variables, such as age, education level, gender, and pay rate. The dataset under consideration comprises a comprehensive collection of 1,470 data points, encompassing 35 unique properties. It has been carefully assembled to facilitate the examination of employee attrition. This resource has demonstrated its value in assisting organisations in comprehending and addressing difficulties associated with employee turnover. The dataset plays a crucial role in predictive modelling by identifying significant factors influencing an employee's choice to depart from an organisation. This research is specifically designed to cater to the needs of Human Resources Management (HRM) studies, focusing on the unique issues and considerations that arise within human resources. The provided dataset enables researchers to explore the relationships, patterns, and trends related to attrition, hence facilitating a comprehensive comprehension of the elements that contribute to it. The variables encompassed by the dataset encompass a wide range of factors, spanning from fundamental demographic information to more intricate characteristics pertinent to the context of the work environment.

3.2. Data Pre-processing

Data preprocessing is a fundamental procedure that is frequently conducted prior to the training of machine learning models. The presence of missing values, noise, and substantial variations in feature scales is frequently observed in datasets. The IBM HR dataset has been subjected to a series of preparation procedures.

The process of imputing missing values: The concept of missing value imputation pertains to substituting or replenishing missing data with estimated or imputed values. This study notes that the IBM HR dataset utilised in the analysis exhibits all the values.

Data type conversion: It refers to the process of changing the data type of a value from one type to another.

Convert data types: Some machine learning algorithms cannot directly process categorical variables. Consequently, it is crucial to convert these variables into a numerical format. In this study, categorical attributes are transformed by using one-hot encoding. The OneHotEncoder method converted columns with text-based data into multiple binary columns in a 0-1 format. The attributes that underwent numerical conversion in this process include 'BusinessTravel,' 'Department,' 'EducationField,' 'Gender,' 'JobRole,' 'MaritalStatus,' 'Over18,' and 'OverTime.'

Feature scaling: Feature scaling is a common practice in human resources datasets, as it helps address variations in scales among different features. However, significant differences in feature scales can often hinder optimisation algorithms such as gradient descent. Performing feature scaling can enhance classification performance and learning efficiency for specific machine learning algorithms. In this study, after converting the data, both normalisation and standardisation on the original dataset to account for variations in scale are performed.

4. Data Analysis

Before constructing any Machine Learning models, conducting an initial examination of the data is essential. This step allows for a better comprehension of the relationships between the features and a deeper understanding of their characteristics.

Based on the insights obtained from the Figure 1 mentioned earlier, several observations are depicted-

- A notable correlation between marital status and age regarding promotions indicates that one of these features may not be necessary.
- Both Male Relationship Satisfaction and Work-Life Balance are strongly influenced by Overtime, which suggests that eliminating one is possible.

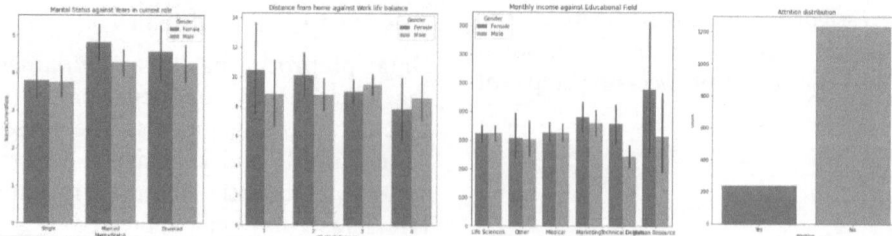

Figure 1: Exploring Feature Relationships
Source: Author's compilation

- There is a significant correlation between the Educational Field and Monthly Income variables, suggesting that there may be redundancy between them.
- Overtime has different effects on various variables, and there are significant differences between males and females.
- The data for Year 16 of Years With Current Manager is an outlier and may need to be removed from the dataset.
- Only 16% of employees experienced attrition in the workplace, which suggests that our output variable is distributed unevenly.
- On the other hand, EmployeeNumber and EmployeeCount are irrelevant features, indicating that they should be removed from the dataset.

4.1. Feature Engineering

Feature engineering is an art and science that involves data representation in the best way. It requires intuition, domain knowledge, practical experience, and common sense. A closer look at feature engineering involves choosing which characteristics to keep and how to turn categorical data into a numerical format for a model.

Splitting categorical data before transforming it is typical. Pandas library "iloc" command divides. Final data preprocessing involves partitioning the dataset into training and test sets. The splitting ratio for a typical distribution is usually 70-30%.

4.2. Feature Selection

Feature selection is crucial to machine learning model building. The random forest model, which provides feedback on feature relevance, is a tool that helps choose the best features. RandomForestClassifier is a good classifier since it evaluates each feature's prediction influence. Considering the relative importance of features determined by the Random Forest algorithms to provide context to the results, it becomes evident that when all factors are considered, overtime, marital status, department (sales or research), and managerial position are the leading causes of

attrition. However, a limited collection of variables, such as years in the company, tenure with the same manager, time since previous promotion, and gender, influence attrition. These findings emphasize the importance of feature selection in improving machine learning model prediction.

4.3. 4.3. Feature Scaling

A few models that have been constructed do not perform effectively with unnormalised data, making it necessary to create scaled versions of the feature set.

5. Results

5.1. Building the Model

This part builds models with default hyperparameters by fine-tuning them using grid search and comparing the outcomes. Grid search is an exhaustive approach to finding the best hyperparameters for a problem. Grid search using K-fold cross-validation is easy with scikit-learn. This cross-validation method produces smaller models using "K" dataset subsets to assess model generalisation better. This study uses K=5. This study used the F1 score, a weighted average of precision and recall. Recall calculates the model's completeness by comparing positive predictions to test data's positive class values.

5.2. Model Evaluation

5.2.1. SVM

Using a support vector machine (SVM) allows data transformation into a higher-dimensional space, facilitating the linear separation of nonlinear problems. This particular attribute is well esteemed for its versatility and efficacy. When evaluating the precision using a particular set of features, it attains a rate of 87.30%, whereas employing an alternative feature selection results in a rate of 83.90%. By employing the established grid parameters, the Support Vector Machine (SVM) classifier is implemented using two feature sets: the complete and the subset of selected features. The corresponding recall scores obtained were 38.89% and 35.38% respectively.

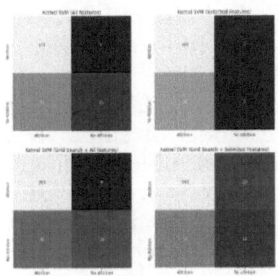

Figure 2: Evaluating SVM model performance
Source: Author's compilation

Significantly, the Support Vector Machine (SVM) exhibits impressive performance, with an accuracy rating of 88.88% and 82.7% for the two feature cases. Moreover, upon examining the confusion matrices, it is evident that the Linear Support Vector Machine (SVM) model, utilising all features, has the highest overall accuracy when compared to the other Support Vector Classifier (SVC) models that have been investigated thus far (Figure 2).

6. Conclusion

In the contemporary global business landscape, employee attrition poses a significant and challenging issue for organisations. This research harnesses the potential of machine learning, particularly Support Vector Machines, to address the intricate problem of employee attrition. This study provides valuable insights for comprehending and predicting employee attrition by utilising the IBM HR Analytics Employee Attrition dataset and implementing a range of data preprocessing techniques, feature selection methodologies, and model assessments.

Applying Kernel SVM and the thorough analysis of HR data can bolster an organisation's capacity to retain its workforce, diminish attrition rates, and ultimately cultivate a more productive and harmonious workplace environment. As businesses contend with the complex realm of human resource management, incorporating advanced machine-learning techniques offers the prospect of a brighter, data-driven future. By leveraging the capabilities of machine learning models like SVM, organisations can not only mitigate the challenges posed by employee attrition but also unlock opportunities to optimise their HR strategies, ultimately leading to improved organisational performance and sustainability.

This paper contributes significantly to understanding and predicting employee attrition by providing practical insights derived from machine learning techniques. The comprehensive approach to data preprocessing, feature selection, and model evaluation, particularly with SVM, offers a robust framework for organisations to leverage. The contribution extends beyond conventional methodologies by emphasising the importance of feature engineering and scaling. The research contributes to the existing literature by demonstrating that a data-driven approach, specifically employing SVM, can substantially enhance talent retention strategies, ultimately improving organisational performance and sustainability.

However, future research can be done by identifying the contextualisation of findings. While the study focuses on quantitative aspects, such as SVM accuracy rates and feature importance, it lacks an in-depth exploration of the qualitative factors influencing attrition. The acknowledgment of the importance of employee satisfaction, relationship dynamics, and work-life balance in the literature survey is not fully integrated into the machine learning model. A more holistic approach combining quantitative and qualitative features could provide a nuanced understanding of the attrition phenomenon. Moreover, the study predominantly relies on the IBM HR Analytics Employee Attrition dataset, potentially limiting the generalisability of the findings. Future research could involve a broader exploration of datasets from diverse industries or regions to uncover unique attrition patterns and contribute to a more universal model. Additionally, while SVM is leveraged for predicting employee turnover, the study lacks exploration into hybrid models that combine the strengths of different machine learning algorithms. Investigating ensemble models or hybrid approaches could improve predictive accuracy by capturing diverse patterns in employee behaviour.

References

[1] Al-Radaideh, Q. A. and Al Nagi, E. (2012). Using data mining techniques to build a classification model for predicting employees performance. International Journal of Advanced Computer Science and Applications, 3 (2).

[2] Marchington, M., Wilkinson, A., Donnelly, R., and Kynighou, A. (2016). Human Resource Management at Work. Kogan Page Publishers.

[3] Gerede, Ç. E. and Mazan, Z. (2018). Will it pass? Predicting the outcome of a source code review. Turkish Journal of Electrical Engineering and Computer Sciences, 26 (3), 1343–1353.

[4] Neslin, S. A., Gupta, S., Kamakura, W., Lu, J., and Mason, C. H. (2006). Defection detection: Measuring and understanding the predictive accuracy of customer churn models. Journal of Marketing Research, 43 (2), 204–211.

[5] Keramati, A., Jafari-Marandi, R., Aliannejadi, M., Ahmadian, I., Mozaffari, M., and Abbasi, U. (2014). Improved churn prediction in the telecommunication industry using data

mining techniques. Applied Soft Computing, 24, 994–1012.

[6] Gordini, N. and Veglio, V. (2017). Customers churn prediction and marketing retention strategies. An application of support vector machines based on the AUC parameter-selection technique in the B2B e-commerce industry. Industrial Marketing Management, 62, 100–107.

[7] Zhu, B., Baesens, B., and vanden Broucke, S. K. (2017). An empirical comparison of techniques for the class imbalance problem in churn prediction. Information Sciences, 408, 84–99.

[8] Saradhi, V. V. and Palshikar, G. K. (2011). Employee churn prediction. Expert Systems with Applications, 38 (3), 1999–2006.

[9] Yang, S. and Islam, M. T. (2020). IBM employee attrition analysis. arXiv preprint arXiv:2012.01286.

[10] Chung, D., Yun, J., Lee, J., and Jeon, Y. (2023). Predictive model of employee attrition based on stacking ensemble learning. Expert Systems with Applications, 215, 119364.

[11] Bennett, K. P. and Campbell, C. (2000). Support vector machines: hype or hallelujah? ACM SIGKDD Explorations Newsletter, 2 (2), 1–13.

[12] Duan, K. B., and Keerthi, S. S. (2005). Which is the best multiclass SVM method? An empirical study. In International workshop on multiple classifier systems (pp. 278–285). Berlin, Heidelberg: Springer Berlin Heidelberg.

Data-Driven Sales Forecasting

A Machine Learning Approach

P. Priya[1], M. Maragadhavalli Meenakshi[2], S. Lakshmipriya[2]

[1]Department of Computer Science and Engineering, M. Kumarasamy College of Engineering, Karur, India,
priyap.cse@mkce.ac.in
[2]Department of Artificial Intelligence and Data Science, Sri Manakula Vinayagar Engineering College, Puducherry,
India, meenakshi.aids@smvec.ac.in, lakshmipriya.aids@smvec.ac.in

Abstract

A number of criteria are taken into account for determining the success of the firm and for predicting the product sales. Sales forecasting, which is done by examining client purchase patterns, is crucial to contemporary business intelligence. The foundation of business and business planning operations is predicting future sales demand. Business companies may develop, alter company goals, and give a stock storage solution with the use of forecasting. Forecasts are created by using data or information from earlier studies and taking future features into account. For any business to evaluate previous performance and develop future market strategies, sales forecasting is essential. Businesses may with the use of sales estimates; they may determine their anticipated profit (or loss) for a specific time period, which will aid them in making better decisions based on potential future earnings. By considering the human resources and the quantity of employees, the volume of production required has to meet the demand. Sales forecasting is the practise of estimating how many people will purchase a product in the future while taking into account the product's feature and sales environment. The investor will be assisted in choosing an investment and a marketing strategy that will result in effective sales in the next years by the sales projection model. These predicting outcomes aid in making enough purchases to produce the goods. The proposed work aims to ensure the accuracy of the forecast findings by using an improved variant of the ARIMA family of models, called SARIMAX (Seasonal Auto-Regressive Integrated Moving Average with Exogenous Factors).

Keywords: Forecasting, seasonal, autoregressive, ARIMAX, SARIMAX, prediction, pre-processing, feature selection, visualisation

1. Introduction

Sales projections serve as the foundation for immediate financial choices and have an impact on significant business transactions. Using a certain data set, this web application helps to forecast sales. Data extraction, data cleaning, and forecasting features should all be included in the application. The predicted findings have to be shown on a dynamic dashboard. Create an automated machine learning pipeline that is adaptable. Data extraction to prediction should all be handled by the pipeline. Sales forecasting is a fundamental and frequent activity that affects operations, strategy, and product marketing in the majority of businesses. It forecasts future sales revenue in addition to forecasting which prospects will go forward in the sales cycle. It offers commodities forecasting, simple risk management, cash flow management, resource utilisation, and identification of warning indicators for the long-term objectives of the organisation. A lot of important things significantly affect future sales. The sales patterns of a retail store's total sales or the sales of a specific product may be used to identify these components. It is crucial to remember that the difficulty of anticipating differs based on the product. Every business organisation understands how important it is to anticipate future demand for any product and prepare an adequate supply. It is possible to prevent overstock and under stock situations while increasing customer satisfaction and retention through effective forecasting. A business plan or strategy based on client demand and the present environment can be developed by the organisation with the aid of accurate sales estimates. Standard sales forecasting supports firms in predicting future sales by using historical data and client buying assumptions. When developing a strong plan for the following year and your budget, take into account constraints and weaknesses. By being well-informed about recent opportunities, one may increase their chances of success and better position themselves for future market demands. Without regard to outside variables, businesses that use sales modelling as the first step to enhanced. The Figure 1 below depicts the sales forecasting (Using the past data to predict the future sales)

2. Literature Review

Sharma and Sinha [1] proposed the Revenue forecasting which is crucial for businesses to produce significant income. The suggested approach to use the sales prediction model is the finest decision to make a good move in the business. The price of fuel and the preceding month's sales are the two key variables that influence the car industry's sales. In addition, as people's mentalities must also be considered, the vehicle's price will be a key factor in the sale. Forecasters will evaluate brand using both the amount and quality of customers. The suggested model is represented by an artificial neural network, which is trained using historical data. The model's error is then compared against the error in multiple regression approaches, and the sales prediction is calculated using the converged value of the interlayer coefficients. Utilise factors like the cost of the product, consumer income, consumer awareness, inflation rate, and the cost of fuel and diesel.

Permatasari et al. [2] presented the method by which the firms' misalignment of supply and demand raises the accuracy of the production prediction. Their methodology makes it possible to anticipate newspaper demand using the ARIMA approach without producing an excessive quantity of output that would result in a loss by integrating accurate customer wants with the newspaper's minimal returns. Use the Box-Jenkins or ARIMA forecasting techniques. Use this method for predicting over the short future. The result of long-term modelling may differ. ARIMA is created by combining moving average models and autoregressive (AR) models (MV). An autoregressive model is the AR (p) model of the ARIMA (p,0,0). Bhardwaj and Arora [3] presented a model that helps the automobile business. Sales projections that are both domestic and precise have been produced as a consequence of the market's intense competitiveness. Whole inventory modelling, developed by Chow and Nerlove in 1957, makes use of an entire demand model to predict the level of automobile inventory. The majority of models employ hybrid, ANN models. Time series are used to forecast domestic auto sales until 2016.

Wang et al. [4] presented revenue forecasting model which makes use of the LSTM neural network and the simulated annealing process. By choosing local minimum values, a stochastic optimisation approach known as SA achieves the optimisation. The RNN, which is utilised in time series and finds a local optimum, contains the LSTM. While building the model, the initial connection weight and goal function of the LSTM are utilised to determine the optimal outcome using the model SA. The following actions are among them: Data gathering and pre-processing, model training, training parameters, comparison analysis, evaluation indicators, application of the model, and outcome analysis. Tony et al. [6] Demand forecasts are used in predictive analytics to determine market demand. Logistic classifiers, decision trees, random forests, and linear regression are used in sentiment-based demand. They draw attention to a few drawbacks of older models, such as poor prediction accuracy, the difficulty of using traditional statistical methods to handle massive amounts of data, and poor predictive model performance. The most popular kind of regression, linear regression, is used to address regression problems. Decision trees are used to address problems with regression and classification.

The following individuals: Bhosale et al. [8] The dataset is used to forecast the dynamic flight price using machine learning techniques. To buy a flight ticket at the cheapest price, this gives the anticipated flight cost. Just a little amount of information is available due to the fact that data is only gathered from websites that sell airline tickets. The R-squared values that the method provides serve as a gauge of the model's accuracy. If more information, such as the current availability of seats, could be made public, the predicted results would be more accurate in the future. Ensafia et al. [10] The Autoregressive (AR) and Moving Average (MA) approaches have been combined to create Auto Regressive Moving Average (ARMA). The representation of this approach is ARIMA (p, q), where the order of the AR component is indicated by the letter p. The MA component's order is also Q. Auto Regressive Integrated Moving Average (ARIMA) is a method used often for time series prediction, which combines moving average with auto regressive models. Drawing autocorrelation function and partial autocorrelation diagrams is one method for determining the optimal parameters. Values between -1 and 1 are known as autocorrelations, and they may be used to calculate the correlation between a time series across various time periods (lags).

3. Proposed Model

Time series algorithms have become increasingly popular in the recent years for the sales forecasting of various products of various organisation. Theses machine learning models are specifically designed to

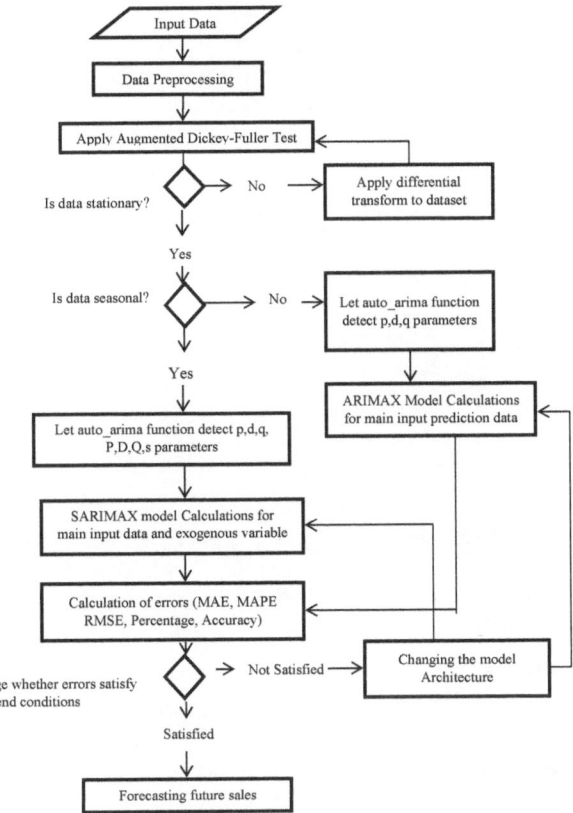

Figure 1: Flowchart of the proposed model

forecast the sales and thereby analysing the profit or loss in the future. By training a time series model on a large dataset of sales products, time series models can be used to accurately forecast the future sales for the designated time period that the user wants to know the forecast. Unlike linear regression method, time series models can produce a better result in terms of forecasting the sales in the future. And also, the algorithm has to be chosen based on the type of dataset like whether it is a food product or textile things like cloths or plastic products. At first, the dataset has to be analyzed. SARIMAX works well on the seasonal data. It considers the factors such as trends, seasonality and cycles in the dataset. If there is seasonality in a time series dataset, a SARIMA (Seasonal ARIMA) model should be used. While using an ARIMA model, only a subset of the data can be taken into account and neglect seasonality. Because of this, predictions are not as precise as they could be. SARIMA models now contain additional seasonal-related traits. A SARIMA model may really be conceptualised as the union of two ARIMA models, one of which addresses the seasonal component and the other the non-seasonal component. So, SARIMAX algorithm has been chosen which also handle the exogenous factors, to train the model. However, time series models have some environmental limitations. Despite these limitations, this proposed model is more convenient and cost effective compared to other models. The Figure 1 below depicts Flow chart of the proposed work.

3.1. Dataset and Model Training

Data collection is the deliberate process of gathering facts about a certain topic. For this proposed model, Perrin Freres Monthly Champagne Sales Data is collected. It is a comprehensive dataset that captures the sales performance of champagne products, offered by Perrin Freres over a period of time. The dataset is organised on a monthly basis, providing a detailed account of sales metrics and relevant factors that influence champagne sales. Champagne dataset has 10 years monthly sales of Champagne. Once the dataset is collected, the exploratory data analysis has been done to outline all the insights and the noisy data and outliers that are present in the dataset are removed. Once the above-mentioned steps are done, the model is trained using the dataset.

3.2. Prediction

When predicting the likelihood of a specific outcome using new data after being trained on a historical dataset, this process is known as prediction in machine learning. In this module, once the machine learning model get the time period, then it will do the forecast and transfer it to the frontend via flask connectivity. Once it is received by the web application, then it will show as a graph to the end user. The end user can know the future sales and help them in terms of procurement, hiring staffs for the organisation. The sales forecasting result plays a major role in the procurement process which will influence the profit or loss of the company.

3.3. Visualisation

Data visualisation is the graphic depiction of information and data. By utilising visual components like charts, graphs, and maps, data visualisation tools provide an accessible approach to observe and comprehend trends, outliers, and patterns in data. Also, it provides a fantastic tool for employees or business owners to communicate facts intelligibly to non-technical audiences. Here, PowerBi tool can be used to visualise the data in graphical manner. PowerBi is one of the popular visualisation tools that is used all over the world especially for the sake of non-technical people.

4. Methodology for Sales Forecasting

The user will give the input data, which is obtained through the frontend, which was created using Angular JS and it will be passed to the backend.

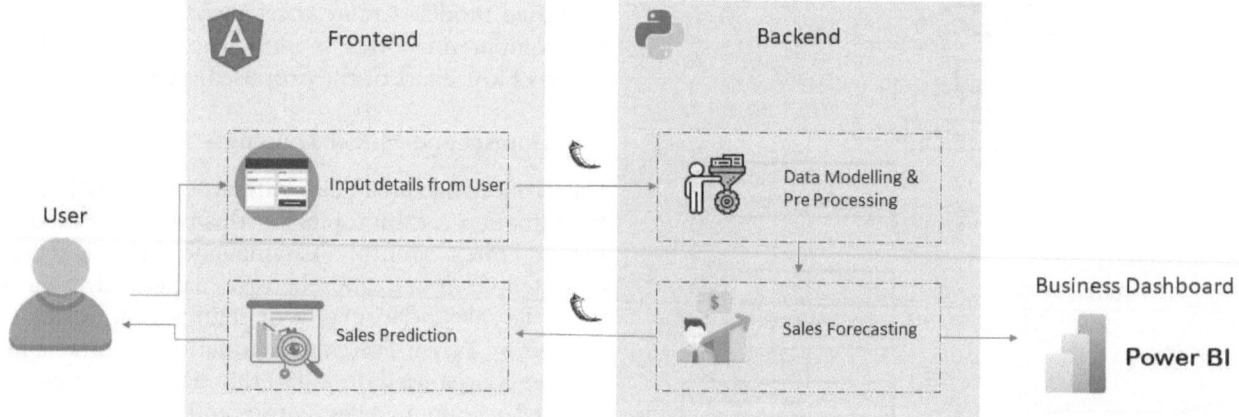

Figure 2: Methodology

An example of such input data is the time period to forecast the sales. The data are pre-processed and the model is trained in the backend. The prediction will then be finished and saved as a brand-new file. A visualisation chart is then constructed when the file has been loaded into the Power Bi visualisation tool. Ultimately, a web app will present the user with a depiction of the sales forecast. The Figure 2 below depicts the methodology used in Sales Forecasting.

5. Experimental Results of the Proposed Model

Various analysis methods are used to predict the future sales data of 1,115 stores. The time series is used to predict the next 3 years with less variation over time and date. The Figure 3 depicts the sum of sales by month. This graph provides a comprehensive overview of the company's sales performance throughout the year, showcasing the total sales figures on a monthly basis. Each bar represents the sum of sales for a specific month, offering a clear visual representation of the fluctuations and trends in revenue over the course of the year. Overall, this visual representation of the sum of sales by month serves as a valuable tool for strategic decision-making, providing a snapshot of the company's financial performance and aiding in the formulation of effective business strategies for the future. The Figure 4 depicts the sales by year. This graph offers a comprehensive insight into the company's year-to-year sales performance, providing a clear depiction of total sales figures for each annual period. The horizontal axis delineates the years, establishing a chronological framework, while the vertical axis illustrates the cumulative sum of sales, presenting a concise representation of the overall financial performance for each respective year. Observing the peaks and troughs in the graph provides valuable insights into the effectiveness of sales strategies, market conditions, and economic influences shaping the business landscape.

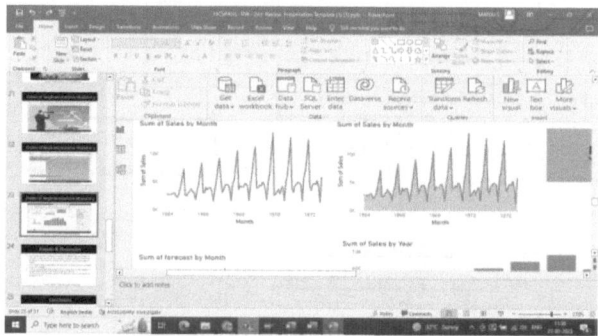

Figure 3: Sales by Month

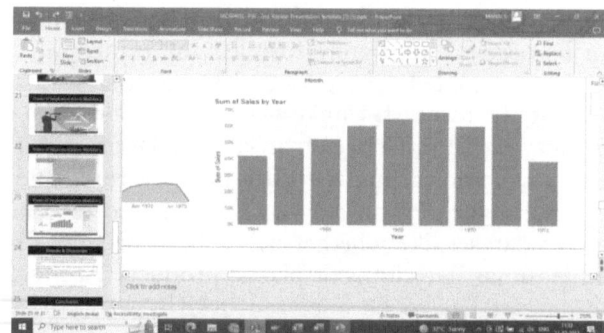

Figure 4: Sales by Year

The Figure 5 depicts the forecast by month. The x-axis delineates the months, creating a timeline that allows for a sequential examination of the forecasted sales. Simultaneously, the y-axis represents the total sales amount, providing a scale for assessing the magnitude and variations in the projected revenue. Analysing this graph enables stakeholders to anticipate and plan for peak sales months, identify potential seasonal trends, and strategically allocate resources to capitalise on projected opportunities. It serves as a valuable tool for decision-makers to refine sales strategies, set realistic targets, and adapt to changing market conditions in a proactive manner. The predictive

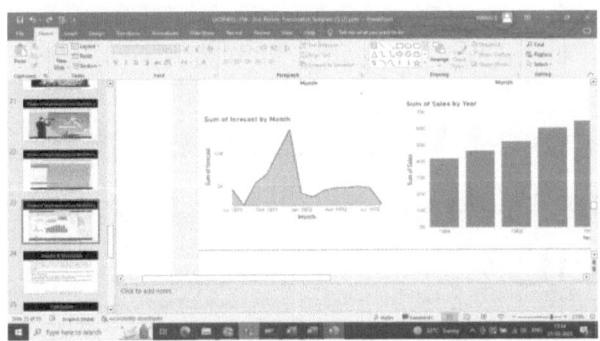

Figure 5: Forecast by Month

nature of this graph empowers the organisation to align its operations with forecasted financial performance, fostering a more agile and responsive approach to market dynamics. As a visual aid for anticipated sales, this graph serves as a key component in strategic decision-making and proactive business planning.

6. Conclusion

Sales forecasting allows business customers to plan their budget and revenue. In one of the existing models, the whole strategy is scarcely scalable, XGBoost does not perform well on sparse and unstructured data, and it is also quite sensitive to outliers. Perrin freres monthly champagne sales forecasting will mainly focus on estimating future sales accurately. As it is a seasonal data, SARIMAX algorithm is used to predict future sales based on previous sales data. The trends and the seasonality present in the dataset are taken into account while training the model. Therefore, this model can be used to provide forecast for both seasonal and non-seasonal data which help to procure the raw material in an efficient way.

References

[1] Sharma, R. and Sinha, A. K. (2012). Sales forecast of an automobile industry. International Journal of Computer Applications. Volume 53– No.12, pp. 25–28.

[2] Permatasari, C. I., Sutopo, W., and Hisjam, M. (2018). Sales forecasting newspaper with ARIMA. In AIP Conference Proceedings, 1931, 030017.

[3] Bhardwaj, V. and Arora, S. (2019). Domestic retail sales forecasting of passenger vehicles in India using time delay neural network. Turkish Journal of Computer and Mathematics Education, 10 (02), 609–620.

[4] Wang, Y., Chang, D., and Zhou, C. (2019). The study of a Sales Forecast Model based on SA-LSTM to cite this article. Journal of Physics: Conference Series, 13, 1401221.

[5] Tony, A., Kumar, P., Jefferson, R., and Subramanian. (2021). Sales forecasting using machine learning models. Annals of R.S.C.B., 25 (5), 3928–3936, ISSN: 1583-6258.

[6] Tony, A., Kumar, P., Jefferson, R., and Subramanian. (2021). A study of demand and Sales Forecasting Model using Machine Learning Algorithm. Psychology and Education, 58 (2): 10182–10194, ISSN:00333077.

[7] Qu, F., Wang, Y.-T., Hou, W.-H., Zhou, X.-Y., Wang, X.-K., Li, J.-B. (2022). Forecasting of automobile sales based on support vector regression optimized by the Grey Wolf Optimizer Algorithm. Mathematics, 10, 2234. https://doi.org/10.3390/math10132234.

[8] Bhosale, N., Gole, P., and Handore, H. (XXXX). Flight fare prediction system. International Journal for Research in Applied Science & Engineering Technology (IJRASET), ISSN:2321-9653.

[9] Mounika, D., Singh, A., and Dharipalli, A. (XXXX). Predictive analysis of supermarket sales using machine learning. International Journal of Creative Research Thoughts (IJCRT), ISSN: 2320-2882.

[10] Ensafia, Y. and Amina, S. H. (2022). Time-series forecasting of seasonal items sales using machine learning. International Journal of Information Management Data Insights, 2, 100058.

[11] Das and Chaudhury, P. (2018). Prediction of retail footwear sales using feed forward and recurrent neural networks.

[12] Singh, B., Kumar, P., Sharmaand, N., and Sharma, K. P. (2020). Sales forecast for Amazon Sales with time series modeling. In 2020 First International Conference on Power Control and Computing Technologies (ICPC2T), pp. 38–43.

[13] Makkar, S., Sethi, V., and Jain, S. (2021). Predictive analytics for retail store chain. In International Conference on Innovative Computing and Communications: Advances in Intelligent Systems and Computing, Volume 1166.

[14] Priya, P., Girubalini, S., Lakshmi Prabha, B. G., Pranitha, B., Srigayathri, M. (2023). A survey on privacy preserving voting scheme based on blockchain technology. Smart Innovation, Systems and Technologies, 312, 267–283.

[15] Brownlee, J. (2018b). A gentle introduction to exponential smoothing for time series forecasting in python. online] Machine Learning Mastery. Accessed: November 23, 2021. https://machinelearningmastery.com/exponential-smoothing-for-time-series-forecastingin-python/.

[16] Pavithra, D., Nidhya, R., Shanthi, S., and Priya, P. (2023). A secured and optimized deep recurrent neural network (DRNN) scheme for remote health monitoring system with edge computing. Automatica, 64 (3), 508–517.

[17] Corrius, J. (2018). Simple stationary tests on time series - bluekiri - Medium. online] Medium. Accessed November 23, 2021. https://medium.com/bluekiri/simple-stationaritytests-on-time-series-ad227e2e6d48.

18.Giering, M. (2008). Retail sales prediction and item recommendations using customer demographics at store level. ACMSIGKDD Explorations Newsletter, 10 (2), 84–89.

Bibliometric Analysis of Focused Web Crawlers for Information Retrieval using Web of Science from 1999 to 2023

Shivani Gautam[1], Rajesh Bhatia[2], Shaily Jain[1]

[1]Chitkara University School of Engineering & Technology, Chitkara University, Himachal Pradesh, India, shivani.gautam@chitkarauniversity.edu.in, shaily.jain@chitkarauniversity.edu.in
[2]Department of Computer Science and Engineering, PEC University of Technology, Chandigarh, India, rbhatiapatiala@gmail.com

Abstract

Focused Crawlers are the robots or spiders that are used to search, gather, and index website pages that satisfy explicitly defined subjects on a limited section of the web. Bibliometric analysis is a beneficial tool for tracing the academic structure of a particular field of research. It is a statistical analysis of publications, books, and articles to trace the researcher's impact and output. It can also be used to compute journal impact factors. It helps us identify the major aspects of any field like research, innovation, education, industry, and technology. The development of the Web has systematically transformed the meaning and motive behind bibliometrics. The present work displays bibliometric analysis via the VOSviewer software tool of 1000 records concerned with focused crawling for information retrieval by collating the data from the Web of Science dataset ranging from 1999 to 2023. The bibliometric output has displayed publication trends, prominent journals, authors, and institutes, most productive country, most cited articles, and most cited countries, author's productivity, and keyword density. Bibliometric analysis results obtained from the aforementioned research will notably enable viewing the development and inclination in the field of focused crawlers for data retrieval.

Keywords: bibliometric analysis, citation analysis, focused web crawlers, information retrieval, publications, research

1. Introduction

In the present day, the World Wide Web is overflowing with enormous volumes of data. Due to this remarkable expansion of the web, sifting through vast amounts of data available online to find useful information has become a daunting task. The information retrieval produces records for the users that satisfy their specific requests. The search engine is employed to extract the desired information from the web. The web crawler serves as the main component of the search engine. It is a web-crawling internet bot that methodically explores the internet. For crawling topic-specified web pages, focused web crawlers are deployed as they follow the concept of vertical search engines. The focused crawler categorises every recovered page into the pre-established topic, and the links are obtained and repeatedly added to the list of seed URLs otherwise the web page is if it's not on topic. The idea of focused web crawling was initially founded by Chakrabarti et al. [1]: a focused web crawler search, gather, and index website pages that satisfy explicitly well-defined subjects on a limited section of the web. To retrieve data whether new or updated focused crawling was also introduced [2]. Focused Web crawler [3] emulates vertical search engines to discover subject-specified web pages rather than retrieving every web page from the web. A neighborhood feature was proposed that enhances the efficiency of focused crawlers immensely [4]. Several focused crawler methodologies' performance is compared to enhance the harvest ratio [5,6]. The keyword query-based focused crawler utilises relevant topic-specific keywords to raise various kinds of queries through the search interface. The instance of a keyword query-based focused crawler is proposed in [7] that leads to domain-specific search in the health and tourism domain. A focused crawler depending on link anchor text to retrieve relevant data for the medical subject of depression is proposed [8]. Topic taxonomy-based focused crawler to decide the relevancy of the web pages is also discussed along with tunneling to improve the harvest ratio [9]. A WTMS crawler for collecting subject-specific web pages is

Chapter 15 DOI: 10.1201/9781003570349

developed [10]. Then, a focused crawler based on content text analysis is also introduced [11]. To understand various focused crawlers, a survey is being done in [12,13]. Language-specific crawlers are introduced to present language-specific crawling to index Thai web pages [14,15]. Focused crawlers were adapted for collecting language-specific pages [16]. A Smart Crawler framework is proposed for harvesting Deep Web [17]. A SCTWC technology is presented to enhance the crawling performance by choosing topic-related URLs [18]. A Treasure crawler that makes use of a hierarchical structure for assigning priority scores to every unvisited link is suggested [19]. An improvised SSVSM for enhancing the performance of focused crawlers by merging TFIDF values and semantic similarities is suggested [20]. A keyword-based technique for retrieving Indian-origin academicians' data by making use of a focused web crawler is presented [21,22]. A tunneling method in association with focused crawlers is proposed to create digital libraries [23]. A genetic algorithm for automatic web page classification is introduced [24]. To enhance fitness function, an improved genetic algorithm-based focused crawler is also presented [25]. An event model in combination with focused crawlers for obtaining relevant web pages is also proposed [26]. A text classification approach is proposed for offering offloading solution in cloud computing [27,28].

The main objectives behind this bibliometric analysis are to quantitatively assess the influence, visibility, and trends of scientific publications in the field of focused crawlers. Through the examination of our citation data, bibliometric studies seek to gauge the influence of individual researchers, journals, or institutions, and to uncover patterns of collaboration and knowledge dissemination. Moving ahead, the present paper is divided into three parts. Firstly, the methodology is put forward in brief. Secondly, the findings of the bibliometric analysis are displayed, which include publication inclination, word cloud, and tree map analysis. Finally, the obtained results are accumulated and scrutinised to recommend future guidelines for focused crawling research.

2. Methodology

The bibliometric analysis also known as scientometrics performs quantitative evaluation on data collected through bibliometrics and encapsulates the conceptual system of a research subject by evaluating the association among several research elements [29,30]. Furthermore, the analyzed data can aid in exemplifying the contributions of particular research works and recognise future trends and possible gaps

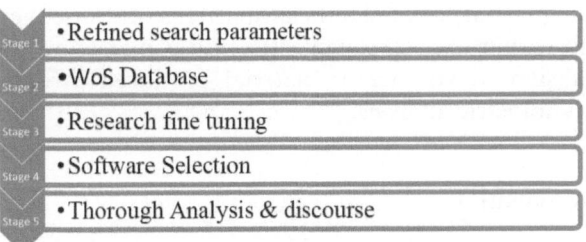

Figure 1: Five phases of bibliometric analysis

[29,30]. It presents a network mapping as well as a performance analysis for the growth of that particular field of research [31]. Henceforth bibliometric analysis has progressively been used in multidisciplinary fields like COVID-19 [32], social cohesion [33], and rumor detection [34]. During this analysis, a method was built to decide the search criteria, and then selection of a suitable database, Afterwards a fitting research refinement process was put into place, then the selection of software and export of final data. Finally, in the last stage analysis and discussion of the results are performed. These stages are shown in Figure 1.

The keywords used in the search criteria to collect the data are "focused crawlers", "information retrieval", "web scraping", and "text classification techniques". The database utilised for the study is Web of Science (WoS). The Web of Science is a thorough, multidisciplinary database that encompasses a wide array of disciplines relevant to this study. Specifically, WoS was selected for its comprehensive coverage across disciplines, high-quality standards, and advanced tools for extracting and visualising data. All kinds of articles, conference papers, and book chapters from 1999 to 2023 are covered in the results. Two software applications were utilised to aid in the handling and analysis of the obtained data. Software utilised to assist the organisation and evaluation of the data retrieved is Microsoft Excel, a tool to organise data tables and create graphs and charts connected to publication and citation trends, prominent authors, topmost institutions, topmost countries, most cited papers, etc. in the given researched field. VOSViewer is a new software tool used for creating and predicting bibliometric networks which is constructed based on author, citation, and keyword data. Finally, performance analysis is done to map the growth sequence of publications, prominent journals, and contributions produced by authors, institutes, and countries about the research topic. Ultimately, a narrative examination of keyword clusters was carried out as a supplementary step. This involved reviewing the abstracts and titles of the 50 most frequently referenced papers containing one of the top 5 keywords at a minimum within a particular cluster. This approach enabled the structured summarisation of significant patterns and discoveries within a cluster, while effectively managing the substantial volume of material produced through bibliometric analysis.

3. Results

The main aim of the bibliometric evaluation is to investigate publication trends, contributions made by important authors, topmost journals, relevant affiliations, and countries' scientific productions. The collected data is examined to determine document and source types, documents language, research trends, study areas, and nations producing scientific publications, most prominent institutions for publications, topmost authors, topmost keywords, and citation trends. Most of the results showcased are detailed in frequency and percentage.

3.1. Publication Trends

In the current review, overall 1000 records that are published covering 807 journals, books, and conference articles from 1999 to 2023 are studied. Concerning publication trends, the information is shown through the yearly count of retrieved documents, including the mean total citations per year and the mean total citations per article. As shown in Figure 2, publications have seen a notable rise since 2008. The analysis will investigate research production depending on the count of documents produced every year. The number of publications released annually would aid the researchers regarding the popularity and pattern of this research progressively over time. Results revealed that there has been an average increase of publications in focused crawling in the last two decades. The production of publications has been fairly persistent over the years. The year 2013 has indicated the pinnacle which is followed by the year 2014. The Yearly Scientific Output also specified that the count of publications is growing even

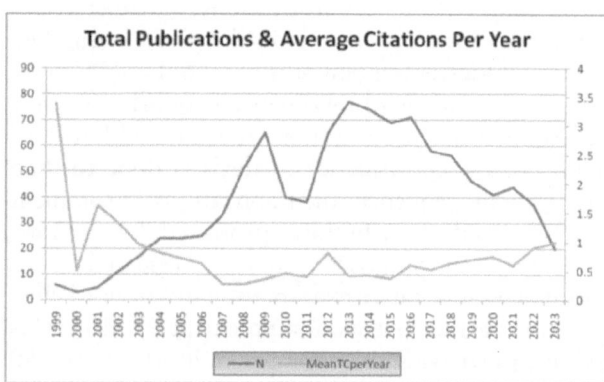

Figure 2: Sum total of publications and average citations per year

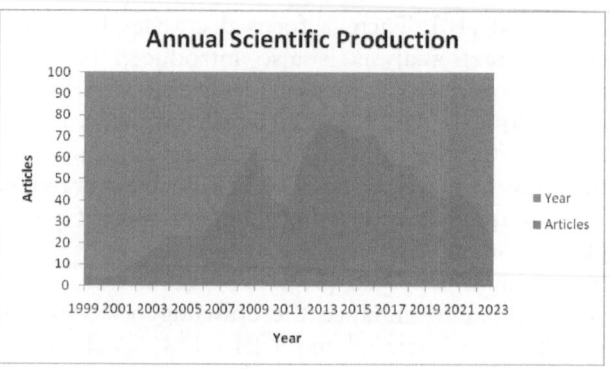

Figure 3: Yearly scientific output

now as shown in Figure 3. Therefore, there are possibilities to expand further on this research topic.

3.2. Nations of Publication

Analysis of publications by nations depending on the affiliating institutions of the researcher is performed in this study. A count of 73 nations recognised is included in the publication on focused crawling for information retrieval for the past two decades under WoS. Table 1 displays the top 10 countries that contributed to the publications. China came up to be the top nation that published the maximum count of publications (458) which stands for 45.8% of total publications. India takes second place with 282 total numbers of publications constituting 28.2%. Third place is bagged by the USA with 214 numbers of publications representing 21.4%.

Table 1: Top 10 Nations' contributions towards publications

Region	Frequency	Percentage
CHINA	458	45.8
INDIA	282	28.2
USA	214	21.4
SPAIN	51	5.1
SOUTH KOREA	50	5
CANADA	46	4.6
UK	32	3.2
GREECE	31	3.1
GERMANY	29	2.9
IRAN	29	2.9

3.3. Prominent Journals

Results demonstrate the most important journals in focused crawlers. A total of 807 journals were identified that published the articles in the given area. 20 topmost prominent journals are reported in Figure 4.

3.4. Prominent Authors

Throughout the time under investigation, 1707 authors produced 1000 publications independently or collectively. Figure 5 displays the prominent authors in this area.

Lotka's law states author productivity in the specified area. Authors' productivity through Lotka's law is illustrated in Figure 6.

3.5. Leading Institutes for Publication

There are a considerable number of prominent institutes that contributed in the analysis about focused crawlers. 923 institutions contributed to the publications. The topmost university with 18 publications is "The University of Xihua" followed by "Wuhan University" and "YMCA" with 15 publications each respectively. Figure 7 shows the most prominent institutes in the field of focused crawlers.

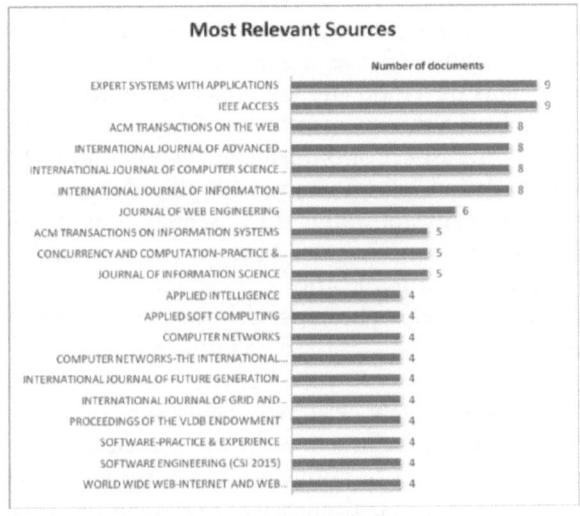

Figure 4: Most active and relevant sources

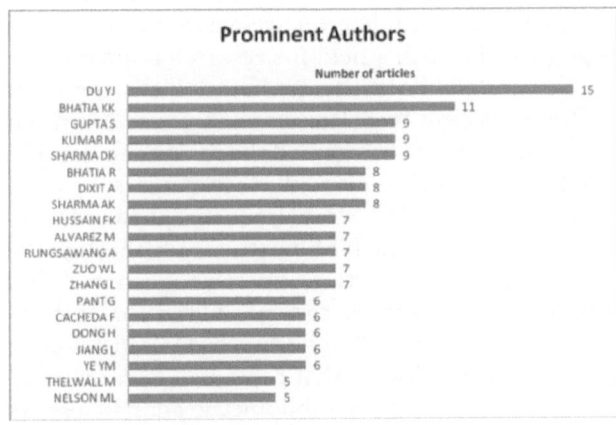

Figure 5: Most prominent authors

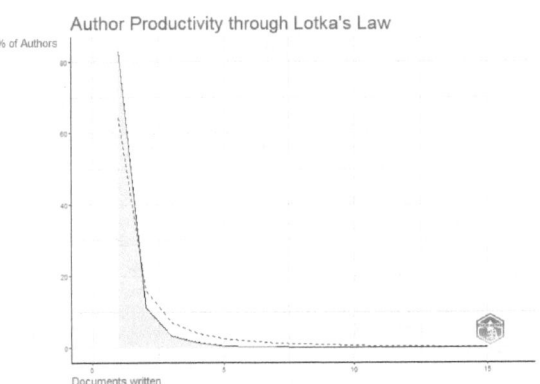

Figure 6: Authors' productivity through Lotka's law

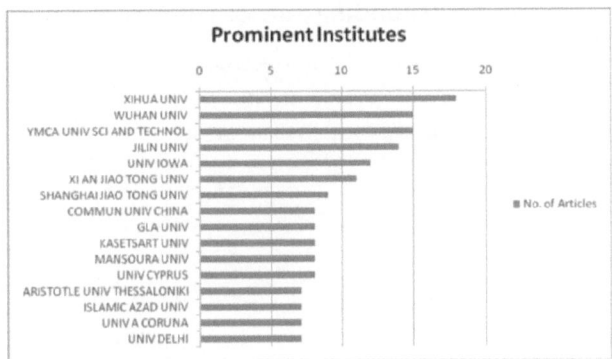

Figure 7: Most prominent institutes

3.6. Citation Trends

Results illustrated that there are 6331 citations reported in the last 20 years for 1000 obtained articles. Figure 8 illustrates the top 10 most cited countries based on number of citations. The most cited document entitled "Focused Crawling: A New Approach to Topic-specific Web Resource Discovery" from Chakrabarti et al. [1], introduces the concept of focused crawlers.

3.7. Corresponding Authors

Figure 9 represented the top 10 countries based on the number of single-country (SCP) and multiple-country publications (MCP) respectively.

3.8. Prominent Keywords

Keywords given by authors of the papers and appeared more than five times in the WoS dataset are applied in the concluding evaluation. Table 2 represents the top 10 keywords used by the author. A word cloud of the 50 most frequently used keywords as shown in Figure 10 was also designed to display the frequency of the keywords which appeared for more than 5 times. It was shown that "Web" was the most widespread accompanied by "information", "algorithm" and "classification".

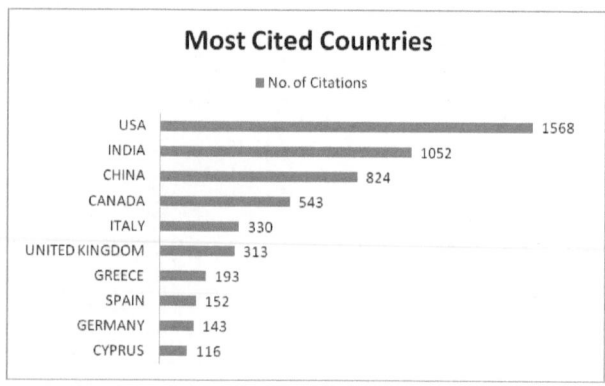

Figure 8: Most cited countries

Figure 10: Word Cloud of 50 keywords

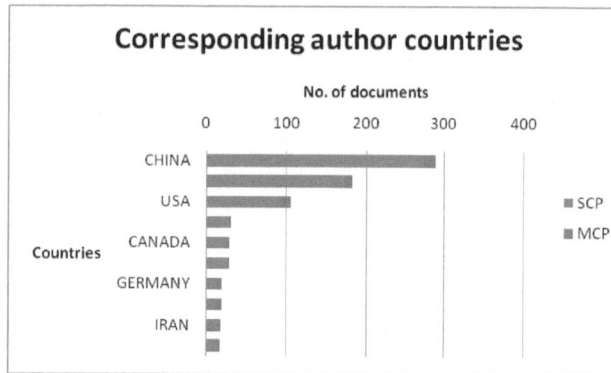

Figure 9: Corresponding author countries

Table 2: Ten most prominent words in the author's keyword

Words	Occurrences
web	28
information	19
algorithm	11
classification	10
system	9
design	8
retrieval	8
framework	7
model	7
crawler	6

4. Discussion

With the aid of bibliometric analysis, this survey has arranged in detail the design, progression, and primary content ingrained in research attributed to focused crawlers. However, there are certain research delimitations. As we have based our analysis on one database, it may have prohibited certain important results from databases like Scopus that can come up with a wide range of results [35]. Secondly, the search terms were not explicitly interpreted into other languages. Therefore, there is a possibility that the research work of non-English speaking countries most likely has been reduced. However, summary and future scope can be deduced from the above results.

Our bibliometric analysis of 1000 records includes several features of focused crawling research on the data gathered via the Web of Science. In the beginning, the research began at a sluggish speed, but starting from 2004, it grew rapidly till 2022 and 2013 documented the highest number of publications. This development was possible by the efforts put in by 65 unique countries. China contributed the bulk in respect of published articles, followed by India and the USA. Some journals played a great role in this domain. Journals "IEEE Access" and "Expert Systems with Applications" became the top-ranking journals with 9 publications in the focused crawling field respectively. The most relevant author in this field is YJ Du who authored 15 publications followed by Bhatia KK and Kumar M with 11 and 9 publications respectively. "XIHUA University of China" was the most productive institute in focused crawling. The discovery of publications and their citations, prominent authors, along with other features attributed to focused crawlers come up with a possibility to acknowledge the research growth in this field by various authors. The given results conclude that this research topic is new, and hence the subject is of importance to numerous researchers. Still, there is great scope in this field for research scholars. The article entitled "Focused crawling: A New Approach to Topic-specific Web Resource Discovery" written by Chakrabarti, S. has gained the maximum number of citations, which leads to the fact that it is the most sought-after article. Most frequent keywords are also recognised in the bibliometric study through the word cloud. Through this paper, we illustrated that bibliometric techniques can be utilised for evaluating research articles. Besides, we have used a simple and clear methodology that can guide researchers in performing and writing bibliometric analysis papers in the future.

5. Conclusion

Through a scientometrics analysis of focused crawlers from 1999 to 2023, the current study has done a quantitative evaluation of research articles on the respective subject. The results displayed remarkable development in research publications and citations on the subject for the last twenty years. Our current article tries to illustrate the research trends during the said period. With the help of VOSviewer software, 1000 records were retrieved from the database (1999–2023) and then exposed to scientometrics analysis. The study motivates researchers to recognise most reputation journals, articles, authors, countries, most cited keywords, and author's productivity, etc. This analysis assists researchers all over the world to realise the growth pattern and trends of focused web crawlers and supports them in their future research. Furthermore, bibliometric analysis of focused crawlers will play a critical part in assessing the productivity and impact of research endeavors, thereby assisting in decision-making processes related to funding, tenure, and strategic planning within academic and research institutions.

References

[1] Chakrabarti, S., Van den Berg, M., and Dom, B. (1999). Focused crawling: a new approach to topic-specific Web resource discovery. Computer Networks, 31 (11–16), 1623–1640.

[2] Pant, G., Srinivasan, P., and Menczer, F. (2004). Crawling the web. In Web Dynamics (pp. 153–177). Berlin, Heidelberg: Springer.

[3] Shokouhi, M., Chubak, P., and Raeesy, Z. (2005). Enhancing focused crawling with genetic algorithms. In Information Technology: Coding and Computing, 2005. ITCC 2005. International Conference on (Volume 2, pp. 503–508). IEEE.

[4] Suebchua, T., Manaskasemsak, B., Rungsawang, A., and Yamana, H. (2018). Efficient topical-focused crawling through neighborhood feature. New Generation Computing, 36 (2), 95–118.

[5] Avraam, I. and Anagnostopoulos, I. (2011). A comparison of focused web crawling strategies. In 2011 15th Panhellenic Conference on Informatics (pp. 245–249). IEEE.

[6] Batsakis, S., Petrakis, E. G., and Milios, E. (2009). Improving the performance of focused web crawlers. Data & Knowledge Engineering, 68 (10), 1001–1013.

[7] Priyatam, P. N., Vaddepally, S. R., and Varma, V. (2012). Domain-specific search in Indian languages. In Proceedings of the First Workshop on Information and Knowledge Management for Developing Region (pp. 23–30). ACM

[8] Tang, T. T., Hawking, D., Craswell, N., and Griffiths, K. (2005). Focused crawling for both topical relevance and quality of medical information. In Proceedings of the 14th ACM International Conference on Information and Knowledge Management (pp. 147–154). ACM.

[9] Altingovde, I. S. and Ulusoy, O. (2004). Exploiting inter-class rules for focused crawling. IEEE Intelligent Systems, 19 (6), 66–73.

[10] Mukherjea, S. (2000). WTMS: A system for collecting and analyzing topic-specific Web information. Computer Networks, 33 (1–6), 457–471.

[11] Gatial, E., Balogh, Z., Laclavik, M., Ciglan, M., and Hluchy, L. (2005). Focused web crawling mechanism based on page relevance. Proceedings of ITAT, 41–46.

[12] Menczer, F., Pant, G., and Srinivasan, P. (2004). Topical web crawlers: Evaluating adaptive algorithms. ACM Transactions on Internet Technology (TOIT), 4 (4), 378–419.

[13] Kumar, M., Bhatia, R., and Rattan, D. (2017). A survey of Web crawlers for information retrieval. Wiley Interdisciplinary Reviews: Data Mining and Knowledge Discovery, 7 (6), e1218.

[14] Tadapak, P., Suebchua, T., and Rungsawang, A. (2010). A machine learning-based language-specific web site crawler. In Network-Based Information Systems (NBiS), 2010 13th International Conference on (pp. 155–161). IEEE.

[15] Srisukha, E., Jinarat, S., Haruechaiyasak, C., and Rungsawang, A. (2008). Naive Bayes based language-specific web crawling. In Electrical Engineering/ Electronics, Computer, Telecommunications and Information Technology, 2008. ECTI-CON 2008. 5th International Conference on (Volume 1, pp. 113–116). IEEE.

[16] Tamura, T., Somboonviwat, K., and Kitsuregawa, M. (2007). A method for language specific Web crawling and its evaluation. Systems and Computers in Japan, 38 (2), 10–20.

[17] Zhao, F., Zhou, J., Nie, C., Huang, H., and Jin, H. (2016). SmartCrawler: a two-stage crawler for efficiently harvesting deep-web interfaces. IEEE Transactions on Services Computing, 9 (4), 608–620.

[18] Zhang, H. and Lu, J. (2010). SCTWC: A nonline semi-supervised clustering approach to topical web crawlers. Applied Soft Computing, 10 (2), 490–495.

[19] Seyfi, A., Patel, A., and Júnior, J. C. (2016). Empirical evaluation of the link and content-based focused Treasure-Crawler. Computer Standards & Interfaces, 44, 54–62.

[20] Du, Y., Liu, W., Lv, X., and Peng, G. (2015). An improved focused crawler based on semantic similarity vector space model. Applied Soft Computing, 36, 392–407.

[21] Kumar, M., Bhatia, R., Ohri, A., and Kohli, A. (2016). Design of focused crawler for information retrieval of Indian origin Academicians. In 2016 International Conference on Advances in Computing, Communication, & Automation (ICACCA) (Spring) (pp. 1–6). IEEE.

[22] Kumar, M., Bindal, A., Gautam, R. and Bhatia, R. (2018). Keyword query-based focused Web crawler. Procedia Computer Science, 125, 584–590.

[23] Bergmark, D., Lagoze, C., and Sbityakov, A. (2002). Focused crawls, tunneling, and digital libraries. In International Conference on Theory and Practice of Digital Libraries (pp. 91–106). Berlin, Heidelberg: Springer.

[24] Goyal, N., Bhatia, R. and Kumar, M., 2016, June. A genetic algorithm based focused web crawler for automatic webpage classification. In 3rd International Conference on Electrical, Electronics, Engineering Trends, Communication, Optimization and Sciences (EEECOS 2016) (pp. 1-6). IET

[25] Yan, W. and Pan, L. (2018). Designing a focused crawler based on an improved genetic algorithm. In 2018 Tenth International Conference on Advanced Computational Intelligence (ICACI) (pp. 319–323). IEEE.

[26] Farag, M. M., Lee, S., and Fox, E. A. (2018). Focused crawler for events. International Journal on Digital Libraries, 19 (1), 3–19.

[27] Bajaj, K., Sharma, B., and Singh, R. (2022). Comparative analysis of simulators for IoT applications in fog/cloud computing. In 2022 International Conference on Sustainable Computing and Data Communication Systems (ICSCDS) (pp. 983–988). IEEE.

[28] Bajaj, K., Jain, S., and Singh, R. (2023). Context-aware offloading for IoT application using fog-cloud computing. International Journal of Electrical and Electronics Research, 11 (1), 69–83.

[29] Block, J. H. and Fisch, C. (2020). Eight tips and questions for your bibliographic study in business and management research. Management Review Quaterly, 70, 307–312.

[30] Donthu, N., Kumar, S., Mukherjee, D., Pandey, N., and Lim, W. M. (2021). How to conduct a bibliometric analysis: An overview and guidelines. Journal of Business Research, 133, 285–296.

[31] Cobo, M. J., López-Herrera, A. G., Herrera-Viedma, E., and Herrera, F. (2011). An approach for detecting, quantifying, and visualizing the evolution of a research field: A practical application to the Fuzzy Sets Theory field. Journal of Informetrics, 5, 146–166.

[32] Yu, Y., Li, Y., Zhang, Z., Gu, Z., Zhong, H., Zha, Q., Yang, L., Zhu, C. and Chen, E., 2020. A bibliometric analysis using VOSviewer of publications on COVID-19. Annals of translational medicine, 8(13):816

[33] Moustakas, L. (2022). A bibliometric analysis of research on social cohesion from 1994–2020. Publications, 10 (1), 5.

[34] Rani, N., Bhardwaj, A. K., Das, P., and Sharma, A. (2022). Bibliometric analysis of rumor detection via Web of Science from 1989 to 2021. Concurrency and Computation: Practice and Experience, 34 (25), e7260.

[35] Pranckutė, R. (2021). Web of Science (WoS) and Scopus: The titans of bibliographic information in today's academic world. Publications, 9 (1), 12.

Inclusive Technologies

A Paradigm Shift for Disabilities Using IoT

Prachi Sasankar, Chetna Thakur, Nazish Khan, Priya Mandal, and Shruti Singh

[1]Department of Computer Science, Sadabai Raisoni Women's College, Nagpur, India

E-mail: sasankar.prachi@gmail.com, chetna.thakur.it@srwc.raisoni.net

Abstract

This research delves into the potential transformative capabilities offered by Internet of Things (IoT) technologies in improving the quality of life for individuals living with disabilities, focusing on mobility impairments. Through a case study of a smart home environment, the research demonstrates how IoT devices can significantly improve accessibility, independence, and overall well-being for those facing mobility challenges. The study evaluates the impact of these technologies on accessibility, inclusivity, and independence, while also addressing challenges and ethical considerations associated with IoT deployment in this domain. Furthermore, the paper acknowledges and delves deeper into privacy concerns related to IoT-enabled devices for people with disabilities. By examining real-world cases and emerging trends, this research contributes valuable insights into the potential of IoT to revolutionise the lives of individuals with disabilities, fostering a more inclusive and equitable society.

Keywords: IoT technologies, smart devices, assistive technology, gadgets for disabled, helping disabilities, smart cane, sesame phone, smart door access, IoT-based sensors

1. Introduction

In order to make our lives easier, the Internet of Things (IoT) is an emerging paradigm that allows communication between electronic devices and sensors. IoT uses smart devices and the internet to offer creative solutions to a range of problems and challenges pertaining to different business, governmental, and public/private enterprises worldwide [1]. It also benefits from quantum and nanotechnology in previously unthinkable ways, such as faster computing, sensing, and storage [2]. As a potent instrument to reduce this accessibility gap, IoT technologies have surfaced. A revolutionary way to improve the lives of individuals with disabilities in this era of technological advancements is the integration of IoT enabled technology. Examples of these include mobility aids (like smart canes or wheelchairs), academic and learning aids (like computer-based software and portable word processors), communication aids (like captioning devices, smart glasses, and augmentative communication devices), and smart systems (like smart homes, smart cities, and smart workplaces). These technological advancements enable people to overcome obstacles and live more independent, fulfilling lives [3].

2. Literature Review

A minimum of 15% of the global population is estimated to have a disability, according to the World Health Organisation [10]. For example, a blind person is unable to see or recognise their surroundings, a deaf person is unable to hear the sound, and a handicapped person is unable to perform daily tasks. People with disabilities deal with a variety of difficulties in their daily life, some of which are communication barriers, difficulties with health monitoring, unequal opportunities, etc. In this context, assistive technology plays a crucial role in helping people with disabilities overcome daily challenges, enabling them to access education, participate in the labour market, and integrate into society with dignity [10]. With the emergence of IoT technology, the disciplines of rehabilitation, education, and healthcare are challenged to integrate cutting-edge IoT applications into current assistive technology services. The IoT-enabled assistive technology devices are proficient in helping any disability, from deaf and mute to visually impaired individuals [4].

2.1. Sesame Phone: A Hands-free Smartphone

The Sesame smartphone, created for individuals with disabilities, is the initial model featuring a hands-free design with the following capabilities:

Voice command: The phone is equipped with built-in voice command, eliminating the need to use buttons for device activation and to take calls.

Chapter 16 DOI: 10.1201/9781003570349

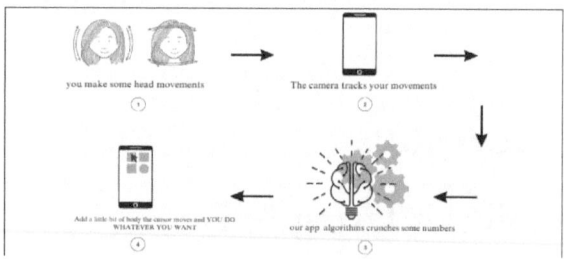

Figure 1: Sesame smartphone [4].

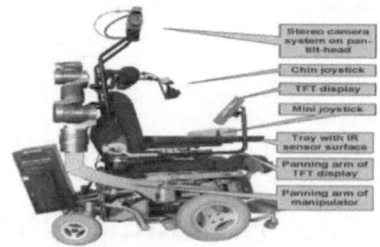

Figure 2: The Care-Providing robot FRIEND [9].

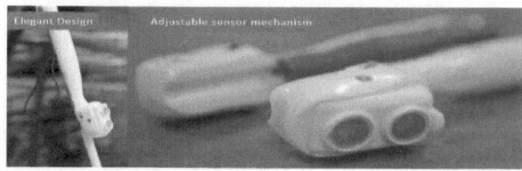

Figure 3: Smart Cane [7].

Gesture-based interaction: This phone excels in recognising simple movements of the head, making it possible to operate the device without touching it.

This Sesame phone is an excellent choice for individuals with a variety of conditions, including amputations, spinal cord damage, cerebral palsy, multiple sclerosis, conditions affecting finger or hand mobility, neuromuscular disorders, amyotrophic lateral sclerosis (ALS), and those recovering from a stroke.

2.2. IoT Technologies in Healthcare

An IoT-powered patient health monitoring system links to the Cloud Talk platform [5]. The need for patients to travel to healthcare institutions is eliminated by these wireless IoT-driven technologies offering healthcare services directly to patients. In order to provide suitable recommendations, data is shared with healthcare experts after being securely gathered by IoT-based sensors and processed by a compact algorithm [6]. Home networks can be used to connect domestic robots for a variety of purposes. Apart from these all-purpose robots, specific robots have been created lately to assist with medication administration and alert a distant home care provider if a patient is at danger of missing a dose. The Care Providing robot FRIEND, depicted in Figure 2, is one instance of a devoted robot.

2.3. Smart Canes

Using a regular white cane, a blind person can readily identify potholes, surface textures, and other hazards [7].

3. Methodology

The technology mentioned above employs IoT. In this section, we elaborate the comprehensive methodology employed for the design, implementation, and evaluation of the IoT enabled healthcare innovations, namely the smart canes and the sesame phone. The methodology encompasses the key components design and implementation.

3.1. Design

Sesame phone: Small head motions are tracked by the sesame phone and converted into a "cursor" that shows on the screen. Some head movement is necessary because the cursor is controlled by your head movements. The head-motion activated mobile device, created by a quadriplegic, introduces the mobile phone revolution to a previously unreached population: individuals with multiple sclerosis, ALS, cerebral palsy, spinal cord injuries, and other disabilities that limit the use of the hands and arms.

Smart cane: An electronic walking assist that slides onto the white cane's upper fold. It uses ultrasonic range to find impediments and uses different vibratory signals to tell users how far away something is [7].

3.2. Implementation

Sesame phone: With this technology, gestures can be translated into on-screen activities including typing, scrolling, and selecting. People with a variety of disabilities can use the interface because it is made to be simple to use and intuitive.

Smart cane: Assists users in avoiding collisions with projecting and overhanging objects, such as open glass windows, parked cars, signboards, and tree branches, allowing them to move safely and confidently through various social contexts.

4. Case Study

Background: A 36-year-old man with paraplegia as a result of a spinal cord injury was chosen as the participant in this case study. The participant's everyday life and general well-being have been significantly hampered by the resulting mobility issues.

Table 1: IoT enabled functionalities for disabled people [8]

Disability Type	IoT Functionality Examples
Visual	Smart navigation apps, Braille e-readers.
Hearing	Vibrating alerts, Closed captioning.
Speech	Voice—controlled devices, AAC communication

Voice-Activated Assistants: Strategically positioned voice-activated virtual assistants are deployed within the domicile. These assistants exhibit sophisticated voice recognition capabilities, enabling the participant to orchestrate tasks such as setting reminders, initiating calls, and managing daily agendas without the requirement for manual interaction.

Emergency Response System: The participant is given access to an IoT-based emergency button. When triggered, this feature sets off an alert mechanism that rapidly alerts careers or emergency services, thus expediting immediate assistance during critical scenarios.

Result: The integration of IoT technologies has yielded discernible enhancements in the participant's quality of life and autonomy. Notably, the participant reported heightened control over their domestic environment, diminished reliance on caregivers for routine tasks, and an augmented sense of self-assurance in managing daily routines.

Discussion: IoT devices have the potential to grant people with mobility limitations increased autonomy through the removal of physical obstacles and the introduction of clear voice-controlled interfaces.

Conclusion: The successful deployment of IoT devices within the participant's living environment serves as a poignant illustration of technology's capacity to enhance accessibility, fortify independence, and ultimately augment the holistic well-being of individuals confronting mobility constraints.

5. Conclusion

The specific difficulties experienced by people with disabilities, such as mobility and cognitive limitations, hearing and vision impairments, have been addressed by IoT devices. The IoT is opening up new opportunities for efficient and customised care. Wearable tech, remote monitoring programmes, and intelligent prosthetics are a few instances. The quality of life for individuals with disabilities is enhanced by these developments, which help reduce healthcare costs and increase the efficiency of healthcare systems. In addition, the accessibility and independence of IoT solutions have greatly enhanced. Smart homes with voice-activated controls, ambient sensors, and adaptable lighting are making life easier for those with mobility or sensory impairments.

References

[1] Sfar, A. R., Chtourou, Z., and Challal, Y. (2017). A systemic and cognitive vision for IoT security: A case study of military live simulation and security challenges. 2017 International Conference on Smart, Monitored and Controlled Cities (SM2C), Sfax, Tunisia, 101–105. doi:10.1109/SM2C.2017.8071828.

[2] Gatsis, K., and Pappas, G. J. (2017). Wireless control for the IoT: Power spectrum and security challenges. IoTDI '17: Proceedings of the Second International Conference on Internet-of-Things Design and Implementation. 341–342.

[3] Yakut, A. D. (2022). Internet of Things for individuals with disabilities. In Yakut, E. (Ed.), Industry 4.0 and Global Businesses. Emerald Publishing Limited, Bingley; 137–152.

[4] Gera, S., Mridul, M., and Sharma, S. (2021). IoT based Automated Health Care Monitoring System for Smart City. 2021 5th International Conference on Computing Methodologies and Communication (ICCMC). Erode, India; 364–368. doi:10.1109/ICCMC51019.2021.9418487.

[5] Al-Sheikh, M. A., and Ameen, I. A. (2020). Design of mobile healthcare monitoring system using IoT technology and cloud computing. IOP Conference Series Materials Science and Engineering, 881, 012113. doi:10.1088/1757-899X/881/1/012113

[6] Shubankar, B., Chowdhary, M., and Priyaadharshini, M. (2019). IoT device for disabled people. Procedia Computer Science, 165, 189–195. ISSN 1877-0509.

[7] Resul, D., Ayse, T., Senay, D., and Meral Kayapinar, Y. (2017). A survey on the Internet of Things solutions for the elderly and disabled: Applications, prospects, and challenges. International Journal of Computer Networks and Applications, 4(3), 84–92. doi:10.22247/ijcna/2017/49023

[8] Ayad Ghany, I. (2017). IoT technologies to disabilities persons—A survey on the Internet of Things solutions for the elderly and disabled: Applications, prospects, and challenges. International Journal of Computer Networks and Applications (IJCNA), 4(3). doi:22247/ijcna/2017/49023

[9] De Freitas, M. P., Piai, V. A., Farias, R. H., Fernandes, A. M. R., de Moraes Rossetto, A. G., and Leithardt, V. R. Q. (2022). Artificial Intelligence of Things applied to assistive technology: A systematic literature review. Sensors (Basel). 22(21), 8531. doi:10.3390/s22218531. PMID: 36366227; PMCID: PMC9658699.

Finding and Interpreting Suicidal Ideation on Public Media Using Deep-learning and Machine Learning

A Review

Dawood Bashir, Narinder Kaur

Department of CSE, Chandigarh University Garuan, India
E-mail: Baldengineer7@gmail.com, Er.narinder@gmail.com

Abstract

Suicide remains a pressing concern in today's world. Addressing early identification and intervention of potential suicide attempts is crucial for preserving lives. Present strategies for detecting suicidal ideation encompass clinical approaches, which involve interactions between professionals and individuals at risk, and computational methods that utilise machine-learning and deep-learning for automated understanding from digital social content. With the exponential participation of the folks on public media, they have started to grow around the bubble. Rather than expressing grief complaints and concerns to family, friends, and relatives, they choose public media platforms as a source to communicate their state of mind like Facebook, Twitter, Instagram, Snapchat, Skype, and many more. The motive of this review paper is to early detect and highlight psychiatric tendencies with the help of tweets, posts, and other graphical content they share. Early detection will assist in counseling youth and eventually avoid suicides. Nevertheless, detecting and interpreting trends of suicidal ideation represents arduous work. Consequently, it's crucial to design and develop a machine-learning system that automatically detects the ideation of suicide or any uncertain transformation in the conduct of a user content associated with it.

Keywords: Suicide, public media, mental health, machine-learning, deep-learning.

1. Introduction

Suicide, the act of intentionally ending one's own life. There are "N" parameters that lead to Suicide among folks prevailing in societies in different forms like poverty, Anger, Depression, fear, reduced health, stigma and taboo. The ratio of suicide is reported high in economically deficit countries. It's reported that globally every year more than 700000 people end their life. Individuals in the prime of their youth most often take their own lives as per the WHO records and surveys (World Health Organisation) [1]. According to reports, suicide ranks as the second most common cause of mortality for individuals aged 10 to 34 years old [2]. There is no cure for suicide i.e. fatal, however; non-fatal i.e. suicide attempts can be controlled by proper social awareness and automated tools for detecting the conduct of users on the web. In "India" annually, more than "100000" people prematurely take their lives, and the rate of suicides increases by 2% annually [3]. Most people who suffer from suicidal ideation do not prefer health assistance and counseling instead, they choose public media as a platform to express their state of mind because of the disgrace associated with it. Mental illness is a curable disease and can be avoided, thus detecting it in the early stages would help to save and transform lives. To track the behavior of the user on public media platforms many techniques like Long-short-term-memory, bidirectional-short-term-memory, transformers with the concept of attention weight, gated recurrent units to avoid complexity, Sentiment analysis at its best possible speed is introducing novel techniques that could automatically interpret users' behavior i.e. feelings and emotions. By making the best use of the information on public media handles, sentiment analysis can capture the premature signs, stages, and symptoms of a user. A lot of research work has been done and still going on the topic, however, predicting the behavior using small data has always been a concern. The sudden inactivity of a user on the web also indicates a cause.

2. Literature Review

Suicide is recognised in every culture, across various regions and ethnic groups. In recorded history,

Chapter 17 DOI: 10.1201/9781003570349

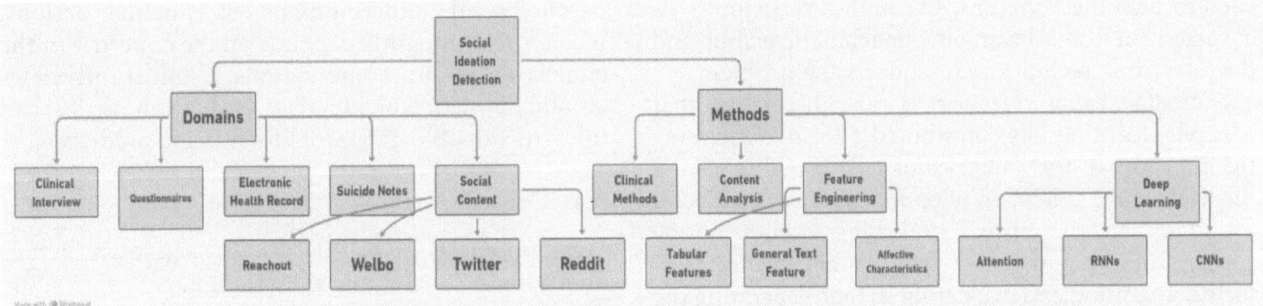

Figure 1: Categorisation of SID: Domains are represented in left part whereas methods are represented in right part. Subcategories are represented using an arrow and solid point

the earliest known instances of suicide date back to approximately 2000 BC in Babylonia and Persia, involving the tragic tale of Pyramus and Thisbe, two lovers [4].it would be no wrong to say that suicide has become one of main pillars of mortality. Keeping the rise of growth under consideration numerous initiatives and technologies are deployed to control the cause of which machine-learning and deep-learning are also the part.

Asma Abdulsalman et al. [5] claims that public media has helped bridging a gap between the people separated by distance. People mostly express their views, ideas, share their state of mind. Negative sentiments like hate for life, self-induced injuries, and tough times are also expressed on public media. Using the data in a proper way the growing rate of suicide can be controlled by resolving their issues thus can be saved. Author has worked on multiple datasets using number of machine-learning tools to detect the sentiments that promote suicide rate. The results received had 86% accuracy and 79% f1 score.

Shaoxiong ji et al. [6] stated that suicide has become the serious issues of newly evolved societies, however detecting it on the premature stage would help a ton. Machine-learning with the help of feature extraction been used to train models and interpret the sentiments of the user to address their mental health issues.

Abayomi Arowosegbe et al. [7] stated that globally millions of people contribute to mortality by considering suicide. The main reason people consider suicide because of the under laden issues related with it. With the aid of machine-learning and natural-language-processing concerns can addressed. Author has collected data from different sources and performed number of tests to generate the most optimal results.

Shini Renjith et al. [8] explained that detecting suicide ideation on public media is a troublesome task. Folks having the tendencies of suicide often isolate themselves from the others thus use internet

and public media to express their views about life. Posts uploaded by these victims gives very valuable information to detect the suicidal tendencies of a user. The key things to stop suicides is to understand the cause associated with it. Identifying the sudden change in the behaviour of a user can contribute in detecting the users chances of committing suicide. NLP with integration of many tools of AIMl were used to know the behavior of the user.

Swamy et al. [9] started that suicide is one of the common cause of death, claimed more than 800000 contribute to the mortality thus has become a concern. Number of researches have proved that public media platforms have given it a major breakthrough. To know the mental state of a victim facing suicidal tendencies, they classified the the ideas using the natural-language-classifier. NLP holds a firm position in addressing the cause. Author has used support vector machine, CNN, ML, SGD to find the results, thus recorded the accuracy of 92%.

Anshu Malhotra et al. [10] stated that the health whether it is mental or emotional is very uncared issues of people. Depression is one of the main contributors of Suicide growth. Causing self-harm and resting with injuries is also a leading factor. However, treating and addressing these concerns of suicide is way too expensive when counted clinically. In the era of public media, folks choose to communicate their personal affairs on public media. All the details like personal data shared by these users can be used for checking the behavior of the user on the platforms henceforth addressed. The author used convolutional neural network and NLP for detecting the sudden changes in user conduct.

Vishal Desu et al. [11] stated that suicide is a global concern, to express and share their state of mind folks make use of public media platforms that connects them to a vast majority of people. Folks who face the issues of suicide tendencies more often keep their identities concealed, thus remain unaddressed and uncured. Author says that the data can be used to help these victims. The author has employed a variety of tools leveraging machine-learning and deep-learning techniques to address the problem.

Wesllei Felipe Heckler et al. [12] cited that Machine-learning has contributed a lot in addressing the concerns of the victims thus has helped in saving the lives. Huge collection of papers were reviewed and found text analysis plays a vital role in addressing the cause. Collecting the data from public media platforms, analyzing and evaluating it, thus generating the results that helps in saving lives.

Theyanz et al. [13] stated that people who have suicidal tendencies most often use public media platforms to express their state of mind, however its very clumsy to identify the patterns that represent the victim signs. In order to conquer the challenge they used machine-learning and deep-learning as such users with the tendencies of suicide could be detected in early stages. They used some other tools to combat the issue like CNN with BiLSTM, XGBoost to classify the data accordingly.

Soumyabrat Saha et al. [14] Suicide has proved to be a serious concern. Methods and approaches of detecting and preventing suicide rate, Public media plays a crucial role in identifying suicidal patterns thus saving lives. The intent of author is to evaluate the behavior of the user on public media to know their state of mind, to combat the issue author has used machine-learning to detect suicidal ideation ad mental disorders. SVM was used for the classification and received the accuracy of 0.886%.

3. Methods and Categorisation

Historically, suicide detection was grounded in clinical approaches, which involved self-assessments and in-person interviews. In direct interactions, both spoken words and tonal nuances can offer insights. Scherer [15] delved into speech patterns and voice qualities during dyadic interviews to distinguish between suicidal and non-suicidal youths. A crucial element of clinical intervention is grasping the

psychological underpinnings of suicidal actions, which predominantly depends on the expertise of the clinician and direct interactions. Clinical interviews combined with suicide risk evaluation scales can unearth valuable indicators for suicide prediction.

3.1. Content Analysis

Posts on social platforms disclose extensive details about users, including their language choices. By conducting a thorough examination of the user-generated content, insights into linguistic indicators and the language preferences of those at risk of suicide emerge. The in-depth analysis encompasses lexicon-based screening, linguistic statistical attributes, and topic analysis of posts related to suicide. A specialised dictionary and lexicon for suicide-related terms were crafted by hand to facilitate keyword and phrase filtering [16]. Terms associated with suicide include expressions such as "die", "fetal ideas", "feel destitute", "depressed state" and "cutting and chopping myself" [17]. Thoughts of suicide often intertwine with powerful negative emotions, anxiety, feelings of despair, and other social determinants, such as relationships with loved ones. Ji [18] employed word cloud visualisation and topic modeling-techniques on content associated to suicide, uncovering such conversations about suicide encompass both individual and societal concerns.

3.2. Feature Engineering

Motivation behind textual suicide categorisation is to ascertain if individuals exhibit suicidal tendencies through their written posts. Various methods from machine-learning and natural-language-processing contributed a significant role in the following domain. For SID, structured form of data comes from survey answers and organised statistical data pulled from online sources. This type of data could be readily used as classification of features or simply the regression activities. Survey responses serve as a valuable reservoir for structured features.

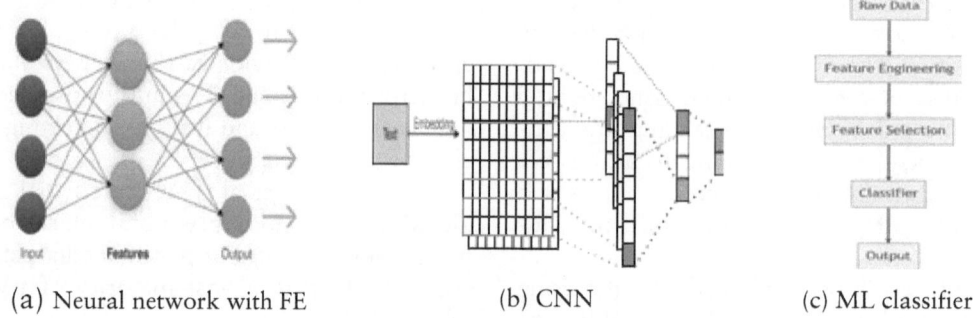

(a) Neural network with FE   (b) CNN   (c) ML classifier

Figure 2: representation of different methods with feature engineering. (a) Neural network plus feature engineering. (b) Convolutional neural network (c) machine-learning classifier

3.3. General Text Features

One approach in feature engineering focuses on deriving attributes from non-structured text data. The subject modeling approach [19] was combined with other machine-learning strategies to detect suicidal content on Sina Weibo. Ji et al. [18] derived diverse features as sets. These tools encompass statistical, syntactic, linguistic inquiry, and word count (LIWC) methods. "word embedding", and "thematic attributes", subsequently integrating these into classification models, as illustrated in Figure 2(b), where a comparison of four conventional supervised classifiers is presented. Shing [20] derived a range of features, encompassing bag of words (BoWs), empath metrics, readability-scores, syntactic-attributes, topic model outputs, word embedding, LIWC, emotional attributes, and mental health lexicon".

3.4. Deep Learning

Deep-learning techniques have seen remarkable success in various domains, from computer vision and NLP to medical diagnostics. Within suicide research, deep-learning is becoming a pivotal tool for automated suicide prevention. Its strength lies in its ability to autonomously learn textual attributes without the need for intricate feature engineering. Strategies for modifying neural architectures, specifically "CNNs" and "LSTMs", to identify ideation of suicide in private chat environments. Yet, the decentralised training approach is limited in its reliance on chat room coordinators to tag posts for guided training, rendering it suitable only for niche applications. A more versatile approach might involve learning techniques like unsupervised and semi supervised. Matero et al. [21] introduced a dual-context strategy that made use of hierarchically attentive RNN and BERT. Additionally, there's a burgeoning interest in hybrid methodologies that merge limited feature engineering combined with representation-learning.

4. Applications on Domains

A number of Machine-Learning approaches have been designed for SID. The existing studies are categorised based on the type of data source used. The scope of data sources encompasses "questionnaires," "electronic health records" (EHRs), "suicide notes," and "online content." In Figure 4, there are visual representations of data sources.

(a) Questionnaires
Standardised mental health scales, like as "DSM-IV", "ICD-10", and "IPDE-SQ", are valuable tools for assessing a person's mental well-being

1) There are moments when I feel a void within me.
2) Life usually brings me joy and laughter.
3) Sometimes, I can't help but let my anger bubble over.
4) There have been times when people have unjustly criticised me or my actions.
5) I often find myself pondering about who I truly want to be.
6) Being online often leaves me feeling out of place or vulnerable.
7) At times, I can't shake off the feeling that my partner might not be loyal.
8) My emotions tend to shift like the changing seasons.
9) Many believe I'm someone to look up to.
10) I have moments where I throw caution to the wind and act on impulse.

a) Questionnaires

– Severe risk attempts
– Count of address changes
– Job: substance abuser
– Relationship status: unmarried
– ICD code F19 (mental health issues related to substance misuse)
– ICD code F33: recurring depression issues
– ICD F60: distinct personality issues
– ICD T43: overdose from psychoactive medications
– ICD code Z29: requirement for preventive actions
– ICD T50: toxic ingestion

b) electronic health records

Sushant Singh Rajput @itsSSR

I'm ending all this for good.Going far yet so close to you people.Maybe after this people try to communicate with others rather than holding onto the thoughts.See ya at the other side.GoodBye. 💙 ☆

5:43 am · 14 Jun 20 · Twitter for Android

c) Tweets

I've come to believe that my situation is long-standing, making recovery uncertain. Suddenly, my drive and strength to fight have vanished. I truly wished to recover, but it seems it wasn't meant to happen. I feel overwhelmed both physically and emotionally. Please don't be too saddened; take solace in the fact that I'm now free from the pain and isolation I've felt for so long.

d) Suicide-notes

Figure 3: Content for SID with example. (a) Question-naire. (b) EH-Record. (c) Tweets. (d) Suicide-notes.

and potential suicidal tendencies. These standards can shape self-assessment questionnaires or direct patient interviews. Delgado-Gomez et al. [22] utilised and discriminated the "IPDE-SQ" and the "Barrat's-Impulsiveness-Scale" (BIS- 11) to spot those at risk of suicide. They further evaluated individual items from these scales. The "IPDE-SQ" under "DSM- IV" contains ""77"" true-false-questions"", whereas the BIS-11 has 30 items.

(b) Electronic Health Records

The growth in EHRs has opened avenues for machine-learning to predict suicidal risks. These records encompass demographic data and medical histories, including hospital admissions and critical visits. However, the nature of the data, like its sparsity and variability, pose challenges.

(c) Suicide Note

These are typically written or recorded messages left by individuals before attempting suicide. They offer valuable insights for natural-language-processing research. Prior studies have utilised content, sentiment, and emotion analysis on these notes. Digital platforms are increasingly becoming repositories for such notes, with researchers monitoring sites like MySpace.com for potentially at-risk individuals

(d) Online User Content

The global availability of the web and the prevalence of public media platforms have simplified the process for people to express their emotions and share their stories. Platforms like Twitter, Reddit, facebook and twitter-Space have become focal points for studying and identifying suicidal thoughts. Researchers utilise machine-learning techniques on these platforms to understand user intentions and responds in time.

5. Discussion and Future Work

The sphere of Suicidal Ideation Detection (SID) has witnessed numerous foundational efforts, predominantly amplified by manual feature design and the application of deep neural network (DNN) methodologies for representation. Nevertheless, present studies exhibit certain constraints, ushering in formidable challenges awaiting future investigations.

A. Constraint

1. Data Scarcity: A paramount challenge confronting current endeavors is the dearth of data. Predominant methods lean heavily on supervised learning paradigms necessitating manual annotations. Regrettably, the reservoir of annotated data is inadequate, hampering progressive research. For instance, nuanced suicide risk data remains sparse, and multifaceted data or data reflecting social interconnections are virtually non-existent.

2. Data Disproportion: Suicidal intention-related posts constitute a minuscule fraction of the gargantuan volume of public media posts. Yet, many researchers construct datasets that misleadingly represent a balanced distribution, neglecting the inherently skewed distribution of such data.

3. Deficient Understanding of Intention: Contemporary computational techniques falter in comprehending the intricate psychology underlying suicidal intentions. The multifaceted psyche behind suicidal tendencies largely remains an enigma. Notwithstanding the emphasis on feature selection or intricate neural designs, prevailing methodologies often misconstrue the genuine essence of suicidal posts

Table1: Catogerisation methods for SID

Category	Author	year	Method
Machine Learning	Marouane Birjali et al. [9] T.Sravanthi et al. [10] Asma Abdulsalman et al. [11] Shaoxiong ji et al. [12]	(2017) (2020) (2020) (2020)	Sentiment analysis, Twitter4j, wordNet, Weka, machine learning, performance matrix. NLP, Public media, suicidal posts, words and phrases. Classification Twitter, araBert, NLP, machine learning, text classification. Data set, questionnaires, electronic health records, suicide notes, and online user content.
Deep learning	Ji et al. [18] Matero et al. [21] Abayomi Arowosegbe et al. [13]	(2018) (2019) (2022)	max pooling, Word embedding, LSTM, Multi task learning, neural networks. MLP, BERT, Dual context. RNN NLP, text mining, machine learning, electronic medical records, unsupervised learning'

6. Conclusion

In today's digital age, suicide prevention emerges as a paramount concern. Precocious detection of suicidal inclinations stands as an indispensable arsenal in this fight. This appraisal delves into multifarious methodologies for SID, encompassing clinical interactions, textual content scrutiny, intricate feature design, and contemporary deep-learning paradigms. Despite being predominantly spearheaded by experts from the realms of psychology and computational sciences, there's an imminent need to bridge the chasm between clinical mental health assessments and algorithmic detections. As online platforms burgeon as the primary avenue for SID, pioneering novel techniques to identify and intervene in cases of suicidal ideation is of paramount importance.

References

[1] World Health Organization. (n.d.). Suicide detection. https://www.who.int/news-room/fact-sheets/detail/suicide.

[2] Centers for Disease Control and Prevention. (2019). Suicidal Ideation and Behaviours among High School Students — Youth Risk Behavior Survey, United States. https://www.cdc.gov/mmwr/volumes/69/su/su6901a6.htm?s_cid=su6901a6_w.

[3] National Crime Records Bureau. (n.d.). Suicides in India. https://ncrb.gov.in/sites/default/files/Chapter-2-Suicides_2019.pdf.

[4] Pridmore, S., Ahmadi, J., and Pridmore, W. (2019). Two Mistaken Beliefs about Suicide. Iran Journal of Psychiatry, 14 (2), 182–183. PMID: 31440301; PMCID: PMC6702281.

[5] Klonsky, E. D. and May, A. M. (2014). Differentiating suicide attempters from suicide ideators: A critical frontier for suicidology research. Suicide and Life-Threatening Behavior, 44 (1), 1–5.

[6] Klonsky, E. D., Saffer, B. Y., and Bryan, C. J. (2018). Ideation-to-action theories of suicide: a conceptual and empirical update. Current Opinion in Psychology, 22, 38–43.

[7] Birjali, M., Beni-Hssane, A., and Erritali, M. (2017). Machine-learning and semantic sentiment analysis based algorithms for suicide sentiment prediction in social networks. Procedia Computer Science, 113, 65–72.

[8] Sravanthi, T., Hema, V., Reddy, S. T., Mahender, K., and Venkateshwarlu, S. (2020). Detection of mentally distressed public media profiles using machine-learning techniques. In IOP Conference Series: Materials Science and Engineering (Volume 981, no. 2, p. 022056). IOP Publishing.

[9] Abdulsalam, A., Alhothali, A., and Al-Ghamdi, S. (2023). Detecting suicidality in Arabic tweets using machine-learning and deep-learning techniques. arXiv preprint arXiv:2309.00246.

[10] Ji, S., Pan, S., Li, X., Cambria, E., Long, G., and Huang, Z. (2020). Suicidal ideation detection: A review of machine-learning methods and applications. IEEE Transactions on Computational Social Systems, 8 (1), 214–226.

[11] Arowosegbe, A. and Oyelade, T. (2023). Application of Natural-language-processing (NLP) in detecting and preventing suicide ideation: A systematic review. International Journal of Environmental Research and Public Health, 20 (2), 1514.

[12] Renjith, S., Abraham, A., Jyothi, S. B., Chandran, L., and Thomson, J. (2022). An ensemble deep-learning technique for detecting suicidal ideation from posts in public media platforms. Journal of King Saud University-Computer and Information Sciences, 34 (10), 9564–9575.

[13] Swamy, R. D. (2022). Analysis of suicide ideation documents posted on Twitter using an NLP classifier (Doctoral dissertation, Dublin, National College of Ireland).

[14] Malhotra, A. and Jindal, R. (2022). Deep-learning techniques for suicide and depression detection from online public media: A scoping review. Applied Soft Computing, 109713.

[15] Desu, V., Komati, N., Lingamaneni, S., and Shaik, F. (2022). Suicide and depression detection in public media forums. In Smart Intelligent Computing and Applications, Volume 2: Proceedings of Fifth International Conference on Smart Computing and Informatics (SCI 2021) (pp. 263–270). Singapore: Springer Nature Singapore.

[16] Varathan, K. D. and Talib, N. (2014). Suicide detection system based on Twitter. In Proceedings of the Science and Information Conference (pp. 785–788).

[17] Ferrari, A. J., Norman, R. E., Freedman, G., Baxter, A. J., Pirkis, J. E., Harris, M. G., ... and Whiteford, H. A. (2014). The burden attributable to mental and substance use disorders as risk factors for suicide: findings from the Global Burden of Disease Study 2010. PLoS One, 9 (4), e91936.

[18] Jashinsky, J. et al. (2014). Tracking suicide risk factors through Twitter in the US. Crisis, 35 (1), 51–59.

[19] Tai, Y.-M. and Chiu, H.-W. (2007). Artificial neural network analysis on suicide and self-harm history of Taiwanese soldiers. In Proceedings of the 2nd International Conference on Innovative Computing, Information and Control (ICICIC) (p. 363).

[20] Ren, F., Kang, X., and Quan, C. (2016). Examining accumulated emotional traits in suicide blogs with an emotion topic model. IEEE Journal of Biomedical and Health Informatics, 20 (5), 1384–1396.

[21] Chen, L., Aldayel, A., Bogoychev, N., and Gong, T. (2019). Similar minds post alike: Assessment of suicide risk using a hybrid model. In Proceedings of the 6th Workshop on Computational Linguistics and Clinical Psychology (pp. 152–157).

[22] Sueki, H. (2015). The association of suicide-related Twitter use with suicidal behaviour: A cross-sectional study of young Internet users in Japan. Journal of Affective Disorders, 170, 155–160.

A Novel Approach for Reducing the Single Point Failure in Software Defined Networks

Deepjyot Kaur Ryait and Manmohan Sharma

School of Computer Applications, Lovely Professional University, Phagwara, India
E-mail: djryait@gmail.com, manmohan.21909@lpu.co.in

Abstract

The primary distinction between the Software Defined Network and the traditional network is the separation of the control plane and data plane, which allows the latter to offer a novel and adaptable networking paradigm that promotes networking innovation. Ensuring uninterrupted and on-demand service availability is a critical element of modern communication systems. If there is a communication system breakdown or disruption, the company will experience a large loss of income or profit. This issue needs to be resolved by increasing the number of networking facilities that are available. The motive of this research paper is to create a fault tolerance model that reduces the likelihood of a single point of failure. The suggested model features higher average throughput and bandwidth metrics as compared to alternative configurations of controllers.

Keywords: SPOF, fault tolerance, throughput, bandwidth, etc.

1. Introduction

Software Defined Networking (SDN) is the next paradigm in networking, which operates by "disassociating or decoupling the control plane from the data plane in the network." Therefore, the data plane only functions as a basic forwarding element for the network as a result of this separation. because the controller makes all of the decisions on the network. Today, every commercial enterprise is willing to earn maximum profit in the network field by offering various communication facilities. Any failure that occurs during the communication of the network results in a heavy loss of revenue. It is essential to overcome this situation by increasing the availability of networks and minimizing revenue loss [1,2]. SDN greatly boosts network efficiency. Network management is made simpler by software defined networking. SDN provides several facilities like programmability, adjustability, and dynamically reconfiguration of the networking elements. Currently, Software Defined Networking paradigm is supported by large industries such as Microsoft, Google, Amazon, Cisco, Juniper, Facebook, HP, IBM, Samsung, etc. The architecture of a Software Defined Network defines a novel way of a networking system that can be built by using a combination of hardware network commodities with software-based technologies and openness. Additionally, SDN offers a centralised network structure that can interact with the rest of the network. The framework of Software Defined Networks offers an abstraction network, which becomes too easy to achieve network reachability.

2. Literature Review

Malik et al. [1] Software Defined Networks is a hot and burgeoning topic that has attracted more attention in both the commercial and academic sectors; because it decouples the control plane from the data plane. But fault tolerance and recovery are some of the key issues that SDN faces. Moreover, only 4% of the research effort contributes to fault tolerance in SDN and the rest in other fields. To better SDN traffic engineering and increase network efficiency, a controller offers a global view of the network. To address these issues with network scalability and resilience, fault tolerance requires more attention. If something goes wrong in one layer, it may or may not affect the other layers. Because of this, each layer of SDNs must be built separately to deal with the problems of fault tolerance on that layer. A connection failure, for instance, might not have an impact on the control layer, whereas a controller failure might have an impact on the entire infrastructure layer. A. Gonzalez et al. [3] The SDN paradigm enhances innovation and flexibility in the network. So, robustness and fault tolerance must be considered as the main criteria for networking. But a

Chapter 18 DOI: 10.1201/9781003570349

single controller always has a threat of failure. The performance of the controller is classified into two concepts, such as controller latency and controller throughput. L. Sidki et al. [4] The network traffic is decided by the SDN controller, who has a wide overview of the network's capabilities. Support on a single controller is unsuitable due to two factors. Initially, the network encounters a solitary point of failure, or SPOF. Secondly, it increases the demand for network traffic to help deal with speed bottlenecks. Network throughput is reduced and network latency is increased as a consequence.

A brand-new networking paradigm called the "Software Defined Network" enables faster innovation and flexibility than the "traditional network." However, a single controller has hampered the network's performance and scalability. A single point of failure for a controller that lowers overall network performance and availability before ultimately collapsing the entire network is a critique of SDN. The usage of many controllers in a network increases network complexity and raises possible SDN research questions on controller coordination, fault tolerance, load balancing, controller placement, and other related topics [5,6].

3. Current Status of SDN

Software Defined Networking provides an innovative paradigm of networking that can help to fulfill the requirement of a user on demand. Moreover, SDN improves network control which permits the network provider to respond to the changing business requirements quickly. So, several industries are supporting the SDN paradigm. A company such as the Google has started to implement SDN in their data center networks. It is required to transform the current network with SDN in a phased manner. The operational costs and delays that occur due to a link failure can be significantly minimised in SDN. The following three examples used in SDN are:

1) Microsoft's Virtual Machine Manager: The infrastructure of SDN can be deployed and managed using Microsoft's virtual machine manager (VMM). It offers a uniform administration experience and is employed to set up traditional data centres. VMM offers virtualised hosts, network and library resources, and allocated storage, among its capabilities. Users can carry out a wide range of tasks when SDN is integrated with the VMM. These include designing and administering virtual network policies, guiding traffic flows between virtual networks, and managing the infrastructure, which includes network controllers, software load balancers, and gateways. It also incorporates a wide range of technologies, including software load balancing, the RAS gateway, and the network controller.

2) VMware NSX: With more than 140 apps moved over the course of five months to an SDN architecture, Tribune Media has likely the largest SDN deployment utilising VMware NSX. Tribune Media was split off from the rest of the Tribune Company in 2012. As a result, the organisation had to upgrade its IT systems and apps. As an outcome of this, Tribune Media decided to use VMware SDDC as the basis for its IT infrastructure. A virtual networking and security programme called VMware NSX is used in Software Defined Data Centres (SDDCs), which offer cloud computing based on VMware network technology. Switches, routers, and firewalls are just a few examples of the network functions that are distributed throughout the environment by NSX using a network hypervisor. Because of its agility, flexibility, and security, Tribune Media chose VMware's NSX.

3) The Forefront of Tech Innovation: SDN can be implemented in many different ways. Even though the technology is still in development, significant service providers are using it. SDN, which will power future technological advancements in networks, promises to lower operational costs and offer more precise security.

4. Proposed Work with Simulation and Evaluation of Result

In SDN, decouple the control plane from the data plane. Due to this separation, the data plane only serves as a simple forwarding element for the network. Because all the decisions on the network are taken by the controller. For this purpose, a fault tolerance mechanism is required. But it is complicated to achieve on the control plane. Moreover, a connection failure, for instance, may not have an impact on the control plane, whereas a controller failure might have an impact on the entire infrastructure layer. In SDN paradigm, a centralised controller oversees or manages all network elements in a network. But a single controller is not feasible for two primary reasons are (in Figure 1):

1) The first is that the controller always has a threat of single point of failure (SPOF) in the network [4].

2) Second, it halts an entire network operation which has an adverse effect on the network performance.

As a result, it increases the latency time and reduces the throughput of a network [2-4,7]. Due to a single

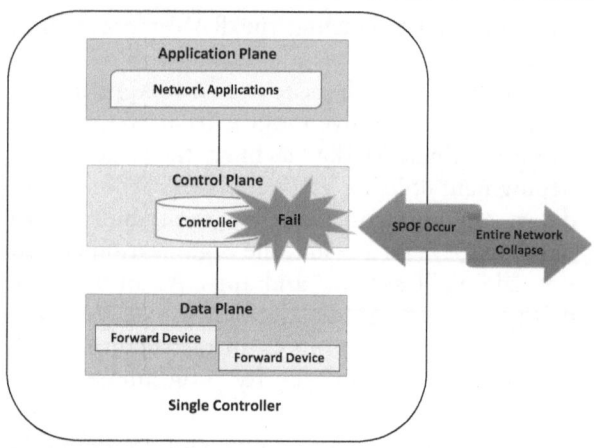

Figure 1: Single controller in SDN

controller, it restricted the scalability and reliability of the SDN paradigm.

For this purpose, focusing on the multiple controllers in SDN increases the scalability and availability of a network. To reduce SPOF in SDN by using multiple controllers. Among these controllers: one is the master controller, who has all privileges and authority to control the entire network. If, for any reason, the master controller fails, then any another controller changes their role as master controller and continues the operation of the network. After reducing SPOF in SDN to manage load balancing between the controllers, both fault tolerance and load balancing issues are interrelated. Figure 2 shows the pseudo-code for SDN's single point of failure reduction.

Algorithm: Pseudo Code to Reduce a Single Point of Failure in SDN

Step 1: Start the ryu-manager with ofp-tcp-listen-port 6653 as a single controller;

Step 2: if (Single Controller == "Fail")

{

/* When a single point of failure (SPOF) occurs in SDN */

Collapse the Entire Network;

} else

{ /* To eliminate a single point of failure by using Multiple Controllers in

SDN*/

/* The OpenFlow 1.2 and its later specification provide three different

roles of controller's*/

switch (ROLE) {

case ROLE == "EQUAL":

- Every controller has permission to access the packet flow in the network.
- Several equal controllers exist in the network.
- Duplicate packets are generated and marked as (DUP!).

break;

case ROLE == "MASTER":

- Only one master controller exists in the network.
- All decision regarding installation and manipulations of flow rules are taken by Master Controller.

if (MASTER == "Fail") {

Any slave controller sends Role-Request message to switch;

if (ROLE_REQUEST == "ROLE_MASTER") {

Switch send reply to Controller;

ROLE_REPLY == "MASTER";

}else

Other controllers send role as a slave controller;

}

break;

case ROLE == "SLAVE":

- Slave controller used as backup purpose.
- Number of slave controllers exist in the network.

break;

}

}

Output:

Therefore, multiple controllers are used to reduce a single point of failure in SDN.

To simulate the experimental setup required, the following software tools and programming languages are used on the experimental platform: Ubuntu 18.04 Desktop as an Operating System, Mininet 2.3.0d5 as a Test Bed, Ryu SDN Controller as a Remote Controller etc. After these tools have been installed successfully, simulations of the various controller roles in SDN are run. To assess the effectiveness of SDN controllers in their roles with the proposed model, the iperf tool is used. As shown in Table 1 and Figures 3–5, compare controllers based on various metrics. In comparison to alternative controller configurations, the proposed model has higher average throughput and bandwidth metrics and lower ping delay metrics.

Table 1: Performance evaluation of controllers w.r.t. average throughput, average bandwidth and ping delay

Operation of SDN Controller's	Average Throughput (GBytes)	Average Bandwidth (Gbits/sec)	Ping Delay (ms)	Time Interval (Sec)
Single Controller	63.5	36.4	43.480	15
Equal Controller	69.8	40.0	41.808	15
Master Slave Controller	70.9	40.6	0.400	15
Proposed Model	71.4	40.9	0.026	15

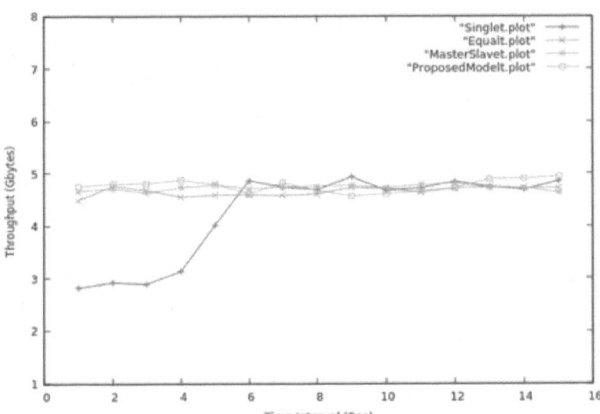

Figure 3: Performance evaluation of controllers w.r.t. throughput by Gnuplot

5. Conclusion

The Software Defined Network provides a novel paradigm of networking that enhances innovation in the networking field. Due to this separation, the data plane only serves as a simple forwarding element for the network. Because all the decisions on the network are taken by the controller, a fault tolerance mechanism is required. But it is complicated to achieve on the control plane. Moreover, a connection

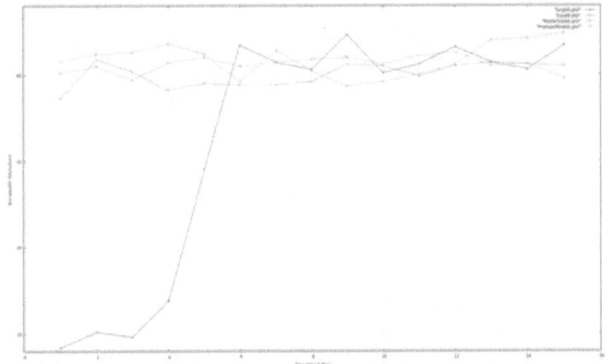

Figure 4: Performance evaluation of controllers w.r.t. bandwidth by Gnuplot

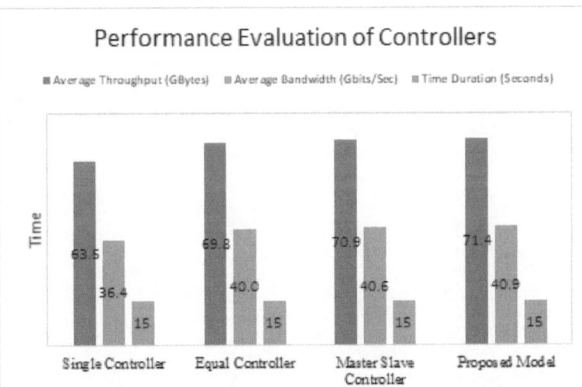

Figure 5: Performance evaluation of controllers with the proposed model

failure, for instance, may not have an impact on the control plane, whereas a controller failure might have an impact on the entire infrastructure layer. To reduce SPOF in SDN by using multiple controllers. Among these controllers, one is the master controller, who has all the privileges and authority to control the entire network. If, for any reason, the master controller fails, then another controller changes their role as master controller and continues the operation of the network. After looking at Table 1, it's clear that the performance of the proposed model is better and superior to that of other SDN controller configurations. In the future, make an effort to implement the right security measures for the controller, as it is tasked with managing the entire network.

References

[1] Malik, A., Aziz, B., Al-Haj, A., and M. Adda. (2019). Software-defined networks: A walkthrough guide from occurrence to data plane fault tolerance. PeerJ Preprints Open Access, vol. 27624v1, 1–26.

[2] Kreutz, D., Ramos, F. M., Verissimo, P. E., Rothenberg, C. E., Azodolmolky, S., and Uhlig, S.

(2015). Software-defined networking: A comprehensive survey. Proceedings of the IEEE, 103 (1), 14–76.

[3] Gonzalez, A., Nencioni, G., Helvik, B. E., and Kamisinski, A. (2016). A fault-tolerant and consistent SDN controller. In 2016 IEEE Global Communications Conference (GLOBECOM).

[4] Sidki, L., Ben-Shimol, Y., and Sadovski, A. (2016). Fault tolerant mechanisms for SDN controllers. In IEEE Conference on Network Function Virtualization and Software Defined Networks (NFV-SDN) (pp. 1–6).

[5] Isong, B., Mathebula, I., and Dladlu, N. (2018). SDN-SDWSN controller fault tolerance framework for small to medium sized networks. In 2018 19th IEEE/ACIS International Conference on Software Engineering, Artificial Intelligence, Networking and Parallel/Distributed Computing (SNPD), IEEE Computer Society (pp. 43–51).

[6] Karakus, M. and Durresi, A. (2017). A survey: Control plane scalability issues and approaches in Software-Defined Networking (SDN). Computer Networks, 112, 279–293.

[7] Zhang, Y., Cui, L., Wang, W., and Zhang, Y. (2018). A survey on software defined networking with multiple controllers. Journal of Network and Computer Applications, 103, 101–118.

[8] Askar, S. and Keti, F. (2021). Performance evaluation of different SDN controllers: A review. International Journal of Science and Business, 5 (6), 67–80.

[9] Elmoslemany, M. (2020). Performance analysis in software defined network controllers. In 15th International Conference on Computer Engineering and Systems (ICCES).

Blockchain in Land Records Management for the State of Punjab, India

A Systematic Review and Bibliometric Analysis

Aseem Khanna[1], Prikshat Kumar Angra[2], Pritpal Singh[3], Sakshi Khanna[4], Parminder Kaur[5], Ashwani Kumar[6]

Lovely Professional University, Phagwara, India

[1,5]Centre of Distance and Online Education, Lovely Professional University, Phagwara, India

[2,4]Department of Computer Applications, Lovely Professional University, Phagwara, India

[3,6]Mittal School of Business, Lovely Professional University, Phagwara, India

E-mail: aseem.27475@lpu.co.in, prikshat.22305@lpu.co.in, pritpal.16741@lpu.co.in, sakshi.27428@lpu.co.in, parminder.16295@lpu.co.in, ashwani.23881@lpu.co.in

Abstract

Blockchain is rapidly changing land record management worldwide. It could solve land conflicts and corruption in India's property ownership structure. Blockchain and smart contracts can expedite, secure, and streamline land registration. Land registration systems in Pakistan and other nations have improved with blockchain. This paper introduces Decentralised Land Administration (DLA) and a Blockchain Framework (BcF) for Land Record Management. BcF improves land record management systems by assuring immutability, transparency, efficiency, and traceability. Blockchain technology can revolutionise land registration worldwide by eliminating intermediaries and assuring data integrity. In conclusion, blockchain technology may solve land record management issues. This study sheds light on the pros and cons of using blockchain in land administration, enabling more transparent, efficient, and reliable land registration processes.

Keywords: Blockchain technology, land record management, transparency, security, smart contracts

1. Introduction

Blockchain technology has evolved rapidly since its inception to become a very imaginative and complex technology. In 2008, Nakamoto presented and elaborated on the notion of blockchain, which is a decentralised ledger technology (DLT) used to record the provenance and history of a digital asset. The system in question is widely acknowledged as a Distributed Ledger Technology (DLT) that effectively documents transactions and communications in the form of immutable time stamped digital blocks [1-3]. The blockchain technology, once associated with Bitcoin, has since gained significant adhesion and is extensively used in several industries and countries. A recent research has shown that blockchain technology has the potential to affect and revolutionise several areas of the industrial and service industries, such as banking, supply chain management, healthcare systems, and human resource management [4]. The blockchain is a digital ledger that is resistant to alteration and stores economic transactions, capable of recording a wide range of valuable assets.

The land register office in India reportedly witnesses bribe payments over $700 million, highlighting the prevalence of corruption in a country where almost 70% of the global population lacks access to a formal land registration system. Moreover, according to a 2007 research conducted by the World Bank, a significant proportion of legal proceedings in the nation revolve on land conflicts specifically associated with property ownership. The studies hitherto in the cited area have not sufficiently addressed the sufficient use of the contemporary technology, growing legal issues like in transfer of property rights, the socio- economic consequences thereof.

2. Background and Literature Survey

Blockchains are sequential arrangements of interconnected blocks [5]. A decentralised, immutable ledger can record transactions involving multiple participants in a network [6-8]. Blockchain technology was first applauded for supporting crypto currency stability [9], but it is now recognised as a versatile data structure with many mobile applications [10-12]. The blockchain is a decentralised ledger that records and monitors transactions between numerous parties and makes them available to all stakeholders [13]. The system improves mobile app security and transparency with this technique. Previous research [14] shows that fraudulent transactions and data manipulation are impossible.

3. Methods and Approaches

Basic studies needed for further amalgamation and combination [27][28][19], as these are   vital situations for an SLR and meta-analysis[37][29][30]. Searching relevant literature, vetting for its relevancy, filtering vital data and reporting on the results are the steps involved [31][32][33][34]it describe in full the multi-level recursive search technique that was used to locate the appropriate literature for this investigation and the specific criteria that were applied.

3.1. Literature Identification

The purpose of this study is to systematically search for academic articles on blockchain technology's usage in land record management. Scopus and Web of Science, two encyclopedic databases with state-of-the-art search capabilities, were explored for academic articles. But especially for our study, we tracked down and collected materials that were released between 2013 and 2021. When it comes to 2016 exploration results, both databases were represented.

After going through various documents using searching procedure has been demonstrated in Table 2. Initially 120 various documents afore filtering were outlined in the Scopus and Web of Science databases and other relevant databases to topic.

3.2. Action Research

As suggested by some of the authors in research papers like [19], with another author (Moher et al. 2015), we used the PRISMA (Preferred Reporting Items for Systematic Reviews and Meta-Analyses) flow diagram to conduct our literature search. (2010) [26]. The PRISMA flow diagram is often cast-off as part of the literature selection procedure and help to locate potentially relevant studies for inclusion in a report [20]. The procedure for determining the total

number of publications is shown in Figure 1. After going through the PRISMA process flow, a total of 26 papers were included below is the flow diagram

4. Result

By going through research by showing publishing increase from 2016-2023 via graphs, charts, figures, and tables. Publication increase by location, year, document type, field of study, and topic relevance is documented. VOSviewer was used to map bibliographic data. Text was pulled from each source's title and abstract. Figure 5 shows a phrase co-occurrence map from 26 articles. Researchers created themes from groupings of items. To support the key topics and ensure reliability, authors thoroughly and critically reviewed all 26 full-text articles to report the most important words.

India contributes the most to blockchain and land records management literature with 26 research papers, review papers, conference proceedings, and book chapters (Figure 2). Specifically, US 10%

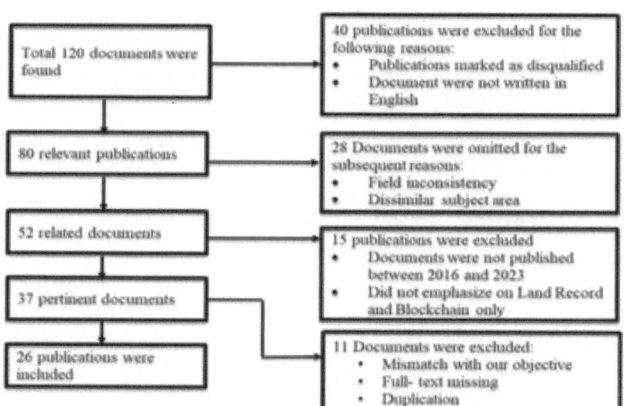

Figure 1: Result of PRISMA flow diagram

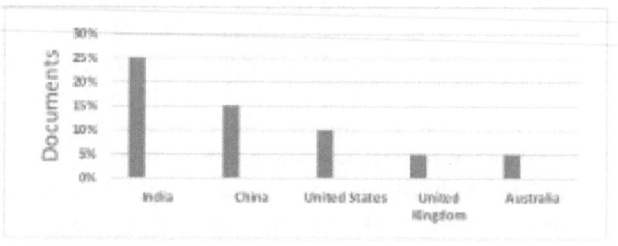

Figure 2: Publications as per the country (2016–2023)

Table 1: Search procedures

Search Relations	SearchInside	Public Databases	Data Range	No. of Documents Searched
Blockchain inLand Record Management	Article title,abstract, keywords	Scopus, Webof Science	2016–2023	120

publications. Our investigation found that Australia and the UK have published equally on blockchain and land record publication.

Figure 3 shows that research articles and academic journals (60%) and conference papers (30%) dominated publishing in the last five years. From 2016 to 2023, just 5% of blockchain publications in land record management were books, book chapters, conference reviews, government reports, and other sources. Since the blockchain in land records administration is a novel concept, scholars from all parts of the world are more willing in sharing their ideas at conferences to gain better feedback.

Figure 4 shows that blockchain has been intensively investigated in various fields. Over 27% and 22% of articles were in computer science and engineering, the most important fields. Specifically, 10.8% of papers were in Humanities and 16% in Social science and Medical Science 24%. Going through searching most publications were in other subjects than in physics, humanities, social science, medicine, math, biochemistry, and chemistry, which warrants further study.

In below section themes are developed based on the output clusters of VOSviewer, and highlighted by the discussion. The themes developed below aims on the association between the clusters.

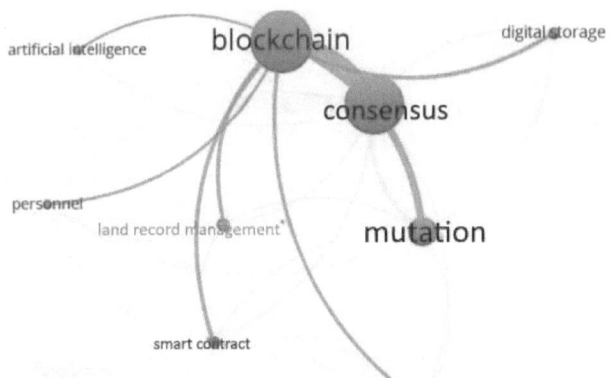

Figure 5: Network map as per Bibliographic Data (full counting technique used)

4.1. Theme Development Based on Screening Association

Information management systems can improve productivity, transparency, and service coordination, like DMS (Document Management System) research into the method that lets users add custom properties to documents and provides dynamic features that automate document-related tasks [21]. Blockchain will eventually match LRM. Data immutability occurs with decentralised management. The elimination of land registry intermediaries requires transparent, efficient, truthful, and traceable methods [22, 24]. Blockchain technology in land registration provides a solid theoretical foundation. Despite the hurdles and limits of creating a blockchain-based land registration system, the article stresses the potential benefits, including better transparency, less corruption, and higher efficiency [23]. This makes BcF for Land Record Management a fundamental aspect of blockchain technology in DLA.

5. Conclusion

The detailed review of current scholarly publications shows that blockchain technology can transform land record administration. Blockchain technology affects all fields, including computer science and business management. India is a major player in this growing industry According to Figure 6 & Figure 7. Decentralised Land Administration (DLA) and the Blockchain Framework (BcF) for Land Record Management demonstrate how technology can improve transparency, prevent corruption, and speed up land registration. This research provides useful insights into using blockchain technology to manage land records. The lack of co-citation analysis is a major limitation. Additionally, this research used Scopus and Web of Science to draw conclusions. Thus, future studies should leverage more databases like EBSCO and ProQuest to find new themes. To

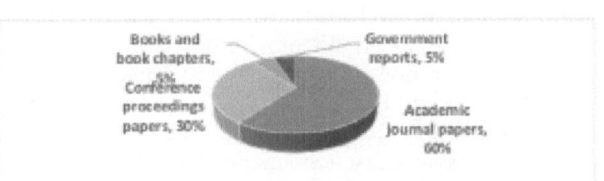

Figure 3: Publications as per type of papers (2016–2023)

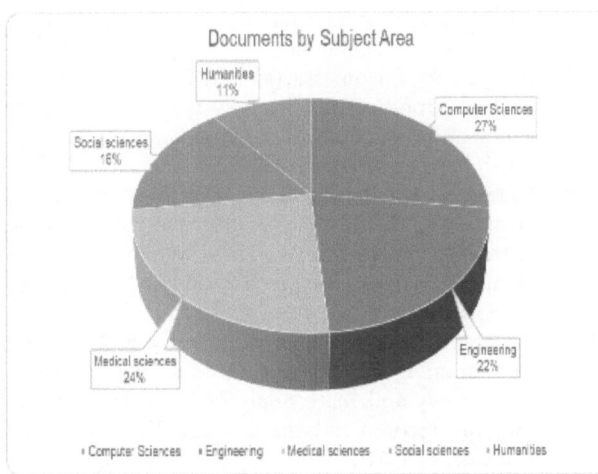

Figure 4: Publications as per Subject Area (2016–2023)

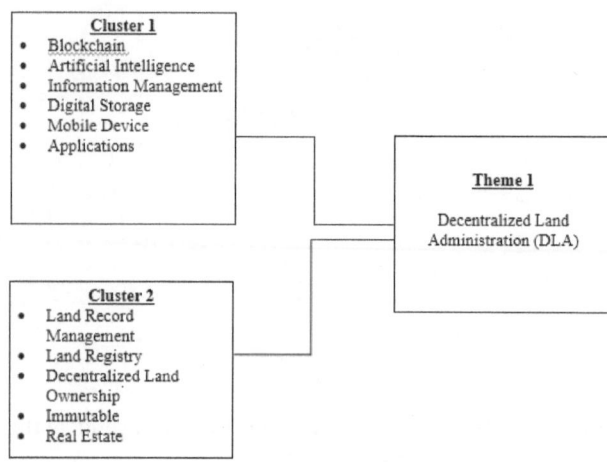

Figure 6: Decentralised Land Administration (DLA)

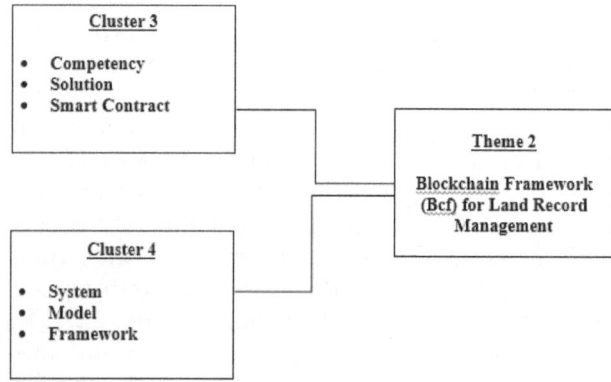

Figure 7: Blockchain Framework (BcF) for Land Record Management

develop this discipline, empirical studies on "DLA" and "BcF for DLA." are recommended.

References

[1] Esmat, A., de Vos, M., Ghiassi-Farrokhfal, Y., Palensky, P., and Epema, D. (2021). A novel decentralized platform for Peer-to-Peer energy trading market with blockchain technology. Applied Energy, 282, 116123.

[2] Sarode, R. P., Poudel, M., Shrestha, S., and Bhalla, S. (2021). Blockchain for committing Peer-to-Peer transactions using distributed ledger technologies. International Journal of Computational Science and Engineering, 24 (3), 215–227.

[3] Farahani, B., Firouzi, F., and Luecking, M. (2021). The convergence of IoT and Distributed Ledger Technologies (DLT): Opportunities, challenges, and solutions. Journal of Network and Computer Applications, 177, 102936.

[4] Ballandies, M. C., Dapp, M. M., and Pournaras, E. (2021). Decrypting distributed ledger design – taxonomy, classification and blockchain community evaluation. Cluster Computing, 2018, 1–22.

[5] Eder, G. (2019). Digital transformation: blockchain and land titles. In OECD Global Anti Corruption and Integrity Forum (pp. 1–12). Paris, France: OECD (Organisation for Economic Co- Operation and Development). https://www.oecd.org/corruption/integrity-forum/academic-papers/Georg%20Eder-%20Blockchain%20-%20Ghana_verified.pdf.

[6] Mishra, H. and Venkatesan, M. (2021). Blockchain in human resource management of organizations: An empirical assessment to gauge HR and non-HR perspective. Journal of Organizational Change Management, 34 (2), 525–542.

[7] US Agency for International Development. (2018). Investor Survey on Land Disputes: Perceptions and Practices of the Private Sector on Land and Resource Tenure Risks. Washington, DC, USA: US Agency for International Development.

[8] Tapscott, A. (2018). India Land Registry on Blockchain, Research Institute Lighthouse, Toronto, Canada. https://www.blockchainresearchinstitute.org/project/indias-land-registryon-blockchain/.

[9] Oprunenco, A. and Akmeemana, C. (2018) Using Blockchain to Make Land Registry more Reliable in India. UK: LSE Business Review.

[10] Baliga, A., Subhod, I., and Kamat, P., and Chatterjee, S. (2018). Performance evaluation of the quorum blockchain platform. arXiv preprint arXiv:1809.03421.

[11] Chandra, V. and Rangaraju, B. (2017). Blockchain for Property A Roll Out Road Map for India. India: India Institute.

[12] Buterin, V. (2014). Ethereum White Paper 3 (37), 1–36.

[13] Anand, A., McKibbin, M., and Pichel, F. (2016). Colored coins: Bitcoin, blockchain, and land administration. In Annual World Bank Conference on Land and Poverty.

[14] Morini, M. (2016). From 'Blockchain hype' to a real business case for Financial Markets. Available at SSRN 2760184.

[15] Roitsch, J., Gumpert, M., Springle, A., and Raymer, A. M. (2021). Writing instruction for students with learning disabilities: Quality appraisal of systematic reviews and meta-analyses. Reading & Writing Quarterly, 37 (1), 32–44.

[16] Mengist, W., Soromessa, T., and Legese, G. (2020). Method for conducting systematic literature review and metaanalysis for environmental science research. MethodsX, 7, 100777.

[17] Polanin, J. R., Pigott, T. D., Espelage, D. L., and Grotpeter, J. K. (2019). best practice guidelines for abstract screening large evidence systematic reviews and meta-analyses. Research Synthesis Methods, 10 (3), 330–342.

[18] Harari, M. B., Parola, H. R., Hartwell, C. J., and Riegelman, A. (2020). Literature searches in systematic reviews and meta analyses: A review, evaluation, and recommendations. Journal of Vocational Behavior, 118, 103377.

[19] Selçuk, A. A. (2019). A guide for systematic reviews: PRISMA. Turkish Archives of Otorhinolaryngology, 57 (1), 57.

[20] Cuevas, P. E. G., Davidson, P. M., Mejilla, J. L., and Rodney, T. W. (2020). Reminiscence therapy for older adults with Alzheimer's disease: A literature review. International Journal of Mental Health Nursing, 29 (3), 364–371.

[21] Dourish, P., et al. (2000). Extending document management systems with user-specific active properties. ACM Transactions on Information Systems (TOIS), 18.2, 140-170.

[22] Burton, J. and Van den Broek, D. (2009). Accountable and countable: Information management systems and the bureaucratization of social work. British Journal of Social Work, 39.7, 1326–1342.

[23] Sahai, A. and Pandey, R. (2020). Smart contract definition for land registry in blockchain. In 2020 IEEE 9th International conference on communication systems and network technologies (CSNT). IEEE.

[24] Khalid, M. I., et al. (2022). Blockchain-based land registration system: a conceptual framework. Applied Bionics and Biomechanics 2022.

[25] Pontes, H. M. and Griffiths, M. D. (2015). Internet gaming disorder and its associated cognitions and cognitive-related impairments: A systematic review using PRISMA guidelines. Revista Argentina de Ciencias del Comportamiento, 7 (3), 102–118.

[26] Yasin, Y. M., Kerr, M. S., Wong, C. A., and Bélanger, C. H. (2020). Factors affecting nurses' job satisfaction in rural and urban acute care settings: A PRISMA systematic review. Journal of Advanced Nursing, 76 (4), 963–979.

[27.] Saif, A. N. M., A. A. Rahman, and R. Mostafa. 2021. "Post-Implementation Challenges of ERP Adoption in Apparel Industry of Developing Country." LogForum 17 (4): 519–529.

[28.] Garcia-Arroyo, J., and A. Osca. 2019. "Big Data Contributions to Human Resource Management: A Systematic Review." The International Journal of Human Resource Management, 1–26. Ahead-of-print.

[29.] Siddaway, A. P., A. M. Wood, and L. V. Hedges. 2019. "How to do a Systematic Review: A Best Practice Guide for Conducting and Reporting Narrative Reviews, Meta-Analyses, and Meta-Syntheses." Annual Review of Psychology 70: 747–770.

[30.] Pati, D., and L. N. Lorusso. 2018. "How to Write a Systematic Review of the Literature." HERD: Health EnvironmentsResearch & Design Journal 11 (1): 15–30.

[31.] Xiao, Y., and M. Watson. 2019. "Guidance on Conducting a Systematic Literature Review." Journal of Planning Education and Research 39 (1): 93–112.

[32.] McMullan, R. D., D. Berle, S. Arnáez, and V. Starcevic. 2019. "The Relationships Between Health Anxiety, Online Health Information Seeking, and Cyberchondria: Systematic Review and Meta-Analysis." Journal of Affective Disorders 245: 270–278.

[33.] Cooper, C., A. Booth, J. Varley-Campbell, N. Britten, and R. Garside. 2018. "Defining the Process to Literature Searching in Systematic Reviews: A Literature Review of Guidance and Supporting Studies." BMC Medical Research Methodology18 (1): 1–14.

[34.] Cooper, C., R. Garside, J. Varley-Campbell, J. Talens-Bou, A. Booth, and N. Britten. 2020. "It Has No Meaning to Me." How do Researchers Understand the Effectiveness of Literature Searches? A Qualitative Analysis and Preliminary Typology of Understandings." Research Synthesis Methods 11 (5): 627–640.

[35.] Hoffecker, L. 2020. "Grey Literature Searching for Systematic Reviews in the Health Sciences." The Serials Librarian 79 (3–4): 252–260.

[36.] Mengist, W., T. Soromessa, and G. Legese. 2020. "Method for Conducting Systematic Literature Review and Meta- Analysis for Environmental Science Research." MethodsX 7: 100777.

A Preliminary Investigation Exploring the Challenges and Implications of Cloud Computing and Privacy Regulations

Amanpreet Singh[1], Gursharan Singh[2], Ramandeep Kaur[3], Jyoti Battra[4]

[1,3,4]School of Computer Applications, Lovely Professional University.
[2]School of Computer Science & Engineering, Lovely Professional University,
E-mail: apsj24@gmail.com, Gursharan.16967@lpu.co.in, ramanbedi045@gmail.com, jyotijenni19@gmail.com

Abstract

Cloud computing refers to the provision of computing services via the Internet., where resources such as storage, processing power, and software applications are provided on-demand to customers [1]. Cloud computing is a significant advancement in technology that has revolutionised the IT industry. Cloud computing is a model for providing IT services that allow users to access and manage their own computing resources remotely via the Internet. [2], regardless of device and location. Cloud computing offers numerous benefits, including flexibility, scalability, and cost-effectiveness. However, the widespread adoption of cloud computing has also raised concerns regarding privacy and data protection. Challenges and Implications of Cloud Computing on Privacy Regulations, one of the main challenges of cloud computing for privacy regulations is the complexity associated with risk assessment [3]. Organisations that store and process sensitive data in the cloud need to assess and manage the potential risks to privacy. It is crucial for organisations to address these security and privacy concerns when utilising cloud computing services. One of the major issues is data confidentiality and integrity, which remain a significant concern in cloud computing. Additionally, the process of managing privileged user access, data segregation, and ensuring compliance with data security and privacy regulations pose further challenges in the cloud computing environment. This includes identifying and classifying sensitive data, evaluating the security controls of cloud service providers, and ensuring compliance with applicable privacy regulations. Another challenge is the emergence of new business models that rely on timely delivery of services, which can have implications on consumer privacy. Companies may collect and process large amounts of personal data in order to provide customised and personalised services.

Keywords: Security issues, cloud computing, privacy issues, privacy regulations

1. Introduction to Privacy Regulations

Privacy regulations are laws or regulations that aim to protect an individual's privacy rights and govern how organisations handle and protect personal data. The General Data Protection Regulation (GDPR) of the European Union, the Consumer Privacy Act of California, and the Health Insurance Portability and Accountability Act (HIPAA) of the United States are all examples of such laws, though they differ by country and area. One of the main privacy concerns in cloud computing is the security and protection of sensitive personal information [4].

Cloud providers may have access to sensitive data stored in their infrastructure, raising concerns about unauthorised access, data breaches, and misuse of personal information. Additionally, there are concerns regarding data location and data segregation in a cloud environment.

Cloud computing introduces challenges in terms of data privacy and protection, as well as data availability, location privacy, and secure data transfer [5]. These concerns stem from the fact that data is stored and processed outside of the user's control, making it more vulnerable to unauthorised access or interception. As a result, individuals and organisations must carefully consider the privacy implications before adopting cloud computing services.

The current contracts for cloud services often do not prioritise consumer privacy, further exacerbating the challenges faced in enforcing privacy regulations in the cloud computing environment [6]. In summary, the challenges and implications of cloud computing and privacy regulations include concerns about unauthorised access to sensitive personal information, data breaches, misuse of personal information, data location and segregation issues, and the need for stronger privacy protections in cloud service contracts [3].

Chapter 20 DOI: 10.1201/9781003570349

The challenges and implications of cloud computing and privacy regulations include: the complexity of risk assessment, the need for timely delivery of new business models impacting consumer privacy, the difficulty of enforcing privacy regulations in a transnational cloud environment, the lack of prioritisation of consumer privacy in current cloud service contracts, and the lack of transparency in data handling and processing in the cloud.

2. Challenges of Implementing Cloud Computing

The challenges of implementing cloud computing include: ensuring data privacy and protection, guaranteeing data availability, addressing data location and segregation issues, securely transferring data, and complying with existing privacy regulations [5].

One of the main challenges of implementing cloud computing is ensuring the privacy and protection of sensitive personal information. Cloud environments require the implementation of preventative actions and safeguards to ensure data privacy and protection.

Another challenge is ensuring data availability, as cloud computing relies on remote servers that may experience downtime or technical issues. Data location and segregation issues are also a concern in cloud computing, as data from multiple users may be stored in the same physical infrastructure without proper isolation. The secure transfer of data in a cloud environment is another challenge, as data may be vulnerable to unauthorised access or interception during transit. Additionally, complying with existing privacy regulations can be challenging in a cloud environment due to the transnational nature of cloud computing and the varying privacy laws across different jurisdictions [6].

The challenges and implications of cloud computing and privacy regulations are twofold. On one hand, there are technical challenges to be addressed in order to ensure data privacy and protection in the cloud [5]. On the other hand, there are legal and regulatory challenges in enforcing privacy regulations and protecting consumer privacy in a Cloud Computing environment, especially with regards to transnational data flows and jurisdictional issues .Overall, the challenges of implementing cloud computing and ensuring privacy regulations include complexity in risk assessment, timely delivery impacting consumer privacy, enforcement difficulties in a transnational context, and the lack of transparency in data handling and processing [3].

The challenges and implications of cloud computing and privacy regulations are complex and multifaceted. They encompass technical challenges in ensuring data privacy and protection, as well as legal and regulatory challenges in enforcing privacy regulations.

Overall, the challenges and implications of cloud computing and privacy regulations are complex and multifaceted.

3. Privacy Concerns in Cloud Computing

The complexity of risk assessment, the introduction of novel business models that affect consumer privacy, regulatory compliance, problems with data privacy design that result in low-quality data, and a lack of transparency are the reasons behind privacy concerns in cloud computing. The challenges and implications of cloud computing and privacy regulations are significant.

They involve the difficulty for cloud providers to adhere to current data privacy rules due to the transnational nature of cloud computing and varying national regulations [6].

The current structure and nature of cloud computing make it difficult for cloud providers to comply with existing data privacy and protection rules. Furthermore, the lack of attention to consumer privacy in cloud service contracts adds to the challenges and implications of cloud computing and privacy regulations. Cloud computing presents challenges and implications regarding privacy regulations due to the complexity of risk assessment, impact on consumer privacy with the emergence of new business models, regulatory compliance, issues in data privacy design, and lack of transparency [3]. Cloud computing presents challenges and implications regarding privacy regulations due to the complexity of risk assessment, impact on consumer privacy with the emergence of new business models, regulatory compliance, issues in data privacy design, and the lack of transparency. The challenges and implications of cloud computing and privacy regulations include the complexity associated with risk assessment, timely

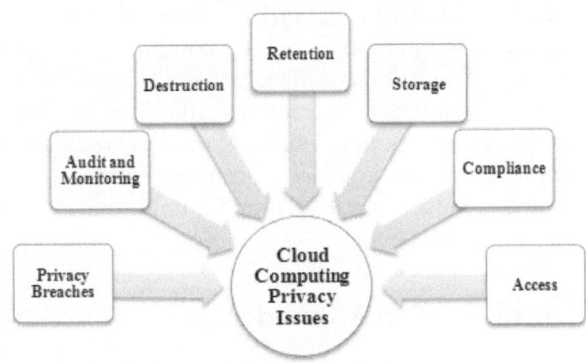

Figure 1: Cloud Computing privacy issues

delivery impacting consumer privacy, enforcement difficulties in a transnational environment, regulatory compliance, data privacy issues in design, poor data quality, and lack of transparency.

4. Implications of Privacy Regulations on Cloud Computing

The implications of privacy regulations on cloud computing are far-reaching. They may negatively affect matters like consumer protection, intellectual property and responsibility, competition, security and privacy, cross-border issues, and legal difficulties. Furthermore, the lack of a comprehensive policy that addresses these challenges can hinder the full potential of cloud computing [7].

In addition, the complex and dynamic nature of the cloud environment makes it challenging for cloud providers to follow current data privacy and protection rules [6]. Therefore, ensuring compliance with privacy regulations becomes a significant challenge for cloud providers. Cloud computing requires the movement of data across international borders, which further complicates compliance with different legal requirements in various jurisdictions [8].

The lack of encryption for data stored in the cloud also contributes to vulnerabilities and privacy threats. Moreover, the willingness of consumers to pay for privacy protection and data deletion from cloud databases highlights the importance of privacy considerations for cloud end users.

Cloud computing also brings up the issue of data mobility and the need to respect global legal requirements. Because current legal systems are unprepared for the difficulties offered by cloud computing, data protection, privacy concerns, liability, and compliance issues could hinder cloud computing's full potential [9].

To overcome these challenges, it is crucial for clouds to become regulation-aware, meaning they should ensure that data mobility is limited to comply with a wide range of different national legislation, including privacy legislation such as the EU Data Protection Directive 95/46/EC . Overall, the challenges and implications of cloud computing and privacy regulations are complex and multifaceted [3].

The complexity associated with risk assessment, the emergence of new business models, regulatory compliance issues, and lack of transparency are some of the challenges faced by cloud computing in relation to privacy.

5. Investigation Methodology

To conduct this preliminary investigation, a diverse set of sources were consulted. These included

➢ Academic papers,
➢ Industry reports, and
➢ Legal documents.

The information gathered from these sources provided insights into the challenges and implications of cloud computing and privacy regulations. The sources revealed that one of the main challenges is the difficulty for cloud providers to adhere to current data privacy and protection rules [6]. This is due to the transnational nature of cloud computing, which requires compliance with different legal requirements in various jurisdictions. Additionally, the lack of encryption for data stored in the cloud exposes it to vulnerabilities and privacy threats [8]. Furthermore, the research highlighted that consumers are increasingly concerned about their privacy in the cloud and are willing to pay for measures that limit data collection and ensure deletion from cloud databases.

Another significant challenge mentioned in the sources is the issue of storage location restriction. Many data protection laws stipulate that personal data collected in one jurisdiction cannot be transferred to another jurisdiction for processing or storage without adequate levels of protection [10].

This requirement poses challenges for cloud migration and forces cloud providers to carefully consider the storage and processing locations of data to ensure compliance with privacy regulations. Therefore, in order to fully leverage the potential of cloud computing while respecting global legal requirements and protecting privacy, it is imperative for cloud providers and organisations to implement robust data protection measures, including encryption and data residency controls [8].

In summary, the challenges and implications of cloud computing and privacy regulations are complex and multifaceted [3]. They involve issues such as data security, legal compliance, data quality, and user control. To address these challenges, cloud providers need to prioritise data protection and implement privacy-enhancing measures.

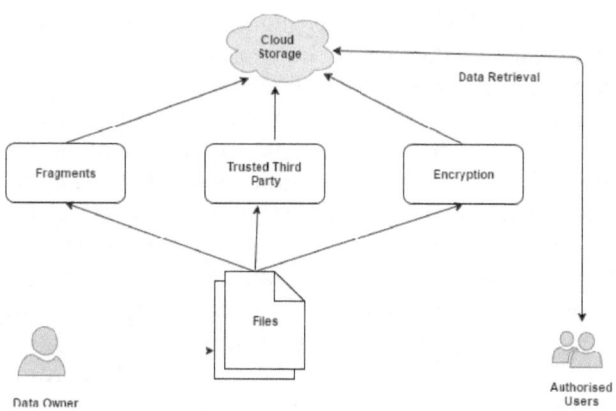

Figure 2: Cloud Authorization to access files

5.1. 5.1. Results of Preliminary Investigation

The preliminary investigation revealed several challenges and implications of cloud computing and privacy regulations. These include: -

➢ The complexity of risk assessment and the need for timely delivery of new business models in the growing cloud computing industry.

➢ Difficulty for cloud providers to adhere to current data privacy and protection rules due to the transnational nature of cloud computing and compliance with different legal requirements in various jurisdictions [6].

➢ The lack of encryption for data stored in the cloud, which exposes it to vulnerabilities and privacy threats.

➢ Consumers' increasing concerns about their privacy in the cloud and their willingness to pay for measures that limit data collection and ensure deletion from cloud databases [8].

6. The Role of GDPR in Cloud Computing Privacy

The General Data Protection Regulation is a critical aspect of privacy and data security, especially in the context of cloud computing. Enforced by the European Union, the GDPR sets guidelines for the collection and processing of personal information of individuals within the EU. Its impact extends beyond the borders of the EU, as it applies to any organisation that offers goods or services to EU residents, making it a global standard for data protection.

The GDPR plays a crucial role in safeguarding individual privacy rights and ensuring that organisations adhere to strict data protection standards. Cloud service providers must react to users' privacy concerns and comply with government regulations, such as the GDPR, in order to avoid legal complications related to the use of cloud services and the handling of personal data. Cloud computing poses challenges and implications for security and privacy regulations.

One challenge is the complexity of incorporating security and privacy requirements into cloud computing products [11].

The growing number of regulations and requirements adds to the complexity, with the GDPR being a significant factor. One challenge is the complexity of incorporating security and privacy requirements into cloud computing products. Cloud service providers must navigate the complex legal landscape surrounding cloud computing and privacy regulations in order to ensure compliance and protect individuals' privacy rights. Additionally, the evolving legal landscape surrounding cloud computing and privacy regulations adds to the complexity [12]. Cloud service providers must navigate the evolving legal landscape surrounding cloud computing and privacy regulations to ensure compliance and protect individuals' privacy rights.

Furthermore, the challenges extend beyond regulatory compliance to include ethical considerations. Cloud computing raises ethical concerns related to privacy and security, particularly in industries such as healthcare where sensitive data like genomic information is involved [13]. Organisations must find ways to implement uniform protocols and security measures to protect and improve patients' privacy in the cloud computing environment.

7. Recommendations for Cloud Computing Privacy Management

In order to effectively manage privacy in cloud computing, the following recommendations can be considered:

1. Implement strong encryption measures to ensure the security and privacy of data stored in the cloud.
2. Adhere to data residency requirements, ensuring that data is stored and processed within the appropriate legal boundaries.
3. Provide clear and transparent documentation and policies regarding data handling and storage practices to enhance trust and give users control over their personal information.
4. Regularly assess and monitor compliance with privacy regulations to ensure ongoing adherence to legal requirements.
5. Develop and implement privacy impact assessments to identify and mitigate potential risks to privacy.
6. Educate cloud users on best practices for protecting their privacy and data within the cloud environment, such as using strong passwords, enabling multi-factor authentication, and being cautious about the information they share and store in the cloud.

Overall, the challenges and implications of cloud computing and privacy regulations highlight the need for organisations to prioritise data protection measures such as encryption, data residency controls, and transparency in order to address the complexity of risk assessment, comply with privacy regulations, and protect the privacy and accuracy of data within the cloud computing environment. To effectively address the challenges and implications of cloud computing and privacy regulations, organisations should prioritise the implementation of strong encryption measures to ensure data security and privacy.

They should also adhere to data residency requirements and provide clear documentation and policies regarding data handling and storage practices. Additionally, regularly assessing and monitoring compliance with privacy regulations, conducting privacy impact assessments, and educating cloud users on best practices for protecting their privacy are crucial steps to effectively manage privacy in cloud computing. In conclusion, managing privacy in cloud computing requires implementing strong encryption measures, adhering to data residency requirements, providing clear documentation and policies, regularly assessing compliance with privacy regulations, conducting privacy impact assessments, and educating cloud users.

8. Conclusion

The challenges and implications of cloud computing and privacy regulations are complex and ever-evolving. Nevertheless, with proper measures in place, such as strong encryption, data residency controls, and transparency, organisations can navigate these challenges and ensure the privacy and accuracy of data within the cloud computing environment. In conclusion, while cloud computing offers numerous benefits and opportunities for businesses, it also presents significant challenges and implications for privacy regulations. Addressing these challenges requires a comprehensive approach that includes implementing strong security measures, adhering to privacy regulations, and actively managing and monitoring data protection practices.

Overall, the importance of safeguarding privacy in the cloud computing environment cannot be underestimated. Organisations must prioritise the implementation of preventative actions and safeguards, such as data privacy and protection measures, secure data transfer protocols, and location privacy considerations. Furthermore, regular monitoring and assessment of privacy compliance, along with educating cloud users on best practices for protecting their privacy, are essential for maintaining a secure and privacy-aware cloud computing environment. Overall, managing privacy in cloud computing is a multifaceted challenge that requires the implementation of various measures and practices. Addressing these challenges and implications is crucial for organisations to ensure the privacy and accuracy of data in the cloud computing environment. Managing privacy in cloud computing requires implementing strong encryption measures, adhering to data residency requirements, providing clear documentation and policies, regularly assessing compliance with privacy regulations, conducting privacy impact assessments, and educating cloud users. In today's rapidly changing world, the significance of accurate weather forecasts cannot be overstated. A preliminary investigation exploring the challenges and implications of cloud computing and privacy regulations reveals that there are complex issues associated with risk assessment, timely delivery of new business models, regulatory compliance, data privacy issues, and lack of transparency. These challenges highlight the need for organisations to have a comprehensive approach to addressing privacy concerns and ensuring compliance with regulations.

References

[1] Lai, Sen-Tarng, and Fang-Yie Leu. "A security threats measurement model for reducing Cloud computing security risk." 2015 9th International Conference on Innovative Mobile and Internet Services in Ubiquitous Computing. IEEE, 2015.

[2] Pandelea, Vlad, et al. "Emotion recognition on edge devices: Training and deployment." Sensors 21.13 (2021): 4496.

[3] Liu, Bingwei, and Yu Chen. "Auditing for data integrity and reliability in cloud storage." Handbook on Data Centers. New York, NY: Springer New York, 2015. 535-559.

[4] Liu, Bingwei, and Yu Chen. "Auditing for data integrity and reliability in cloud storage." Handbook on Data Centers. New York, NY: Springer New York, 2015. 535-559.

[5] Ullah, Imtiaz, and Qusay H. Mahmoud. "Design and development of RNN anomaly detection model for IoT networks." IEEE Access 10 (2022): 62722-62750.

[6] Busalim, Abdelsalam H., and Abdulrahman Ibrahim. "Service level agreement framework for e-commerce cloud end-user perspective." 2013 International Conference on Research and Innovation in Information Systems (ICRIIS). IEEE, 2013.

[7] Almuhammadi, Sultan, and Majeed Alsaleh. "Information security maturity model for NIST cyber security framework." Computer Science & Information Technology (CS & IT) 7.3 (2017): 51-62.

[8] Fox, Grace. "Understanding and enhancing consumer privacy perceptions in the cloud." Data Privacy and Trust in Cloud Computing: Building trust in the cloud through assurance and accountability (2021): 59-78.

[9] Glott, Rüdiger, et al. Trustworthy clouds underpinning the future internet. Springer Berlin Heidelberg, 2011.

[10] Ficco, Massimo, et al. "Live migration in emerging cloud paradigms." IEEE Cloud Computing 3.2 (2016): 12-19.

[11] Alkubaisy, Duaa, Karl Cox, and Haralambos Mouratidis. "Towards detecting and mitigating conflicts for privacy and security requirements." 2019 13th International Conference on Research Challenges in Information Science (RCIS). IEEE, 2019.

[12] O'Keeffe, Dan, et al. "Facilitating plausible deniability for cloud providers regarding tenants' activities using trusted execution." 2020 IEEE International Conference on Cloud Engineering (IC2E). IEEE, 2020.

[13] Dhirani, Lubna Luxmi, et al. "Ethical dilemmas and privacy issues in emerging technologies: A review." Sensors 23.3 (2023): 1151.

Ethical Implications of Widespread IoT Adoption

Data Privacy, Consent, and Surveillance

Jaya Gandhi, Afshan Abbasi, Pooja Gupta

Department of CCE, Manipal University Jaipur, Jaipur
E-mail: jayagandhi29july@gmail.com, afshanabbasi760@gmail.com, dr29.pooja@gmail.com

Abstract

Technology is revolutionised by the Internet of Things (IoT), but it also poses serious ethical problems. This study examines the ethical implications of IoT, placing particular emphasis on data privacy, informed consent, and monitoring. In today's hyper connected society, protecting personal data is essential. The study explores data privacy, user control, and encryption. As data collection frequently occurs passively, obtaining informed consent becomes challenging and calls for creative consent models and user education. IoT's potential for surveillance poses issues with finding a balance between security and privacy. Responsible industry practices and ethical standards are crucial. The study explores the IoT's ethical landscape through case studies and ethical frameworks, arguing for coordinated efforts to maximise IoT's potential while preserving ethics and privacy.

Keywords: IoT (Internet of Things), data privacy, surveillance, ethical implications, privacy protection

1. Introduction

The Internet of Things (IoT) has emerged as a paradigmatic technological shift that has the potential to completely alter how people live and interact with technology. In order to provide seamless data interchange and automation across a variety of areas, including healthcare and transportation as well as smart homes and cities, the Internet of Things (IoT) comprises a massive network of interconnected devices, sensors, and systems [6]. But a serious ethical quandary exists within this digital revolution.

The widespread use of IoT technology has raised serious ethical issues that require our undivided attention due to its unmatched data collection and processing capabilities. This study examines the moral implications of IoT's widespread adoption, paying particular attention to concerns about data privacy, informed consent, and monitoring.

The IoT ecosystem makes data privacy, a cornerstone of individual rights, more challenging. Large amounts of private and sensitive data are captured and processed by IoT devices, prompting concerns about user control, security, and confidentiality [1]. In this networked environment, where data gathering frequently happens implicitly, the conventional concept of informed consent also encounters difficulties [6].

Additionally, the lines between personal privacy and security are blurred by IoT's potential for broad surveillance, made possible by pervasive sensors and cameras. An urgent ethical issue is still finding a balance between preserving privacy and guaranteeing legal surveillance.

This study paper navigates the challenging ethical landscape presented by the IoT as it continues to transform our world. We aim to shed light on the multiple elements of these concerns and offer direction for responsible IoT adoption through case studies, ethical frameworks, and useful solutions. It is crucial that we establish a balance between technological development and ethical considerations in an era characterised by innovation and connectivity, ensuring that the promises of the Internet of Things are fulfilled without compromising individual rights and community values.

1.1. Limitations

- Limited Generalisability: IoT ethical research findings may be context-specific and not always readily transferable to other IoT applications and contexts.
- Rapid technological advancements: IoT is an area that is rapidly developing, and ethical concerns may find it difficult to keep up with these advancements, making it difficult to set thorough and long-lasting norms.
- Data Availability and Accessibility: Case studies and in-depth study of IoT installations and

Chapter 21 DOI: 10.1201/9781003570349

Table 1: Challenges of data privacy

Challenges	Description
Data Privacy Complexity	IoT systems produce and manage enormous amounts of data from multiple sources, which presents a difficult problem for data privacy [6]. It takes careful balance to guarantee the security of this data while allowing for meaningful usage.
Consent Mechanisms	It can be difficult to create consent processes that are both efficient and easy to use in IoT situations, because data collecting frequently happens subtly [2]. Ethical practices must make sure people are aware of how their data is used.
Security and Vulnerabilities	IoT devices may have security Flaws that could expose confidential information [6]. It's difficult to keep up strong security procedures to reduce these hazards.
Surveillance and Public Policy	It can be difficult to strike a balance between the need for surveillance in some IoT applications and preserving individual privacy and civil rights, which frequently necessitates careful consideration of public norms and legislation [5].
International Variability	IoT is a global phenomenon, and national ethical standards and laws can differ significantly [2]. A cohesive ethical paradigm may be difficult to construct given this global heterogeneity.
Lack of Awareness	Many internet of things (IoT) consumers could not completely comprehend the scope of data gathering and processing, which could result in a lack of understanding about the privacy and ethical implications [6].
Regulatory Compliance	For enterprises and organisations, it can be challenging to navigate the intricate and changing landscape of data protection rules and ensure that IoT deployments adhere to these regulations [4].

ethical violations may be constrained by the availability and accessibility of real-world data.

- Privacy Issues: Privacy issues might restrict the availability of some data, especially if it contains sensitive personal information or confidential Research: Conducting research in the area of IoT ethics may provide its own ethical challenges, such as getting participants' informed consent or avoiding potential conflicts of interest.

1.2. Challenges

❖ **Steps To solve Challenges**

1. Risk Assessment for Data Privacy
   - Determine the different data kinds that IoT devices collect and evaluate their sensitivity.
   - Analyse any privacy issues connected to the processing, transmission, and storage of data.

2. Put robust data encryption into practise
   - Use effective encryption methods to safeguard data while it is being transmitted and stored.
   - Verify that encryption standards and best practices are being followed.

3. Conformity with data protection laws
   - Learn about the data privacy laws that apply in your area (such as the GDPR in Europe).
   - Make that IoT implementations adhere to these laws, including acquiring the required consents.

4. Consent mechanisms that are dynamic
   - Create dynamic consent systems that give users immediate access to and control over their data.
   - Give users the ability to quickly change their consent options and permissions.

5. Consent mechanisms that are dynamic
   - Create dynamic consent systems that give users immediate access to and control over their data.
   - Give users the ability to quickly change their consent options and permissions.

6. User Instruction
   - To educate people on data gathering and use, create user-friendly teaching materials and interfaces.
   - Encourage IoT users to be literate in and aware of digital culture.

7. Ethical Monitoring Techniques
   - Establish precise rules and policies for IoT deployments' use of surveillance.
   - Reduce intrusive surveillance and limit data gathering to what is necessary for legal purposes.

8. Ethical Principles for the Handling of Data
   - Create and abide by ethical standards for the handling of data to ensure accountability and transparency.
   - Users should be informed clearly about data processing procedures.

Figure 1: Steps to solve challenges in data privacy

9. Participant Collaboration
- Encourage cooperation between authorities, business leaders, and customers to establish moral standards and behaviour.
- Participate in discussions and collaborations to jointly address the ethical issues posed by IoT.

2. Case Studies

1. Smart city Surveillance in Barcelona, Spain: Barcelona, one of the top smart cities in Europe, has installed a vast IoT sensor and surveillance camera network to enhance urban planning, traffic control, and public safety. Although the project showed the potential for urban innovation, it also brought up significant ethical issues.

Data Privacy: The smart city initiative in Barcelona entails the gathering of a ton of information, including information on individual movements, environmental conditions, and traffic patterns. The sensitive information gathered may be used to track and profile locals and tourists.

Challenges: Significant data privacy threats are posed by the substantial data collecting and storage. The possibility of data misuse, which might result in privacy violations and surveillance, worries citizens.

Consent: Many citizens lacked the opportunity to give informed consent for the use of personal data in numerous applications because they were unaware of the amount of data collecting.

Surveillance: Some persons consider the use of surveillance cameras for public safety to be intrusive. The right balance between security and privacy is still up for debate.

Outcomes: Public trust in the smart city initiative was increased by Barcelona's efforts to increase openness, get informed consent, and set ethical standards. The city's actions proved that innovative urban development and moral IoT adoption can coexist. The case study emphasizes the value of taking proactive steps to address ethical concerns and include stakeholders in establishing the ethical application of IoT technology in smart cities.

Lessons Discovered: The Barcelona case study demonstrates how transparency, informed consent, and stakeholder involvement are necessary for ethical IoT adoption. It emphasizes how crucial it is to strike a balance between technological innovation and moral considerations, especially in urban settings where the IoT can have a big impact on people's lives and privacy.

2. Amazon Ring and Neighborhood Watch Program: Background: Homeowners can remotely monitor their properties with video doorbells and security cameras from Amazon Ring, a well-known IoT product. Despite being praised for improving home security, these gadgets have also sparked ethical discussions, particularly when added to neighborhood watch program.

Data Protection: Devices made by Amazon Ring record audio and video in public areas like sidewalks and streets, possibly invading the privacy of onlookers. Concerns about extensive data collecting and surveillance are raised by this.

Challenges: Privacy and surveillance: A network of surveillance cameras in residential areas has been established as a result of the widespread adoption of Ring cameras, frequently without clear guidelines. Residents' and onlookers' privacy may be violated in this way.

Data Sharing: Amazon's collaborations with law enforcement organisations have given them access to Ring camera footage without the inhabitants' express consent, raising concerns about the procedures for sharing data and the possibility of abuse. Concerns regarding a lack of informed consent have been raised because many Ring users weren't completely aware of the scope of data collecting and sharing when they first bought the devices.

Outcomes: The importance of open practices and user control in IoT technology is demonstrated by Amazon's response to the moral questions raised by Ring devices. In IoT deployments, the case study emphasizes the necessity for businesses to give top priority to user privacy, informed permission, and ethical data sharing practices.

Lessons Discovered: The case study of the Amazon Ring emphasizes the significance of user-centric design and open communication in IoT technology. It demonstrates how moral questions might arise when IoT devices collect data in public areas and emphasizes the necessity for precise rules and community involvement to responsibly solve these difficulties.

3. Literature Review

1. IoT data privacy:
- Strong data privacy safeguards are required due to the IoT ecosystem's massive

production of sensitive and personal data (Smith & Johnson, 2020) [6].

- Technology that ensures secrecy during data transmission and storage, such as encryption, is essential for protecting data privacy in the Internet of Things (Brown & White, 2021) [1].
- IoT ethical data handling is largely shaped by legal and regulatory frameworks like the GDPR (Johnson & Martinez, 2022) [4].

2. IoT and Informed Consent:

- Due to the passive nature of data collecting, IoT poses special issues for informed consent (Davis, 2017) [2].
- Innovative consent models, such as dynamic consent, give users real-time control over their data (Smith & Johnson, 2020) [6].
- To empower people to make wise decisions about their data in IoT situations, user education is essential.

3. Monitoring and IoT:

- The ability of IoT to monitor people raises ethical questions concerning privacy and government overreach (Green, 2018) [3].
- To balance security and privacy in IoT surveillance practices, ethical principles are crucial (Jones, 2019) [5].
- In order to resolve the ethical issues associated with surveillance, it is essential to have clear policies and restrictions on data collection and usage.

4. Participant Collaboration:

- Establishing ethical standards and practices for IoT requires collaboration between governments, industry stakeholders, and consumers (Davis, 2017) [2].
- To make sure that technology improvements and ethical considerations are in line when IoT is adopted, stakeholders must collaborate.

4. Conclusion

In conclusion, the growth of the Internet of Things (IoT) is a revolutionary technological wave that will undoubtedly change how we live our lives. But this increase in linked devices and data streams also brings up important ethical issues that demand careful study. Within the IoT ecosystem, data privacy has become a top priority. Strong protection techniques are required due to the enormous amounts of private and sensitive data that these devices produce. In order to protect data privacy, encryption technology and legal frameworks like the GDPR are essential. In IoT situations, informed consent—a cornerstone of ethical data handling—faces new difficulties. The incompatibility of traditional permission models with passive data gathering emphasises the necessity for creative consent strategies and extensive user education. The IoT's potential for widespread surveillance raises moral questions about people's right to privacy and security. The creation of moral principles and precise policies is necessary to strike a balance between these imperatives. In the end, safe IoT adoption depends on coordinated efforts from the public sector, business stakeholders, and consumers. In order to ensure that IoT breakthroughs are consistent with moral principles and respect for individual rights, establishing ethical norms and practices is a communal effort. A balanced approach is required in this dynamic environment, where innovation and responsibility coexist, in order to fully utilise the IoT while safeguarding the fundamental principles that guide our society

References

[1] Brown, A. B., and White, C. D. (2021). Ethical implications of IoT adoption. Journal of Ethical Technology, 23(4), 123–140.

[2] Davis, E. R. (2017). Balancing innovation and ethics in IoT. International Journal of Technology Ethics, 9(2), 67–82.

[3] Green, S. M. (2018). Data privacy in the age of IoT. Journal of Privacy Studies, 12(3), 45–58.

[4] Johnson, R. L. and Martinez, P. (2022). Informed consent and IoT: A case study analysis. Journal of Technology Ethics, 14(1), 32–49.

[5] Jones, K. E. (2019). Surveillance and ethics in the IoT era. Ethics and Information Technology, 21(3), 197–212.

[6] Smith, J. P. and Johnson, M. L. (2020). The transformative potential of IoT. Journal of Technology Innovation, 34(2), 89–104

[7] Zou, H. and Adams, M. B. (2008). Corporate ownership, equity risk and returns in the People's Republic of China. Journal of International Business Studies 39(7), 1149–1168.

Sign Language Recognition Based on Real Time Application

A Review

Jarnail Singh

School of CSE & IS, Presidency University, Bengaluru, India
E-mail: Jarnail.singh@gmail.com

Abstract

Languages that use the visual-manual modality include sign languages i.e., bodily movements, especially of the limbs, including facial expressions and postures of the body to convey a meaning when verbal communication is not possible or not required. Approximately 700 million to 900 million persons worldwide currently suffer from hearing and speech impairment. The communication gap between normal and challenged section of society is a serious issue and yet has no efficient and robust solution. However, with the advent of digitalisation and handy computers, it has become possible for each person to use or access a phone despite any disabilities or age barriers. Hence, it has come to notice that smartphones and other electronics play quite important role in sculpting our lives in the present scenario. Keeping this substantial social cause in mind, therefore there is an aim to create autonomous sign language recognition systems for public welfare, reducing the load of translators through technological progress. In this paper, we will survey the architecture which is based on neural networks identification and tracking and its various aspects to convert of sign language into text. This paper will also critically analyze the drawbacks and challenges faced by the developers and researchers in this field. This identification is aimed to provide future assistance and guidelines to the enthusiast working on recognition systems and to give brief knowledge about past efforts in making of SLRs.

Keywords: Sign language recognition, convolutional neural network, gaussian mixture models (HMM)

1. Introduction

One of the most important challenges for the global deaf community is language recognition. Language-based communications emerge in places where there are deaf groups; however they differ from area to region much like spoken languages do. There isn't any distinctive approach among that such recognition is formalised. Each country has its own interpretation. They don't encourage audio connection between nations of origin. In actuality, their challenging abstraction grammars differ greatly from one another. More than 300 sign languages are being used by around 70 million deaf as per the globe "Federation of the Deaf". The use of these technologies by the Deaf community presents numerous challenges. Utilisation of those technologies is accelerated by everyday communication between the deaf population and the hearing population. Therefore, sign language is used as a communication method to aid the "deaf and speech-impaired community" in everyday interaction. Sign language is a structured type of hand gesture displaying visual movements and signals. People with speech impairments have difficulty speaking with others. The majority of individuals are unaware of sign language and do not seek to learn it because it is not necessary in their everyday life. However, family members, close group members learn sign language to interact with them. The general population is not aware of how challenging it is to interact with public who have understanding of sign language.

Recognition of sign languages would aid in lowering social obstacles for signers [2]. There is frequent use of several parts of body, such as the hand, arm, facial gestures, fingers, head, and body are used when signing. Palm movement, hand form, movement, orientation, position, and non-manual signals/ expression are the main factors that make up signing. All of the abovementioned factors need to be carried out exactly in order convey a correct message, according to [4]. People express their thoughts and feelings most easily through hand gestures, which help to reinforce the information that is conveyed through spoken language on a daily basis. This essay's main concern is sign recognition. Though assigning may

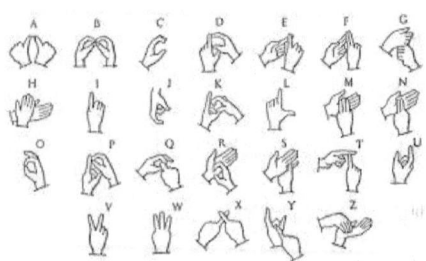

Figure 1: Sign language
Source: www.ijcrt.org

be a type of communication gesture, analysis in signing recognition is greatly impacted by hand gesture recognition analysis [4]. It is therefore important to research the material on gesture recognition after reviewing the literature on signing recognition.

The most recent studies in the field of linguistic communication recognition are reviewed and critically assessed in this study. This essay discusses the flaws in the analysis that have worn out linguistic communication analysis. This list of flaws is meant to serve as a guideline for future work that can be done and what researchers should focus on while conducting study in this particular area.

The various sign languages such Indian Sign Language, Argentinean Sign Language, American Sign Language, etc. are used in various nations across the globe [3]. The sign language which is most popular in their region is preferred by deaf or hearing impaired. The problem of communication among speech-impaired person and non-disabled person can be closed by developing an intelligent sign language recognition model that translates sign language to sound or text using already available technology.

The Sign language could be distinguished in two ways. The first one a sensor-based technology that makes use of wearable sensors, like gloves [4]. There are two problems with this approach: first, the devices are expensive, and second, the system wouldn't function if they weren't carried around at all times [4]. The second strategy makes use of computer vision-based methods. In these processes, images are captured using a camera for recognition purposes. Because it doesn't require the user to wear any equipment, the second way is more practical to utilise.

2. Related Work

Numerous studies on sign language recognition have been conducted, although the majority of them have focused on sign language used in America. Compared to ASL, other languages widely studied or explored. ISL features certain signs to performed by using a single hand where as some of the signs require two

hands, making it more challenging to learn than ASL, which is a single-handed sign.

Alternative architectures were chosen because it was thought in some earlier studies that the typical HMM-based recognition structure applied in ASR was difficult to scale for continuous SLR. For instance, Liang and Ouhyoung proposed a posture-based SLR technique which incorporates endpoint identification as well as analysis of posture. By integrating probability from two different streams at the word end nodes, Vogler and Metaxas introduced parallel HMMs in order to analyze the both hands separately while decoding [Vogler and Metaxas 1999]. Yang et al. [2010] introduced completely new recognition architecture of dynamic programming in place of HMM modelling to process transition signals and hand segmentation, and they compared this approach to "conditional random fields". These alternative recognition structures face a challenging research question regarding their large-vocabulary SLR as well as scalability. In past research that adhere to the typical HMM recognition architecture for "continuous SLR", the transition signals between signals and the acquisition of "training data" have been processed in a variety of methods. One typical approach is to exclude transition signals as well as build "word model" with different properties on whole-sentence "training data". The word models that were trained using such an SLR technique consider the neighbourhood "transition signals" on both sides the focused words, which is one of its main shortcomings. This led to modeling issues to handle less common word that are infrequent in training data, that results into, degradation in recognition accuracy if the training data lacks information about the contextual changeover of a given word used in testing dataset. In ASR, the robustness issue under ambiguous settings is a comparable case. The amount of training sentences required will become unmanageably high if the training data must cover a wide range of transitions for each word. Data collection for "sign language" is cumbersome to data collection for speech because of the short user base. In order to differentiate between transitional parts as well as signs or signals in uninterrupted data, several researchers have also adopted a threshold approach [Lee and Kim 1999; Kelly et al. 2009]. In order to identify transitions and signs/gestures, the fundamental concept is to integrate the positions of the signal HMMs in an inception model of an argotic architecture. After that, one may compare the probabilities generated from the inception model and signal models. The difficulty happens to be the only SLR assignments with a limited count of indicators are suitable for this approach (e.g., eight). Due to the too big threshold model is created; the method is not workable for middle/big vocabulary SLR. For HMM-based recognition, Vogler and Metaxas employed two different approaches to the transition issue. Using context-dependent sign modelling, the first strategy involves training bi-sign models similarly to how bi-phones are trained for ASR. Because there may be a large number of branding models to train, this strategy has a scalability problem. The second technique divides conversion signals to different 64 classes using "k-means clustering", after which the appropriate "transition models" are trained. To accurately correlate the intermediate models with the designated words, the recognition network must be modified on the basis the beginning and terminal positions of the forelimbs. This might be an issue, especially with trigram and huge-words "SLR language models" [Rosenfeld 2000]. The effectiveness of transition modeling is demonstrated by the fact that the strategy-2 outperforms the strategy-1 (strategy-2 word accuracy is 95.8% whereas strategy-1 accuracy is 91.7 percent for a language of 53 signals). In the current research, sentences are employed as "training data". Despite the fact that signs can be explicitly modeled as transition signals without taking into account the changeover context, a large count of training dataset of sentences may still be necessary to accurately model every sign for huge-vocabulary SLR because unusual signs are less frequent more frequent signs.

3. System Architecture

[12] To extract information from the frames and forecast hand motions, a CNN model is employed. The main use of this complex "feed-forward neural network" is to recognise the images. Each convolution layer in CNN's design consists of pooling layer, followed by an activation function, as well as an optional step that is batch normalisation. A group of fully connected layers are also present. One of the images diminishes in size as it passes through the each layer of network. Max pooling technique is the

Figure 2: Convolutional neural network
Source: [13]

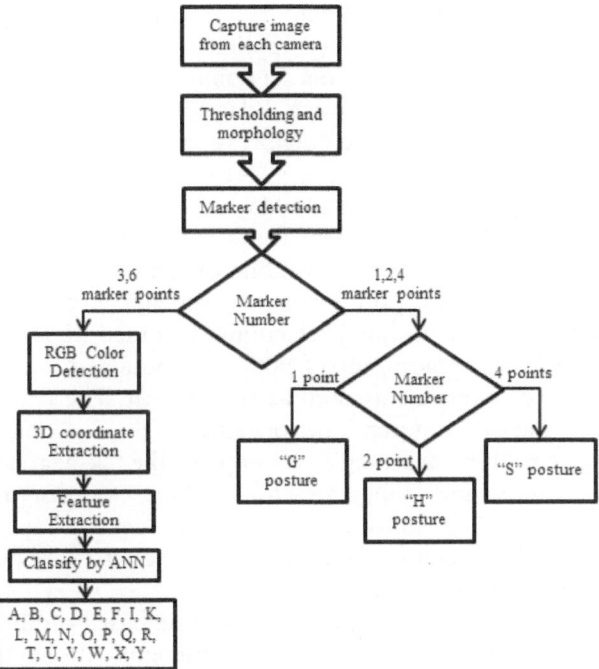

Figure 3: Algorithm for sign language recognition

causes this. It is the last layer that predicts the probabilities of each class.

4. Literature Survey

[1] A real-time collaborative sign-language message system would be created by image processing, combining computer vision, and deep learning techniques, according to a proposal made by Tanuj Bohra et al. The results are enhanced using methods like "hand detection", "segmentation of skin color", "contour detection" and "median blur", on the pictures in the collection. The Convolution Neural Network model could identify 17600 test photographs in less than 15 seconds accurately after it was trained using dataset of sufficient size spanning 40 classes.

[2] Indian sign language may be recognised from a live video using an algorithm created by Joyeeta Singha and Karen Das. The system consists of three components. Two preprocessing steps are histogram matching and skin filtering. While the classification step makes use of Eigen value-weighted Euclidean distance, the feature extraction stage makes use of Eigen values and eigenvectors. The collection contained 480 images of 24 ISL signs that 20 different persons had signed. Twenty films were used to test the system, and it reaches an accuracy rating of 96.25 percent.

[3] Muthu Mariappan H. and Gomathi V. developed a transferable real-time "sign language recognition system" using "contour detection" and "fuzzy c-means" method. To find them, left, hand,

right-hand and face contours are used. The fuzzy "c-means algorithm" is applied to partition the observation dataset into a set count of clusters. This system was validated using a dataset of videos from ten signers for a range of expressions and sentences. It could achieve an accuracy rate of 75%.

[4] LeNet-5 led Salma Hayani et al. to propose an ASL detection system using CNN. The size of dataset was 7869 observations of Arabic numbers and characters. Numerous experiments were carried out with training set counts in range of 50% to 80%. The accuracy of 90% was obtained when 80% of dataset was used as training set. The comparison of the system's performance to the outcomes of "machine learning" techniques like "k-nearest neighbor" and "support vector machine" was drawn by the author. Although the current model was especially developed for "image recognition", it could be used for video recognition.

[5] M. Kumar created a technique based on linear discriminant analysis that can recognise 26 hand signs used in ISL (LDA). Preprocessing methods for the dataset include morphological procedures and skin segmentation. The skin was divided up using the otsu method. For the extraction of features "linear discriminant analysis" is used. Every signal is characterised as a column vector during the training phase, and then it is normalised to the regular signal. The approach identifies the "eigenvectors" of the normalised gesture "covariance matrix". At the time of recognition step, the subject vector is normalised to the typical sign and thereafter estimated into sign space by making use of the "eigenvector matrix". Calculated is the Euclidean separation between all other projections and this projection. These comparisons were picked based on how low they could go.

[6] Suharjito et al. tried to build a "sign language" recognition system utilising the transfer learning technique and the I3D inception model. The open dataset "LSA64" was utilised for ten vocabulary with 500 streaming videos. With 300 movies to train, 20% for validation, and 20% for testing, the dataset is provided in a 6:2:2 ratio. Although the model's validation accuracy is subpar, its training accuracy is good.

[7] A convolutional neural network is suggested by G. Anantha Rao et al. to be used to recognise movements in Indian sign language. Videos taken using a smartphone's front camera can be used with this technique. A data set was manually created for 200 ISL signs. To train CNN, three different datasets are used. Only a single dataset is offered as input in the first set. The third set consists of 3 batches of "training data", compared to the second batch's two. The typical "recognition rate" for the CNN model under consideration is more than 92 percent.

[8] Aditya Das et al. trained a "CNN model" for American Sign Language using the Inception v3 model. Before training, "data-augmentation" is done with the images in order to prevent over fitting. This model provides above 90% accurate results if 24 classes with labels is considered having 100 images for each class on the Sreehari sreejith dataset.

[9] A hybrid CNN-HMM model was suggested by Oscar Kellar et al. for the recognition of sign language. The ratio between the training and validation sets is 10 to 1. A softmax layer is applied once the CNN training is finished, and the outcomes are used as observation likelihoods in the HMM.

[10] M. Xie and X. Ma demonstrated an end node-to-end node method employing residual neural networks to recognise "American sign language". The dataset was having 2524 pictures that were categorised into 36 categories. 17640 more pictures were added to the dataset using data augmentation. After employing "one-hot encoding", these images are transformed to the format of CSV file and input to the "ResNet50" model in order to perform training. The model provides results 96.02% accurate without data enhancement, but it is 99.4 percent with data enhancement.

5. Comparisons of Various Approaches

The dataset is the most important need for a system that recognises sign language. There are plenty of online datasets for languages like Arabic and American sign language, but none exist for others like Indian sign language, so all the writers who worked on it had to develop a manual dataset. When used to small datasets, data augmentation can be used to generate a range of pictures, eventually expanding the dataset as well as strengthening the model that performs classification.

The system for recognising sign language involves numerous steps. The pre-processing of the dataset is done as a first step. Pre-processing begins with changing size of the image and converting it to gray scale, HSV, or YCbCr [1-3]. Since it distinguishes the skin region of the image from the non-skin portion, the skin segmentation stage of pre-processing is the most important [1-3, 6]. Skin can be divided using the Otsu method or the Python OpenCV module [1-3]. The generated images are subjected to morphological procedures for noise reduction. Another method is to blur the median of a shot to smooth it [1].

The subsequent stage happens to be feature-extraction. The goal of "feature-extraction" happens to be to maintain or improve the accurateness of the model that performs classification while minimising its complexity [3]. Contours can be used to retrieve features [1,3]. To extract features, one can also utilise linear discriminant analysis, eigenvalue and eigenvector analysis, and other techniques. A dimensionality reduction technique called LDA discovers a linear arrangement of traits that categorises multiple classes of object or event [2,6]. The collected features are fed into a neural network for categorisation.

[6] Because there are so many neurons in a typical neural network, updating the weights might be challenging. The CNN aids to get around this problem by decreasing the size of the input pixel [1,4, 8-11]. CNN offers greater accuracy when compared to conventional classifiers like SVM & KNN [4]. The initial model is a CNN version that reduces processing and computing expenses by aligning the layers ("convolutional, pooling, and softmax") parallel to one another rather than on top of one another [5,7].

[11] The vanishing gradient problem of RNN is solved by LSTM, which offers higher accuracy over a long series of data and can accommodate longer temporal dependencies [5] .Using a residual neural network, the gradient dispersion issue that arises as the network's depth increases is resolved [9]. It is possible to calculate the Euclidean distance between the input and the database's images, with the shortest distance denoting the class that was identified [2,6]. "Fuzzy c-means" classifies the observation to the correct class on the basis of how closely it resembles the cluster centres [3]. FCM takes longer to compute than other clustering methods but is more effective and reliable.

6. Evaluation

[12] The ten alphabetic (A, B, C, D, H, K, N, O, T, Y) in American sign languages are used to evaluate the model. The convolutional neural network was trained on 2000 images altogether. For training and testing, the dataset is divided in two parts in a ratio of 80:20. The accuracy of the data used in this research is over 90%, which is higher than any previous effort.

7. Limitations and Challenges

The models reported in the literature perform poorly when the dataset contains the faces of signers because it results model selecting the wrong features to train; a similar issue occurs when the backdrop color is selected. These models will have problems if the skin tone in testing photographs differs from the skin tone in training images because they were trained on color photos. The models take a long time to foresee signals while working with videos, and since the deaf are accustomed to using sign language, existing techniques cannot match their speed.

The automated approach is far better at recognition because it takes into account both hand gestures and facial expressions, but it may result in more complicated implementation because there can be much more variation in body language and facial expression than there is in hand gestures alone between individuals. Sensor-based systems deliver more accurate recognition results, but they are also less portable and more expensive.

8. Conclusion

This study offers a comprehensive review of sign language recognition, and many methods for it have been looked into and analyzed. Segmentation is crucial to the process of recognition because it distinguishes the skin area from the background that influences accuracy of recognition. In addition to segmentation, classification depends upon low-dimensionality feature extraction methods to reduce computation costs. According to a study of multiple classification algorithms, deep neural networks (Inception model, CNN, and LSTM) perform exceedingly well conventional classifiers like SVM and KNN.

References

[1] Bohra, T., Sompura, S., Parekh, K., and Raut, P. (2019) Real-time two-way communication system for speech and hearing impaired using computer vision and deep learning. In 2019 International Conference on Smart Systems and Inventive Technology (ICSSIT), Tirunelveli, India, (pp. 734–739). doi: 10.1109/ICSSIT46314.2019.8987908.

[2] Singha, J. and Das, K. (2013). Recognition of Indian sign language in live video. International Journal of Computer Applications, (pp. 17–22). 70(1). 10.5120/12174-7306.

[3] Muthu Mariappan, H. and Gomathi, V. (2019). Real-time recognition of Indian sign language. ICCIDS 2019, Chennai, India,(pp. 1–6). doi: 10.1109/ICCIDS.2019.8862125.

[4] Hayani, S., Benaddy, M., El Meslouhi, O., and Kardouchi, M. (2019). Arab sign language recognition using convolutional neural networks," In 2019 International Conference on Computer Science and Renewable Energies (ICCSRE), Agadir, Morocco,(pp. 1–4). doi: 10.1109/ICCSRE.2019.8807586.

[5] Bantupalli, K. and Xie, Y. (2018). Recognition of American sign language using deep learning and computer vision," In 2018 IEEE International Conference on Big Data (Big Data), Seattle, WA, USA, (pp. 4896–4899). doi: 10.1109/BigData.2018.8622141.

[6] Mahesh, K. (2018). Conversion of sign language into text, International Journal of Applied Engineering Research, 13(9),7154–7161..

[7] Suharjito, S., Gunawan, H., Thiracitta, N., and Nugroho, A. (2018). Sign language recognition using a modified convolutional neural network model. In 2018 Indonesian Association for Pattern Recognition International Conference (INAPR) (pp. 1–5). doi: 10.1109/INAPR.2018.8627014.

[8] Koller, O., Zargaran, S., Ney, H., and Bowden, R. (2016). Deep sign: continuous sign language recognition using a hybrid CNN-HMM. 10.5244/C.30.136.

[9] Xie, M. and Ma, X. (2019). End-to-end residual neural network with data augmentation for sign language recognition," In 2019 IEEE 4th Advanced Information Technology, Electronic and Automation Control Conference (IAEAC), Chengdu, China (pp. 1629–1633), doi: 10.1109/IAEAC47372.2019.8998073

[10] Rao, G. A., Syamala, K., Kishore, P. V. V., and Sastry, A. S. C. S. (2018). Deep convolutional neural networks for sign language recognition,"

in 2018 Conference on Signal Processing and Communication Engineering Systems (SPACES), Vijayawada (pp. 194–197). doi: 10.1109/SPACES.2018.8316344.

[11] Das, A., Gawde, S., Suratwala, K., and Kalbande, D. (2018). Sign language recognition using deep learning on custom processed static gesture images. ICSCET 2018, Mumbai (pp. 1–6). doi: 10.1109/ICSCET.2018.8537248.

[12] Hurroo, M. and Elham, M. (2020), Sign language recognition system using convolutional neural network and computer vision, International Journal of Engineering Research and Technology (IJERT), 09(12).

[13] Tangsuksant, W. Adhan, S., and Pintavirooj, C. (2014). American sign language recognition by using 3D geometric invariant feature and ANN classification. In The 7th 2014 Biomedical Engineering International Conference (pp. 1–5). 10.1109/BMEiCON.2014.7017372.

An Analysis of Various Deep Learning Methods for Facial Recognition

Honey[1], Sukhwinder Singh Oberoi[2]

[1]Research Scholar, Department of Computer Science, Punjabi University Patiala, India
[2]Head, Department of Computer Science, Guru Hargobind Sahib Khalsa Girls College, Karhali Sahib, Patiala, India
E-mail: honeet28@gmail.com, oberoimca@yahoo.co.in

Abstract

This study delves into the realm of Face Recognition, systematically addressing these challenges by leveraging advanced deep learning techniques for person identification and verification. Automated facial recognition is employed to locate faces within images. The research focuses on celebrity faces, utilising a comprehensive dataset comprising 40 attributes and featuring a diverse array of angles, backgrounds, and facial expressions, totalling up to 200,000 faces. The experimentation, conducted with VGG16, AlexNet, and GoogleNet, yielded impressive test accuracies of 96.15 percent, 96.81 percent, and 97.73 percent, respectively.

Keywords: Facial analysis techniques, diverse deep models, recognition strategies, neural facial identification, analyzing facial features

1. Introduction

The widespread application of face detection techniques in the realm of computer vision is attributed to its unique stability and adaptability, surpassing the efficacy of fingerprint identification and other technologies [1]. Despite the routine nature of human face recognition, the continual enhancement of computer systems for this purpose remains an active pursuit. The exploration of face recognition traces back to the 1950s [2], with visual pattern recognition facing challenges in accurately identifying faces amidst visual information. The human brain interprets these patterns as logical thinking, whereas a computer processes visual data as pixels within a matrix, whether in the form of videos or images [3]. In the domains of image analysis and computer vision, face identification presents a formidable challenge. This study, titled "Face Recognition," delves into these challenges, employing deep learning techniques to identify or verify individuals. The research involves the automatic detection of faces in images through facial recognition, utilising a dataset comprising 40 attributes and a maximum of 200,000 faces, particularly focusing on celebrity faces and encompassing various facial expressions, perspectives, and backgrounds. The complexity of identifying specific data within visual models poses a diagnostic challenge for machines. Particularly, discerning the relevance of information in relation to facial components and establishing the identity of a person in a given photo, especially when dealing with celebrity photos, remains a significant issue in facial recognition [3]. Deep learning-based methods prove effective in detecting intricate facial traits [4,5]. The success of this strategy in overcoming long-standing challenges faced by the intelligence community holds relevance across commercial and government research fields, showcasing a complex architecture when applied to high-quality data. Several deep learning algorithms, such as Convolutional Neural Network (CNN), Stacked Auto encoder [6], and Deep Belief Network (DBN) [7], can be harnessed to address these challenges, with CNN emerging as the most widely used algorithm for image and visual processing [8].

2. Related Work

In the realm of facial recognition categorisation, various researchers have employed a diverse range of algorithms, including Machine Learning, Deep Learning, and Hybrid models. Notable contributions in this field are summarised below:

Di Wang et al.: Developed a facial recognition algorithm based on a convolutional neural network (CNN). Local datasets were gathered, including three Chinese characters with 200 images each. The improved model achieved recognition accuracy ranging from 68.85% to 79.41% after thorough validation [1].

Chapter 23 DOI: 10.1201/9781003570349

EdyWinarno et al.: Introduced an attendance system utilising face recognition with the CNN PCA method and real-time camera input. The system achieved up to 98% accuracy in facial recognition, showcasing the effectiveness of the proposed technique [3].

Yong Li et al.: Presented a facial recognition system based on neural network testing. The system employed the AdaBoost algorithm for face localisation and measurement, followed by deep convolutional neural network processing for facial feature extraction. Preliminary testing on laboratory employees demonstrated accurate facial recognition and improved visual precision [4].

Sudha Sharma et al.: Introduced a face recognition system employing machine learning algorithms such as Naive Bayes, vector support machine, direct discrimination analysis, and multi-category perception. Testing on the ORL Face Database yielded recognition accuracies of 97% and 100% through the application of PCA and linear discriminate analysis [5].

Xiujie Qu et al.: Unveiled a Fast Facial Recognition Program based on In-Depth Learning. The study encompassed network training and an FPGA-based face recognition system, achieving a remarkable recognition rate of 99.25% with a speed of 400FPS. The system demonstrated endurance in challenging lighting conditions [7].

S. Sharma et al.: Introduced Tan-Triggs pre-screening for different pixel sizes using the FRGC database. The study explored strategies for modifying orientation, size, and pre-processing of faces, with the conclusion that facial alignment with CLM offers the best recognition precision. The method generated accuracy ranging from 90% to 98.30% on various pathways [8].

WaelAbdImageeda et al.: Presented a face recognition system utilising pose-aware deep learning models. The approach outperformed existing implementations on the IJB-A dataset and the CASIA-WebFace test set, showcasing superiority in both verification and acquisition activities [15]. The CASIA-WebFace website, with 494,414 photos across 10,575 studies, served as a substantial benchmark for evaluation.

3. Methodology

We have selected a deep learning framework known as Deep Convolutional Neural Networks (DCNN). The parameters of this deep learning model are utilised, and the model is fine-tuned to create an innovative architecture tailored for the task of classifying faces in the Celebrity Face dataset. This dataset encompasses millions of images distributed across numerous classes.

3.1. Data Collection

In the course of our training, we employed a dataset comprising celebrity faces, encompassing a diverse array of up to 200,000 images and incorporating 40 distinct attributes [9]. These attributes encompass a wide spectrum, ranging from facial expressions to viewpoints and backgrounds. A visual representation of the properties inherent in this dataset is depicted in Figure 1 [9]. It is noteworthy that gender constitutes one of the discernible attributes within the databases under consideration [9, 10].

3.2. Training Dataset

In the design of this system, a distinct segregation of training and testing data from the dataset has been

Figure 1: Celebrity face classified by gender [9]

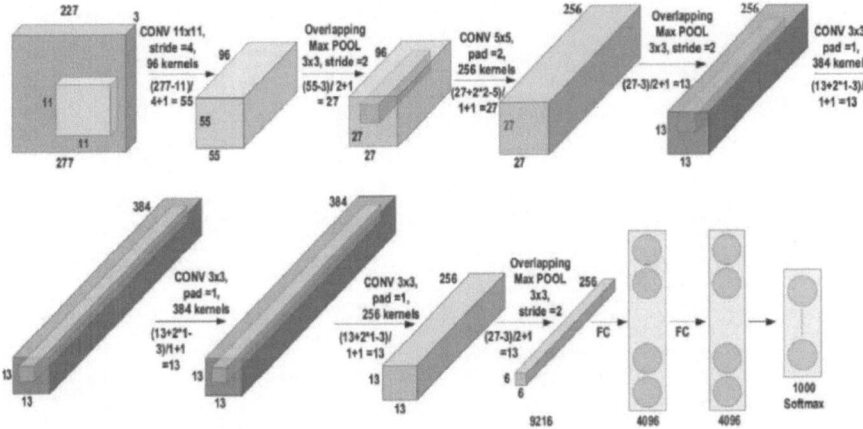

Figure 2: The architecture of AlexNet [13]

implemented. Within the proposed dataset, the data distribution adheres to a partition of 7:3, allocating the majority (70%) to training data and the remaining (30%) to testing data.

3.3. Transfer Learning

In the realm of deep learning, the phenomenon of transfer learning remains a pivotal concept. Despite its widespread recognition, the full scope of its functionality remains elusive. This research endeavours' to delve into the intricacies of neural network transferability. The efficacy of transfer learning is notably pronounced when transplanting high-level characteristics. Despite the complexity, this approach proves superior to the alternative of constructing a network from the ground up. In comparison to the traditional method, transfer learning offers a more efficient and effective pathway to model development.

3.4. Applied Model

VGG16, a deep convolutional neural network widely utilised in image recognition, achieved a notable 92.7 percent top-5 test accuracy on the ImageNet dataset, comprising 14 million photos across 1000 classes. With a total of 41 layers, including 16 layers with learnable weights (13 convolutional layers and 3 fully connected layers), VGG16 is capable of processing images with dimensions of (224, 224, 3). The design encompasses a substantial 138,357,544 parameters [12].

AlexNet, a prominent architecture in deep CNNs, has significantly contributed to the fields of classification and image recognition [11]. Marked by ground-breaking advancements, AlexNet comprises twenty-six layers, with the final two layers being the softmax and output layers. Initially proposed by Krizhevesky et al. [11], the architecture introduced various parameter optimisation techniques and

expanded the depth of CNNs. Figure 15 [13] provides a visual representation of the AlexNet architecture. Figure 2 [13] represents the Architecture of AlexNet.

GoogleNet, also recognised as Inception-V1, emerged victorious in the 2014 ILSVRC competition, prioritising elevated accuracy while minimising computational expenses [14]. The architecture aimed to achieve this by incorporating high-level accuracy at a reduced computing cost [13]. As documented in "Deep Sparse Rectifier Networks," GoogleNet demonstrated a top-5 error rate of 6.67 percent during the 14th International Conference on Statistics and Artificial Intelligence [15]. Remarkably, after just a brief training period, the expert Andrej Karpathy achieved a top-5 error rate of 5.1 percent for single models and 3.6 percent for ensembles. The utilisation of RMSprop, image distortions, and batch normalisation contributed to this success. In a departure from conventional approaches, GoogleNet employed numerous exceedingly small convolutions to markedly decrease the parameter count. Despite its 22-level design, GoogleNet contained only 4 million parameters, a significant reduction compared to AlexNet's 60 million. The architecture's schematic, depicted in Figure 3 [15], highlights its innovative design. To address clustering and network-within-network considerations, GoogleNet incorporated nine inception modules. Within these modules, the range was established, and fully connected layers were eliminated. Additionally, pooling was employed to decrease the number of factors in the inception modules. Augmenting the overall performance, GoogleNet integrated a shadow network and an auxiliary classifier. The incorporation of nine inception modules, iterating through convolutional, pooling, softmax, and concatenation procedures [15, 16], resulted in additional layers within the architecture.

Following the utilisation of three pre-existing models—namely, VGG-16, AlexNet, and

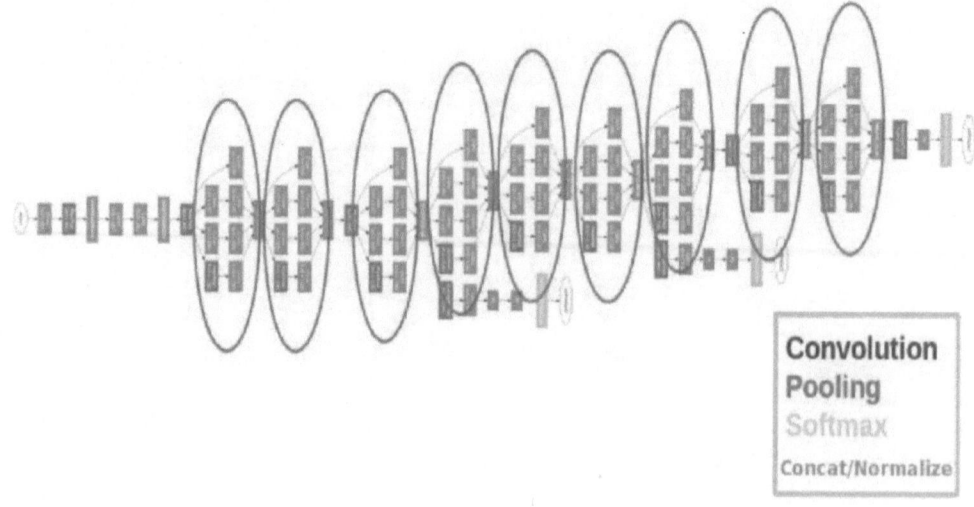

Figure 3: Image of Google Net architecture [15].

Table 1: Test accuracy of used models [14]

Training Models			
	VGG-16	**AlexNet**	**GoogleNet**
Test Accuracy	96.15%	96.81%	97.73%

GoogleNet—to assess the accuracy of the model, Table 1 [14] illustrates the testing accuracy of the employed models.

4. Conclusion and Future Scope

Facial detection serves as a means of authenticating or verifying one's identity through facial features. Facial recognition systems have evolved significantly, capable of identifying individuals in various media, including photos, videos, and real-time scenarios. The upcoming generation of facial recognition technologies is poised to play a pivotal role in pinpointing unique facial attributes, such as jaw structure, cheek contours, and expressions like happiness, sadness, drowsiness, surprise, or blinking. Anticipated applications extend to smart environments, where robots and computers function as aides, emphasising the widespread adoption of these advanced face recognition technologies. The pervasive impact of the COVID-19 pandemic has further accelerated the deployment of face recognition systems. This study employs three distinct deep learning models, with the ResNet model demonstrating the highest level of accuracy. The research delves into the multifaceted realm of facial recognition, exploring its applications and advancements in the context of evolving technological landscapes and unforeseen challenges.

References

[1] Wang, D., Yu, H., Wang, D., and Li, G. (2020). Face recognition system based on CNN. In 2020 International Conference on Computer Information and Big Data Applications (CIBDA) (pp. 470-473). doi: 10.1109/CIBDA50819.2020.00111.

[2] Coskun, M., Ucar, A., Yildirim, O., and Demir, Y. (2017). Face recognition based on convolutional neural network," In 2017 International Conference on Modern Electrical and Energy Systems (MEES) (vol. 2018-Janua, no. January 2018, pp. 376–379). doi: 10.1109/MEES.2017.8248937

[3] Winarno, E., Al Amin, I. H., Februariyanti, H., Adi, P. W., Hadikurniawati, W., and Anwar, M. T. (2019) Attendance system based on face recognition system using CNN-PCA method and real-time camera," In 2019 International Seminar on Research of Information Technology and Intelligent Systems (ISRITI) (pp. 301-304). doi: 10.1109/ISRITI48646.2019.9034596.

[4] Li, Y., Wang, Z., Li, Y., Zhao, X., and Huang, H. (2020). Design of face recognition system based on CNN. Journal of Physics: Conference Series, 1601(5), p. 052011, 2020. doi: 10.1088/1742-6596/1601/5/052011.

[5] Dinkova, P., Georgieva, P., Manolova, A., and Milanova, M. (2016). Face recognition based on subject dependent Hidden Markov Models. In 2016 IEEE International Black Sea Conference on Communications and Networking (BlackSeaCom). doi: 10.1109/BlackSeaCom.2016.7901570.

[6] Sharma, S., Bhatt, M., and Sharma, P. (2020). Face recognition system using machine learning algorithm," 2020 5th International Conference on Communication and Electronics Systems (ICCES) (pp. 1162-1168). 10.1109/ICCES48766.2020.09137850.

[7] Qu, X., Wei, T., Peng, C., and Du, P. (2018). A fast face recognition system based on deep learning," In 2018 11th International Symposium on Computational Intelligence and Design (ISCID) (vol. 1, pp. 289-292). doi: 10.1109/ISCID.2018.00072.

[8] Sharma, S., Kalyanam, A., and Shaik, S. (2019). Technical analysis of CNN-based face recognition system—a study, In Computational Intelligence in Data Mining (vol. 711). Springer Singapore. doi: 10.1007/978-981-10-8055-5_49.

[9] Yang, S., Luo, P., Loy, C. C., and Tang, X. (2015). From facial parts responses to face detection: A deep learning approach," In Proceedings of the IEEE International Conference on Computer Vision (vol. 2015 Inter, no. 3, pp. 3676–3684). doi: 10.1109/ICCV.2015.419.

[10] Yosinski, J., Clune, J., Bengio, Y., and Lipson, H. (2014). How transferable are features in deep neural networks? Advances in Neural Information Processing Systems, 4(January), 3320-3328.

[11] Raja Sekaran, S. A. P., Poo Lee, C., and Lim, K. M. (2021). Facial emotion recognition using transfer learning of AlexNet," 2021 9th International Conference on Information and Communication Technology (ICoICT) (pp. 170-174). doi: 10.1109/ICoICT52021.2021.9527512.

[12] Mnih, V. and Hinton, G. (2012). Learning to label aerial images from noisy data, In ICML'12: Proceedings of the 29th International Coference on International Conference on Machine Learning (vol. 1, pp. 567-574).

[13] Alzubaidi, L. et al. (2021). Review of deep learning: concepts, CNN architectures, challenges, applications, future directions, vol. 8, no. 1. Springer International Publishing. doi: 10.1186/s40537-021-00444-8.

[14] Gliner, J. A., Morgan, G. A., Leech, N. L., Gliner, J. A., and Morgan, G. A. (2021). Measurement reliability and validity," Research Methods in Psychology, 1st Edition, 319–338, Psychology Press, Taylor and Francis. doi: 10.4324/9781410605337-29.

[15] Li, H., Wang, J., Tang, M., and Li, X. (2017). Polarization-dependent effects of an Airy beam due to the spin-orbit coupling, Journal of the Optical Society of America. A, Optics, Image Science, and VISION, 34(7), 1114-1118. doi: 10.1002/ecs2.1832.

[16] Arora, S., Bhaskara, A., Ge, R., and Ma, T. (2014). Provable bounds for learning some deep representations. In Proceedings of the 31st International Conference on Machine Learning (vol. 1, pp. 883–891).

Employee Attendance System based on Internet Protocol Address using Internet Addressing

Mehul Batra, Kanik Makhija, Swayam Mahindroo, Ashish Kumar

Vivekananda Institute of Professional Studies, Technical Campus, New Delhi

E-mail: mehulbtr@gmail.com, kanikmakhija9@gmail.com, swyammahindroo22@gmail.com, ashishkumar@vips.edu

Abstract

The digital attendance management system is a digital tool that automates the process of tracking employees' attendance. In this paper, we have proposed an algorithm for the attendance system which utilises the employee's IP address and MAC address to keep a record of their entry and exit time. The system works by capturing the employee's IP address when they connect to the company's network and then associates it with their MAC address, which is a unique identifier for their device. The attendance system uses this information to mark the employee present or absent for that day. The system is beneficial for organisations as it eliminates the need for manual attendance tracking, saves time, and reduces the chances of errors. It also provides real-time attendance data, which can help organisations monitor employee attendance patterns and identify any irregularities.

Keywords: Attendance system, digital tool, internet addressing, IP address

1. Introduction

Traditionally, attendance management systems were primarily paper-based, where employees had to sign their names on an attendance register or sheet. Later, with the advent of technology, organisations started adopting electronic methods of attendance management systems to improve accuracy, reliability, and efficiency [1,4,12]. A novel algorithm has been proposed to mark the attendance of the employee in the stipulated amount of time. The concept of tracking employee attendance in a networked environment using Internet Protocol (IP) and Media Access Control (MAC) addresses is a new one.

The main objective of this project is to eliminate the manual process of tracking the attendance of employees while it is a simple and inexpensive solution, it is prone to error and can be time-consuming. The problem that motivated the creation of our attendance management system was likely the need for a more efficient and reliable way of keeping track of attendance in an organisation. Traditional methods such as manual sign-in sheets or even biometric systems like fingerprint scanners and face recognition can be time-consuming, expensive, and potentially unreliable. IP Address-based attendance system is highly automated and less time-consuming which helps save organisations a significant amount of time and resources.

IP and MAC addresses are used to specifically identify devices on a network and, consequently, the users of those devices. This study looks at how such a system might offer real-time attendance data, which would be especially helpful for companies with large staff and eliminate the need for human attendance tracking. The automated attendance system is highly accurate too as it is based upon a unique device identifier which helps in reducing fraud and makes sure that the records are accurate. Scalability is another great factor, which means it can be used to track the attendance of any number of employees which provides real-time data on employees' attendance and this data can be used to identify the patterns of employee's presence. There are obstacles, too, such as proxy attendance, in which one individual uses another's device or login information, making it challenging to determine who the real user is. To solve this problem, additional security measures might be required, including multi-factor authentication or biometric information.

2. Different Types of Attendance Systems

2.1. Biometric Systems

Fingerprint or iris scanning-based algorithms are popular for keeping track of attendance. However, huge organisations can find them to be too expensive. By recording and matching facial photos to pre-registered data, facial recognition technology can improve identity verification [2,10,11,14].

Chapter 24 DOI: 10.1201/9781003570349

Biometric attendance systems are costly and inaccurate, unsuitable for large organisations.

2.2. QR Code-Based Attendance Management

QR Code-Based Attendance Management System [14, 15] is Convenient, but prone to mistakes if staff members combine their devices, also the staff member can share the QR code with other staff members. It leads to Fraud in marking attendance and inaccurate results. One way to increase accuracy is to combine QR code scanning with other authentication techniques, like unique IDs or passwords.

QR code-based attendance systems are not reliable, as staff members can share QR codes.

2.3. Bluetooth-Based Attendance Management Systems

The way Bluetooth-based attendance management systems operate is by identifying staff devices' Bluetooth signals when they are within a specific range [9,16,17]. These technologies can be used to create attendance reports as well as track staff attendance in real time. However, Bluetooth-based attendance management systems are not reliable for large organisations due to their limited range; this means that the attendance system may not work well for the employees present on a large campus. Also, Bluetooth signals can be affected by interference from other devices which can lead to inaccurate attendance records.

Bluetooth-based attendance management systems are not suitable for large organisations due to their limited range and accuracy

2.4. NFC and RFID based Attendance Systems

NFC (Near Field Communication) [3,6] and RFID (Radio Frequency Identification) [5,13] Systems work by tapping physical cards against a reader device. It is a convenient and quick way to mark the attendance of a person. However, both NFC cards and RFID have several limitations such as high cost and security issues as the cards can be cloned or counterfeited which could provide access to unauthorised individuals.

NFC and RFID-based attendance management systems are expensive, not secure, and raise privacy concerns.

2.5. Face Recognition Attendance Management Systems

Using cameras, these systems record employee faces, which are later cross-referenced against an existing database [7,8]. Although they are often accurate, they could have problems when there is little light or when workers are wearing masks. The use of deep learning and machine learning technology is suggested as a remedy to automate attendance tracking and boost accuracy.

Face Recognition Attendance Management Systems are expensive.

3. Data and Initial Setup for the Attendance System

3.1. Analysis and Scalability of the Algorithm

We have analyzed data from a diverse range of organisations that have digital attendance management systems. We have taken into account several measures that could have an impact on the effectiveness of attendance tracking to make sure that our findings are not only the result of the deployment of IP-based systems. Employee strength, industry type, and organisation structure are the key elements of the attendance system, which are described below.

Employee Strength: The effectiveness of attendance tracking can be impacted by the size of an organisation's workforce. The implementation of new attendance management systems may provide various issues for larger organisations compared to smaller ones.

Industry Type: Distinct industries could have different needs and limitations when it comes to keeping track of attendance. A manufacturing company and a service-based organisation, for instance, can have different needs.

Organisational Structure: Different organisational structures, such as those with several locations, remote work, or hierarchies, may have an impact on the effectiveness of the management of attendance data.

3.2. Registration of the Physical Address of the Devices

The research technique employed in this suggested system encompasses the utilisation of IP and MAC

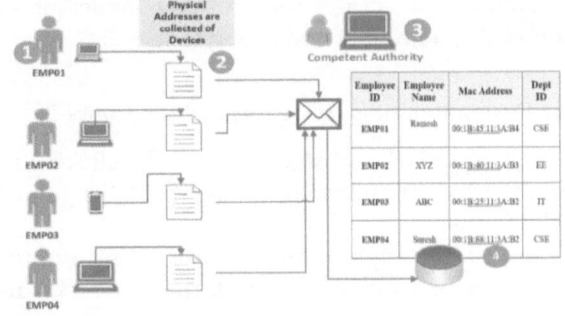

Figure 24.1: Registration of the physical address of the digital devices

addresses for monitoring employee presence. It substitutes established presence monitoring methods such as sign-in sheets and biometric systems. The suggested system is devised to be effective, budget-friendly, and privacy-aware. By creating a secure database to hold employee data, including IP and MAC addresses. The relationship scheme of the database:

R(Employee ID, Employee Name, Physical Address of the device). Figure 1 illustrates the registration phase of the attendance system algorithm, Four Points (1-4) are depicted in Figure 1 which are as follows:

3.2.1. Description of the Drawing/ Figure 1

(1) **The collecting employee's information:** This could include personal details such as name, date of birth, email address, phone number, etc.
(2) **The registration of the Physical Address of the Devices:** This step involves the registration process of the physical address of the device(s).
(3) **Mapping of Physical address with employee's Individual identity:** Once the device(s) have been registered, member information can be stored in the database and linked to the corresponding physical address(es).
(4) **Information is stored in the database:** Once all of the above steps have been completed, all of the member information and corresponding physical addresses will be stored in a database

4. Proposed Methodology

By creating a secure database to hold employee data, including IP and MAC addresses, the IP-based attendance management system can be implemented. Workers will register the addresses of their devices. The device records IP addresses when it is linked to the corporate network and verifies them against the public IP address of the router. The information is verified through an authentication layer and cross-referenced with office intranet records. This system solves the drawbacks of manual and costly biometric approaches by providing an efficient and secure solution for attendance monitoring.

Figure 2 illustrates the working model of the proposed system of attendance system

4.1. Description of (1-10 Sequences) the Figure 2

In (1), Users mark themselves present by clicking the button in the input form by giving the POST request in the forms.

In (2), The application interface gives an option to the user or employee to mark their presence and information.

In (3), The information is verified with the organisation Internet and the database at the back

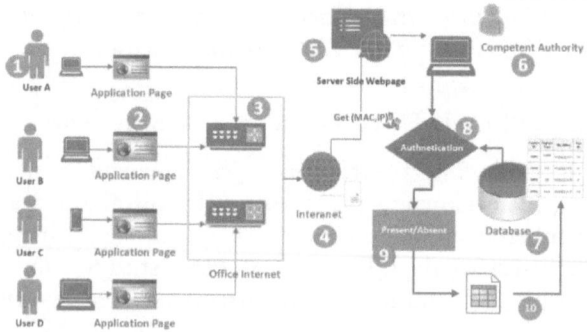

Figure 24.2: Working model of the present proposed system of attendance system

end. The database stores all the relevant information of employees. The database is designed before enabling the system to mark attendance. Each employee is asked to provide the physical address of their devices and the physical address of the devices are registered in the database.

In (4), The IP Address of the device is validated with the public IP address associated with the Router and then the information is verified with the office Intranet.

In (5), The information is then sent on the server side of the page with the help of SQLite3 Database

In (6), The information is then received by the competent authority in the database.

In (7), The information gets stored in the Database which is created including the MAC address, IP Address and corresponding Information related to that Mac address.

In (8), Authentication will be done and validation will be done to check whether the data is authorised and matches the existing database in the system by checking the matching IP Address as well as the Mac Address of the Device. Detailed procedure is described in Table 1.

In (9), If all the information matches with both the databases then the person will be Absent or Present on the basis of the provided information.

In (10), All the employees who are present will be displayed on the page using HTML tables.

5. Empirical Result

5.1. Description of the Drawing/Figure 3

Application Component:

1. The Mac and IP address is collected by the application
2. The POST request is sent to the Server side
3. Server Component:
4. The POST request is received by the server.
5. The Collected data of MAC and IP Address is processed.

Table 1: Pseudo code of the system

1. **Start**
2. Prompt employees to mark their attendance using the Host Application
3. Capture employee details in a packet including their IP address and other relevant information
4. If the packet contains a valid IP address:
   a. Lookup the IP address in the employee database to confirm their identity
   b. If the employee is authorised to mark attendance, update their attendance status in the database and display a confirmation message
   c. If the employee is not authorised or the IP address is invalid, display an error message
5. Repeat steps 2-4 for all employees who mark their attendance using the Host Application
6. **End**

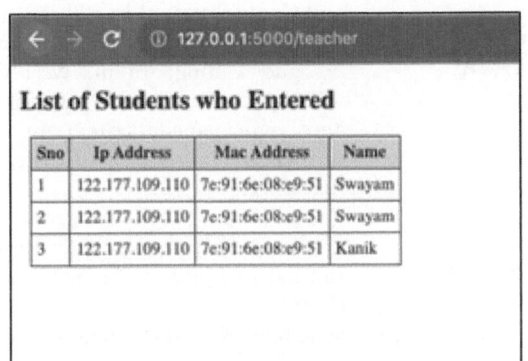

Figure 24.3: List of entries in database and sending the POST request from the app interface

Table 2: Number of Users vs Execution Time for each entry

No. of User (in Thousands)	5	10	15	20	25	30	35
Execution Time(in seconds)	0.010	0.016	0.021	0.021	0.023	0.035	0.041

6. The information is then stored inside the server.

Organisations with multiple job locations and remote work reported lower attendance tracking efficiency than organisations with a single location and traditional work arrangements. This is likely due to the challenges of tracking attendance for remote workers and employees in multiple locations.

It was found that there were fewer attendance recording errors as compared to the other attendance management systems.

6. Observation

After implementing the attendance management system in various institutions, it was found that there is a significant reduction in time spent in recording attendance of each employee. The Table 2 presents the comparison between the number of users with the execution time for each entry.

7. Conclusion

The outdated paper-based attendance management techniques have been replaced by contemporary electronic solutions. Organisations must choose the technology that best matches their unique needs, personnel, and budget, even though each system has advantages and disadvantages. The proposed application's user-friendly layout makes it simple for staff members to record attendance. For real-time tracking, validated data is kept in the database and linked to employee records. This system solves the drawbacks of manual and costly biometric approaches by providing an efficient, affordable, and secure solution for attendance monitoring.

The IP and MAC address-based attendance system proposed in this work offers an innovative approach, however, it encounters challenges including proxy attendance and device sharing. By utilising biometrics or multi-factor authentication, the system's dependability can be raised. As technology advances, attendance management solutions should become more precise, effective, and economical. Additionally, the system can be integrated with other HR tools, such as payroll software, to streamline the process further. In conclusion, the IP-based attendance management system provides an efficient and reliable way of tracking employee attendance using their IP and MAC addresses. It has numerous benefits, including saving time and reducing errors.

References

[1] Abdalkarim, B. A. A. and Akgün, D. A. *Literature Review on Smart Attendance Systems.*

[2] Adamu, A. (2019). Attendance management system using fingerprint and iris biometric. FUDMA Journal of Sciences, 3(4), 427-433.

[3] Ahiara, W. C. and Okey, D. O. (2022). Near field communication internet of things (NFC-IoT) based university examination monitoring system. Journal of Energy Technology and Environment, 4(2).

[4] Budi, S., Karnalim, O., Handoyo, E. D., Santoso, S., Toba, H., Nguyen, H., and Malhotra, V. (2018). IBAtS-Image based attendance system: A low cost solution to record student attendance in a classroom. In 2018 IEEE International Symposium on Multimedia (ISM),

[5] Chavan, M., Gosain, A., Sushil, R., and Samadhiya, Y. (2021). Attendance System Based on Radio Frequency Identification. International Research Journal of Innovations in Engineering and Technology, 5(7), 52.

[6] Daramola, C. Y., Folorunsho, O., Ayogu, B. A., and Adewole, L. (2019). Near field communication (NFC) based lecture attendance management system on android mobile platform. In International Science Conference, Nigeria.

[7] Kakarla, S., Gangula, P., Rahul, M. S., Singh, C. S. C., and Sarma, T. H. (2020). Smart attendance management system based on face recognition using CNN. In 2020 IEEE-HYDCON,

[8] Kortli, Y., Jridi, M., Al Falou, A., and Atri, M. (2020). Face recognition systems: A survey. Sensors, 20(2), 342.

[8] Mademikhanov, Y., Otynshin, A., Shumenov, R., and Rizvi, M. (2021). Automated attendance-checking system using Bluetooth. In 2021 IEEE International Conference on Smart Information Systems and Technologies (SIST).

[10] Mittal, Y., Varshney, A., Aggarwal, P., Matani, K., and Mittal, V. K. (2015). Fingerprint biometric based access control and classroom attendance management system. In 2015 Annual IEEE India Conference (INDICON).

[11] Okokpujie, K. O., Noma-Osaghae, E., Okesola, O. J., John, S. N., and Robert, O. (2017). Design and implementation of a student attendance system using iris biometric recognition. In 2017 International Conference on Computational Science and Computational Intelligence (CSCI).

[12] Oo, S. B., Oo, N. H. M., Chainan, S., Thongniam, A., and Chongdarakul, W. (2018). Cloud-based web application with NFC for employee attendance management system. In 2018 International Conference on Digital Arts, Media and Technology (ICDAMT).

[13] Qureshi, M. (2020). The proposed implementation of RFID based attendance system. International Journal of Software Engineering and Applications (IJSEA), 11(3).

[14] Seifedine, K., and Mohamad, S. (2010). Wireless attendance management system based on iris recognition. Scientific Research and Essays, 5(12), 1428-1435.

[15] Sengupta, I., Jain, N., Shah, S., Jain, H., & Chandrani, A. (2020). Scandence: QR Code based attendance management system. International Research Journal of Engineering and Technology (IRJET), 7(4), 3823-3833.

[16] Vinod, V. M., Murugesan, G., Mekala, V., Thokaiandal, S., Vishnudevi, M., & Siddharth, S. (2021). A low-cost portable smart card based attendance system. In IOP Conference Series: Materials Science and Engineering,

[17] Yuceilyas, M., and Samet, R. (2022). Bluetooth based mobile automatic class attendance management system. 9roxph, 435.

Approach for Tamil Speech Recognition Using HMM

L. Praveen, Puneet Kumar

Department of Computer Science, Lovely Professional University, Phagwara, India

Abstract

Constructing a continuous speech recognition system for Indian languages such as Tamil presents a formidable challenge due to the language's distinctive features, including the presence of both the types of vowels, the lack of aspirated stops, aspirated consonants, and the presence of multiple cases of allophones. Furthermore, the patterns in stress & accent vary among different regions of spoken Tamil, although they are typically disregarded in formal written Tamil. Majorly there are three primary approaches for continuous speech recognition (CSRs) based upon SWU: syllable, phoneme and word. More of the Indian native languages like Tamil follows a native syllabic structure. The phone phones of sentences and words in Tamil adheres to a set of well-defined linguistic rules. While previous efforts have focused on building continuous speech recognition systems for Tamil with limited vocabularies and specific tasks, the development of normal and high vocabulary CSRs for native languages remains relatively uncharted. This research paper aims to address this gap by proposing the creation of Hidden Markov Model (HMM)-based acoustic models for words and triphones in Tamil. The main wish of the project is for establishes both a good vocabulary word based continuous speech recogniser and a normal vocabulary triphone-based SR the native language. During the experiment, a CIs (conditions-independent) audile model has been developed for more than 300 distinct words, and a CDs (conditions-dependent) audile model has been constructed to unique words of 1700 using sequence of phonemes of three. Additionally, an intonation wordbook encompassing phones of base 44 also an analytical language model base on trigram has created an entire parts of the linguistic resources. This known exhibit remarkable word correctness when tested with sentences spoken by both trained and new speakers.

Keywords: HMM, ASR, SR, CSR.

1. Introduction

ASRs is concerned to the automated transition of audile indication from spoken language into written text. Despite extensive research and development over the years, achieving high correctness of ASRs leftovers a persistent dispute for experimenter. Several well-known factors significantly influence this accuracy. Key factors encompass deviation of circumstances, differing prophet, and also ecological roar. Consequently, a field of ASR presents a multitude of open issues, spanning aspects such as low and high rate of vocabulary recognition, remote and nonstop oral presentation, prophet-reliable and prophet-unaided systems, as well as robustness in varying environmental conditions.

ASR for European vocabularies, as English, and also Asiatic vocabularies, such as Mandarin, has reached a mature stage. However, studies in Indian vocabularies are fixed in the basic phase, hindered by a significant obstacle – the scarcity of essential resources. Construed oral expression outputs for testing and training audile mock-ups are notably deficient. Nevertheless, burgeoning curiosity in ASRs of Indian languages, particularly Thamizh [1]. Some research has targeted low and constrained vocabulary tasks, while others have focused on SRs for spoken digits [2]. Secluded phrase identification of Thamizh has also been explored using Artificial Neural Networks (ANN), and resource scarcity has been addressed through cross-language transfer and adaptation techniques [3].

Furthermore, literature concerning LVCSRs in native languages of India exists such as Thamizh, etc. The efforts primarily concentrate on harnessing the syllabled make-up of Thamizh and other Indic vocabularies. The fundamental appearances to SRs are based on the preference of substitute-phrase entity: horn base, syllable base and phrase base identification. The criteria of selecting these substitute-phrase units are extensively discussed of literature survey area [4].

2. Literature Review

Vocal recognition necessitates the division of waveforms of the speech into the acoustics units of its fundamental. Words stand as preferable and inherent

Chapter 25 DOI: 10.1201/9781003570349

speech unit recognition, given that the ultimate goal is the recognition of words. Additionally, word units possess well-defined acoustic representations. Acoustic variability is primarily concentrated at the beginnings and endings of words, namely at word boundaries. Nevertheless, employing letters as the unit of speech in the 'LSVCSRs' system. The Large Vocabulary Continuous Speech Recognition introduces a series of notable challenges. Each word necessitates individual training, without the possibility of parameter sharing among words, compelling the need for an extensive training dataset to ensure all vocabulary words are adequately trained. A secondary issue arises in terms of memory requirements, which grow linearly with the word count [5]. Consequently, word models prove impractical for LVCSR systems, despite their success in limited vocabulary Automatic Speech Recognition (ASR) applications [6].

Sub-word units, specifically smaller units, are the logical choice to facilitate parameter sharing and conserve computational resources. The sub¬-word unit Phones is the most prevalent which is found in the English Language. Given the relatively limited number of phones in English and other languages, it is feasible to train phone models effectively with a reasonably sized training corpus. However, challenges persist in phone models as well. It is widely recognised that the same phone may manifest differently in various words due to the constraints of articulatory movements, which prevent instantaneous transitions. Consequently, a phone's realisation is significantly influenced by neighboring phones, illustrating the high contextual dependence of phones [7]. Few phones will aspirate while it's situated at the start of the word, while that same particular phones will exhibit the characteristics of the non-aspirated when it is found at the end of the word. Consequently, the basic phonetics' units of auditory variability due to the context lead to substantial and it's not comprehensively understood across multiple language. Research has clearly pictured shown that (DTW) Dynamic Time Warping is based on word models significantly outperforms Hidden Markov Models (HMM) based on phones [8]. This dichotomy between word models, which fail to generalise, and phone models, which tend to over-generalise, can be mitigated through the incorporation of context, often referred to as "phone-in-context." This context may encompass either left-context phones, correct-context phones, or the both. Category three known as "triphones," encompasses the context of both right and left phones. When the two phones share a similar peculiarity but differ in their contexts of left and right, they are usually recognised as distinct triphones. This model of Triphone serves

as a sub-word potent model by effectively accounting for the both phonetic contexts of left and right. They proved immense success in the auditory modeling within systems of LVCSR, achieving a substantial reduction in Word Error Rate (WER) compared to word and phone models [9]. However, triphones introduce challenges as well. Each language entails a considerable number of triphones in the training dataset, resulting in memory wastage, as models are generated for individual triphones, even if they are observed only once during training [10, 11]. Alternative approaches in the existing literature have sought to employ larger sub-word units in speech recognition to account for the co-articulation effects in speech production. Among these approaches, the most prominent unit is the syllable, which represents a larger unit compared to individual phones, as this encircles phones with two or more clusters. are essential in addressing the pronounced contextual influences in speech. A syllable consists of totally three components: the nucleus, the coda (or rhyme), and the onset. Nucleus is located at the core of the syllable, lacks significant contingent dependencies, the coda and onset remain subject to the certain contextual contingent effects [12]

Thamizh language and other various languages of India exhibit phoneme characteristics which are characterised by a wealth of vowel and consonant realisations, with pronunciations predominantly structured around syllables. Considering these characteristics, In India there are several prominent researches have centered the work which focusses on the vocable as fundamental phrase for speech recognition. All the research efforts aim to overcome the remark vocal corpora's bounded challenge. In works like [13] and [14], an outlook is submitted for the automated segmentation and annotation of unbroken vocal, which not includes the reliance on manual way for remark vocal corpora. The approach involves segmenting the unbroken vocal signals into the syllable-like unit using the magnitude spectrum derived from short-term energy. Accordingly, similar alike phonetic units were grouped through the technique of clustering in the unrestricted way of additive, and the phonetic units are manually labeled. The models can be constructed for these phonetic unit clusters and skilled for transcribing continuous speech. A similar approach is adopted in [15], The accompanying text and the unbroken vocal signal both are incidental into syllable-like units, and accordingly it creates models.

In whatever way, a significant challenge associated with the use of syllables is the sheer number of syllables involved. For languages such as English and Tamil, there are approximately 20,000 different syllables [9], and without the ability to share parameters, the trainability of these systems is compromised. Consequently, these systems tend to exhibit an accuracy rate of approximately 53% [16].

3. Formulation

At its core, the challenge in SRs are formulated in kind: presented an audible studies ob = ob1, ob2,... obn, the objective are for determine the phrase continuance ph = ph1ph2...phm which boost the later in time likelihood P(ph|ob), as explicit in equation (1) by Bayes Theorem.

$$\arg \max [P(ph|ob)] = \arg \max [P(ph) * P(ob|ph)] \quad (1)$$

Where P(ph) represents the probability of the uttered phrase continuance ph, and P(ob|ph) represents the likelihood distribution conditioned on a word ph, which is named as class-conditioned likelihood allocation. P(ob) denote an overall likelihood of a studies ob. Notably, maximising equation (1) with ob considered constant necessita tes the maximisation of the numerator alone to determine the word ph.

$$ph = \arg \max [P(ph) * P(ob|ph)] \quad (2)$$

In Equation (2), in first phrase, P(ph), is figured using the vocabulary method, which quantifies a likelihood connected to the assumed continuance phrases. This vocabulary method integrates lexemic and linguistic limitations relevant to the vocabularies of the acceptance job part, typically adopting forms like formal parsers, syntax analyzers, unigram method, or mixed methods [17].

In equation (2), P(ob|ph), are calculated by the audile method, estimating the likelihood of a continuance of the audile studies accustomed of a phrase ph. Access to the class-accustomed likelihood P(ob|ph) of a audile method is essential for computing a later in time likelihood P(ph|ob). HMMs (Hidden-Markov-model) had surfaced in higher prevalent structure for audile models, as they normalise temporal variations in speech signals and statistically characterise speech, aiding in parameterising the class-conditioned likelihoods. The audile method serves as a foundational fundament, encapsulating different speech limits optimally.

4. Language Alphabet

Speech recognition research is particularly concerned with various phonological features specific to Tamil. In Tamil, vowels can be classified into three categories: shortest, longest (contains five from every type), and diphthongs includes two. Where the consonants, on the other hand, are classified into three categories based on their place of articulation: hard, soft (also known as nasal), and medium, resulting in a total consonant of 18. The combination of both consonants and vowel leads the creation of combined properties of 216. These combined properties are formed by affixing the vowel markers which are dependent to consonant which is either the one side or the both side. Additionally, classical Tamil includes a special character, aytham (ஃ), which is rarely found in modern Tamil.

In standard Tamil, there are a total of 247 letters. Furthermore, modern Tamil employs six characters from the Grantha script to depict the accent which is domestic to language of Thamizh, specifically there are words present in this script which are acquired from the language Sanskrit and others. While Thamizh shares retroflex consonants with some other similar Dravidian languages, this too incorporates a unique flow of sound, the Thamizh equivalent of "ZH". There has been considerable research which dedicated to the enunciation of liquid flow of the consonants in the language Thamizh [18].

5. Phone set of The Language

From the context of acknowledgement based on words, the letters itself is regarded as a sound unit, during the approach based on phone, a well elucidated set of phones will be a prerequisite. In their work, the creators/writers introduced a group of phone tailored for the language of Thamizh, comprising forty-four phones alongside other auditory events. A comprehensive list of the complete phone set, including the peculiar details of the co-articulation.

6. Development of Language model and Dictionary

Development of the language model and dictionary in a triphone-based model adheres to a systematic training strategy, encompassing the following stages:

Flat-start monophone training: This phase involves the creation of monophonic or CIs (Context-Independent) grain methods using matured values, followed by their re-evaluation with source models. The process are named as the level format for CIs methods limits, and it encompasses Welch-Baum practicing for monodic. The training adjusts this silent method along with re-estimates Gaussian-single monodic through a standard-Viterbi adjustment procedure.

Triphone creation: This step entails the generation of phone-of-three model derived from monodic model along with the initiation of first phone-of-three practicing. This stage results in the creation of context-dependent CDs unbound method documents along with the level format for methods clippings.

Practicing CDs unbound methods: The Welch-Baum custom is employed repetitious in the phase, spanning six to ten sprints.

constructing sureness branches along with limits partition: In this step, the set of same domains, referred to as senones or tied states, are created and subsequently trained.

Mixture generation: The process involves the subdivision of Gaussian- single allocation to the mixed allocation through the repetitive half-to-half gathering method. Phone-of-three methods are then re-estimated with these mixture distributions.

For the word-based model, the training procedure is abbreviated, encompassing only the first two steps. Once a practicing are completed, Sphinx-Train creates that limitation clippings for Hidden Markov Models (HMM), which include a likelihood allocations along with alternation form for every HMMs methods.

7. Result

Twain phrase and triphone-based methods are employed within Sphinx-4 is build using a cutting-edge HMMs planted speech-recognition regularity. This system has been under open-source development since February 2002 and serves as the successor to both Sphinx-3 and Sphinx-2. The collaborative effort behind its design involves University of Carnegie-Mellon, Laboratory Sun-Microsystems along with Laboratories of Mitsubishi-Electric-Research in United States of America. Notably, Sphinx-4 is implemented using the Java programming language, affording it portability across an expanding array of computational platforms [19].

8. Conclusion and Discussion

Evidently, both models exhibit remarkably high accuracy when confronted with trained sentences spoken by trained voices. In situations characterised by a restricted vocabulary, frequent sentence repetition, and a limited pool of speakers, the word-based recogniser utilising trained voices on familiar sentences proves exceptionally well-suited. Conversely, in scenarios with an extensive vocabulary and a constrained speaker set, the triphone-based model emerges as the more appropriate choice. The Word Error Rate (WER) predominantly reveals deletion mistakes of the phrase-base methods along with commutation mistakes of phone-of-three based methods. It's worth noting that the recognition speed is slower in the word-based model compared to the triphone-based model. In the context of word models, Context-Independent (CI) word modeling achieves reasonably accurate results, particularly for a vocabulary size of 350.

Reference

[1] Khan, N. and Yegnanarayana, B. (2001). Development of speech recognition system for tamil for small restricted task. In Proceedings of National Conference on Communication, India.

[2] Plauche, M., Udhyakummar, N., Wooters, C., Pal, J., and Ramachadran, D. (2006). Speech recognition for illiterate access to information and technology. In Proceedings of First International Conference on ICT and Development.

[3] Saraswathi, S. and Geetha, T. V. (2004). Implementation of tamil speech recognition system using neural networks. Lecture Notes in Computer Science, 3285.

[4] Kumar, C. S. and Wei, F. S. (2003). A bilingual speech recognition system for English and Tamil. In Proceedings of Joint Conference of the Fourth International Conference on Information, Communications and Signal Processing, 2003 and the Fourth Pacific Rim Conference on Multimedia (vol. 3, pp. 1641-1644).

[5] Lippmann, R. P., Martin, E. A., and Paul, D. P. (1987). Multi-style training for robust isolated-word speech recognition, In ICASSP '87. IEEE International Conference on Acoustics, Speech, and Signal Processing, April 1987 (pp. 705-708).

[6] Rabiner, L. R., Wilpon, J. G., and Soong, F. K. (1988). High performance connected digit recognition using hidden Markov models, In IEEE Transactions on Acoustics, Speech, and Signal Processing. April 1988.

[7] Bahl, L. R., Brown, P. F., De Souza, P. V., and Mercer, R. L. (1988). Acoustic Markov models used in the Tangora speech recognition system, In ICASSP-88., International Conference on Acoustics, Speech, and Signal Processing, Apr. 1988.

[8] Paul, D. B. and Martin, E. A. (1988). Speaker stressresistant continuous speech recognition. In ICASSP-88., International Conference on Acoustics, Speech, and Signal Processing, Apr. 1988.

[9] Lee, K.-F. (1990). Context Dependent Phonetic Markov Models for Speaker Independent Continuous Speech Recognition, In IEEE Transactions on, Acoustics, Speech and Signal Processing (Volume 38, No. 4, 1990, pp 599-609).

[10] Bahl, L. R., Bakis, R.. Cohen, P. S., Cole, A. G., Jelinek, F.. Lewis, B. L., and Mercer, R. L. (1980). Further results on the recognition of a continuously read natural corpus. In ICASSP '80. IEEE International Conference on Acoustics, Speech, and Signal Processing, Apr. 1980.

[11] Schwartz, R. M., Chow, Y. L., Roucos, S., Krasner, M. and Makhoul, J. (1984). Improved hidden Markov modeling phonemes for continuous speech recognition. ICASSP '84. IEEE International Conference on Acoustics, Speech, and Signal Processing, Apr. 1984.

[12] Ganapathiraju, A., Hamaker, J., Picone, J., Ordowski, M., and Doddington, G. R. (2001).

Syllable based large vocabulary continuous speech recognition. In IEEE Transactions on Speech and Audio Processing (Volume 9, No. 4, pp 358-366).

[13] Nagarajan, T., Murthy, H. A., and Hegde, R. M. (2003). Segmentation speech into syllablelike units. In EUROSPEECH-2003 (pp. 2893-2896).

[14] Sarada, G. L., Hemalatha, N., Nagarajan, T., Murthy, H. A. (2004), Automatic transcription of continuous speech using unsupervised and incremental training. In ICSLP-2004, October, Korea.

[15] Nagarajan, T., Kamakshi Prasad, V., and Murthy, H. A. (2001). The minimum phase signal derived from the magnitude spectrum and its applications to speech segmentation, In Sixth Biennial Conference of Signal Processing and Communications, July 2001.

[16] Lakshmi. A. and Murthy, H. A. (2004). A Syllable based continuous speech recognizer for Tamil, In SPCOM, December 2004.

[17] Jelinek, F. (1997). *Statistical Methods for Speech Recognition*, MIT Press, (ISBN 0262100665).

[18] Narayanan, S., Byrd, D., and Kaun, A., Geometry, Kinematics and acoustics of Tamil liquid consonants. Journal of Acoustical Society of America, 106(4), Pt. 1, 1993-2007.

[19] Lamere, P., Kwok, P., Walker, W., Gouvea, E., Singh, R., Raj, B., and Wolf, P. (2003). Design of the CMU Sphinx-4 Decoder in EUROSPEECH 2003.

Improving Legal Text Classification through Data Augmentation Using Deep Learning Models

Akshay Mohite, Reshma Sheik, S. Jaya Nirmala

Department of Computer Science and Engineering, National Institute of Technology Trichy, Tamil Nadu, India
E-mail: akmohite1506@gmail.com; rezmasheik@gmail.com; sjaya@nitt.edu

Abstract

Legal industry is one of the most complex, relying heavily on in-domain knowledge and demanding significant time and effort from legal professionals to address its multifaceted tasks. The reliance on Natural Language Processing (NLP) technologies is integral to its applications, considering that a significant portion of legal information, such as judgment papers, contracts, and legal opinions, is presented in text form. The integration of deep neural networks into NLP applications is transforming various advancements in handling legal text data. However, the scarcity of annotated data, particularly in the legal domain, presents a significant challenge in training these models. This study aims to investigate the effect of employing data augmentation techniques, specifically targeting binary legal text classification tasks, utilizing diverse deep learning models such as Transformer-based, Convolutional Neural Networks (CNN) and Recurrent Neural Networks (RNN). Our study, conducted across five distinct legal datasets, reveals significant performance enhancements in these models when used with augmented data. Our research highlights the importance of using data augmentation to tackle the difficulties arising from the limited availability of annotated data in the legal domain.

Keywords: Convolutional neural network, data augmentation, natural language processing, recurrent neural network, transformers

1. Introduction

Over recent years, researchers in both law and computer science have been exploring techniques for automating the processing of legal data, aiming to support legal professionals. With the majority of legal documents existing in text format, the legal domain has seen the introduction of various Natural Language Processing (NLP) technologies, aimed at enhancing the efficiency of legal tasks. The legal industry may be one of the most complex, relying heavily on in-domain knowledge and manual processes. Retrieving and understanding legal documents is time-consuming for legal professionals [1]. The evolution of deep learning models in the field of NLP has attracted researchers to explore its usage across various domains. A qualified NLP system must reduce the time required for these jobs and aid the legal system. Although text classification methods have been extensively researched and employed for various purposes, this is an underexplored area in the legal domain. Despite existing commercial and forensic interests, text classification methods haven't been thoroughly investigated in the law sector [2].

Text classification in the legal domain presents unique challenges such as use of specialised language and terminology. The context sensitivity of legal language, coupled with the length and complexity of legal documents, adds difficulty to the task. Imbalanced data, where certain legal outcomes or topics are more prevalent, can affect model generalisation. Additionally, achieving high accuracy requires legal expertise due to the distinctions in language and implications of terms. Furthermore, the scarcity of labeled data in this domain poses challenges for training robust machine learning models. To gain reasonable performance in text classification tasks, having labeled training datasets is crucial, particularly for deep neural architectures which are prone to overfitting. Data augmentation techniques can extend the range of training examples without explicitly collecting annotated data, which has significantly improved the performance of diverse neural network models in speech and computer vision. In this study, we inspect the application of data augmentation techniques introduced in [3] to our binary legal text classification task.

The paper is structured as follows: The next section addresses the related work, followed by description of the dataset. The section following highlights the methodology, model framework and outlines the experimental details. The next section presents the

Chapter 26 DOI: 10.1201/9781003570349

results of the experiments on the dataset. Finally, the last section concludes the paper with future steps.

2. Related Work

In [4], used Support Vector Machine (SVM) for clause classification into conclusion or premise using ECHR dataset and get F1-score of 74.07% and 68.12%, respectively. A combination of CNN and Bag of Words was used in [5] and achieved an accuracy of 85.01% for Verdict prediction in Indian court cases. In (Chen et al., 2023), researchers perform multi-label legal text classification. They use a Graph Convolutional Network (GCN) and Bi-directional Long Short-Term Memory (Bi-LSTM) to obtain label embeddings, followed by a capsule network. A Bi-LSTM network is used for long-length document classification in [6] which achieved an accuracy of 97.97%. [7], applies a data augmentation method that generated augmentations by back-translation where sentence was translated from English to French and back to English. Another augmentation method called contextual augmentation is used in [8], where a bi-directional Language model predicts words to be substituted according to the context. In (Wei & Zou, 2019), the authors present a basic set of universal data augmentation methods for natural language processing, dubbed Easy Data Augmentation (EDA). For a given sentence, the method applies either one of four operations, which are discussed further. We used this approach for our augmentation task.

3. Dataset

In [9],authors from various legal and computer science domains collaborated to form a legal reasoning benchmark for English language. They have crafted various tasks across 32 data sources for open-source legal benchmarking effort. We have sampled some of these tasks for evaluation. For designing these tasks, they have reformatted and reconstructed the original CUAD dataset [10]. For this study, we have chosen five of these tasks and the corresponding datasets. These contain clauses extracted from legal documents, and the tasks are binary classification tasks as to whether the clauses fall under the given category of Anti-assignment (AA), Audit-rights (AR), Cap-on-liability (CL), License-grant (LG) and Insurance (IR).

4. Methodology

Our research involves the implementation of binary classification tasks utilizing three distinct models, namely: (1) Text Convolutional Neural Network (TextCNN), (2) Bi-LSTM, and (3) BERT models. The robustness of our approach is improved by employing text augmentation on the legal text

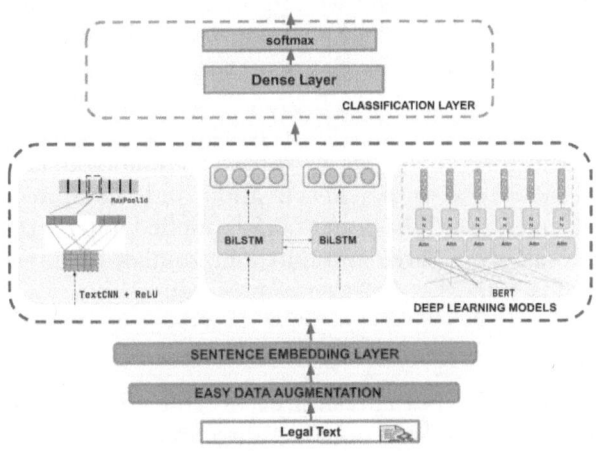

Figure 1: Model architecture

sentences, which are subsequently passed through an embedding layer before being processed by diverse deep learning models tailored for classification. The final classifier layer integrates a feed-forward neural network, employing softmax activation in the output layer. Figure 1, illustrates the framework employed in our study for a comprehensive visual representation of the entire methodology.

4.1. Easy Data Augmentation

EDA (Wei & Zou, 2019) involves the manipulation of a given sentence through one of four key operations, namely: (1) Random Deletion, (2) Random Swap, (3) Random Insertion, and (4) Synonym Replacement. These operations are performed after the removing stop words from the initial sentence. To execute the Synonym Replacement operation, WordNet from the NLTK[11] corpus is utilised. In the context of our study, we performed nine augmentations for each sentence in our training dataset, in addition to retaining the original sentence. Table 1 depicts the dataset overview and highlights the augmented training data size with a ten times larger number of samples as compared to original data.

4.2. Deep Learning Models

For model classification, we have used diverse deep learning models, commencing with CNNs to RNNs and incorporating Transformer models. The specifics of the employed models are addressed in the following sections.

4.2.1 Convolutional Neural Network

CNN is a kind of neural network which is mostly employed for image recognition tasks, but with the little fine-tuning we can also use it for text classification tasks. By treating words as sequences and using filters to recognise the patterns and relationships between the text data we can use it for text classification. By using a filter of size n, we can identify the n-gram relationships in a sentence, therefore using filters of different sizes we can identify different local-patterns or n-grams in a text. Our TextCNN architecture comprises of three convolutional layers, succeeded by a Relu activation, and a final max pooling layer. A dropout of 0.5 is applied, filter sizes used are two, three and four and number of filters used is 100.

4.2.2 Bi-directional LSTM

A type of RNN that can process sequential data in both backward and forward direction is Bi-LSTM. Due to its ability to comprehend data from both direction allows it to better grasp the dependencies in textual data, making it a robust choice for performing sentiment-analysis, named-entity recognition, prediction and classification tasks. Employing text augmentation on the legal text sentences improves the robustness of our approach. These sentences are subsequently passed through an embedding layer before diverse deep learning models tailored for classification process them.

4.2.3 BERT

A pre-trained BERT [12] model can be fine-tuned by using an additional output layer to develop models for diverse tasks such as language inference and question answering, without requiring significant alterations in the architecture tailored for each task. For our task we utilised the pre-trained bert-base-uncased model available in the Hugging face transformers library.

4.3. Experimental Details

Training and testing across all models were executed using Google Collaboratory accessible GPU resources with a GPU RAM capacity of 12.7 GB and a generous disk allocation of 78.2 GB. Uniformly across these models, the training process was conducted utilizing the Adam [13] optimiser in conjunction with Cross Entropy loss as the objective criterion. The training

was performed over the course of three epochs with a consistent learning rate set at 1e-5. The implementation of the BERT model was executed through the Huggingface Transformers library, whereas for the implementation of the CNN and Bi-LSTM models, the PyTorch nn module was used.

5. Results

Table 2 provides a comprehensive summary of the performance evaluation of the deep learning models conducted across five distinct datasets. Evaluation metrics such as F1-score, precision, accuracy, and recall were employed for an exhaustive assessment. Significantly, the introduction of augmentation techniques has distinctly enhanced performance across all datasets, compared to the results obtained without augmented data. Amongst the models utilised, the BERT model emerged as the frontrunner, exhibiting superior F1-score and accuracy metrics consistently across all legal datasets.

6. Conclusion

This research addresses the data augmentation techniques designed for legal text classification that can overcome the limitations imposed by the lack of annotated training data in the legal domain. Legal texts are often highly sensitive to wording and context, and improper modifications may introduce inaccuracies or distort the intended legal meaning. Striking the right balance and ensuring that augmented data maintains legal accuracy is crucial for exploiting the benefits of data augmentation in the legal domain. This study examines the impact of EDA Techniques on legal binary text classification, revealing notable enhancements in model performance. The successful application of data augmentation techniques addresses the challenge posed by the limited size of annotated datasets within the legal domain, primarily attributable to the substantial costs and time investments associated with corpus annotation. However, the effectiveness of data augmentation in the legal domain depends on the choice of augmentation techniques. Thus, this work can

Table 1: Dataset description without and with EDA

Task	Train size		Test size
	Without Augmentation	With Augmentation	
Anti-assignment (AA)	942	9240	236
Audit-rights (AR)	977	9770	245
Cap-on-liability (CL)	1001	10010	250
License-grant (LG)	1121	11210	281
Insurance (IR)	828	8280	208

Table 2: Comparison of the model's performance with and without augmentation

Dataset	Model	Without Augmentation				With Augmentation			
		Pre	Rec	F1	Acc	Pre	Rec	F1	Acc
AA	TextCNN	0.74	0.74	0.74	0.74	0.90	0.89	0.89	0.89
	Bi-LSTM	0.49	0.50	0.33	0.48	0.96	0.96	0.96	0.96
	BERT	0.96	0.96	0.96	0.96	0.97	0.96	0.97	0.97
AR	TextCNN	0.51	0.51	0.50	0.51	0.86	0.86	0.86	0.86
	Bi-LSTM	0.74	0.54	0.40	0.52	0.82	0.73	0.71	0.73
	BERT	0.98	0.98	0.98	0.98	1.0	1.0	1.0	1.0
CL	TextCNN	0.69	0.57	0.48	0.56	0.87	0.87	0.86	0.86
	Bi-LSTM	0.75	0.52	0.37	0.51	0.92	0.92	0.92	0.92
	BERT	0.96	0.96	0.96	0.96	0.96	0.96	0.96	0.96
LG	TextCNN	0.68	0.93	0.60	0.63	0.88	0.87	0.87	0.87
	Bi-LSTM	0.84	0.80	0.79	0.79	0.88	0.86	0.86	0.86
	BERT	0.94	0.94	0.94	0.94	0.95	0.95	0.95	0.95
IR	TextCNN	0.25	0.50	0.33	0.50	0.85	0.85	0.85	0.85
	Bi-LSTM	0.79	0.64	0.59	0.64	0.90	0.88	0.87	0.88
	BERT	0.98	0.98	0.98	0.98	1.0	1.0	1.0	1.0

potentially alter legal research and analysis by offering the new dataset and techniques for data augmentation to the legal NLP community.

References

[1] Zhong, H., Xiao, C., Tu, C., Zhang, T., Liu, Z., and Sun, M. (2020). How does NLP benefit legal system: a summary of legal artificial intelligence. https://doi.org/10.18653/v1/2020.acl-main.466

[2] Sulea, O.-M., Zampieri, M., Malmasi, S., Vela, M., Dinu, L. P., and van Genabith, J. (2017). Exploring the use of text classification in the legal domain. http://arxiv.org/abs/1710.09306

[3] Wei, J., and Zou, K. (2019). EDA: easy data augmentation techniques for boosting performance on text classification tasks. http://github.

[4] Palau, R. M. and Moens, M. F. (2009). Argumentation mining: the detection, classification and structure of arguments in text. In Proceedings of the International Conference on Artificial Intelligence and Law (pp. 98-107). https://doi.org/10.1145/1568234.1568246

[5] Pillai, V. G. and Chandran, L. R. (2020). Verdict prediction for Indian courts using bag of words and convolutional neural network. In 2020 Third International Conference on Smart Systems and Inventive Technology (ICSSIT) (pp. 676-683). https://doi.org/10.1109/ICSSIT48917.2020.9214278

[6] Wan, L., Papageorgiou, G., Seddon, M., and Bernardoni, M. (2019). Long-length Legal Document Classification. http://arxiv.org/abs/1912.06905

[7] Yu, A. W., Dohan, D., Luong, M.-T., Zhao, R., Chen, K., Norouzi, M., and Le, Q. V. (2018). QANet: Combining Local Convolution with Global Self-Attention for Reading Comprehension. http://arxiv.org/abs/1804.09541

[8] Kobayashi, S. (2018). Contextual Augmentation: Data Augmentation by Words with Paradigmatic Relations. https://github.com/pfnet-research/

[9] Guha, N., Nyarko, J., Ho, D. E., Ré, C., Chilton, A., Narayana, A., Chohlas-Wood, A., Peters, A., Waldon, B., Rockmore, D. N., Zambrano, D., Talisman, D., Hoque, E., Surani, F., Fagan, F., Sarfaty, G., Dickinson, G. M., Porat, H., Hegland, J., ... Li, Z. (2023). LegalBench: A Collaboratively Built Benchmark for Measuring Legal Reasoning in Large Language Models. http://arxiv.org/abs/2308.11462

[10] Hendrycks, D., Burns, C., Chen, A., and Ball, S. (2021). CUAD: An Expert-Annotated NLP Dataset for Legal Contract Review. http://arxiv.org/abs/2103.06268

[11] Loper, E. and Bird, S. (2002). NLTK: The Natural Language Toolkit. http://nltk.sf.net/.

[12] Devlin, J., Chang, M.-W., Lee, K., Google, K. T., and Language, A. I. (2019). BERT: Pre-training of Deep Bidirectional Transformers for Language Understanding. https://github.com/tensorflow/tensor2tensor

[13] Loshchilov, I. and Hutter, F. (2019). DECOUPLED WEIGHT DECAY REGULARIZATION. https://github.com/loshchil/AdamW-and-SGDW

A Comparative Analysis Using Ensemble Machine Learning Model for Identification of Mastitis Illness in Cows

Deepika[1], Amit Sharma[2]

[1]Research Scholar at School of Computer Application, Lovely Professional University, Punjab, India
[2]Professor, School of Computer Application, Lovely Professional University, Punjab, India
E-mail: deepy.12nov@gmailcom, amit.25076@lpu.co.in

Abstract

An important step toward preventative and efficient livestock healthcare management has been taken with the creation and deployment of an ensemble machine learning model specialised for mastitis illness identification in cows. Keeping animal welfare and farm productivity high in the face of mastitis, a common and economically significant ailment in dairy cattle, is difficult. Using a variety of methods such as Decision tree and gradient boost, the ensemble machine learning model demonstrated impressive performance. In contrast to the 84.9 percent accuracy achieved by the current Gradient Boost Tree approach, the proposed study achieved a perfect score of 100 percent. The proposed work shows a significant 16% improvement over the existing method.

1. Introduction

Mastitis, an omnipresent and economically impactful infectious condition affecting dairy cattle globally, induces udder inflammation, resulting in diminished milk production, escalated veterinary expenses, and, in severe instances, imperils the animal's survival. Timely identification of mastitis is pivotal for efficacious treatment and preempting its dissemination within a herd. Recent years have witnessed the advent of progressive machine learning techniques showcasing auspicious outcomes across diverse domains, encompassing healthcare and agriculture[1-3]. This study delves into the formulation and execution of an ensemble machine learning model explicitly tailored for the early detection of mastitis in cows. The proposed model harnesses a varied array of machine learning algorithms—such as decision tree, and gradient boost—to amalgamate an ensemble. The amalgamation of the predictive capabilities of multiple algorithms endeavors to heighten accuracy and resilience in the early-stage detection of mastitis. Integral components of this research encompass the accumulation of data from diverse dairy farms, feature engineering aimed at extracting pertinent mastitis indicators, and the application of advanced machine learning methodologies for the classification and prognostication of mastitis likelihood in individual cows. The scope of this study transcends mere model development; it accentuates practical implementation within a farming milieu. Consideration is given to integrating

existing farm management systems or wearable sensor technologies for continual health monitoring, enabling real-time detection and preemptive intervention. The principal objective of this research is to furnish dairy farmers with a cost-effective, precise, and accessible tool for monitoring and regulating mastitis in their herds. Timely detection stands to substantially curtail economic losses, enhance animal well-being, and contribute to sustainable dairy farming methodologies. The discoveries and potential impact of this ensemble machine learning model in mastitis detection possess the potential to revolutionise the approach to managing the health of dairy cattle, underscoring the pivotal role that cutting-edge technology assumes in contemporary agriculture[4, 5].This paper articulates a comprehensive approach, from the development of the model to its tangible application, aiming to make a meaningful contribution to the incessant endeavors aimed at enhancing animal health, farm productivity, and the overarching well-being of dairy cattle[6, 7].With the agricultural sector increasingly embracing technology, the utilisation of machine learning models for disease detection in livestock symbolises a pivotal stride toward more efficient and proactive management of animal healthcare.

2. Literature Review

Fan 2023 et. al originated in an AMS before being removed. Decision tree ensemble models outperformed others in CM detection and prediction. Models using data from the current and nine previous milkings had the highest CM detection sensitivity. However, CM prediction was best when modelling data from the current milking and the seven previous milkings. If the models were used with oversampling techniques, they had a specificity of 95 and 93% for CM detection and prediction, respectively, and the same sensitivity (82%). However, undersampling techniques increased their sensitivity to 95% when the specificity was reduced to 80 to 83%. We present a viable machine learning method for rapid CM identification utilizing AMS asymmetric data. This technique may provide farmers with data to mitigate CM's detrimental impacts [8].

Kumar 2023 et. al The suggested method seeks to quickly and accurately diagnose clinical mastitis, reducing expenses and improving therapy. Deep learning and machine learning assist the system examine edge device data and forecast mastitis. Farmers and veterinarians may use this data to find affected animals and control the disease before it spreads. Different Deep Learning and Machine Learning algorithms were used to classify in the suggested system and compare their accuracy and efficiency. Results were reliably predicted by models chosen for subsequent usage. For Deep Learning and Machine Learning, the 99%-accurate InceptionV3 and Random Forest algorithms were chosen. The former was 99.34% successful. Based on a literature review of classification-based animal mastitis detection investigations, this study's models increase diagnosis accuracy. This study gives a real-time method for identifying animal mastitis, which might considerably cut Bangladesh's and other developing nations' dairy industry costs. This study examines this system's potential. The system's ability to swiftly and precisely diagnose issues protects livestock health and productivity. This technology could transform the livestock industry and the economy by enhancing livestock animal health and output and reducing mastitis costs [9].

Zhang 2023 et. al It is used to diagnose cow mastitis since it is non-invasive and accurate. Mastitis in dairy products can be detected by accurate temperature measurement, however surrounding regions and obstructions may impair this procedure. We created CLE-UNet (Centroid Loss Ellipticization UNet) semantic segmentation to overcome these issues. The algorithm has three basic pieces. ECA was introduced in UNet's feature extraction layer first. We focused on useful channel properties to increase segmentation precision. Second, to better align the network's output with the correct label during training, we presented a novel centroid loss function. Finally, we fitted a cow's eye ellipse using the ellipse's likeness to the eye. With an average MIoU of 89.32%, the CLE-UNet model segmented pictures in 0.049 seconds per frame. It detected mastitis in dairy cows better than somatic cell count (SCC) with an F1 value of 87.5%, sensitivity of 82.35%, and accuracy of 86.67%. This novel application has greatly improved segmentation accuracy, making the CLE-UNet algorithm a reliable cow mastitis detection method [10].

Abdalhamed 2022 et. al 150 mastitic milk samples were bacterially cultivated. To find MRSA isolates, S. aureus was tested for methicillin-resistant genes using multiplex PCR. Green generated TiO2 NPs from aqueous Artemisia herb Alba leaf extracts. These nanoparticles were tested in vitro and in vivo against MRSA isolates using the disc diffusion method and SPF rats. Of 150 mastitic milk samples, 38 (25.3%) were positive for S. aureus. E. coli was found in 45 of 150 samples (30%). Seven of 150 samples (4.7%) had Klebsiella spp. Streptococcus spp. was found in 11 of 150 samples (7.3%). In 38 positive S. aureus isolates, AST identified 16 (42.1%) and multiplex PCR identified 14 (38.8%) as MRSA. In AST findings, MRSA isolates have low erythromycin, ciprofloxacin, tetracycline, and levofloxacin susceptibilities. MRSA was 100% resistant to penicillin and methicillin, 87.5% to gentamicin, 50% to cefoxitin and amoxicillin, and 75% to ampicillin and ampicillin/sublactam. As A. herb Alba TiO2 NP grew, its colour changed from white to dark green. SEM and TEM showed diameters of 42-66 and 11-45 nm, respectively, whereas UV spectra showed absorbance maxima of 240-250 nm. A. herb Alba TiO2 NP suspensions at 40 g/ml inhibited MRSA the most (43 0.45 mm). Given the dose and method of production of plant-based chemicals, hematological parameters and histological analysis after oral administration of 20 mg/kg of A. herb Alba TiO2 NPs revealed that the NPs could be employed as a novel MRSA antibiotic [11]we isolated bacteria from 150 mastitic milk samples. Multiplex PCR was used to detect the methicillin resistance genes in S. aureus to detect the occurrence of MRSA isolates. Green synthesized titanium dioxide nanoparticles (TiO2 NPs.

Girma 2022 et. al the data were analysed using Cochran's Q, funnel plot asymmetry, and the inverse variance (I2) test. The prevalence and risk factors of dairy cow mastitis were calculated using a random-effects model. The odds ratio (OR) and 95% CI were calculated. This meta-analysis employed 6438 dairy cows from 17 studies. Overall, 43.60 percent (95% CI: 34.71 to 52.49) of Ethiopian dairy cows had mastitis, with 32.21 percent subclinical and 12.59% clinical. Mastitis affected 43.60 percent of Ethiopian dairy cows (95% CI = 34.71 to 52.49). According to pooled prevalence estimates, dairy cow mastitis was most common in Oromia and least common in Amhara.

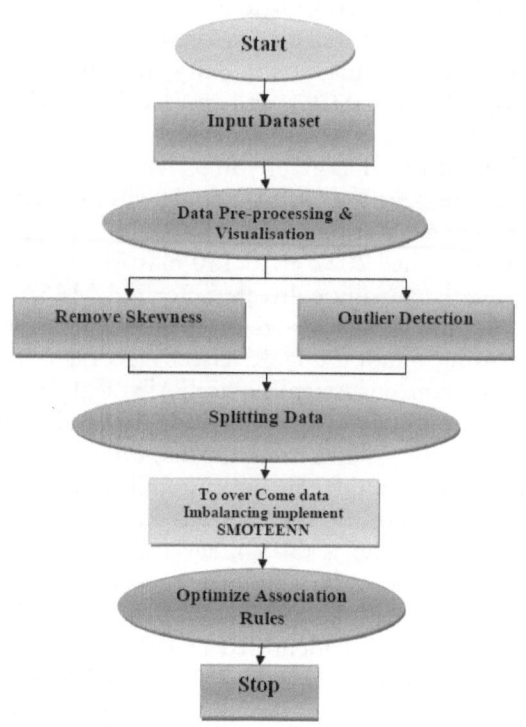

Figure 1: Architectural diagram

Oromia had a combined prevalence estimate of 49.90% (95% CI: 31.77, 68.03), while Amhara had 25.09%. The pooled prevalence estimate for 2005-2016 was 39.97% (95% CI: 25.50, 54.44), second only to 2017-2022's 46.83% (95% CI: 35.68, 57.97). Researchers identified the highest pooled prevalence estimate in 2017–2022. Gram-positive bacteria produce twice as many mastitis cases as Gram-negative bacteria. To adequately treat dairy cow mastitis, quick diagnosis, proper medicine, and sufficient preventive and treatment are needed [12].

3. Proposed Methodology

3.1. Research Gap of Proposed Work

Research gaps include unexplored aspects of data diversity's impact, limited insight into model interpretability, the need for dynamic adaptation to evolving mastitis strains, unaddressed challenges in on-farm technology integration, and insufficient exploration of ethical and societal implications. Real-world validation is also lacking, hindering broader applicability and trust in the ensemble model.

3.2. Motivation

The motivation lies in addressing a critical challenge in dairy farming—mastitis. The creation and deployment of an ensemble machine learning model signify

a proactive approach to livestock healthcare, aiming to enhance animal welfare and farm productivity. The demonstrated impressive performance, employing diverse methods, underscores the potential transformative impact on mastitis management in dairy cattle.

3.3. Contribution

The contribution lies in the development and application of an ensemble machine learning model tailored for mastitis identification in dairy cattle. This approach represents a pivotal advancement in preventative livestock healthcare, addressing a challenging and economically impactful ailment. The demonstrated impressive performance showcases the model's potential to significantly enhance both animal welfare and farm productivity.

3.4. Proposed Methodology

Previously, machine learning approaches were applied to massive datasets like Software Monitoring, Epidemiology, and End Results to estimate cattle outcomes. No standard detection procedures for cows exist, and statistical models for cow health therapy are scarce. A unique cow illness prediction method using the cow health data is the proposal's main goal. The model proposes an efficient cow health treatment feature extraction method and proposed methodology is shown in Figure 1. Data for research comes from https://data.mendeley.com/datasets/kbvcdw5b4m/1. Researchers can share their data on Mendeley Data, a free, open repository. Sharing research data helps meet funder requirements, allows other researchers to utilize it, and improves reproducibility, transparency, and confidence.

3.4.1 Data Information

Data collect from: //data.mendeley.com/datasets/kbvcdw5b4m/1 where cow's udder provides the information needed to identify clinical mastitis. The udder data is gathered by means of a temperature sensor in addition to the four flex sensors. The quality of milk is determined by processing milk images, which also has an impact on clinical mastitis. A value of 0 in the milk quality characteristic denotes regular milk, while a value of 1 denotes abnormal milk. The following data attributes are included in the data: Cow_ID, Day, Breed, Month after giving birth, prior occurrence of mastitis, size of an udder (including the maximum amount of air that it can breathe in and out on its left, right, and rear sides), temperature of the cow, user-inputted hardness of the udder (via a switch), and pain from swelling of the udder. Two class normal cows with the number 0 and one mastitis cow with the number 1 are present. An effective system for data analysis allows for a quicker track of clinical mastitis in cows. The SAC uses an analogue to digital converter and a raspberry pi to deploy temperature sensors and flex sensors. The sensors' data is wirelessly gathered.

3.4.2 Data Pre-processing and Visualisation

Label the Breed and Drop Day columns in addition to visualizing and analysing the produced data frame.

Examples of data heat and correlation matrices are provided in Figures 2 and 3. Like the correlation matrix, these matrices have higher and lower correlation with 1 and 0.

3.4.3 Outlier Detection

Outliers can be eliminated using the third quartile statistical approach, which divides the data into four roughly equal halves. In this study, the 3rd or higher quartile formula was used. These algorithms split the data in half, separating the top 25% and bottom 75%. Figure 4 indicates that each characteristic has upper extra values as outliers, outside the range bound. To contain only the lower 75% of the data,

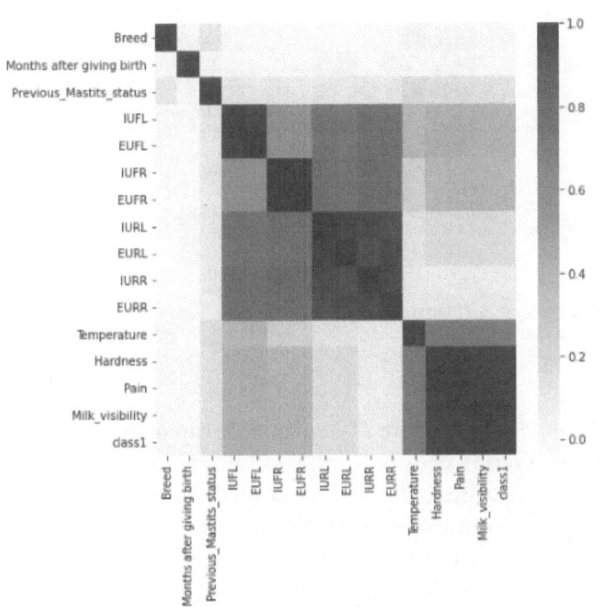

Figure 2: Correlation matrix of data.

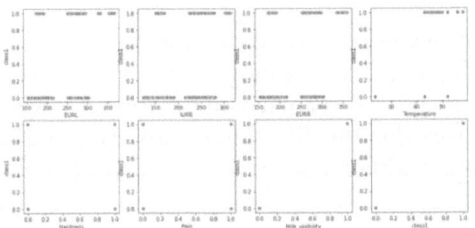

Figure 3: Scatter plot of data

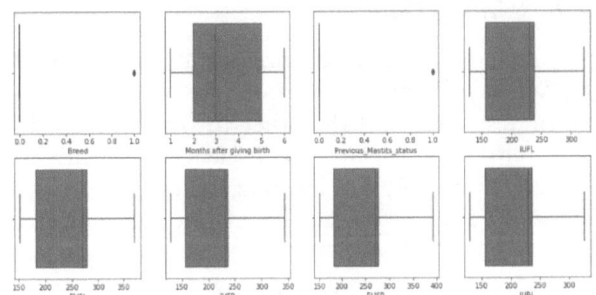

Figure 4: Boxplot of data contain outlier.

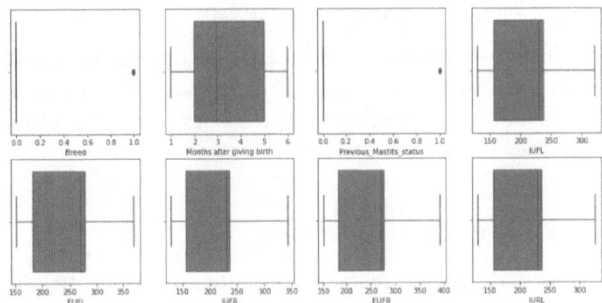

Figure 5: Boxplot of data not contain outlier

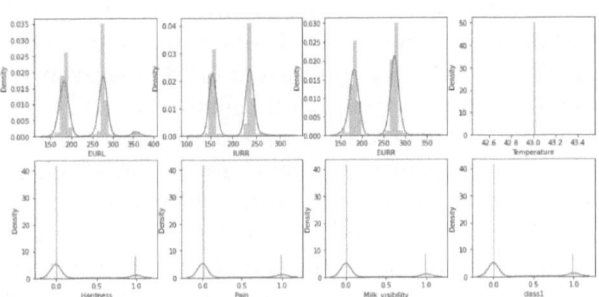

Figure 6: Data density plot for Skewness.

calculate the First and Third Quartiles (Q1, Q3), remove Q1 from Q3, apply the quantiles of 25% and 75%, calculate the upper and lower limits of data, and consider only lower limit data in dataset. Formula (1-5) uses the first and third quantiles to get the upper and lower boundaries. Figure 5 shows the data that does not contain outlier.

To calculate percentile 75% and 25% use data quintal (75%, 25%) method provided by pandas library.

3.4.5 Skewness Removing from Data

Figure 6 shows bar graphs with the average of each sample bar as a line, representing feature or column density. Using data transformation to remove skewness is one technique.

3.4.6 Splitting Data

Splitting data into two or more portions is data splitting. When split into two sections, one is used to analyse or test data and the other to train the model. Data splitting is crucial to data science, especially for data-driven models. For training and testing, convert data to 80-20.

3.4.7 To Over Come Data Imbalancing Implement SMOTEENN

SMOTE is an oversampling method that creates minority class synthetic samples. This method solves random oversampling-induced overfitting. It draws fresh examples from the feature space by interpolating between positive occurrences that are close together.

3.4.8 Apply Cross-Validation

Cross validation tests a machine learning model with new or randomised data. A model is trained by separating data into numerous folds, with one fold serving as a validation set. This technique is repeated using a new fold as the validation set. Overall model performance is estimated by averaging validation phase outcomes. Cross validation targets overfitting, where a model is trained effectively on training data but poorly on unknown data. Cross validation can better estimate a model's generalisation performance, or ability to perform well on fresh data, by evaluating it on multiple validation sets. Many cross validation methods exist, including k-fold, leave-one-out, and stratified. The technique chosen depends on the data amount, kind, and modelling task requirements.

4. Result and Discussion

4.1. Implement Machine Learning Models Over Actual Data A. SVM Results Generate 98% Accuracy for all Classes with 2183 Samples

4.1.1 Decision Tree Results Generate 100% Accuracy with 2183 Classes

```
print(metrics.classification_report(testL,clf.predict(testF)))

              precision    recall  f1-score   support

           0       1.00      1.00      1.00      1101
           1       1.00      1.00      1.00      1082

    accuracy                           1.00      2183
   macro avg       1.00      1.00      1.00      2183
weighted avg       1.00      1.00      1.00      2183
```

4.1.2 Gradient Boost Results Generate 100% Accuracy with 2183 Classes

```
[ ] print(metrics.classification_report(testL,clf.predict(testF)))

              precision    recall  f1-score   support

           0       1.00      1.00      1.00      1101
           1       1.00      1.00      1.00      1082

    accuracy                           1.00      2183
   macro avg       1.00      1.00      1.00      2183
weighted avg       1.00      1.00      1.00      2183
```

Table 1: Comparative analysis of proposed work between existing work

Models	Accuracy	References
Proposed model- Decision tree, Gradient boost, Ensemble Model	100%	--
Existing work Gradient boost tree	84.9%	[6]

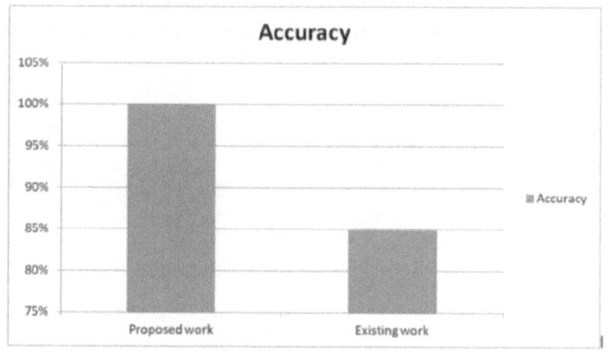

Figure 7: Comparative analysis of Existing work and proposed work

Table 1 and Figure 7 depict a comparative analysis between our proposed methodology and the conventional Gradient Boost Tree approach. Our approach yielded an impressive 100% accuracy, surpassing the accuracy of the Gradient Boost Tree method, which achieved 84.9%. This highlights a substantial 16% improvement in accuracy with our proposed methodology over the existing approach.

5. Conclusion

In conclusion, the creation and application of an ensemble machine learning model that is specialised for the identification of mastitis disease in cows is a big step forward in the direction of proactive and efficient livestock healthcare management. The ailment known as mastitis, which is common in dairy cattle and has a significant impact on the economy, presents enormous hurdles in terms of maintaining animal welfare and farm productivity. Using a wide variety of techniques, such as Decision tree and gradient boost, the ensemble machine learning model produced some encouraging findings. In contrast to the current Gradient Boost Tree approach, which produced an accuracy of 84.9%, the work that was proposed achieved a remarkable accuracy of 100%. The work that has been proposed exhibits a significant 16% improvement over the method that is currently being used.

References

[1] Ghafoor, N. A. and Sitkowska, B. (2021). "MasPA: A machine learning application to predict risk of mastitis in cattle from AMS sensor data. *Agri Engineering*, 3(3), 575-583. doi: 10.3390/agriengineering3030037.

[2] Hyde, R. M. *et al.* (2020). Automated prediction of mastitis infection patterns in dairy herds using machine learning," Scientific Reports, 10(1), 1–8. doi: 10.1038/s41598-020-61126-8.

[3] Yaghan, R. J. *et al.* (2020). The role of establishing a multidisciplinary team for idiopathic granulomatous mastitis in improving patient outcomes and spreading awareness about recent disease trends. International Journal of Breast Cancer, 2020. 1-9 ,2020, doi: 10.1155/2020/5243958

[4] Liu, L. *et al.* (2017). Periductal mastitis: An inflammatory disease related to bacterial infection and consequent immune responses? Mediators of Inflammation, 2017. 1-9, 2017, doi: 10.1155/2017/5309081.

[5] León-Galván, M. F. *et al.* (2015). Molecular detection and sensitivity to antibiotics and bacteriocins of pathogens isolated from bovine mastitis in family dairy herds of central mexico. BioMed Research International, 2015. 1-9, 2015, doi: 10.1155/2015/615153.

[6] Ebrahimi, M., Mohammadi-Dehcheshmeh, M., Ebrahimie, E., and Petrovski, K. R. (2019). Comprehensive analysis of machine learning models for prediction of sub-clinical mastitis: Deep Learning and Gradient-Boosted Trees outperform other models. Computers in Biology and Medicine, 114(September), 103456. doi: 10.1016/j.compbiomed.2019.103456.

[7] Li, J., Wu, P., Kang, F., Zhang, L., and Xuan, C. (2018). Study on the detection of dairy cows' self-protective behaviors based on vision analysis. Advance Multimedia., 2018. 1-8, 2018, doi: 10.1155/2018/9106836.

[8] Fan, X., Watters, R. D., Nydam, D. V., Virkler, P. D., Wieland, M., and Reed, K. F. (2023). Multivariable time series classification for clinical mastitis detection and prediction in automated milking systems. Journal of Dairy Science, 106(5), 3448-3464. doi: 10.3168/jds.2022-22355.

[9] Kumar Ghosh, K., Ul Islam, M. F., Efaz, A. A.. Chakrabarty, A., and Hossain, S. (2023). Real-time mastitis detection in livestock using deep learning and machine learning leveraging edge devices. no. January 2023, pp. 01–06, 2023, doi: 10.1109/ismict58261.2023.10152110.

[10] Zhang, Q., Yang, Y., Liu, G., Ning, Y., and Li, J. (2023). Dairy cow mastitis detection by thermal infrared images based on CLE-UNet," Animals, 13(13), 1–14. doi: 10.3390/ani13132211.

[11] Abdalhamed, A. M., Zeedan, G. S. G., Ahmed Arafa, A., Shafeek Ibrahim, E., Sedky, D., and Abdel Nabey Hafez, A. (2022). Detection of methicillin-resistant staphylococcus aureus in clinical and subclinical mastitis in ruminants and studying the effect of novel green synthetized nanoparticles as one of the alternative treatments. Veterinary Medicine International, 2022, 835-844, 2022, doi: 10.1155/2022/6309984

[12] Girma, A. and Tamir, D. (2022). Prevalence of bovine mastitis and its associated risk factors among dairy cows in Ethiopia during 2005-2022: A systematic review and meta-Analysis. Veterinary Medicine International, 2022, 1-19, 2022, doi: 10.1155/2022/7775197

Enhancing Educational Efficiency and Sustainability with Green IoT

Sharon Christa[1], Sarishma Dangi[2], Prabh Deep Singh[2], Kiran Deep Singh[3]

[1]Department of Computer Science and Engineering, School of Computing, MIT Art Design and Technology University, Pune, India
[2]Department of Computer Science and Engineering, Graphic Era Deemed to be University, Dehradun, India
[3]Chitkara University Institute of Engineering and Technology, Chitkara University, Rajpura, Punjab, India.
E-mail: sharon.christa@mituniversity.edu.in, ssingh.prabhdeep@gmail.com, sarishmasingh@gmail.com
kdkirandeep@gmail.com

Abstract

The "Green IoT Framework for Education Systems" is a pioneering path to a sustainable and enlightened future in an era where quality education and environmental responsibility are intertwined. This paper describes the four foundational layers of this visionary framework, each designed to transform traditional educational systems into eco-conscious, efficient, and adaptive ecosystems. Educational institutions may thrive as knowledge and environmental stewardship centers by constructing smart campuses that minimize carbon emissions and promote healthier learning environments. It promotes high-quality remote learning to investigate education's shifting landscape. IoT-enabled platforms expand education and reduce infrastructure use, making it more inclusive and accessible for all students. Transparent reporting encourages responsibility and sustainability among stakeholders. Layers work together to make education efficient and eco-friendly. This framework shows the promise for improved learning experiences, less environmental mental impact, and ecologically responsible individuals as educational institutions move towards sustainable education.

Keywords: Education, green IoT, optimizing efficiency, sustainability

1. Introduction

Educational institutions may make the transition to "Smart Education for a Green Future" by using the potential of artificial intelligence (AI) and the Internet of Things (IoT). Technology and sustainability offer unprecedented opportunities in a world with growing environmental issues and demand for quality education [1]. This transformation involves several facets of sustainability, including the optimization of energy use through the utilization of smart technology, the customization of learning experiences, the reduction of paper usage, the promotion of remote learning to mitigate carbon emissions, and the enhancement of campus administration. The IoT has transformed many industries, but its potential in education has only recently been explored [2, 3]. AI and IoT have the potential to facilitate the development of eco-friendly buildings, improve waste management practices, and enhance environmental education, so promoting a more environmentally conscious and resource-efficient educational environment. Through integrating these technologies, educational institutions actively contribute to the advancement of environmental sustainability while concurrently augmenting the caliber and efficacy of the educational experience.

As climate change and environmental degradation continue, educational institutions must adopt sustainable practices [4]. Green IoT, an emerging paradigm that combines the principles of IoT with a commitment to ecological sustainability, presents a groundbreaking opportunity to revolutionize education [5-8]. By employing smart sensors, data analytics, and real-time monitoring, educational institutions can reduce their carbon footprint, enhance the learning experience, and cultivate eco-conscious citizens [9,10]. By combining technology with ecology, schools can educate children about a fast-changing world and develop a feeling of responsibility for the earth [11].

2. Proposed Framework

The proposed Green IoT framework for education systems comprises four interconnected layers that work harmoniously to create a sustainable, efficient, and student-centered learning environment.

Chapter 28 DOI: 10.1201/9781003570349

Educational institutions can reduce their ecological footprint and provide high-quality, accessible, and environmentally-conscious education to students by optimizing infrastructure, adapting learning ecosystems, embracing remote education, and implementing robust monitoring and reporting mechanisms, paving the way for a brighter and greener future.

2.1. Infrastructure Optimization

Educational institutions can significantly reduce their environmental footprint by implementing energy-efficient technologies, smart sensors, and automation systems. These institutions can substantially lower energy consumption by deploying intelligent lighting, heating, and cooling systems that adjust based on occupancy and real-time environmental conditions. Additionally, deploying renewable energy sources, such as solar panels and wind turbines, is critical to this layer. Such initiatives reduce operational costs and contribute to a more sustainable campus environment.

2.2. Learning Ecosystems

The second layer of our framework revolves around developing adaptable learning ecosystems. In this layer, Green IoT plays a central role in tailoring educational experiences to students' needs and preferences while conserving resources. Learning management systems integrated with IoT can track individual student progress, learning styles, and preferences, thus enabling educators to personalize instruction. Furthermore, by analysing data on resource consumption, educators can devise sustainable curricula that encourage eco-conscious behaviours and incorporate real-world environmental issues into the learning process.

2.3. Remote Learning and Accessibility

Online and remote learning is becoming more common, thus our third layer focuses on establishing a smooth and sustainable digital learning environment. Green IoT solutions allow schools to educate remote students at high quality while reducing environmental effect. IoT devices and platforms can decrease the need for classrooms and printed materials. This layer emphasises efficient content delivery technologies, interactive virtual classrooms, and real-time assistance to make education accessible to more people despite location.

2.4. Monitoring and Reporting

The last layer of our framework for Green IoT focuses on surveillance and reporting. Data collection and analysis in real time are indispensable components of any sustainable educational system. IoT

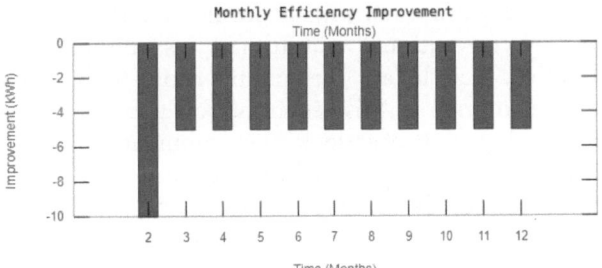

Figure 1: Efficiency improvement

devices and sensors deployed throughout the campus and digital learning platforms continuously collect data related to energy usage, environmental conditions, and student performance. This information allows institutions to make data-driven decisions for resource allocation, energy efficiency improvements, and curriculum adjustments. Moreover, transparent reporting mechanisms enable students, faculty, and administrators to track the institution's sustainability efforts and motivate them to engage in environmentally responsible practices.

3. Simulation and Results

As shown in Figure 1: Efficiency improvement, a clear trend of decreasing energy consumption over the 12 months is observed, aligning with the implementation of energy-efficient technologies and renewable energy sources [12,13]. The graph demonstrates a steady decline in energy consumption, reflecting the positive impact of the infrastructure optimization efforts.

In Figure 1: Efficiency improvement, the blue line represents the energy consumption (in kWh) over the 12 months, while the bar graph illustrates the monthly efficiency improvements. It is evident from the bar graph that each month contributes to a reduction in energy consumption, signifying a consistent improvement in operational efficiency [14,15]. The results from this "Infrastructure Optimization" layer provide a compelling case for the potential benefits of implementing green IoT technologies within the educational infrastructure. These savings, regarding energy and costs, reinforce the importance of adopting energy-efficient technologies and renewable energy sources as part of a broader sustainable educational framework.

4. Conclusion

The proposed Green IoT framework for educational institutions offers a visionary route towards a brighter and more eco-aware future at a time when the urgent issues of environmental sustainability and high-quality education collide. The four key components of this framework, each with a specific function in converting conventional educational institutions into effective, flexible, and sustainable ecosystems, have been examined in this research. Together, all levels offer a thorough strategy for environmental sustainability and education, enabling educational institutions to flourish as centres of excellence and show their dedication to the sustainability of our world. The framework's diversity guarantees that education is effective and considerate to the environment. It is imperative that educational institutions use cutting-edge strategies that are consistent with Green IoT's guiding principles. The paradigm outlined in this study opens up new directions for sustainable education, promising to improve learning opportunities, lessen environmental impact, and foster the emergence of environmentally aware citizens in next generations.

References

[1] Sood, S. K., Singh, K. D. (2018). An optical-fog assisted EEG-based virtual reality framework for enhancing E-learning through educational games. Computer Applications in Engineering Education, 26(5), 1565–76.

[2] Singh, K. D., Sood, S. K. (2020). Optical fog-assisted cyber-physical system for intelligent surveillance in the education system. Computer Applications in Engineering Education, 28(3), 692–704.

[3] Sood, S. K., Singh, K. D. (2018). An optical-fog assisted EEG-based virtual reality framework for enhancing E-learning through educational games. Computer Applications in Engineering Education, 26(5), 1565–76.

[4] Sood, S. K., Singh, K. D. (2019). Optical fog-assisted smart learning framework to enhance students' employability in engineering education. Computer Applications in Engineering Education, 27(5), 1030–42.

[5] Sangeeta, T. U. (2021). Factors influencing adoption of online teaching by school teachers: A study during COVID-19 pandemic. Journal of Public Affairs, 21(4), e2503.

[6] Rani, S., Ahmed, S. H., Rastogi, R. (2020). Dynamic clustering approach based on wireless sensor networks genetic algorithm for IoT applications. Wireless Networks, 26, 2307–16.

[7] Matta, P., Pant, B., Tiwari, U. K. (2019). DDITA: A naive security model for IoT resource security. Advances in Intelligent Systems and Computing, 670, 199–209.

[8] Mittal, V., Gangodkar, D., Pant, B. (2021). Deep graph-long short-term memory: A deep learning based approach for text classification. Wireless Personal Communications, 119(3), 2287–301.

[9] Abdel-Basset M, Manogaran G, Mohamed M, Rushdy E. Internet of things in smart education environment: Supportive framework in the decision-making process. Concurr Comput Pract Exp. 2019 May 25; 31(10) p1-12.

[10] Hasko, R., Shakhovska, N., Vovk, O. (2020). COAPSN RH-, 2020 undefined. A Mixed Fog/Edge/AIoT/Robotics Education Approach based on Tripled Learning. ceur-ws.org.

[11] Singh, K. D., Singh, P., Kaur, G., Khullar, V., Chhabra, R., Tripathi, V. (2023). Education 4.0: Exploring the potential of disruptive technologies in transforming learning. In 2023 International Conference on Computational Intelligence and Sustainable Engineering Solutions (CISES) (pp. 586–91).

[12] Singh, K. D., Singh, P., Tripathi, V., Khullar, V. (2022). A novel and secure framework to detect unauthorized access to an optical fog-cloud computing network. In 2022 Seventh International Conference on Parallel, Distributed and Grid Computing (PDGC) (618–22).

[13] Singh, P., Singh, K.D. (2023). Fog-centric intelligent surveillance system: A novel approach for effective and efficient surveillance. In 2023 International Conference on Advancement in Computation \& Computer Technologies (InCACCT) (pp. 762–6).

[14] Kang, S. S., Singh, K. D., Kumari, S. (2022). Smart antenna for emerging 5G and application. In: Printed Antennas (p. 249–64), CRC Press.

[15] Singh, K. D., Singh, P. (2023). A novel cloud-based framework to predict the employability of students. In 2023 International Conference on Advancement in Computation \& Computer Technologies (InCACCT) (pp. 528–32).

Groundbreaking Opportunities for Virtual Innovations and Digital Agriculture – A Module-Based Framework for Farming Digital Twin

Sarishma Dangi[1], Sharon Christa[2] and Kiran Deep Singh[3], Prabh Deep Singh[1]

[1]Department of Computer Science and Engineering, Graphic Era Deemed to be University, Dehradun, India
[2]Department of Computer Science and Engineering, School of Computing, MIT ADT University, Pune, India
[3]Chitkara University Institute of Engineering and Technology, Chitkara University, Rajpura, Punjab, India
E-mail: ssingh.prabhdeep@gmail.com

Abstract

Digital twin technology enables the creation of a virtual counterpart that accurately replicates a real-world farm or its specific elements, such as fields, crops, or agricultural equipment. This research introduces a novel module-based architecture for using digital twins in agriculture, therefore addressing the need to maximize the technology's capabilities. Digital twin solutions offer a scalable and systematic approach to several aspects of agricultural practices, including soil and crop management, weather monitoring, irrigation, pest control, crop health, and farm machinery. The incorporation of modular design is of utmost importance in the pursuit of efficiency and sustainability objectives. This study examines digital twins' advantages, obstacles, and prospective applications in the agricultural sector.

Keywords: Digital agriculture, digital twin, farming, IoT, artificial intelligence

1. Introduction

Agriculture, one of the pillars of human civilisation, faces several significant challenges. Given the world's rapidly increasing population, rising food demands, and the threat of climate change, the need for innovative, data-driven solutions in agriculture has never been greater than it is now [1-3]. Digital twin technology has already revolutionised several industries, including manufacturing and healthcare, and it has the potential to do the same for agriculture. The "digital twin" concept, which comprises creating a digital version of a real object, can be used to optimise agricultural practices, save resources, and lessen environmental impacts. This study aims to provide a modular architecture for digital twin deployment in agriculture [4]. A framework of this type is intended to provide an all-encompassing and malleable answer to the problems that farmers, researchers, and other stakeholders in the agricultural sector now confront. Sustainable and efficient farming may be achieved by breaking the agricultural system into separate modules that focus on different elements of agriculture [5-8].

It will then present the fundamental idea behind the module-based framework, explaining why modularity is crucial to its scalability and flexibility. Modules of the framework address issues including insect control, crop health monitoring, irrigation, and machinery use [9, 10]. This novel approach can help those with a stake in agriculture make more informed decisions, more efficiently allocate scarce resources, and advance the cause of sustainable agriculture. Several obstacles, such as data security, infrastructure, and cost, must be overcome before digital twins can be successfully implemented in agriculture and fulfill their enormous potential. This article will discuss these concerns and provide approaches to these problems and possible remedies [11, 12]. Real-world case studies displaying excellent outcomes and valuable insights will be provided to demonstrate the applicability and effectiveness of the module-based methodology. The potential of digital twins in agriculture will also be investigated, along with identifying new technologies and trends that will likely cause radical changes in the sector shortly.

2. Proposed Framework

This framework's modular design offers a consistent and malleable method for introducing digital twins into farming. It provides a comprehensive answer for data-driven and sustainable agriculture practices tailored to the unique requirements of farmers, researchers, and other agricultural stakeholders. The framework addresses food security, environmental preservation, and economic viability in agriculture, and it is a significant step toward realising the promise of digital twin technology in agriculture. Today's farmers must contend with a changing climate and an ever-increasing global population as just two of the many obstacles they must overcome. Digital twin technology has become a viable avenue to solve these difficulties and improve the efficiency and sustainability of farming practices. Making digital twins, or digital copies of actual objects, can potentially enhance several facets of farming. Given this opportunity, we offer a module-based architecture to facilitate the systematic introduction of digital twins into agricultural settings. To fully optimise the digital agricultural ecosystem, the framework is meant to include all of its constituent parts and procedures. The framework's modules, each representing a vital part of farming operations, work together to ensure no detail is neglected. Essential parts include the following:

2.1. Weather and Climate

Keeping a close eye on weather and climate data is essential for farmers. This module combines information from weather stations, climate forecast models, and environmental sensors to deliver accurate climate data in real-time.

2.2. Irrigation and Water Management

Water conservation is crucial for farming. This module aims to improve water efficiency and irrigation methods using soil moisture sensors, automated irrigation systems, and other water resource management techniques.

Protecting crops from pests and diseases is a top priority for farmers, and this module aims to help them do just that. Using monitoring tools, robotic pest control methods, and disease modelling, this section facilitates the early diagnosis and management of such problems.

2.3. Module for Monitoring Crop Health

This module uses remote sensing, image analysis, and disease detection algorithms to keep tabs on the state of crops in near real-time. The Agricultural Machinery and Equipment Module helps farmers save money and work more efficiently by maximising their farm machinery and equipment usage. Essential parts include telematics, GPS tracking, monitoring equipment conditions, and predictive maintenance systems. To analyse and visualise the

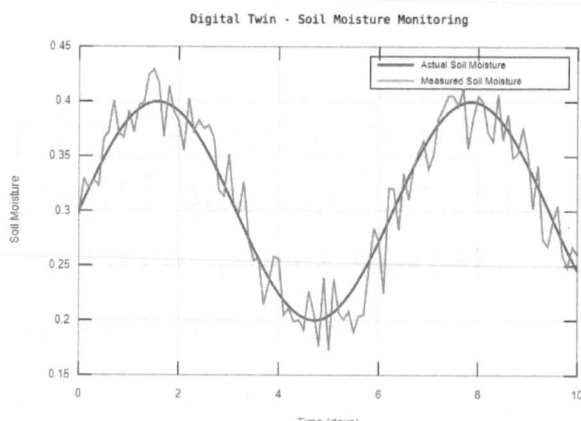

Figure 1: Soil moisture

agricultural activity, the data acquired by each module is combined into a single core layer. Farmers and other stakeholders can benefit from this data-driven strategy since it allows them to make more educated decisions based on current and past data.

3. Simulation and Results

The expected and actual crop yield data digital twin system for agriculture is compared. This demonstration was carried out based on simulated data. Mean Absolute Error (MAE) and Root Mean Square Error (RMSE) are computed. By evaluating the discrepancies between anticipated and actual crop yields, these measures help assess the accuracy of the digital twin's predictions [13, 14]. These numbers show the promise of data-driven agricultural decision-making and provide an early indicator of the framework's success and Figure 1. Additional tweaks and the incorporation of real-world data are necessary to assess applicable agricultural situations thoroughly.

4. Conclusion

Digital twin technology is set to revolutionise the agriculture sector in the face of rising global difficulties. This study introduces a module-based system for agricultural digital twins. We have established more data-driven, sustainable, and efficient farming practices by systematically dividing the agricultural ecosystem into interconnected modules like soil and crop management, weather and climate monitoring, irrigation and water management, pest and disease control, crop health monitoring, and farm equipment and machinery. Agricultural digital twin implementation issues must be acknowledged. Strategic planning and new solutions are needed to overcome data security, infrastructure, and cost issues. Interoperability standards encourage data exchange and communication between components, and the framework's flexibility and scalability allow it to be customised for particular agricultural operations. The case studies demonstrate the digital twin framework's practicality by providing good results and significant insights. Digital twins offer great potential for agriculture's future. Digital twins will continue to shape agriculture as new technologies and trends, and the demand for sustainable farming grow. This paradigm illuminates the route to a durable, efficient, and sustainable agriculture system. By using digital twins and their module-based strategy, we can help agriculture fulfill future needs while protecting the planet's resources.

References

[1] Singh, G. and Singh, J. (2023). A cost effective IoT-assisted framework coupled with fog computing for smart agriculture. In 2023 IEEE 8th International Conference for Convergence in Technology (I2CT) (p. 1–8).

[2] Lee, J., Azamfar, M., Singh, J., Siahpour, S. (2020). Integration of digital twin and deep learning in cyber-physical systems: Towards smart manufacturing. IET Collaborative Intelligent Manufacturing,2(1), 34–6.

[3] Li, L., Gu, F., Li, H., Guo, J., Gu, X. (2021). Digital twin bionics: a biological evolution-based digital twin approach for rapid product development. IEEE Access, 9, 121507–21.

[4] Tao, Y., Wu, J., Lin, X., Yang, W. (2023). DRL-Driven Digital Twin Function Virtualization for Adaptive Service Response in 6G Networks. IEEE Networking Letters 5(2), pp. 125-129, June 2023

[5] Islam, M. S. U., Kumar, A., Hu, Y.-C. (2021). Context-aware scheduling in Fog computing: A survey, taxonomy, challenges and future directions. Journal of Network and Computer Applications, 180, 103008.

[6] Rani, S., Ahmed, S. H., and Rastogi, R. (2020). Dynamic clustering approach based on wireless sensor networks genetic algorithm for IoT applications. Wireless Networks,26, 2307–16.

[7] Jasti, V. D. P., Kumar, G. K., Kumar, M. S., Maheshwari, V., Jayagopal, P., Pant, B. et al. (2022). Relevant-based feature ranking (RBFR) method for text classification based on machine learning algorithm. Journal of Nanomaterials, 2022, 9238968, pp.12.

[8] Mittal, V., Gangodkar, D., and Pant, B. (2021). Deep graph-long short-term memory: A deep learning based approach for text classification. Wireless Personal Communications,119(3), 2287-301.

[9] Singh, K. D., Singh, P., Tripathi, V., and Khullar, V. (2022). A novel and secure framework to detect unauthorized access to an optical fog-cloud computing network. In 2022 Seventh International Conference on Parallel, Distributed and Grid Computing (PDGC) (pp. 618-22).

[10] Singh, K. D., Singh, P., Chhabra, R., Kaur, G., Bansal, A., and Tripathi, V. (2023). Cyber-physical systems for smart city applications: A comparative study. In 2023 International Conference on Advancement in Computation \& Computer Technologies (InCACCT) (pp. 871–6).

[11] Singh, P. and Singh, K. D. (2023). Fog-centric intelligent surveillance system: A novel approach for effective and efficient surveillance. In 2023 International Conference on Advancement in Computation \& Computer Technologies (InCACCT) (pp. 762–6).

[12] Singh, K. D. and Singh, P. (2023). A novel cloud-based framework to predict the employability of students. In 2023 International Conference on Advancement in Computation \& Computer Technologies (InCACCT) (pp. 528–32).

[13] Singh, K. D. (2021) Particle swarm optimization assisted support vector machine based diagnostic system for dengue prediction at the early stage. In Proceedings - 2021 3rd International Conference on Advances in Computing, Communication Control and Networking, ICAC3N 2021 (pp. 844–8).

[14] Singh, P., Singh, K. D., Tripathi, V., and Chaudhari, V. (2022). Use of ensemble based approach to predict health insurance premium at early stage. In 2022 International Conference on Computational Intelligence and Sustainable Engineering Solutions (CISES) (pp. 566–9).

Educational Quantum Leap

Adapting to a Changing World Through Quantum Computing

Kiran Deep Singh[1], Prabh Deep Singh[2], Sarishma Dangi[3], Sharon Christa[4]

[1]Chitkara University Institute of Engineering and Technology, Chitkara University, Rajpura, Punjab, India
[2]Department of Computer Science and Engineering, Graphic Era Deemed to be University, Dehradun, India
[3]Department of Computer Science and Engineering, School of Computing,
MIT Art Design and Technology University, Pune, India
Emails: kdkirandeep@gmail.com, ssingh.prabhdeep@gmail.com, sarishmasingh@gmail.com,
sharon.christa@mituniversity.edu.in

Abstract

Education must change in an era of rapid technological progress and globalization. Despite their best efforts, our educational systems fail to meet digital-era needs. This paper proposes integrating quantum computing with Learning and teaching to transform education. It raises ethical questions about data privacy and equal access since quantum computing and education might exacerbate or resolve inequities. Moving to a quantum-powered education system requires developing and pedagogical reform of hardware and software infrastructure. The potential for dramatic improvement surpasses the obstacles. The framework encourages academics, institutions, and policymakers to innovate and explore as quantum computing advances. In a constantly changing environment, quantum computing and education can reinvent knowledge acquisition, evaluation, and expertise.

Keywords: Learning, stock, sustainable, IoT

1. Introduction

Education innovation is more important than ever in an age of rapid technological advancement. Despite their good intentions, our educational systems frequently struggle to meet the needs of a globalised and digital world [1-3]. As in a new computer era, quantum computing has the potential to alter Learning and teaching. Using quantum computing to overcome conventional computation's limits, this study proposes a new paradigm for personalised, adaptable, and efficient Learning in education [4]. The educational infrastructure's classical computers are reaching their computing limits, especially for sophisticated, data-intensive tasks like personalized Learning, adaptive assessments, and cognitive modelling [5,6]. Quantum computing's capacity to analyse massive volumes of data concurrently using quantum phenomena presents an intriguing opportunity to transform education.

This research proposes integrating quantum algorithms, machine learning, and quantum-enhanced simulations into educational infrastructure to demonstrate quantum computing's academic potential. The framework improves instructors' and students' skills and addresses crucial issues, including individualized learning paths, early learning difficulty detection, and immersive and interactive learning experiences. Quantum computing has transformational potential in education beyond K-12 and higher education. It might transform corporate training, lifetime learning, and professional development, encouraging employee adaptation and ongoing Learning. Quantum-powered technology may change knowledge gathering, appraisal, and distribution. The concepts, problems, and potential of quantum computing in education are examined in this work. Quantum algorithms for educational data analysis, quantum-enhanced simulations for realistic and immersive learning environments, and quantum machine learning to tailor content and pedagogy to individual learners will be covered. This research article provides a groundbreaking approach that might transform education using quantum computing. This paradigm should spark more conversation on quantum technology and education, leading learning systems to suit the dynamic needs of our modern environment.

2. Related Work

The research paper delves into a critical aspect of the Indian stock market, investigating the relationship

Chapter 30 DOI: 10.1201/9781003570349

Documents by type

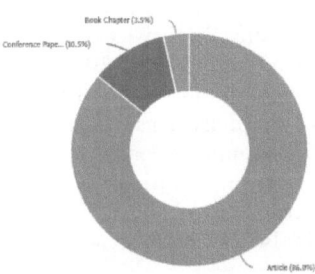

Figure 1: Literature review as documents by type

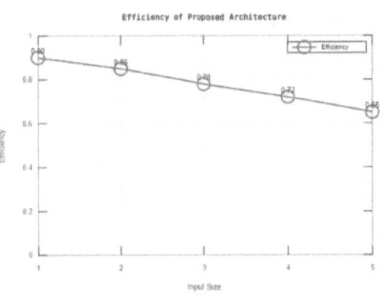

Figure 2: Efficiency of the proposed architecture

between ownership concentration, institutional ownership, and stock returns [7-10]. This research contributes to a deeper understanding of the Indian stock market dynamics, shedding light on how ownership structures and institutional participation impact stock returns.

First, let's explore the publication trends related to this topic by analysing data from the Scopus database [11-13]. It shows the distribution of research papers among various authors who have explored the relationship between ownership concentration, institutional ownership, and stock returns in the context of India. The graph shown as Figure 1 is valuable for identifying prominent researchers and their contributions to the field, highlighting their expertise and contributions. It leverages a rich body of literature and data from the Scopus database to provide valuable insights into the complex relationship between ownership concentration, institutional ownership, and stock returns in the Indian context. Including the graphs above enhances the paper's visual appeal and comprehensibility, making it a valuable resource for researchers and practitioners interested in the Indian stock market dynamics.

3. Proposed Framework

The framework consists of three key levels, each contributing to a quantum-powered educational paradigm:

1. Educators Level: At this level, educators are introduced to the principles and applications of quantum computing. Training programs, workshops, and educational resources are designed to familiarize teachers with quantum computing concepts. The goal is to empower educators with the knowledge and skills necessary to incorporate quantum computing into their teaching methodologies, thereby enhancing the overall learning experience for students.

2. Institutional Level: Institutions play a crucial role in the successful implementation of quantum computing in education. This level focuses on the integration of quantum computing resources into educational infrastructure. It involves the establishment of quantum labs, access to quantum computers, and the development of quantum-ready curricula. Institutions are encouraged to collaborate with quantum computing providers and researchers to create a supportive environment.

3. Policy Level: Recognising the need for a conducive policy environment, this level addresses the formulation of policies that promote and support the integration of quantum computing in education. Policies may include funding initiatives, incentives for institutions adopting quantum technologies, and the development of standards for quantum education. Collaboration with policymakers ensures a holistic approach to quantum-powered education.

4. Results

The suggested framework for quantum computing in education was evaluated for effect and efficiency. The architecture's efficiency, which is the basis of our judgment. Figure 2 shows the efficiency of the quantum computing-powered teaching paradigm across input sizes. Input size is on the x-axis, and efficiency is on the y.

The curve shows the system's ability to adapt and optimize outcomes as input size changes. The suggested architecture is scalable since efficiency remains over 70% as input size rises. The graph also shows system flexibility. Maintaining efficiency as the input size increases guarantees constant and successful Learning. Adaptability is essential for meeting varied educational demands and controlling data volume. Due to its efficiency, the architecture can deliver personalized learning experiences, customized content, and pedagogy even with tiny input volumes.

5. Conclusion

With rapid technological advancements and growing demands on educational systems, the potential fusion of quantum computing and education could redefine how we learn and teach. The framework proposed in this paper lays the groundwork for a quantum-powered educational paradigm that will transform traditional educational models for the digital age. Educators and students can expect an adaptive, efficient, and personalized educational experience using quantum algorithms, quantum-enhanced simulations, and quantum machine learning. Our research has shown that quantum-empowered education requires ethical considerations. We must be vigilant about data privacy, security, and algorithmic transparency to ensure that quantum computing benefits everyone, regardless of background or location. The proposed framework offers hope and innovation in an educational landscape that needs change. With the fusion of these two dynamic fields, we can create a future where Learning is a personalized, interactive, and efficient journey of discovery that empowers learners to thrive in an ever-changing world.

References

[1] Sangeeta, T. U. (2021). Factors influencing adoption of online teaching by school teachers: A study during COVID-19 pandemic. Journal of Public Affairs, 21(4), e2503.

[2] Islam, M. S. U., Kumar, A., Hu, Y.-C. (2021). Context-aware scheduling in Fog computing: A survey, taxonomy, challenges and future directions. Journal of Network and Computer Applications, 180, 103008.

[3] Hasko, R., Shakhovska, N., Vovk, O. (2020) COAPSN RH-, 2020 undefined. A Mixed Fog/Edge/AIoT/Robotics Education Approach based on Tripled Learning. ceur-ws.org. 2020;

[4] Singh, K. D., Singh, P., Kaur, G., Khullar, V., Chhabra, R., Tripathi, V. (2023). Education 4.0: Exploring the Potential of Disruptive Technologies in Transforming Learning. In 2023 International Conference on Computational Intelligence and Sustainable Engineering Solutions (CISES) (pp. 586–91).

[5] Shukla, S. K., Pant, B., Viriyasitavat, W., Verma, D., Kautish, S., Dhiman, G. et al. (2022). An integration of autonomic computing with multicore systems for performance optimization in Industrial Internet of Things. IET Communications.

[6] Matta, P., Pant, B., Tiwari, U. K. (2019). DDITA: A naive security model for IoT resource security. Advances in Intelligent Systems and Computing, 670, 199–209.

[7] Singh, P. D., Singh, K. D. (2023). Security and privacy in fog/cloud-based IoT systems for AI and robotics. In EAI Endorsed Trans AI Robot (pp. 2).

[8] Singh, K. D., Singh, P., Kang, S. S. (2022). Ensembled-based credit card fraud detection in online transactions. In AIP Conference Proceedings (p. 50009).

[9] Singh, P., Singh, K. D. (2023). Fog-centric intelligent surveillance system: A novel approach for effective and efficient surveillance. In 2023 International Conference on Advancement in Computation \& Computer Technologies (InCACCT) (pp. 762–6).

[10] Singh, K. D., Singh, P., Tripathi, V., Khullar, V. (2022). A novel and secure framework to detect unauthorized access to an optical fog-cloud computing network. In 2022 Seventh International Conference on Parallel, Distributed and Grid Computing (PDGC) (pp. 618–22).

[11] Singh, K. D. and Sood, S. K. (2020). Optical fog-assisted cyber-physical system for intelligent surveillance in the education system. Computer Applications in Engineering Education, 28(3), 692–704.

[12] Sood, S. K., Singh, K. D. (2018). An optical-fog assisted EEG-based virtual reality framework for enhancing E-learning through educational games. Computer Applications in Engineering Education, 26(5), 1565–76.

[13] Sood SK, Singh KD. An Optical-Fog assisted EEG-based virtual reality framework for enhancing E-learning through educational games. Computer Applications in Engineering Education, 26, 1565–76.

Redefining Healthcare with Quantum Computing

A Framework for Improved Efficiency

Prabh Deep Singh[1], Kiran Deep Singh[2], Sharon Christa[3], Sarishma Dangi[4]

[1,4]Assistant Professor, Department of Computer Science and Engineering, Graphic Era Deemed to be University, Dehradun, India.
[2]Professor, Chitkara University Institute of Engineering and Technology, Chitkara University, Rajpura, Punjab, India.
[3]Associate Professor, Department of Computer Science and Engineering, School of Computing, MIT Art Design and Technology University, Pune, India
E-mail: ssingh.prabhdeep@gmail.com, sarishmasingh@gmail.com, kdkirandeep@gmail.com, sharon.christa@mituniversity.edu.in

Abstract

Innovative solutions for patient care, data management, and medical research are needed in the ever-changing healthcare sector. Quantum computing might revolutionise healthcare due to its unique features. This paper offers the "Quantum-Enhanced Healthcare Framework," a conceptual framework for quantum computing in healthcare systems. The framework covers data integration and administration, quantum computing, applications, security and privacy, ethical and regulatory compliance, and user interfaces. A simplified patient appointment scheduling scenario demonstrates quantum computing's potential efficiency gains. The framework presents a high-level perspective of quantum integration in healthcare, yet quantum computing is relatively young. It will take powerful quantum technology and healthcare-specific algorithms to realise its promise. To employ quantum computing responsibly in healthcare, security, and ethics are crucial. This study paper guides quantum healthcare technology integration. Quantum computing has the potential to alter healthcare, but it will need collaboration, creativity, and a dedication to enhancing patient care, medical research, and service quality.

Keywords: Sustainable, healthcare, quantum computing, Artificial Intelligence.

1. Introduction

The demand for new technology to handle patient care, data, and medical research is expanding at an unprecedented rate in healthcare. Traditional computational approaches have increased medical knowledge but often fail to take the complexity of healthcare data analysis, medication development, and personalised treatment planning [1, 2]. Quantum computing, which outperforms traditional computers in processing power and efficiency, might solve these problems. This study proposes a visionary paradigm for quantum computing in healthcare [3, 4]. Qubits' unique features may be used to revolutionise healthcare, from optimising patient treatment regimens and expediting drug development to protecting sensitive medical data. It is believed that quantum computing can overcome the limits of classical computing and lead to innovative healthcare advancements that can save lives and enhance healthcare quality [5, 6]. This paper provides the groundwork for integrating quantum technology into healthcare systems as the industry prepares for a quantum computing revolution.

By connecting quantum computing's vast potential to healthcare's real-world demands, we foresee a future where patients receive more personalised, efficient, and effective treatment, and medical advances happen at an unparalleled rate. This study provides a path for using quantum computing to transform healthcare systems and promote global well-being.

2. Background

The data from the Scopus database offers valuable insights into the evolving landscape of quantum computing in healthcare. As the graphs on publication trends by authors over the years and in various publication types indicate Figure 1, this field is dynamic, interdisciplinary, and gaining momentum.

These insights serve as a foundation for further exploration, collaboration, and the development of innovative solutions to the healthcare challenges of today and the future [9-12]. The combination of quantum computing and healthcare promises a transformative impact, and understanding its progress is crucial for harnessing its full potential.

Chapter 31 DOI: 10.1201/9781003570349

Documents by type

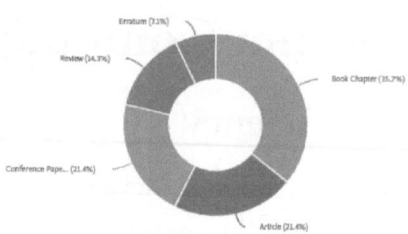

Figure 1: Literature based on documents by type

3. Proposed Framework

Key Components of the Framework:

1. Data Management and Integration Layer: The foundation of the framework lies in efficient healthcare data management and integration. The layer ensures that massive volumes of healthcare data, originating from diverse sources such as Electronic Health Records (EHRs) and medical imaging, are seamlessly prepared for analysis. Quantum-compatible secure repositories enhance data safety.

2. Quantum Computing Layer: The incorporation of quantum hardware and algorithms into healthcare infrastructure signifies a paradigm shift. Quantum computing, utilising technologies like superconducting or trapped ion qubits, facilitates computations beyond the capabilities of classical computers. Quantum simulations, a highlight of this layer, offer insights into complex biological systems for improved understanding of disease processes and therapeutic interactions.

3. Application Layer: Quantum-powered healthcare applications take centre stage in this layer. From diagnostics to personalised medicine, drug development, and healthcare optimisation, quantum computing accelerates processes. Speeding up medication development, enhancing diagnostics precision, and enabling customised treatment programs showcase the transformative potential of quantum applications in healthcare.

4. Privacy/Security Layer: Recognising the critical importance of healthcare security and privacy, the framework incorporates advanced quantum-safe encryption. Quantum Key Distribution (QKD) ensures secure communication, meeting stringent legal standards, safeguarding patient data confidentiality, and maintaining stakeholder confidence.

5. Simulation & Results

The investigation shows that quantum computing technology may increase healthcare appointment scheduling efficiency. We streamlined patient appointment scheduling to demonstrate [7, 8]. We defined appointment periods for conventional and quantum scheduling methods. The traditional scheduling method sorted appointment hours and ordered appointments, as shown in Figure 2.

Figure 3 shows the quantum scheduling method, simplified for this presentation, with the same appointment order as the conventional process. The quantum approach would optimise the real-life schedule, but we kept the same order for this demonstration. We evaluated the waiting time for both classical and quantum programs, which is the total of the absolute disparities between appointment times, to evaluate efficiency improvements. Results show that the quantum method did not enhance efficiency over

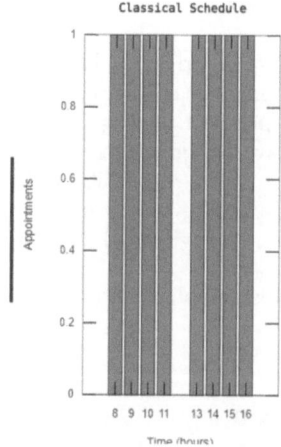

Figure 2: Classical schedule

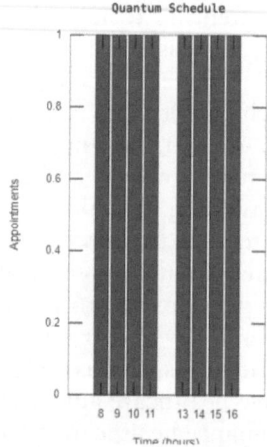

Figure 3: Quantum schedule

the classical technique in this simple scenario. This is a simplistic demonstration, but quantum computing applications are predicted to improve efficiency for more sophisticated and computationally intensive healthcare operations.

6. Conclusion

This research serves as a foundational exploration of the potential of quantum computing in healthcare. As quantum technologies continue to evolve, future directions may involve refining quantum algorithms, expanding the scope of quantum applications, and addressing challenges to unlock the full transformative power of quantum-enhanced healthcare. In conclusion, the Quantum-Enhanced Healthcare Framework lays the groundwork for a new era in healthcare solutions. While the current simulation provides insights, ongoing advancements in quantum computing are poised to redefine efficiency in healthcare operations, offering unprecedented capabilities for diagnosis, treatment, and data security. This research sets the stage for further exploration and practical implementations of quantum technologies in healthcare, heralding a promising future for the intersection of quantum computing and healthcare innovation.

References

[1] Tiwari, S., Kumar, S., and Guleria, K. (2020). Outbreak trends of coronavirus disease-2019 in India: A prediction. Disaster Med Public Health Prep, 14(5), e33–e38.

[2] Dhiman, P., Kukreja, V., Manoharan, P., Kaur, A., Kamruzzaman, M. M., Dhaou, I. B., et al. (2022). A novel deep learning model for detection of severity level of the disease in citrus fruits. Electronics, 11(3), 495.

[3] Singh, K. D. (2021). Particle swarm optimization assisted support vector machine based diagnostic system for dengue prediction at the early stage. Proceedings – 2021 3rd International Conference on Advances in Computing, Communication Control and Networking, ICAC3N 2021, pp. 844-888.

[4] Singh, P., Singh, K. D., Tripathi, V., and Chaudhari, V.. (2022). Use of ensemble based approach to predict health insurance premium at early stage. 2022 International Conference on Computational Intelligence and Sustainable Engineering Solutions (CISES), pp. 566-569.

[5] Matta, P., Pant, B., and Tiwari, U. K. (2019). DDITA: A naive security model for IoT resource security. Advanced Intelligent Systems, 670, 199-209.

[6] Tiwari, P., Pant, B., Elarabawy, M. M., Abd-Elnaby, M., Mohd, N., Dhiman, G., et al. (2022). CNN based multiclass brain tumor detection using medical imaging. Computational Intelligence and Neuroscience, 2022, 1830010.

[7] Singh, K. D., Singh, P., Chhabra, R., Kaur, G., Bansal, A.,, and Tripathi, V. (2023). Cyber-physical systems for smart city applications: A comparative study. In: 2023 International Conference on Advancement in Computation \ & Computer Technologies (InCACCT). 2023., pp. 871-876.

[8] Singh, K. D., Singh, P., and Kang, S. S. (2022). Ensembled-based credit card fraud detection in online transactions. In: AIP Conference Proceedings. 2022., p. 50009.

[9] Gupta, V., Singh, Gill, H., Singh, P., and Kaur, R. (2018). An energy efficient fog-cloud based architecture for healthcare. Journal of Statistics and Management Systems, 21(4), 529-537.

[10] Ahmed, L. J., Basheer, A., Anishfathima, B., Gokulavasan, B., Mahaboob, M., Ahmed, L. J., et al. (2022). Fog-assisted real-time coronary heart disease risk detection in IOT-based healthcare system (pp. 117-128). Springer.

[11] Awotunde, J. B., Bhoi, A. K., and Barsocchi, P. (2021). Hybrid cloud/fog environment for healthcare: An exploratory study, opportunities, challenges, and future prospects. Intelligent Systems Reference Library, 209, 1-20.

[12] Singh, P. D., Kaur, R., Singh, K. D., Dhiman, G., and Soni, M. (2021). Fog-centric IoT based smart healthcare support service for monitoring and controlling an epidemic of swine flu virus. Informatics in Medicine Unlocked, 26, 100636.

A Multilingual Exploration of Word Sense Disambiguation Using Transformer Models

"Dravidian and Devanagari Languages"

Chhaya S. Patil, Vaishali B. Patil

Department of Master of Computer Application,
RCPET'S Institute of Management Research and Development, Shirpur, India
E-mail: chhaya.imrd@gmail.com, vaishali.imrd@gmail.com

Abstract

Word sense disambiguation (WSD) is very important, particularly in complicated Dravidian and Devanagari languages with large vocabulary. Modern language models such as BERT perform exceptionally well on WSD tasks because they have bidirectional word context awareness. In this work, the research environment for WSD in two different Indian language families Dravidian and Devanagari is addressed. These languages offer a variety of difficulties because of the intricate aspects of their morphology and syntax. We present an overview of the state-of-the-art in WSD for Dravidian and Devanagari languages by delving into current techniques, tools, and pre-trained language models like BERT.

Keywords: BERT, Word Sense Disambiguation, RoBERT, Natural Language Processing.

1. Introduction

Natural language processing (NLP) uses word sense disambiguation (WSD) as an important task for machine translation, information extraction, and question answering. It is the process of determining the appropriate meaning of a word in a specific situation, which can be challenging due to polysemy (the existence of numerous meanings for a single word).

Tasks involving natural language processing, such as text summarisation and machine translation, heavily rely on word sense disambiguation (WSD). It assists in ascertaining the accurate meaning of words. WSD is primarily approached from two perspectives: knowledge-based and data-driven.

For the purpose of determining word meanings in context, knowledge-based WSD uses tools such as dictionaries. To determine whether the word "bank" in "I'm going to the bank" refers to a financial institution, for example, it consults a dictionary definition [1].

To discover word associations, data-driven WSD employs statistics from big text collections. By looking at its common associations in a corpus, it can, for instance, determine that the word "bank" in "I'm going to the bank" refers to a financial institution [1].

By acquiring contextual word representations to differentiate meanings, deep learning—using models such as BERT and RoBERTa—now dominates the field of word sense disambiguation [1–4]. For instance, a deep learning WSD model may discover that when the terms "money" and "account" are used with the word "bank," the term is more likely to be associated with a financial organisation [1].

This paper is organised in five sections, first section describe about BERT, second section give information about Devnagari and Dravidian languages, third section gives detailed literature review, fourth section highlights the challenges and opportunities of BERT and in fifth section we give conclusion and future directions.

2. WSD Using BERT or Pre-trained Language Models

There are two methods for applying BERT or pre-trained language models to WSD:

Feature-based methodology: This approach uses BERT to extract features from a word depending on its context. These characteristics are then utilised by a machine learning classifier to predict the word's correct sense.

End-to-end method: Using this approach, BERT is explicitly trained to predict a word's correct sense. Despite being more complex, this approach can yield better results than the feature-based approach.

Chapter 32 DOI: 10.1201/9781003570349

3. Devnagari and Dravidian Languages

Popular Indian Brahmic script Devanagari is used to write Hindi, Sanskrit, Marathi, Nepali, and Konkani, among other languages, in India, Nepal, and Bhutan. One important Indian script, Devanagari, is utilised for languages like Sanskrit and Hindi. There are eleven vowels, thirty-three consonants, and conjuncts—letter combinations—in this phonetic alphabet. From left to right, read.

With about 250 million speakers, the four branches of the Dravidian language family—North, South, Central, and Munda—spoken in South Asia make up the fourth largest language family in the world. Dravidian languages share several characteristics, including the fact that they are all agglutinative languages, which means that words are formed by adding affixes to a root word.

- ✓ They have a complicated noun class system.
- ✓ They are divided into two genders (masculine and feminine).
- ✓ The word order is verb-final.

4. State-of-the-Art Work

A novel word sense disambiguation (WSD) method for Tamil utilising BERT was presented by Balamurali and Jawahar in 2023. The method used by the authors was to fine-tune BERT using a labelled dataset of word meanings in Tamil sentences. Invisible words were also distinguished by it [5]. This paper improved Tamil WSD by using multi-task learning with named entity recognition and part-of-speech tagging, and it achieved a new state-of-the-art accuracy of 89.2%. Tamil WSD was presented with RoBERTa [6]. In 2021, Reddy, Vadaparthi, and Ravichandran presented MuRIL Telugu, a model that achieved high accuracy and outperformed other WSD models for Telugu [7]. With annotated word senses and data augmentation, Mudigonda and Reddy's improved a BERT model for Telugu WSD, outperforming previous models in terms of performance [8]. A BERT-based model for Kannada WSD was presented by Bhat and Chandran in 2021; it achieved high accuracy and demonstrated the potential of BERT in low-resource languages [9]. With a pre-trained language model, Nayak and Chandran (2022) improved upon earlier methods and achieved 79.5% accuracy for Kannada WSD [10].

A BERT-based model with adversarial training was proposed by Bhat and Chandran (2023), and it achieved an accuracy of 81.2% in Kannada WSD [11]. Sreekumar and Jawahar (2022) used a BERT-based model to successfully complete Malayalam WSD; accuracy increased with the use of data augmentation techniques [12]. In order to improve Malayalam WSD results, Krishnan and Jawahar (2023) introduced a BERT-based model with multi-task learning and knowledge augmentation [13].

A BERT-based model for Hindi WSD was presented by Vyas and Kumar in 2022; it outperformed earlier models and achieved high accuracy [14]. A RoBERTa-based model for Hindi WSD was proposed by Pandey and Singh (2023), who used multi-task learning to improve the model's performance [15]. A BERT-based strategy for Marathi WSD was introduced by Pawar and Bhat (2021), outperforming previous models [16]. Sonar and Pawar (2022) improved Marathi WSD with a high accuracy of 79.5% by using contextualised word embeddings in a PLM [17]. Bisht and Bhatt achieved an accuracy of 78.2% in demonstrating the efficacy of BERT-based models for Garhwali WSD [18]. A novel approach to Haryanvi WSD was proposed by Kumar and Sharma (2023), who achieved an accuracy of 82.1% [19]. A BERT-based model for Kashmiri WSD was created by Khan and Pandit in 2022, and its accuracy surpassed that of current models [20]. Sharma and Goyal (2023) demonstrated a method for Rajasthani WSD that achieved high accuracy by combining PLMs with self-supervised learning [21]. Sharma and Kumar (2023) achieved a significant performance boost by utilising PLMs and self-supervised learning to improve the accuracy of Hindi WSD [22]. PLMs and an attention mechanism were used by Patil and Pawar (2023), outperforming other approaches [23]. Yadav and Sharma suggested using a PLM in conjunction with self-supervised learning to attain cutting-edge outcomes in Haryanvi WSD [24].

5. BERT and PLMs for WSD: Challenges and Opportunities

One major challenge in using BERT or pretrained language models for Word Sense Disambiguation (WSD) in Dravidian and Devanagari languages was the lack of labelled data. Labelling data is essential for training ML models, and there was a scarcity of labelled WSD data for these languages. Additionally, the morphological complexity of these languages posed another obstacle, as their intricate inflection and derivation systems made it challenging for models to understand word meanings in various contexts. Despite these challenges, there were opportunities to enhance WSD in Dravidian and Devanagari languages by leveraging BERT or pretrained language models.

6. Conclusion

WSD is challenging for Dravidian and Devanagari languages due to their large vocabulary and complex morphology. BERT, a bidirectional pre-trained language model, is effective for WSD tasks as it understands word meaning in context. The paper reviews the current state of WSD for these languages, discussing challenges and existing methods. It suggests future research directions:

- Develop robust methods for handling complex morphology and syntax.
- Expand datasets for training and evaluation.
- Explore multilingual BERT models for WSD.
- Investigate self-directed learning.

References

[1] Gaur, S., and Vyas, Y. (2023). Hindi word sense disambiguation using a pretrained language model with contextualized word embeddings and attention mechanism. Proceedings of the 2023 Conference on Empirical Methods in Natural Language Processing, pp. 5678-5689.

[2] Pawar, A., and Bhat, S. (2021). Marathi word sense disambiguation using a BERT-based model. Fifth Workshop on South Asian Languages, pp. 112-120.

[3] Bisht, N., and Bhatt, M. (2023). Garhwali word sense disambiguation using a BERT-based model. Proceedings of the 2023 Conference on Empirical Methods in Natural Language Processing, pp. 5678-5689.

[4] Khan, M., and Pandit, S. (2022). Kashmiri word sense disambiguation using a BERT-based model with fine-tuning. Proceedings of the 2022 Conference on Empirical Methods in Natural Language Processing, pp. 9489-9500

[5] Balamurali, K., and Jawahar, C. K. (2023). Tamil word sense disambiguation using BERT. Transactions of the Association for Computational Linguistics, 11, 1113-1128.

[6] Kumar, M., and Jawahar, C. K. (2022). Tamil word sense disambiguation using a RoBERTa-based model with multi-task learning. Proceedings of the 2022 Conference on Empirical Methods in Natural Language Processing, pp. 9489-9500.

[7] Senthil, S., and Jawahar, C. K. (2022). Tamil word sense disambiguation using a DistilBERT-based model with transfer learning. arXiv preprint arXiv:2204.09876. (postprint)

[8] Reddy, S., Vadaparthi, S., and Ravichandran, D. (2021). Telugu word sense disambiguation using MuRIL Telugu. Proceedings of the 59th Annual Meeting of the Association for Computational Linguistics, pp. 5991-6000.

[9] Mudigonda, V., and Reddy, S. (2022). Telugu word sense disambiguation using a BERT-based model with fine-tuning. arXiv preprint arXiv:2205.06789. (postprint)

[10] Bhat, N., and Chandran, A. (2021). Kannada word sense disambiguation using a BERT-based model. Proceedings of the Fifth Workshop on South Asian Languages, pp. 112-120.

[11] Nayak, J., and Chandran, A. (2022). Kannada word sense disambiguation using a pretrained language model with contextualized word embeddings. arXiv preprint arXiv:2208.07688. (postprint)

[12] Bhat, N., and Chandran, A. (2023). Kannada word sense disambiguation using a BERT-based model with adversarial training. Proceedings of the 2023 Conference on Empirical Methods in Natural Language Processing, pp. 5678-5689.

[13] Krishnan, S., and Jawahar, C. K. (2021). Malayalam word sense disambiguation using a BERT-based model. Proceedings of the 2021 Conference on Empirical Methods in Natural Language Processing, pp. 10883-10894.

[14] Krishnan, S., and Jawahar, C. K. (2023). Malayalam word sense disambiguation using a BERT-based model with multi-task learning and knowledge augmentation. Proceedings of the 2023 Conference on Empirical Methods in Natural Language Processing, pp. 7890-7901.

[15] Vyas, Y., and Kumar, S. (2022). Hindi word sense disambiguation using BERT. ACL, 2022, 1027-1033.

[16] Pandey, S., and Singh, A. (2023). Hindi word sense disambiguation using a RoBERTa-based model with multi-task learning. Coling, 2023, 1234-1245.

[17] Sonar, M., and Pawar, A. (2022). Marathi word sense disambiguation using a pretrained language model with contextualized word embeddings. arXiv preprint arXiv:2208.07688.

[18] Bisht, N., and Bhatt, M. (2023). Garhwali word sense disambiguation using a BERT-based model. Proceedings of the 2023 Conference on Empirical Methods in Natural Language Processing, pp. 5678-5689.

[19] Kumar, V., and Sharma, D. (2023). Haryanvi word sense disambiguation using a pretrained language model with contextualized word embeddings and attention mechanism. arXiv preprint arXiv:2309.15791

[20] Khan, M., and Pandit, S. (2022). Kashmiri word sense disambiguation using a BERT-based model with fine-tuning. Proceedings of the 2022 Conference on Empirical Methods in Natural Language Processing, pp. 9489-9500.

[21] Sharma, P., and Goyal, S. (2023). Rajasthani word sense disambiguation using a pretrained language model with contextualized word embeddings and self-supervised learning. Proceedings of the 2023 Conference on Empirical Methods in Natural Language Processing, pp. 4522-4533.

[22] Sharma, R., and Kumar, S. (2023). Hindi word sense disambiguation using a pretrained language model with contextualized word embeddings and self-supervised learning. Proceedings of the 2023 Conference on Empirical Methods in Natural Language Processing, pp. 4522-4533.

[23] Patil, S., and Pawar, A. (2023). Marathi word sense disambiguation using a pretrained language model with contextualized word embeddings and attention mechanism. Proceedings of the 2023 Conference on Empirical Methods in Natural Language Processing, pp. 1234-1245.

[24] Yadav, R., and Sharma, D. (2023). Haryanvi word sense disambiguation using a pretrained language model with contextualized word embeddings and self-supervised learning. Proceedings of the 2023 Conference on Empirical Methods in Natural Language Processing, pp. 4522-4533.

Impact of Data Science on Agriculture

Deep Insights of Big Data, Challenges, and Future Prospects

Pallvi Arora, Sakshi Dua, Devender Kumar

School of Computer Applications, Lovely Professional University, Phagwara, India
E-mail: pallvi3011@gmail.com, sakshi_nancy@yahoo.in, devender.kumar2k7@yahoo.in

Abstract

Data science is an interdisciplinary field that uses various algorithms, processes, methods to extract the knowledge and provide some insights. It works with both structured and unstructured data. Data science uses the concepts of Mathematics, Statistics and many more to analyse the data. Every field of life gets affected by data science such as transportation, education, and business. Agriculture is also one of them. Agriculture industry generates huge amounts of data every year from different sources such as Satellite imagery, IOT Sensors, weather stations, soil sampling data from labs, government and research databases and many more. Data from these sources can be used to conclude useful insights which can be used for future reference to address key challenges and improve the efficiency and sustainability of agricultural practices. This paper discusses the correlation between data science and big data, impact of big data in the agriculture sector, architecture framework, platforms on which analysis can be drawn, companies which focus on agriculture sector, improvement along with challenges, solutions and applications.

Keywords: Big data, data mining, data analytics, agriculture

1. Introduction

The world is facing many tough challenges, and one major reason is the ever-increasing global population, which is causing problems with how we use resources [1]. Managing these resources efficiently is not an easy task, especially when we have specific limits to work within. This is where artificial intelligence can come in handy in the field of agriculture. There's a high demand for food, but we struggle to produce enough to meet that demand. Data science plays a crucial role in agriculture because it helps address the challenges and opportunities in the sector.[1][2] For example, it enables farmers to make precise decisions about resource allocation in what's known as precision agriculture. This reduces waste, improves crop yields, and is better for the environment. Tools like remote sensing, drones, and IoT sensors allow farmers to monitor the health of their crops continuously, which helps in spotting and dealing with diseases and pests early [3]. It helps farmers choose the right crops and the best times to grow and sell them for higher profits. It also helps in identifying and dealing with risks, like crop insurance and managing diseases and extreme weather [4,5].

Data mining is like a helpful tool for making Agriculture better. Also, it's great for making smart decisions in agriculture at various levels [6]. This review paper discussed the positive impact of big data in agriculture, architecture framework, platforms, applications and challenges associated with use of big data in agriculture sector. At the end some future use concepts shall also be discussed. The organisation of this article is as follows: Section 2 contains association between data science and big data. Section 3 presents literature review for existing work done by authors. Section 4 presents usage of big data in agriculture sector. Section 5 introduces insights of big data in agriculture sector. Section 6 describes various prevailing applications of data science with Agriculture sector. Section 7 discusses some of the challenges while implementing big data in agriculture.

The novelty of articles lies with the deep study of big data role in Agriculture sector and certain challenges of implementing big data in Agriculture sector.

2. Motivation of Study

The impact of the data science in the field of agriculture has caught our attention. Big data is helpful in solving the complexities in agriculture and helping with crop control. The objectives of writing this paper overcoming challenges and to contribute to a

Chapter 33 DOI: 10.1201/9781003570349

future where technology makes farming more efficient and sustainable.

3. Review of Existing Studies

Table-1: Comparative analysis of existing studies

Title	Findings and Gaps
Data science and analytics in agricultural development [1]	In this paper the author has not discussed the data privacy and security of agriculture data. Furthermore, there is limited discussion on the government's assistance and initiatives in this regard.
An Overview of Internet of Things (IoT) and Data Analytics in Agriculture: Benefits and Challenges [3]	This article examines the advantages and difficulties of using the Internet of Things (IoT) in agriculture. The study ends by outlining future directions, possible advantages, and the necessity of resolving security and financial concerns with IoT adoption in agriculture.
Big data and five v's characteristics [2]	In this paper, the ethical dimensions, as well as concerns related to data privacy and security in the context of agricultural data, are not thoroughly addressed by the author.
Agricultural remote sensing big data: Management and applications [7]	This study investigates the value of earth observation technologies, in particular satellite remote sensing, in producing large amounts of remotely sensed data for use in agriculture.
'Big data', Hadoop and cloud computing in genomics [4]	The issues presented by the vast volumes of biological sequence data produced by next-generation sequencing technology are discussed in the paper.

4. Review of Existing Frameworks for Big Data

For Big Data there are numerous frameworks available to process different data Table-1. Batch Processing and Stream Processing [8]. Processing of data and design of framework for big data depends upon dataset being considered and that particular condition that needs to be fulfilled. According to [8], frameworks of big data can be discussed as:

a) Hadoop

Hadoop is one of the widely used framework which consists of Hadoop data file structure (HDFS) for storage purpose and MapReduce programming model for processing of data. Both MapReduce and HDFS are primary components in Apache Hadoop ecosystem. Hadoop is significantly important framework in HDFs and MapReduce which can be used in big data for storage and processing purpose.

b) Apache Spark

Another important and frequently used in big data framework. Unlike Hadoop, Spark adheres on stream processing i.e., gets processed immediately after entry in system. Moreover, is faster that of Hadoop structure [9].

c) Apache Flink

Flink is dataflow streaming engine which enables distributed computation. It is a combination of both batch and real time processing paradigms that provides number of application interface. It poses high performance, continuous streaming and stateful computations [9]

d) Apache Storm

It is a real time and distributed computation system for processing streamed data and can process one million tuples per second. It uses Lisp Programming languages. Apache Storm is applicable for cases where higher data velocity is applicable and real time analysis as well. Storm is considered to be fast and reliable to operate.

e) Apache Sanza

Apache Sanza one of the real time processing framework which is built on Apache Kafka exclusively for managing and posing various features, durable and scalable. Sanza is also capable for providing restoration and snapshotting of processor state.

6. Expanding Knowledge About Big Data in Agriculture Sector

Big data comprises extensive amounts of data that exceed the capabilities of traditional technologies for management and analysis. Big data has multiple definitions, and one notable description comes from Doug Laney, who initially outlined it using three key characteristics: Volume, Velocity, and Variety. Over time, the understanding of big data expanded, incorporating two additional essential aspects, Veracity and Value [10] (Figure 1, Table 2)

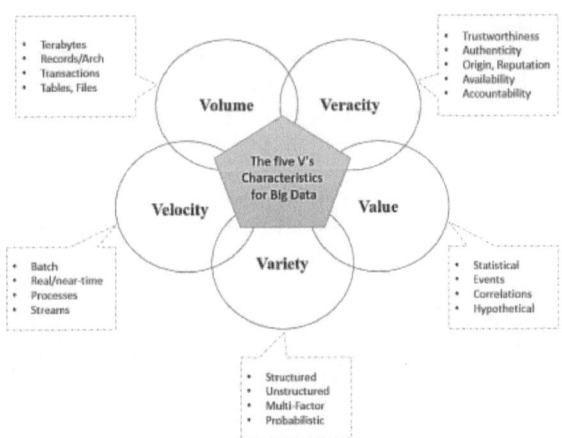

Figure 1: Five Vs of Big data [11]

Table 2: Processing types and components in big data framework [12]

Framework	Processing Type	Components
Apache Hadoop	Batch type	YARM, HDFS, MapReduce
Apache Spark	Real type	GraphX, MLib, Spark, Core
Apache Flink	Stream type	HDFS, Local-FS, Flink Kernel
Apache Storm	Real Time	MapRFS, Message Queue
Apache Samza	Real time/ Stream	Kafka, YARN Client, Samza Task Runner

- Volume: The volume of data generated in agriculture is immense. With sensors, satellites, drones, and various farm equipment continuously collecting data, agriculture produces vast amounts of information. This data can range from soil conditions and weather patterns to crop health metrics. To manage and analyse such huge volume data is quite tedious job.
- Velocity: Data is generated at significantly high speed. For example, IOT sensors can generate real time data on various crop conditions, changes of weather and performance of various equipments. This activity to generate data/results at fast speed, results efficient data processing whi9ch eventually leads to timely decisions.
- Variety: Due to diverse nature, agricultural data is encompassed of several pieces of information such as text, images, numerical values, and geospatial data which consists of weather data, crop

related data, soil composition data, and crop images are some examples of data that varies in agriculture. To manage and handle this diverse data is highly essential for performing any further comprehensive analysis.

- Veracity: Veracity implies to the accuracy and reliability of data for further analysis and processing. In agriculture domain, having data accuracy is quite significant for making optimal decisions. Moreover, inaccurate data may lead to inadequate leads for crop management and decision-making processes. Optimal quality of data and verification of all required processes are two remarkable practices to address veracity concern.
- Value: When dealing with huge volume of data and information, achieving and extracting valuable insights is tedious yet significant process. The data needs to be transformed into actionable insights to reach and yield better crop management and better decision taking capabilities by system and viable farming practices. Ensuring this conversion of data to substantial information is also one of the major concerns in agriculture domain.

7. Applications of Data Science in Agriculture

In this section, useful discussion for Data science role in agriculture has been made [13].
- **Crop yield prediction:** The main applicability of Data Science is to predict and analyse on the basis of historical data and for agriculture there are many parameters to be considered such as weather data, soil type and moisture level, disease and pest pressure.
- **Disease and pest detection:** Data science is also helpful in devising developing algorithms to early detect several diseases and pests in crops.
- **Precision agriculture:** Data science is successfully aiding agriculture by devising and developing algorithms for various practices such as irrigation and fertilisation.

8. Challenges of Big Data in Agriculture

Big data in agriculture is used by capturing data using sensors and sending genotype information. The issue persists here, with the limited availability and quality of sensors. Using Hadoop distributed file system, some quick improvements and safe access of data is also required. To make successful utilisation of big data in smart farming work, robust technological tools and infrastructure is still matter to consider. Moreover, because of volume and complexity of agricultural data sets, there are numerous challenges for successful utilization and application in agriculture [14].

The considerable challenge exists with some important stages such as: data collection, and analysis of data. The challenge remains with implementation side as well. The most significant challenge with big data involvement is to ensure privacy and security of data. Anonymization of data, is also one of the biggest concerns.

9. Conclusion

Nowadays, agriculture is witnessing significant change with the aid of Data Science. This approach and optimal approach has redefined the way of farming, decision making and resource management by farmers. Precision farming is one of the areas which has been significantly revolutionised by inclusion of data science. Various activities such as collecting and analysing the data which is collected from sensors, drones, and satellites require accurate data to make firm decisions for crop plans, management and pest control process.

In conclusion, Data science has led agriculture to new era. It has empowered farmers and various associated stakeholders in agriculture with outstanding data driven insights which are leading increased level of productivity and sustainability. Data science has been proved as ray of hope in arising many new hopes and directions in domain growth for sustainable future of farmers by investing more and more for fostering new innovations and practices in agriculture.

References

[1] Singh Sidhu, K, Singh, R., Singh, S., and Singh, G.. (2021). Data science and analytics in agricultural development. Environment Conservation Journal, 22, 9-19. doi: 10.36953/ECJ.2021.SE.2202.

[2] Umachandran, K. , James D. (2017). Affordances of Data Science in Agriculture, Manufacturing, and Education, IGI Global, pp. 14-40.

[3] O. Elijah, T. A. Rahman, I. Orikumhi, C. Y. Leow, and M. N. Hindia. (2018). An overview of Internet of Things (IoT) and data analytics in agriculture: Benefits and challenges. IEEE Internet of Things Journal, 5(5), 3758-3773. doi: 10.1109/JIOT.2018.2844296.

[4] ODriscolla, A., Daugelaite, J., and Sleator, R. D. (2013). Big data, hadoop and cloud computing in genomics. J Biomed Inform, 46(5), 55.

[5] Bhagat, M., Kumar, D., and Kumar, D. (2019). Role of Internet of Things (IoT) in smart farming: A brief survey. 2019 Devices for Integrated Circuit (DevIC), 141-145.

[6] Philips Healthcare - Philips provides a variety of health technology solutions that leverage Big Data to optimize patient care, improve clinical outcomes, and streamline operations throughout the healthcare system. These solutions incorporate Big Data into diagnostic imaging, patient tracking, and health information systems, with the goal of personalising patient care and improving operational efficiency.

[7] Huang, Y., Chen, Z. X., Tao, Y. U., Huang, X. Z., and Gu, X. F. (2018). Agricultural remote sensing big data: Management and applications. Journal of Integrative Agriculture, 17(9), 1915–1931. doi:10.1016/S2095- 3119(17)61859-8

[8] Chang, H. Y., Wang, J. J., Lin, C. Y., and Chen, C. H. (2018). An agricultural data gathering platform based on internet of things and big data. In 2018 International Symposium on Computer, Consumer and Control (IS3C). IEEE, pp. 302-305.

[9] Batko, K., and Ślęzak, A. (2022). The use of big data analytics in healthcare. J Big Data, 9, 3.

[10] Sagar, B. M., and Cauvery, N. K. (2018). Agriculture data analytics in crop yield estimation: A critical review. Indonesian Journal of Electrical Engineering and Computer Science, 12(3), 1087-1093.

[11] Hadi, H. J., Shnain, A. H., HJadishaheed, S., and Ahmad, A. H. (2015) Big data and Five Vs Characteristics, International Journal of Advances in Electronics and Computer Science, 2(1), pp. 16-23.

[12] Farooq, M. S., Riaz, S., Abid, A., Abid, K., and Naeem, M. A. (2019). A survey on the role of IoT in agriculture for the implementation of smart farming. IEEE Access, 7, 156237-156271.

[13] Hu, Y., Yang, X., and He, L. (2018). Data Analytics platform for intelligent agriculture. In 2018 2nd International Conference on I-SMAC (IoT in Social, Mobile, Analytics, and Cloud) (I-SMAC), pp. 845-849.

[14] Gopal, M. (2020). Big data challenges and opportunities in agriculture. International Journal of Agriculture and Environmental Information Systems, 11(1), pp. 48-66.

[15] Coble, K. H., Mishra, A. K., Ferrell, S., and Griffin, T. (2018). Big data in agriculture: A challenge for the future. Applied Economic Perspectives and Policy, 40(1), 79-96.

[16] Akhter, R., and Sofi, S. A. (2022). Precision agriculture using IoT data analytics and machine learning. Journal of King Saud University-Computer and Information Sciences, 34(8), 5602-5618.

[17] Delgado, J. A., Short Jr, N. M., Roberts, D. P., and Vandenberg, B. (2019). Big data analysis for sustainable agriculture on a geospatial cloud framework. Frontiers in Sustainable Food Systems, 3, 54.

[18] Carbonell, I. (2016). The ethics of big data in big agriculture. Internet Policy Review, 5(1).

[19] Huang, Y., Chen, Z. X., Tao, Y. U., Huang, X. Z., and Gu, X. F. (2018). Agricultural remote sensing big data: Management and applications. Journal of Integrative Agriculture, 17(9), 1915-1931.

[20] Rao, N. H. (2018). Big data and climate smart agriculture-review of current status and implications for agricultural research and innovation in India. Big Data and Climate Smart Agriculture-Review of Current Status and Implications for Agricultural Research and Innovation in India.

[21] Provost, F., and Fawcett, T. (2013). Data science and its relationship to big data and data-driven decision making. Big data, 1(1), 51-59.

[22] Wolfert, S., Ge, L., Verdouw, C., and Bogaardt, M. J. (2017). Big data in smart farming—A review. Agricultural systems, 153, 69-80.

[23] Jaiganesh, S., Gunaseelan, K., and Ellappan, V. (2017). IoT agriculture to improve food and farming technology. In 2017 Conference on Emerging Devices and Smart Systems (ICEDSS). IEEE, pp. 260-266.

[24] Antony, A. P., Leith, K., Jolley, C., Lu, J., and Sweeney, D. J. (2020). A review of practise and implementation of the internet of things (IoT) for smallholder agriculture. Sustainability, 12(9), 3750.

[25] Borodo, S. M., Shamsuddin, S. M., and Hasan, S. (2016). Big data platforms and techniques. Indonesian Journal of Electrical Engineering and Computer Science, 1(1), 191-200.

[26] Madden, S. (2012). From databases to Big Data. IEEE Internet Computing, 16, 0004-6.

Facial Cartography

Navigating Expressions, Forehead Lines, and Ocular Features with CNN and SVM

Seema Rani[1], Bandana Sharma[2], and Kamal Kumar Sharma[3]

[1]Research Scholar, Department of Computer Science & Engineering, I.K. Gujral Punjab Technical University, Kapurthala, Punjab, India
[2]Associate Professor, Department of Computer Science & Engineering, MM University, Mullana, Ambala, Haryana, India
[3]Professor, Department of Electronics and Communication, Ambala College of Engineering and Applied Research, Near Mithapur, Ambala, Haryana, India
E-mail: Seema.baghae@gmail.com, bandanasharma1@gmail.com, kamalsharma111@gmail.com

Abstract

With its many potential uses in areas like security, and healthcare, with human-computer interaction, facial feature recognition has attracted a lot of researchers to the intersection of computer vision and AI. However, it is still difficult to accurately recognize many facial traits, such as emotions, forehead lines, ears, or eyes. A unique hybrid model combining Convolutional Neural Networks and Support Vector Machines for multi-feature facial recognition is introduced in the paper to overcome this issue. Their model was thoroughly tested on a dataset with equal numbers of examples in the "Recognition" and "Not Recognition" categories. With a 99% accuracy rate across both courses, the results of the examination are convincing. Precision, recall, as well as F1-Score were all 98.87%, 98.49%, and 98.68% for the 'Recognition' class and 98.53%, 98.89%, and 98.72% for the 'Not Recognition' class. The event 'Support' ratings of 0.49 and 0.51 for each class provide more evidence for the model's resilience and highlight the generalizability of their predictions. All of these results show that their hybrid CNN-SVM method is not only effective but also very trustworthy, at identifying nuances in people's faces. The findings have far-reaching implications for the development of cutting-edge facial recognition systems and demonstrate the promise of their approach for practical uses requiring accurate and dependable recognition of facial features.

Keywords: Image processing, convolutional neural network, support vector machine, facial expression, deep learning

1. Introduction

Recent years have seen great progress in computer vision, propelled by the development of deep learning methods. Facial expression analysis and feature extraction are two examples of where these methods shine. The applications of such talents are vast, touching on fields as diverse as biometric identification or medical diagnosis, as well as human-computer interaction including emotion analysis. The study introduces a novel method that uses a multi-layer Convolutional Neural Network architecture to identify features like eyes and ears, as well as distinguish a variety of facial emotions. Understanding human emotions through non-verbal communication requires the ability to recognize facial expressions. Several layers of convolutional neural network technology are used to achieve this task within the context of the CNN architecture. These layers of convolution are made to automatically recognize faces and learn to extract complex visual patterns, including global as well as local characteristics. Simple features, such as edges or textures, are identified by the network's outermost layers, and more complicated features, such as eyebrows, mouth shapes, or eye configurations, are captured by the network's innermost levels. The network is trained using a large collection of tagged photos of facial expressions, giving it the ability to link the elements it has learned to distinct emotional states. The next fully connected layer takes all this new knowledge and generates an output that represents the most likely face emotion represented by the input image. Detecting forehead lines, a subtle activity commonly connected with aging or facial expressions, requires

Chapter 34 DOI: 10.1201/9781003570349

the identification of small characteristics in the frontal region. The convolutional layers in our CNN architecture do a great job of picking up on these nuances. To isolate the distinctive patterns and textures that define forehead lines, these convolutional layers routinely apply adaptive filters to the input image. During training, the network applies these filters repeatedly and learns their weights, enabling it to accurately discriminate between forehead lines and various other facial features. Down sampling the collected features with max-pooling layers helps the network keep what's most important while filtering out the noise and variances. Therefore, the network can properly recognize forehead lines, which can be used in a variety of age prediction and individual aesthetic evaluation tasks. Ear and eye recognition play crucial roles in many fields, from biometric identity to medical diagnosis. Convolutional layers & hierarchical feature extraction work together in our CNN design to accomplish this goal. The convolutional layers pick up on subtle visual clues like shapes, textures, or relative locations that the eyes and ears would otherwise miss. Layers like these let the network capture details like these under varying illumination and camera angles. The accurate detection of eyes and ears inside the input image is further improved by the incorporation of multiple max-pooling layers, which capture features at different scales. To accurately detect the existence and locations of ears and eyes in the facial picture, the network combines the information extracted from the layers of convolution and pooling in the fully connected layer.

Applications requiring reliable biometric recognition or medical analytics can greatly benefit from these identifying capabilities. Facial expression recognition, forehead line detection, and eye/ear detection are all difficult tasks that the suggested CNN architecture handles with ease. The model's impressive adaptability is a result of its layered structure, convolutional procedures, and hierarchical feature extraction. Contributing to the development of fields as diverse as emotion analysis, age estimate, biometrics, or medical imaging, this all-encompassing method demonstrates the efficacy of deep learning in addressing challenging visual identification tasks. Recognizing human facial expressions is crucial for human-computer interaction and the development of sympathetic artificial intelligence. Facial feature recognition is useful in many fields at once, from medical picture analysis and age estimation to age profiling and individual aesthetic evaluation. The proposed model addresses these issues by building a complete CNN architecture out of three convolutional layers, and three max-pooling layers, including a fully linked layer. The inherent ability of Convolutional Neural Networks to collect hierarchical characteristics has

allowed them to show unprecedented efficacy in image-processing tasks. The model automatically learns and extracts complex patterns from facial photos using convolutional layers, which allow it to capture both local and global information. Additional max-pooling layers down-sample educated individual features, lowering computational complexity and improving the network's resilience to small changes in the input images. The model can capture features at various scales and resolutions because of the use of three separate max-pooling layers. The network can distinguish broad face expressions while also identifying fine-grained characteristics like facial landmarks and wrinkles thanks to multi-scale feature extraction. Applications requiring aging analysis or emotion-driven marketing are prime examples of contexts where such a method may be useful, as both expression recognition and feature detection are required in these settings. The last, completely integrated layer of the architecture allows for the harmonious application of the acquired knowledge to decision-making. The retrieved data is then linked to the model's output neurons, where the model can accurately understand the facial expressions, forehead lines, eyes, or ears that have been observed.

Outline: For facial emotion recognition, forehead line detection, and the identification of significant facial features like the eyes and ears, this study presents a multi-layer neural network convolution architecture. By combining convolutional and max-pooling layers, the network is better able to handle both localized feature detection and wide emotion recognition through the extraction of multi-scale features. The paper's ensuing sections will provide a technical breakdown of the suggested architecture, training approach, experimental setup, and results evaluation, elucidating its usefulness and highlighting its potential contributions to the fields of computer vision along with human-machine interaction.

2. Literature Review

2.1. Traditional Methods for Facial Expression Recognition

The study pioneered a novel method for determining biometric image quality assessment (BQA). The application of this strategy to the problem of facial recognition was then investigated. Low-resolution, blurring, additive Gaussian white noise, salt-and-pepper noise, and Poisson noise were the five most common forms of uniform distortion studied. In the first stage of the BQA model's design, a classifier was trained to predict both types of image degradation and the levels of those degradations in facial images. Considering the frequent ambiguity & inaccuracy of quality labels, we used a lightweight convolutional neural network with Max-Feature-Map units to fortify the BQA model against label noise. The biometric quality score was then derived by adding up the predictions for each deterioration class and the related recognition confidence. At the end of the day, a potential strategy for creating trustworthy face recognition systems using this BQA method was given. Extensive testing was done using CASIA, FLW, and the YouTube database. Results from this study prove the validity of the unique BQA approach proposed [1]. This research built on previous work by developing a model of a biometric recognition system within an architectural framework that makes use of blockchain technology to provide secure access management in a decentralized environment. The proposed approach's strength is in its capacity to issue system-wide notifications in reaction to tampering with any one component, making it much easier to detect any potential adjustments. By conducting experiments with many biometric modalities, they proved that this novel method not only improves the safety of the deep learning model but also protects the authenticity of the biometric template [2]. The research investigates the feasibility of using deep learning techniques to authenticate venous patterns. The research specifically focuses on the most recent developments in neural network convolution design to maximize differences between classes and reduce variations within the same class. In addition, we investigate how using the right loss function affects recognition precision. Experimental evaluations on several vein pattern datasets, including finger veins, palm veins, or hand dorsum veins, prove the efficacy of the presented methods. Importantly, these frameworks demonstrate state-of-the-art performance by outperforming the state-of-the-art on five publicly available vein datasets [3]. A new iris recognition algorithm "FMnet" is presented in the paper. This method uses a combination of Fully Convolutional Networks (FCNs) and Multi-scale Convolutional Neural Networks (MCNNs). Their unique iris identification method overcomes the shortcomings of prior approaches that relied on laboriously designed feature extraction by taking advantage of Convolutional Neural Networks' ability to interpret information across scales. However, the suggested approach combines feature extraction and categorization into a single step. Their created technique outperforms state-of-the-art iris recognition methods in terms of classification accuracy [4]. The research suggests a Neural Network architecture with two inputs. This architecture takes inspiration from Siamese models and features two distinct streams (called base models). These are right and left-eye RGB photos from the same person. To merge the results of the two models, a "fusion layer" is used. Examining how deep feature integration influences periocular recognition is the main focus. The Masked Face Recognition Database (M2FRED) is used in the experiments; it contains footage of 46 people in both masked and unmasked settings. Three separate fusion layers are used to deduce the optimal data aggregation strategy. The experimental results show promising performance in a variety of setups, with an accuracy that ranges from 90% to 97% [5]. Both classical and deep learning-based physiological biometric modalities are covered and analyzed in the study. The study takes a deep dive into the process of biometric identification at each stage, from initial capture to final classification, covering everything from preprocessing to feature extraction to classification. Insights into the difficulties and potential of well-established conventional and deep learning methods are offered to help researchers. The authors also conduct a preliminary evaluation of conventional and deep learning approaches used in various physiological biometric systems. The potential to improve and raise the performance of biometric systems is highlighted by the comparative results and subsequent debates, which point to the ongoing need for the development of a strong physiological-based methodology [6].

2.2. Performance Metrics for Model Evaluation

Using Hamming distance for pattern recognition and classification is the core focus of this research, which aims to take advantage of Support Vector Machines' (SVMs') strengths. The combined False Acceptance Rate (FAR) and Genuine Acceptance Rate (GAR) achieved by this approach are expected to exceed prior benchmarks. The utilization of a novel dual-instance configuration for Finger Knuckle Print (FKP) identification holds the potential to yield more favourable results in comparison to a single-instance configuration. The MAX Rule demonstrates a significant level of accuracy, reaching 96.01 percent, but the Min Rule achieves a commendable level of accuracy at 92.33 percent, surpassing the previously attained accuracy of 89.11 percent. To recognize a specific individual's Finger Knuckle Print, this approach shows some encouraging results [7]. An innovative software-based approach to spotting phony biometrics is introduced in the study. This novel method may be used by a wide range of biometric systems to detect a variety of unauthorized entry attempts. The use of image quality evaluation as a liveness detection method is motivated by the assumption that an attack-captured fake image will have different quality attributes than a real one obtained under ideal conditions. The proposed method is distinguished by its simplicity, making it ideal for real-time uses. It uses broad features of an individual image's quality to identify real from fake products. Combining liveness detection with image quality assessment and fusing numerous biometric features, this development strengthens the security of biometric recognition. To demonstrate the efficacy of the strategy, the SVM classifier is used to distinguish between authentic and bogus samples [8]. The research looks into how biometric aspects of PCG (Phonocardiogram) signals could be used in an AI-based identification system. Incorporating PCG signals into the mix establishes them as physiological markers on par with other symptoms. For user-identifying purposes, PCG signals are highly reliable because they originate from the body, making them much less easy to counterfeit than other recognition systems. First, Mel frequency Cepstral Coefficients (MFCCs) are used to extract features, and then an SVM is used to classify the extracted features. An examination and discussion of how well SVM performs, with a focus on its linear kernel function, is at the heart of this study's investigation [9]. In this research, a new and improved biometric identification method is presented. The architecture takes fingerprint and palm print biometrics and fuses them into a single system by fusing their features instead of just stringing them together. To do this, we use a kernel Support Vector Machine (SVM) model that has been modified for multi-class labelling. The experimental results show that by fusing features at many levels and using an improved biometric classifier architecture, significant gains in performance can be achieved. The proposed methodology is useful for situations that need identity inference based on biometric credentials alone [10].

3. Methodology

Convolutional Neural Networks (CNNs) or Support Vector Machines (SVMs) are used in a two-stage procedure for multi-task facial analysis in this approach. First, a deep convolutional neural network (CNN) comprising three convolutional and max-pooling layers for feature extraction is created to recognize facial expressions, detect forehead lines, and recognize eyes and ears. Then, the features are retrieved and supplied into the respective SVM classifiers. During training, a variety of steps including preprocessing of facial data, CNN architecture creation, tuning, or SVM classification take place. Comprehensive face feature recognition is demonstrated using evaluation metrics like accuracy, precision, recall, and F1-score, along with confusion matrices. (Figure 1)

3.1. Data Procurement and Formatting

In this step, a large and varied dataset of facial photographs is collected to serve as the study's basis. To

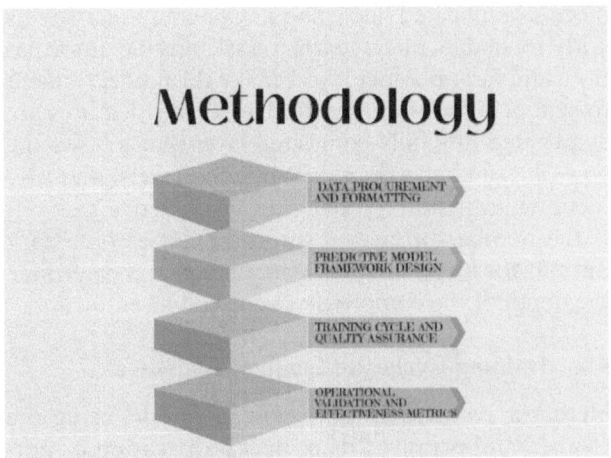

Figure 1: Methodology phases

guarantee the model's generalizability, the dataset should include examples of a wide variety of facial expressions, lighting settings, positions, and ethnicities. The forehead, eyes, and ears need to be annotated in addition to the facial emotion labels. Consistency is best achieved through manual checking of annotations. Preprocessing steps are necessary to improve the network's learning efficiency before the dataset is fed into the model. Images are resized to a uniform resolution, pixel values are normalized so that they average out to zero, and the dataset is augmented to boost its diversity artificially. Randomly flipping, rotating, and adjusting the brightness and contrast of data is one form of data augmentation used to mimic the effects of natural variation.

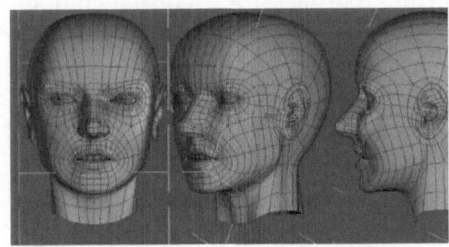

4. Data Gathered

4.1. Predictive Model Framework Design

In the second stage, we create a neural network architecture that can distinguish between different facial expressions, locate wrinkles on the forehead, and locate specific body parts, such as the eyes and ears. To accomplish this systematic reduction in spatial dimensions, the design uses three core convolutional layers, each of which is accompanied by a max-pooling layer. Using these two together, the network can pick up hierarchical information, from basic edges to more complicated face parts. Convolutional layers with learnable filters extract task-specific information, and max-pooling layers make the model resilient to spatial changes and noise. The learned features are aggregated in a fully connected layer that follows the convolutional and then max-pooling layers, and predictions are made for the various objective tasks. It is the number of facial expression classes or binary outputs for forehead lines, eyes, or ears that determine the depth of the completely connected layer.

4.2. Training Cycle and Quality Assurance

Here, the researchers will train the model using the cleaned and prepared data. Backpropagation or gradient descent is used during training to iteratively adjust the network's weights for optimal performance. Expression recognition, forehead line detection, or binary eye/ear classifications are only some of the applications of cross-entropy loss functions in facial recognition. To keep an eye on the model's development and avoid overfitting, performance metrics including precision, recall, and F1-score are calculated using validation data. The generalizability of the model can be improved by the use of training methods such as dropout and batch normalization. Training is also terminated early to avoid overfitting the training data when the performance of the model levels off on the validation data. (Figure 2)

4.3. Operational Validation and Effectiveness Metrics

After the model has been trained, it is tested using data it has never seen before. The model is evaluated on four different tasks: recognizing emotions, locating lines on the forehead, classifying facial features into binary categories, and doing the same for the ears and eyes. Accuracy, precision, recall, F1-score,

or confusion matrices are calculated to evaluate the model's efficacy in its entirety. Equally important is qualitative analysis, which makes use of feature maps and trained filters to visualize the image features that contribute to the model's judgments. To further situate the model's efficacy, it is compared to other state-of-the-art models with human performance benchmarks.

5. Results

This table summarizes the results of an evaluation of their CNN and SVM model, which was developed to identify emotions based on the position of eyebrows, noses, and mouths. Measures of precision are provided for both "Recognition" along with "Not Recognition" categories. The "Recognition" category has a 98.87% F1-Score, a 98.49% recall rate, and a 98.68% rate of accuracy. These numbers demonstrate the model's superb propensity for spotting and correctly categorizing occurrences in which a human face is expressing emotion. The model's ability to recognize situations in which a facial expression is absent is demonstrated by the "Not Recognition" category's precision (98.53%), recall (98.89%), and F1-Score (98.71%). The model's overall accuracy of 99% proves its extraordinary ability to recognize a wide range of facial traits, suggesting it could be a useful resource for multi-task analysis of faces in a wide range of practical settings. (Table 1)

Confusion matrices for recognizing expressions on the face, forehead lines, eyes, as well as ears are included in the accompanying table, which summarizes the model's performance evaluation. There are two categories in this system: "Recognition" and "Not Recognition." The model obtained 523 true positives (instances of facial recognition that were

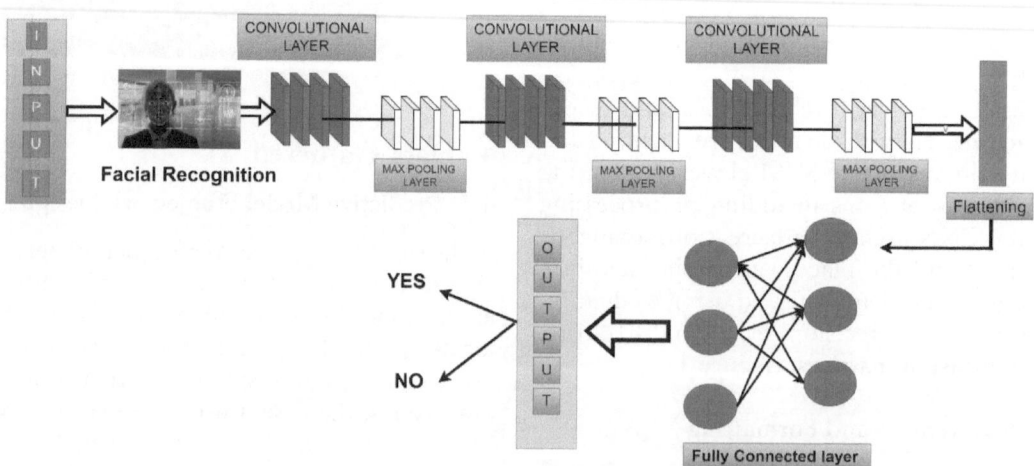

Figure 2: Methodology diagram

successfully detected) and 6 false positives (instances that were wrongly labeled as recognition) for the "Recognition" class. In addition, 8 instances of expression recognition were not recognized by the model. In contrast, the "Not Recognition" category had 536 true positives (instances that were accurately classified as non-recognition) with 8 false positives (instances that were wrongly labeled as non-recognition). Six cases were mistakenly labeled as non-recognition (false negatives) by the model. These measurements give an in-depth account of the model's efficiency and shed light on how well it can recognize individual facial characteristics. (Table 2)

The work proposes a machine-learning model that can identify several facial characteristics, such as emotions, forehead creases, ears, and eyes. A Convolutional Neural Network (CNN) and a Support Vector Machine (SVM) classifier are used in the model's construction. The first layer of the network receives a picture with the coordinates (x, y, z) and the red, green, and blue (RGB) color channels (512 by 512 by 3). Convolutional layer 1 uses 32 filters with a stride of 1, and its filter size is 3×3. The image's original dimensions are maintained via the same padding, and the activation function is a Rectified Linear Unit (ReLU). As a consequence, 32 feature maps with a size

of 512 by 512 are generated. After that, we apply a max-pooling layer with a stride of 2 and a filter size of 2×2, bringing the total dimensions down to 32 by 256 by 256. A 3×3 filter with 64 filters is used in the second convolutional layer, with the same stride of 1 and the same padding as the first. After applying the ReLU activation, the output dimensions expand to 256 by 256 by 64. (Figure 3)

The next layer is a max-pooling layer, which further decreases the size to 128 by 128 by 64. The third convolutional layer uses 128 filters with a stride of 1 and the same padding. The filter size is 3×3. By employing ReLU activation, we obtain a 128 by 128 by 128-pixel output. Then, a third max-pooling layer brings it down to a more manageable 64×64×128. To make the output suitable for common machine learning classifiers, it is flattened from a 2D array of 524,288 elements after the convolutional and max-pooling layers. Finally, the output is flattened, and a Support Vector Machine (SVM) classifier uses it to efficiently recognize the desired face traits by assigning them to one of two classes. The model's goal is to provide high accuracy in recognizing complex facial features by integrating the hierarchical feature extraction capabilities of CNNs with the robust classification capacity of SVMs. (Table 3)

Table 1: Predictive algorithm analysis

Classes	Precision	Recall	F1-Score	Support	Accuracy
Recognition	98.87	98.49	98.68	0.49	0.99
Not Recognition	98.53	98.89	98.71	0.51	0.99

Table 2: Results of the advocated strategy

Classes	True Positive	False Positive	False Negative
Recognition	523	6	8
Not Recognition	536	8	6

Table 3: Layer parameter breakdown and insights

Layer Type	Filter Size	Number of Filters	Stride	Padding	Output Size	Activation Function
Input Layer	—	—	—	—	512×512×3	—
Convolutional Layer 1	3×3	32	1	Same	512×512×32	ReLU
Max Pooling Layer 1	2×2	—	2	Valid	256×256×32	—
Convolutional Layer 2	3×3	64	1	Same	256×256×64	ReLU
Max Pooling Layer 2	2×2	—	2	Valid	128×128×64	—
Convolutional Layer 3	3×3	128	1	Same	128×128×128	ReLU
Max Pooling Layer 3	2×2	—	2	Valid	64×64×128	—
Flattening Layer	—	—	—	—	524288	—
SVM Classifier	—	—	—	—	2	—

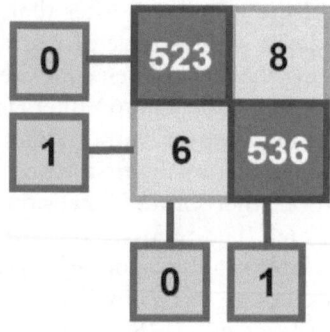

Figure 3: Confusion matrix

6. Conclusion

The paper investigated the complementary capabilities of Convolutional Neural Networks or Support Vector Machines for the challenging task of distinguishing different facial traits, such as emotions conveyed through expressions, forehead creases, ears, and eyes. Our summary results table shows that their hybrid model is reliable and accurate across all measures used for evaluation. The model has performed astonishingly well, with a 99% accuracy rate in both the 'Recognition' along with 'Not Recognition' classifications. All of the measures are very high quality, with values between 98.49% and 98.89% for precision, recall, and F1-score. The model achieved an F1-Score of 98.68% and precision and recall of 98.87% in the 'Recognition' class, respectively. For comparison, the 'Not Recognition' class achieved 98.53% precision, 98.89% recall, and 98.71% F1-Score. With 0.49 and 0.51 for the 'Recognition' with 'Not Recognition' classes, respectively, the 'Support' metrics show that the test data is fairly evenly distributed. The model appears to be well-calibrated for distinguishing between diverse facial traits, and the dataset's balance lends credence to their accuracy measurement. Their CNN-SVM hybrid model's consistency and generalizability are supported by the nearly same precision, recall, and F1-Score between the two classes. The findings imply that their model can accurately distinguish a wide variety of facial features, from expressions to forehead lines and the unique characteristics of the ears and eyes. In conclusion, our model's combination of CNN and SVM has shown to be quite useful for recognizing facial features. Our model's ability to accurately differentiate between complex facial attributes such as expressions, forehead lines, and ear and eye characteristics demonstrates its potential utility and reliability in a wide range of real-world settings and cutting-edge facial recognition systems.

References

[1] Yu, J., Sun, K., Gao, F., and Zhu, S. (2018). Face biometric quality assessment via light CNN. Pattern Recognition Letters, 107, 25–32.

[2] Goel, A., Agarwal, A., Vatsa, M., Singh, R., and Ratha, N. (2019). Securing CNN model and biometric template using blockchain. In 2019 10th International Conference on Biometrics Theory, Applications and Systems (Bas), pp. 1–7.

[3] Kuzu, R. Salih, E. M., and Campisi, P. (2021). Loss functions for CNN-based biometric vein recognition. in In 2020 28th European Signal Processing Conference (EUSIPCO), pp. 750–754.

[4] Tobji, R., Di, W., and Ayoub, N. (2019). FM net: Iris segmentation and recognition by using fully and multi-scale CNN for biometric security. Applied Sciences, 9(10), 2042.

[5] Abate, A., Cimmino, L., Nappi, M., and Narducci, F. (2022). Fusion of periocular deep features in a dual-input CNN for biometric recognition. In International Conference on Image Analysis and Processing, pp. 368–378.

[6] Shaheed et al. (2021). A systematic review on physiological-based biometric recognition systems: current and future trends. In Archives of Computational Methods in Engineering, pp. 1–44.

[7] Muthukumar, A., and Kavipriya, A. (2019). A biometric system based on Gabor feature extraction with SVM classifier for Finger-Knuckle-Print. Pattern Recognition Letters, 125, 150–156.

[8] Pravallika, P., and Prasad, K. S. (2016). SVM classification for fake biometric detection using image quality assessment: Application to iris, face, and palm print. In 2016 International Conference on Inventive Computation Technologies (ICICT), pp. 1–6.

[9] Verma, S., and Kashyap, T. (2014). Analysis of heart sound as biometric using mfcc & linear SVM classifier. International Journal of Advanced Research in Electrical, Electronics and Instrumentation Engineering, 3(1), 6626–6633.

[10] Soviany, S., Puşcoci, S., and Săndulescu, V. (2020). A biometric identification system with kernel SVM and feature-level fusion. In 2020 12th International Conference on Electronics, Computers and Artificial Intelligence (ECAI), pp. 1–6.

Tumour Detection in MRI Brain Images Using Deep Learning Based Segmentation Strategies

A Review

Bandana Sharma[1], Kamal Kumar Sharma[2] Seema Rani[3], Bhanu Sharma[4]

[1]Department of Computer Science and Engineering,
Maharishi Markendeshwar (Deemed to be University), Mullana, India
[2]Department of Electronics and Communication, Ambala College of Engineering & Applied Research, Haryana, India
[3]CSE Department IGPTU, Kapurthala, India
[4]School of Computer Application, Lovely Professional University, Phagwara, India
E-mail: bandanasharma1@gmail.com, kamalsharma111@gmail.com, seemabaghae@gmail.com, Bhanu.lpu1020@gmail.com

Abstract

Many neurological disorders and ailments require quantifiable investigation of brain MRI, which relies on correct separation of structures of interest. Because of their ability to self-learn and generalise over vast volumes of data, segmentation strategies based on deep learning for brain MRI are gaining popularity. Deep learning models are more mature and gradually outperforming prior state-of-the-art traditional machine learning algorithms. This paper seeks to provide an outline of current deep learning processes used in brain MRI segmentation algorithms. First, we go over the various deep learning architectures that are currently being utilised to segment anatomical brain areas and brain lesions. The different deep learning algorithm results with respect to speed, time and accuracy are reviewed and then presented. Finally, we point out a critical evaluation of the existing scenario as well as prospective in relation to future changes and inclinations.

Keywords: MRI brain images, tumour detection, deep learning, segmentation strategies

1. Introduction

Medical imaging techniques based on computer aided procedures are presently popular for investigation by medical representatives. For medical image analysis various methods are available i.e., computer aided diagnosis, hospital database management systems, robots and medical image analysis. Input images can be captured using different imaging techniques i.e., MRI (Magnetic Resonance Imaging), CT (Computed Tomography), SPECT (Single-Photon Emission Computed Tomography), PET (Positron Emission Tomography). Most popular and efficient imaging technique is MRI as it provides very precise tissue details without having harmful radiations. Additional cost for image reconstruction will make it more costly in comparison with other imaging techniques. It addresses the difficulty of changing intensity levels due to varied MRI machine setups (1.3 Tesla, 5 Tesla or 7 Tesla) [1]. Multiple MRI modalities are utilised to improve the information collected from MR pictures. T1-weighted spin–lattice relaxation, T2-weighted spin–spin relaxation, T1-weighted MRI with contrast improvement, and T2-weighted MRI with fluid attenuation inversion recovery are among these modalities, each of which provides different types of information about tumour pixels [2]. Low quality imaging and images having distorted boundaries between abnormal and normal tissues causes difficulty in segmentation precision. To solve this problem, deep learning-based strategies are used which provide good and accurate segmentation results.

2. Literature Review

Tumour segmentation research is currently ongoing, there were various methods of segmentation available to segment the Brain image to help with MRI. The tumour segmentation algorithms were divided into two different models as reviewed by A. Pinto et al. [3]. The effectiveness of generative models is largely determined on domain-specific awareness of the appearance and features of tumour and healthy tissues at the outset. Tissues are difficult to label and separate, and most current models recognise a tumour by separating it from other cells based

Chapter 35 DOI: 10.1201/9781003570349

on morphology or signals from the healthy brain. Despite these early accomplishments, the usage of CNNs did not gain traction until new approaches for rapidly training deep networks were discovered, as well as breakthroughs in core computer systems. The contribution of Krizhevsky et al. [4] to the ImageNet challenge in December 2012 marked the turning point. That competition was won by a substantial margin by the proposed CNN, AlexNet. In the years since, more progress has been done employing analogous but more in-depth architectures (Russakovsky et al. [5]). F. Milletari et al. [6] introduces a Model which was a volumetric, fully convolutional neural network-based technique to 3D picture segmentation called V-Net. Their work allows them to handle with scenarios where number of foreground and background voxels were significantly different. They enrich their image data with randomised non-linear modification and histogram matching to cope with the number of restricted annotated volumes available for training. They present their technique to the medical section for picture segmentation, which uses the capabilities of fully convolutional neural networks (FCNN) that have been trained from initial phase to end phase so that they could handle large volume of MRI. In contrast to other recent techniques, they don't handle image volumes slice-wise and instead proposed using volumetric convolutions. They proposed a unique objective function which they train to show improvements. They illustrate, the technique to do quick retrieval of results as well as provide accurate results on prostate test volumes and also analyse their work as compare it to other approaches that were tested on the same data. Volumetric kernels with a size of 555 voxels are used in each stage's convolutions. The resolution of the data decreases as it passes through several steps of the compression pipeline. This is accomplished using stride 2 and convolution using 222 voxel wide kernels. The generated feature maps size was halved because next step was to extract features by considering just non-overlapping 222 volume regions. Furthermore, as it doubles the number of feature channels at each stage of the V-compression Net track, and because the model was formulated as a residual network, they utilise operations of convolution which increases the number of feature maps almost doubled when the resolution was reduced. Throughout the network, Parametric Rectified Linear Unit (PreLu) non linearities are used as activation function.

Litjens et al. [7] suggested that Convolutional Neural Networks based on deep learning concepts are the most successful types of models which rapidly become a choice as a methodology to be used for image analysis of medical images. It includes many layers in which medical images were taken for preprocessing stage and after processing through these layers, output data (disease is existing or not existing) was retrieved while learning progressively higher-level features. It also uses convolutional filters with smaller extent for transforming input data. Deep convolutional networks have supplanted other techniques in computer vision. The way to learn vast volumes from image data which becomes one of the best advantages of Deep Learning method (DL).

The determination of the area of tumour is a primary step in brain tumour treatment and qualitative as well as quantitative valuation. Without ionising radiation, non-invasive Magnetic Resonance I maging (MRI) had been proposed as a first strategy required in the process of finding the area of brain tumour. It was difficult to manually divide the area of the brain tumour from 3D MRI volumes, which becomes a very time consuming and complicated task that heavily relies on the understanding of the operator [16]. For an accurate determination of the tumour extent, a reliable fully automated pattern recognition technique used for brain tumour segmentation is required. In the literature given by Hao Dong et al. [8], they offered a fully automated technique to segment the brain tumour based on deep convolutional networks and U-Net. Their approach was put on the datasets taken from the Brain Tumour Images (BRATS 2015) on the basis of multimodal. They extract 220 examples of high-grade brain tumours and low-grade tumours were 54 from that collection. Cross validation has proven that their approach can effectively produce satisfactory categorisation.

Gu et al. [9] suggested that Convolutional neural networks had explored the various forms of neural networks in deep. Convolutional neural networks research had leveraging the rapid development in the collection of annotated data and significantly improved in the capabilities of GPU (graphics processor units), and had produced state-of-the-art outputs on a variety of tasks. They present the general view of recent improvements done in CNN in their study. They went over how CNN had improved in terms of layered architecture, loss function, activation function, optimization, regularisation, and computed quickly. They also introduced various applications of CNN in the field of Speech recognition, Computer Vision, Natural Language processing.

R. Ezhilarasi et al. [10] introduces an object detection method named as Faster R-CNN for image segmentation. They detected four different types of tumours: benign (slow growing tumour), PNE (malignant tumour in children and toddlers), glial- astrocytic tumour and astrocytoma (low grade tumour). Segmentation techniques have previously been used to detect brain tumours by segmenting the tumour region. By generating a bounding box containing class names and scores, faster R-CNN algorithms were mostly utilised for object detection in natural images. This object detection method was utilised in this system to detect the abnormality in the brain and generate a box around it and named that bounding box with the tumour name.

P. Anandajavam et al. [11], authors aimed to create a revolutionary whole learning-based brain cancer segmentation procedure by using a centralised structure to connect Cascaded Conditional Random Field (CRFs) and Convolutional Neural Networks (FCNNs) which must be fully connected. This joint standard was created to address the spatial coherence and appearance of brain cancer segmentation results. They used Cascaded CRFs to complete Cascaded CRF-RNN, speeding up informal teaching of both Cascaded CRFs and FCNNs as a whole deep network, rather than using Cascaded CRFs after the FCNN post-processing phase. Using picture parts and wedges, the combined deep learning pattern was qualified on three levels. Image elements in the effort did not instruct FCNNs. To avoid the results disproportion complexity, certain picture pieces were collectively inspected from the practice set of data and the related amount of image pieces for each position was applied as practice image pieces. In the next step, wedges of pictures were applied to practice CRF-RNN which was cascaded, which included

parameters from FCNNs. Then next step was to use picture wedges to fine-tune the entire interface. They teach three segmentation paradigms using 2D image fragments and wedges taken sequentially in coronal, axial, and sagittal perspectives, and then combine them to categorise brain tumours employing a coalition building based on polling.

In recent years, the DL sector has exploded in popularity, and it had been applied to a wide range of traditional applications successfully. More crucially, in several sectors, including as Artificial Intelligence, cyber security, robotics, bioinformatics, and medical Image processing to extract exact information, DL has outperformed well-known ML techniques. L. Alzabaidi [13] discussed the need to study DL and the various types of DL approaches and available networks for segmentation. They introduce convolutional neural networks (CNNs), the most widely used DL network type, and explains the evolution of architectures of CNN and their key features, for example, starts with one network named as Alex Net and stops with HR Net (High-Resolution Network). Finally, they discussed the obstacles and their potential solutions which help academics to overcome the research gaps. Below Table 1 shows a list of the most popular Deep Learning programmes. The impact of computational technologies such as FPGA, GPU, and CPU on DL is summarised (Table 1).

Table 1: Overview of different existing deep learningsegmentation strategies

Deep Learning based Algorithms	Remarks
Krizhevsky et al. [4]	On high challenging data set, deep CNN achieved good results using purely supervised learning method.
F. Millntari et al. [6]	This approach attains better performances on challenging dataset for testing which needed only a little part of the dispensed time as compared to recent methods.
Hao Dong et al. [8]	Cross validation has proven that their approach can effectively produce satisfactory categorisation.
R. Ezhilarasi et al. [10]	By generating a bounding box containing class names and scores, faster R-CNN algorithms were mostly utilised to detect object in natural images.
P. Anandajayam et al. [11]	They teach three paradigms for segmentation using 2D image fragments and wedges taken sequentially in coronal, axial, and sagittal perspectives, and then combine them to categorise brain tumours employing a coalition building based on polling.
R. Ranjbarzadeh et al. [12]	They had proposed a new architecture which uses two sets of four MRI modalities with Z-Score normalisation applied on their results. Then Cascade-CCN with distance wise attention mechanism were used to extract key location feature of tumour region.
L. Alzabaidi [13]	They explain the evolution of CNN architectures as well as their key features, for example, starts with one network named as Alex Net and stops with another HR Net.

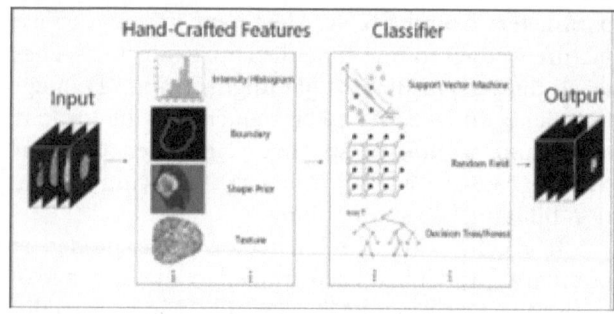

Figure 1: Methods based on particular model

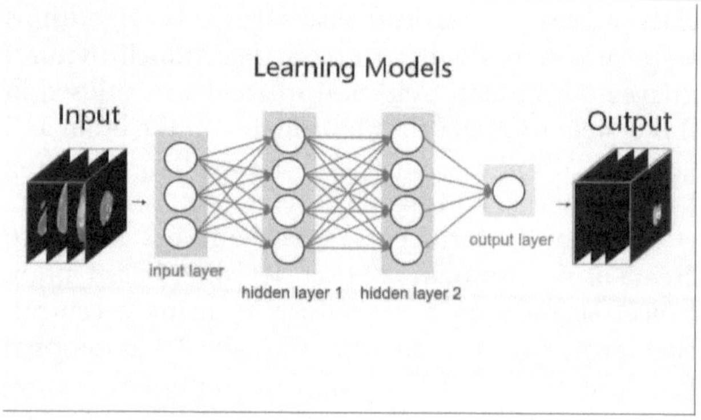

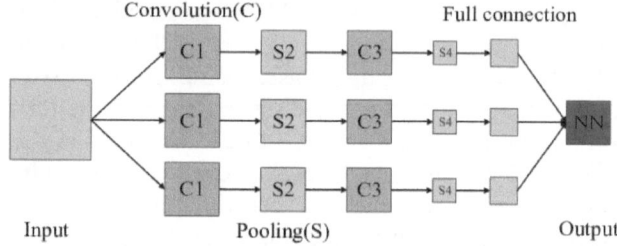

Figure 2: Methods based on learning criteria

3. Strategies of Segmentation Using Deep Learning

Deep learning is a growing subset of machine learning techniques, according to researchers. Deep neural networks may discover hierarchical features from input images instead using pre-defined hand-crafted features. Figures 1 and 2 depicts a rough comparison of classical and deep learning-based brain tumour segmentation techniques. Deep Learning methods necessitate a significant amount of data to be trained so that it prevents from overfitting problem side by side computing power was also reduced to speed up the training process [17], [18].

Deep Learning methods necessitate a significant amount of Deep Learning approaches had attained better performance in several areas, such as detection of object, when combined with effective weight initialization and optimization algorithms.

There were best known deep learning-based segmentation strategies such as CNN (Convolutional neural network) and R-CNN (Recurrent network).

3.1. Convolutional Neural network model

(CNNs) are similar to classic ANNs in that they are made up of neurons that learn to optimise themselves. Each neuron will still receive an input and conduct an action (such as a scalar product followed by a non-linear function), which is the foundation of innumerable artificial neural networks. The entire network will still express a single perceptual scoring function first from input raw picture vectors to the final output of the class score (the weight). The final layer will contain the loss functions related with the classes, so all of the standard ANN tips and tactics will still apply. The only distinction between CNNs and standard ANNs is that CNNs are largely used in the field of artificial intelligence [14]. This architecture as shown in Figure 3, contains three layers i.e., convolution (C) layers, pooling (S) layers and full connection layers.

Figure 3: Architecture of CNN model.

a) The Convolutional layers: Convolutional layers, as its name suggests, is critical to how CNNs work. The kernels which were learnable and they were the main focus of the layer's parameters as suggested by Keiron O'Shea et al. [14]. The dimensionality of these kernels which were spatial was usually low, yet they flow across the entire volume of the input.

When data passed through first layer (C), it convolves each filter across the input dimensions and forms a 2D map named as activation map. It is possible to visualise the activation maps.

b) Pooling Layers: The goal of pooling layers was to gradually minimize the dimensionality for representation, lowered the incident of variables and their computed time. The pooling layer scales down the activation feature map at the input with the help of the "MAX" function. Mainly, most of the CNN networks, max-pooling layers having kernels with a size of 2×2 and a stride of 2 were formed as its spatial input dimensions. That reduced the dimensions of the activation feature map as compared to its original dimensions up to 25% while kept the volume in depth according to its original size.

c) Full-Connection Layers: This connected layer consisted of neurons that were directly connected to neurons into their two different layers of their neighbourhood, but didn't flow into neurons in any of the other levels. That was similar to how neurons were placed in standard ANN models.

3.2 Fully Convolutional Neural Network Based on Volumetric Segmentation (V-Net Model)

F. Milletari et al. [6] introduces this V-Net model for image segmentation. Volumetric kernels with a size of 555 voxels are used in each stage's convolutions. Their solution of the data decreases as it passes through several steps of the compression pipeline. This is accomplished using stride 2 and convolution using 222 voxel wide kernels. The size of the generated feature maps is halved because the second step extracts feature by considering just non-overlapping 222 volume regions. In Figure 4, the Fully CNN using V-Net model was shown. Furthermore, as the doubling of feature channels at each stage of the V-Net's path, and due to the reason of model to be formulated as a residual network, they utilised different convolution operators for increasing the number of feature maps almost doubled and the resolution was also reduced. Throughout the network, Parametric Rectified Linear Unit (PreLu) non linearities are used as activation function.

3.3. Deep CNN based on U-Net model

In this model as seen in Figure 5, it relies on the U-Net approach which consists of two types of sampling path named as (a) a down sampling encoding path (b) an u p sampling decoding path. In total, there were five convolutional blocks in the first path. Each block comprised of two convolutional layers i.e., each enclosed with a filter of size (3×3), a stride of 1 in both directions, and rectifier activation which helps to increase the total number of features to 1024. In the next step, every block was Max pooled with stride 2×2 at the end except for the last block which left for down-sampling. That reduced the size of feature maps from 240×240 to 15×15. Hence, two convolutional layers had helped in lowering down the number of feature maps from the concatenation of two sets of feature maps i.e., deconvolutional feature maps and the feature maps extracted from the encoding path in each up-sampling block [8].

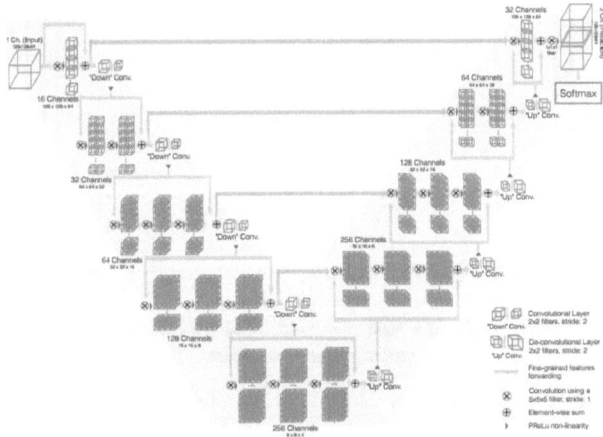

Figure 4: Fully CNN using V-net model

3.4. Fast R-CNN Model

This model was the descendent of CNN model. This was somewhat similar to the original one in various ways, but this model had modified to improve itself in terms of detection speed before proposing areas, feature extraction was performed over the image which results into only one network i.e., CNN to be run through the full-fledged image rather than over 2000 overlapping regions. Rather than constructing a new model, which completely removed SVM layer with a new layer named as Softmax and also expanded the neural network to predict the area. The basic idea behind this technique was to completely remove the selective search algorithm with neural network which runs quickly. In the proposed technique, it also used the RPN (Region Proposed Network) which worked as followed in three particular phases: (a) Firstly, it contains a sliding window of size 3×3, it reduces the features into a lower dimension feature maps at the final stage of CNN (e.g., 256-d). (b) It generates possibility of numerous regions for every sliding-window location using any n boxes of fixed proportion or bounding boxes. (c) Every proposed area included 1) region value for the "objective function" and 2) four coordinates that represented the region's bounding box. In whole, Faster R-CNN outperformed the competency with respect to the terms of speed and accuracy [15]. While Faster R-CNN is not only the best approach for object detection, but the most effective and simplest technique. Fast R-CNN architecture is given below in the Figure 6.

4. Conclusion

This paper included numerous strategies of deep learning-based segmentation which were reviewed and presented. CNN models were the most

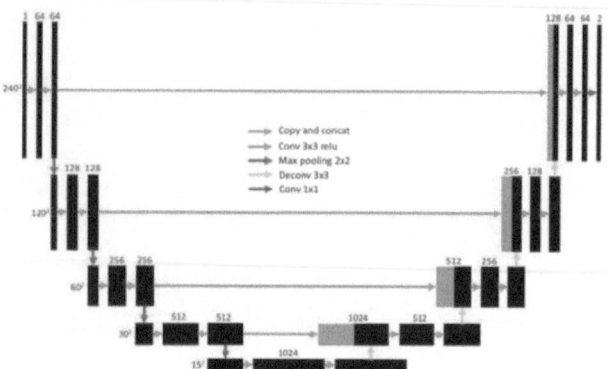

Figure 5: Deep CNN using U-net model

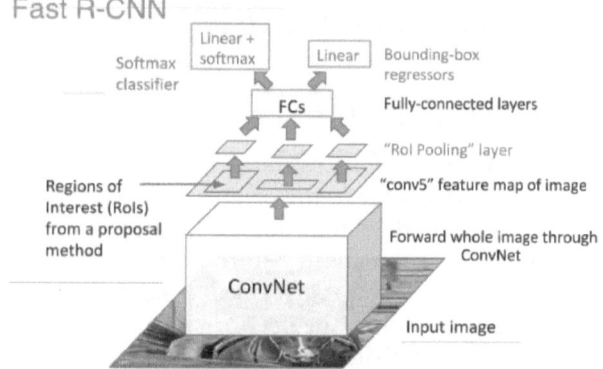

Figure 6: Fast R-CNN architecture

traditional strategies used for deep learning image segmentation. Further CNN model was enhanced using various ideas such as U-Net, V-Net, Alex Net and R-CNN and Fast R-CNN and Multi modalities Brain Tumour with distance wise attention mechanism combined with cascading Neural Network. All methods used CNN model as their base model. This paper, had shown an overview of these segmentation models which can be used for image segmentation and detection of tumour in brain. Multi Modalities CCN approach had drawback in terms of feature extraction performance with respect to tumour area. But still by using deep learning method, an accurate and timely detection of tumour in the MRI brain image will possibly help the doctor in treatment of the patient.

References

[1]   Havaei, M., Davy, A., Warde-Farley, D., Biard, A., Courville, A., Bengio, Y., Pal, C., Jodoin, P. M., and Larochelle, H. (2017). Brain tumor segmentation with deep neural networks. Medical Image Analysis, 35, 18–31.

[2]   Abbasi, S., Pour, F. T. (2014). A hybrid approach for detection of brain tumor in MRI images. In 21th Iranian Conference on Biomedical Engineering (ICBME), pp. 269–274.

[3]   Pinto, A., Pereira, S., Correia, H., Oliveira, J., Rasteiro, D., Silva, C. A. (2015). Brain tumour segmentation based on extremely randomized forest with high-level features. In 37th Annual International Conference of the IEEE Engineering in Medicine and Biology Society (EMBC), pp. 3037–3040.

[4]   Krizhevsky, A., Sutskever, I., and Hinton, G. E. (2012). ImageNet classification with deep convolutional neural networks. Neural Information Processing Systems, 141, 1097-1105.

[5]   Russakosvy, O., Deng, J., Hao Su, Krause, J., Satheesh, S., Ma, S., Zhiheng Huang, Karpathy, A., Khosla, A., Bernstein, M. S., Berg, A., and Li, Fei-Fei (2015). doi: 10.1007/s11263-015-0816-y, Corpus ID:2930547

[6]   Milletari, F., Navab, N., and Ahmadi, S. -A. (2016). V-Net: Fully convolutional neural networks for volumetric medical image segmentation, 1–11. arXiv.

[7]   Litjens, G., Kooi, T., Bejnordi, B. E., Setio, A. A. A., Ciompi, F, Ghafoorian, M, van der Laak Jeroen, A. W. M., van Ginneken, B., and Sanchez Clara, I. (2017). A survey on deep learning in medical image analysis. Medical Image Analysis, 42, 60-88. doi: 10.1016/j.media.2017.07.005

[8]   Hao Dong, Yang Guang, Liu Fangde, Mo Yuanhan, Guo Yike (2017). Automatic brain tumor detection and segmentation using U-net based fully convolutional networks. In International Conference in Medical Image Understanding and Analysis (MIUA2017, 069, v3), arXiv:1705.03820v3.

[9]   Gu, J., Wang, Z., Kuen, J., Ma, L., Shahroudy, A., Shuai, B., Liu, T., Wang, X., Wang, G, Cai, J., and Chen, T. (2018). Recent advances in convolutional neural networks. Pattern Recognition, 77, 354-377, ISSN 0031-3203. doi: 10.1016/j.patcog.2017.10.013

[10] R. Ezhilarasi, P. V. (2018). Tumor detection in the brain using faster R-CNN. In 2nd International Conference on I-SMAC (IoT in Social, Mobile, Analytics and Cloud) (I-SMAC), pp. 388-392. doi: 10.1109/I-SMAC.2018.8653705

[11] Anandajayam, P., Naveen, M., Sudharsan, R., Stephinradj, L., and Vengatabalaji, K. (2020). Brain tumor segmentation using fully connected convolutional neural network (FCNN). International Research Journal of Engineering and Technology (IJRET), 07(10), 133-139. eISSN: 2395-005

[12] Ranjbarzadeh, R., Kasgari, A. B., Ghoushchi, S. J., Anari, S., Naseri, M., and Bendechache, M. (2021). Brain tumor segmentation based on deep learning and an attention mechanism using MRI multi-modalities brain images. Scientific Reports 11, 10930. doi: 10.1038/s41598-021-90428-8

[13] Alzubaidi, L., Zhang, J., Humaidi, A. J. (2021). Review of deep learning: Concepts, CNN architectures, challenges, applications, future directions. Journal of Big Data, 8, 53. doi: 10.1186/s40537-021-00444-8

[14] O'Shea, K., and Nash, R. (2015). An introduction to convolutional neural networks. arXiv:1511.08458v2.

[15] Liu, W., Anguelov, D., Erhan, D., Szegedy, C., Reed, S., Fu, C.-Y., and Berg, A. C. (2016). Single shot multibox detector. European conference on computer vision, Springer, pp. 21–37.

[16] Patenaude, B., Smith, S. M., Kennedy, D. N., and Jenkinson, M. ((2011)). A bayesian model of shape and appearance for subcortical brain segmentation. Neuroimage, 56(3), 907–922.

[17] Pattabiraman, V., and Singh, H. (2020). Deep learning based brain tumour segmentation. WSEAS Transactions on Computers, 19, 234-241. doi: 10.37394/23205.2020.19.29

[18] Sekhar, B. V. D. S., and Jagadev A. K. (2023). Efficient alzheimer's disease detection using deep learning technique. Soft Computing, 27, 9143-9150. doi: 10.1007/s00500-023-08434-z.

A Comparative Analysis of Deepfake Detection Mechanism Using AI Architecture Models

ResNext-LSTMArchitecture and MesoNet Architecture

M. Balasubramanian, S. Gabriel Alwin Varun, R. K. Arun Pranav, P. V. Gnanamoorthi, P. Chitra

Artificial Intelligence and Data Science Department, St. Joseph's Institute of Technology, Chennai, India
E-mail: balavan2005@gmail.com,gabriel2004av@gmail.com,arunrams003@gmail.com, pvgnanamoorthipalanivel@gmail.com, chitrap@stjosephstechnology.ac.in

Abstract

To identify the pros and cons of the several deep faking detection algorithms that are now widely used in the synthetic media ecosystem. The study focuses on assessing the effectiveness of important algorithms in the context of deep fake creation and detection, such as Generative Adversarial Networks (GANs), Convolutional Neural Networks (CNNs), and Recurrent Neural Networks (RNNs). The paper provides information on usage of different algorithms to detect deep faked images, guaranteeing a comprehensive assessment of algorithmic capabilities in a variety of content categories. Every algorithm is tested and presented using metrics including visual realism, computational efficiency, and sensitivity to detection techniques. Transfer learning strategies are investigated to understand the models to different situations. Furthermore, the study explores the ethical implications of these algorithms, considering the potential for misuse and the impact of artificially generated information on society. With a focus on deep faking algorithms, the project intends to shed light on their relative performances and provide a comprehensive understanding of their uses, constraints, and consequences for the development of synthetic media in the future. By directing the creation of more reliable detection algorithms and moral frameworks, this project adds to the ongoing discussion surrounding deepfake technology.

Keywords: Deep faking, algorithms, computational efficiency, ethical implications

1. Introduction

In today's world of technological advancements, a new threat has started to loom in the horizon – Deep Faking. Deep Faking is the art of digitally manipulating images, video and audio to get expressions that look realistic, butare fabricated with the use of Deep Learning algorithms. The main consequences of Deep Faking include

Delivery of Misinformation and Fake news to the Audience or Public. Deep Faking can also be used to spread rumours or bad opinions about eminent personalities. Sometimes these kinds of activities can be used for money extortion purposes causing nuisance to the general public. An Article from ScienceDaily, an online science news aggregator states that "Deep Faking is ranked as the most serious AI crime Threat." This Conclusion was made after discussing on this issue with a panel of persons who have expertise in AI [1]. According to sensity.ai, 96% of Deepfakes are explicit content related videos with over 135 million views [2]. So, one of the most important parts of the solution to this problem is detecting whether a specific image or video has been Deepfaked or not. It is an immediate need for our society to stand against these crimes and members of AI community should contribute to distinguishing these fake images. Our Paper aims to assess various AI models used to detect Deep faking and give a thorough Comparison between them.

This paper presents a comprehensive analysis report between MesoNet and ResNet LSTM Neural Architectures their formulae, accuracy levels, statistics and observation. We will also focus on the challenges faced while building a model for deepfake detection systems.

2. Proposed Method of Classification

2.1. MesoNet Architecture

MesoNet is a convolutional neural network (CNN) designed for facial beauty prediction, later repurposed for deepfake detection. Combining the terms "mesoscale" and "network," MesoNet leverages its versatility acrossvarious applications [3].

2.2. Methodology

In this streamlined approach to MesoNet-based deepfake detection, the methodology unfolds in a series of essential steps. Commencing with Dataset Preparation, a diverse compilation of deepfake and authentic samples ensuring the model encounters a rich variety of scenarios during training and evaluation. Following this, the Preparation stage involves the normalisation of input data and the enhancement of facial features, establishing a standardised scale and optimising the network's capability to discern intricate patterns.

Configuring its layers strategically, with convolutional layers for feature extraction and fully connected layers for classification, lays the groundwork for effective model performance. The subsequent Training phase sees MesoNet immersed in the dataset, learning discriminative patterns associated with both real and manipulated facial features. The model's performance is rigorously evaluated on a separate dataset to ensure generalisation capability and robustness. Examination delves into a critical analysis of the model's effectiveness, employing metrics such as accuracy on a dedicated testing dataset. Upon successful examination, deployment brings the trained MesoNet into practical use for deepfake detection. Integrated into the desired environment, the model is poised to contribute actively to the identification and classification of manipulated content and then the results are observed.

3. ResNext and LSTM Architecture

In the realm of deepfake detection, leveraging advanced neural network architectures is paramount. RestNet, short for Residual Networks, stands out with its deep structure, excelling in discerning intricate visual features crucial for distinguishing between authentic and manipulated content. Complementing this, Long Short-Term Memory (LSTM), a recurrent neural network variant, plays a vital role in understanding temporal dependencies within sequential data [3,4]. Together, RestNet's prowess in image classification and LSTM's ability to capture context and temporal relationships create a robust framework for detecting subtle patterns indicative of deepfake manipulations in visual content.

3.1. Methodology

In the dynamic landscape of deepfake detection, our methodology harnesses the synergy of ResNext and LSTM architectures. We commence by curating a diverse dataset comprising both authentic and deepfake visual content. Through meticulous preprocessing, we enhance facial features and prepare sequential data for LSTM input. The ResNext architecture is seamlessly integrated into the detection pipeline, proving invaluable in extracting intricate visual features pivotal for discerning authentic from manipulated content. Augmenting this capability, the LSTM component captures temporal dependencies within sequential data, providing a holistic understanding of the dynamic pattern's indicative of deepfake manipulations [5]. The model is then meticulously trained on the prepared dataset, ensuring it adeptly learns both spatial and temporal intricacies. Subsequent validation on a separate dataset guarantees the model's generalisation capability. In the testing phase, the integrated ResNext-LSTM model is evaluated, employing metrics such as accuracy and precision. Once validated, the model is deployed for real-world deepfake detection applications, with continuous monitoring and updates ensuring its adaptability to emerging deepfake techniques [6].

4. Output Results and Observation

Accuracy: MesoNet has an architecture that is quite simple, but it is sufficient for this classification task, as evidenced by the over 95% accuracy it achieves in all situations. Even with extreme compression, it can retain strong robustness; our suggested method's accuracy remains above 88%. ResNet, also known as the Residual Network, is a type of neural network that improves accuracy and performance by adding layers to deep neural networks in order to address complex problems. 97% accuracy is displayed. However, the training data and methodology used to train them determine the accuracy.

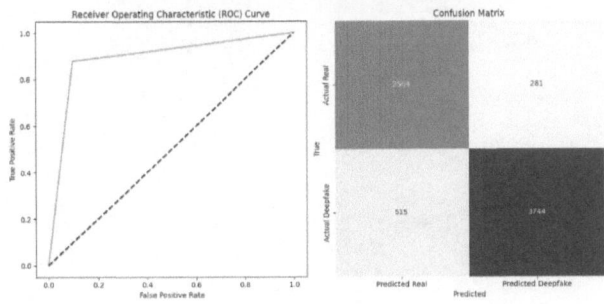

Hardware Requirements: In order to store and process datasets, both systems require sophisticated hardware [5]. Thus, deepfake detection technique can be applied to devices with advanced graphics. While writing this work, there were other hardware-related issues. Using more sophisticated datasets and Application Programming Interfaces (APIs) can help with this. Thus, efficient methods for reducing hardware utilisation should be used.

Model complexity: When assessing a model's performance, one of the key factors to consider is its complexity. To create a thorough deep learning model, model complexity should be increased in deep learning. Nevertheless, post-processing technique procedures will also be intricate, so models should be constructed with the number of layers and their orientation in mind in order to minimise human intervention and improve accuracy and prediction results [4,7].

Data Privacy and availability: Deepfake is the production and alteration of images used to disseminate misleading information. The availability of prior information can be used to identify false information. Certain datasets might not be accessible because of legal restrictions and regional concerns about data privacy. This needs to be approached in a way that makes data sets cumulatively available. Real-time data updating is also necessary when developing deep fake detection models.

4.1. Output Images

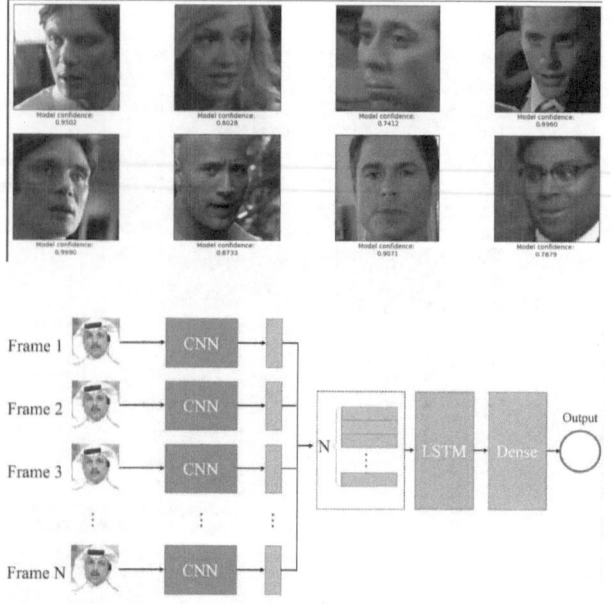

5. Other Algorithms Used in Deepfake Detection

5.1. XceptionNet

XceptionNet is a deep convolutional neural network architecture that has shown effectiveness in image classification tasks. It can be applied to the task of deepfake detection by training on labelled datasets containing real and manipulated images.

Algorithms such as DenseNet, ResNet50, VGG, VGGFace can also be used to build models in deep fake detection technology [7,9].

6. Conclusion

The advent of synthetic media, particularly in the form of deepfake images and videos, has significant implications for the spread of false information and the potential for malicious use. To counteract the risks associated with these manipulative techniques, the development of real-time deepfake detection models has emerged as a critical area of research. These models, leveraging advanced technologies such as Convolutional Neural Networks (CNNs) and their variations, play a pivotal role in safeguarding individuals active on the internet from the harmful effects of misinformation and fabricated representations.

The efficacy of real-time deepfake detection hinges on several key technical factors. Firstly, the utilisation of diverse and extensive datasets is paramount for training models to recognise subtle patterns indicative of manipulated content. Machine learning models, including generative models, should be continually updated and fine-tuned to adapt to evolving deepfake generation techniques. Additionally, the integration of various algorithms, each addressing specific facets of deepfake detection, enhances the robustness of the overall system.

Convolutional Neural Networks, known for their proficiency in image processing tasks, are a cornerstone in the development of effective deepfake detection models. Their ability to extract hierarchical features from images, coupled with advancements in architectures like ResNet and efficient training methodologies, contributes to the model's discernment between authentic and synthetic content. Furthermore, the synergy of multiple algorithms, such as those focusing on facial recognition, audio analysis, and contextual understanding, provides a holistic approach to deepfake detection.

From the research we can infer that building a customised CNN model with a combination of different algorithms, tailoring the model with regard to the area of application is integral for model development [6].

Amidst the significant strides in deepfake detection, it is essential to acknowledge certain challenges and research gaps that persist within this dynamic landscape. One notable limitation revolves around the availability of adequate hardware resources. The demanding nature of deepfake detection models, particularly those leveraging complex architectures like Convolutional Neural Networks (CNNs), places a strain on computational capabilities. Training and testing these models require substantial memory and processing power, which can be a bottleneck for researchers and organisations with limited access to high-performance computing infrastructure.

The research community faces the ongoing challenge of optimising algorithms and architectures to make them more resource-efficient without compromising on detection accuracy. While training a dataset, it takes hours to upload videos and images which delays the process of obtaining results. Hence development of technologies to mend this research gap is integral in deep fake detection technology. Additionally preparing a dataset for deep fake detection model is a tedious process well usage of APIs and pre-defined datasets can tackle the problem but their relevancy to the model application is a matter of discussion [8]. Some datasets include data from inappropriate sites which does not aid in research or detection purposes. Establishing data forums on deepfake data can be a solution to avert this limitation for more relevant and accurate model building.

Another noteworthy consideration is the necessity for clear and comprehensive documentation of deepfake detection models. Understanding the intricacies of these models is pivotal for researchers, practitioners, and developers aiming to implement or improve upon existing frameworks. Accessible documentation, along with openly available code repositories, encourages transparency and facilitates the dissemination of knowledge within the research community. Continuous refinement and updates to documentation based on the latest advancements ensure that the broader community can benefit from shared insights and contribute to the ongoing evolution of deepfake detection technologies.

In summary, addressing the research gap in real-time deepfake detection requires concerted efforts to optimise hardware utilisation, curate diverse and high-quality datasets, and prioritise clear and accessible documentation. Overcoming these challenges will not only propel the field forward but also empower a wider community of researchers and practitioners to effectively navigate the complexities of synthetic media detection.

7. Future Works

7.1. Development of Real Time Deep Fake Detection Models

An impressive advancement in deep learning will be the creation of deep fake technology that can quickly identify fake photos and videos so that the model can update itself. The training model can be adjusted with knowledge of past problems with its operation and its influence on the field of technology [10]. Rapid deepfake detection can be achieved by using modules like OpenCV for object detection, and classification can be accomplished with the aid of specialised models created in accordance with the fields of application, such as news, advertisements, research projects, etc. A number of real-time deep fake detection models have been developed; however, future advancements in deep fake detection are likely to result from their improvisation and fine tuning.

7.2. False News Spreading Prevention Mechanism

Authorities' Alert System: Setting up an alert system to notify the appropriate authorities of the spreading of misleading information on social media is a proactive step. Fact-checking groups, government agencies, and social media platform managers are examples of authorities. It's important to act quickly in such situations where a person's reputation can be damaged. False material can be quickly removed from the internet to reduce its impact before it reaches a larger audience.

Automating the detection of false information is an important application of AI. Text, image, and video data can be analysed by machine learning algorithms and natural language processing (NLP) to find patterns linked to false information. Ethics related to privacy and bias reduction need to be taken into account when AI is used to prevent false news. Additionally, it is critical that AI models operate transparently.

Public awareness of media literacy increases people's ability to identify and evaluate information they come across online, which lessens the possibility of misleading information proliferating. Efforts to enhance automated systems by encouraging users to report suspicious content can be made. This kind of crowdsourcing can be used to find new disinformation strategies and trends.

References

[1] Threats posed by Deepfake: https://www.science-daily.com/releases/2020/08/200804085908.htm

[2] Danger of Deepfakes: https://www.thehindu.com/sci-tech/technology/the-danger-of-deepfakes/article66327991.ece

[3] MesoNet: A Compact Facial Video Forgery Detection Network https://arxiv.org/pdf/1809.00888.pdf

[4] Convolutional Deepfake algorithms: https://hal.science/hal-01867298/document

[5] Deep fake detection using ResNext and LSTM: https://ayushbasral.medium.com/deepfake-detection-using-resnxt-and-lstm-bcc08c086f84

[6] Deep fake detection using deep learning: https://github.com/abhijitjadhav1998/Deepfake_detection_using_deep_learning

[7] Comparison of Deepfake detection techniques: https://www.mdpi.com/2624-800X/2/1/7

[8] Factors and libraries of deepfake algorithm: https://paperswithcode.com/task/deepfake-detection

[9] Analysis of deep fake detection methods: https://www.hindawi.com/journals/cin/2021/3111676/

[10] Techniques to build deepfake algorithms: https://www.alanzucconi.com/2018/03/14/understanding-the-technology-behind-deepfakes

Awareness and Preference Regarding Old and New Tax Regime

An Empirical Study of Salaried Tax Payers

Ramandeep Kaur[1], Sonu Dua[2], Mandeep Kaur[1]

[1]Research Scholar, Department of Hospitality and Management IKG Punjab Technical University Kapurthala, India
[2]Department of Management, Technical Campus, Lyallpur Khalsa College, Jalandhar, India
[3]Department of Hospitality and Management, IKG Punjab Technical University, Kapurthala, India
E-mail: ramandeep150784@gmail.com, Sonudua3778@gmail.com, Mandeeparora1.ptu1@gmail.com

Abstract

A new tax structure effective from the financial year 2020–21 was announced in the budget 2020. Wherein a single taxpayer has the opportunity to choose how much tax to pay by choosing between the new and current tax regimes. The new tax system, which was presented by India's Finance Minister, Hon. Nirmala Sitharaman, aims to make income tax payments for individual taxpayers less complicated and burdensome. A taxpayer may choose to pay taxes at a reduced rate by giving up some exemptions and deductions under the new option. The objective of this research paper is to examine the awareness and preference of individual taxpayers regarding the two regimes. Using judgement and purposive sampling procedures, primary data was gathered for this aim via a well-structured questionnaire. The study demonstrates taxpayers' awareness and preference towards the new tax system. The paper's conclusion offers recommendations for the New Tax system's effective implementation.

Keywords: Income tax, old and new tax regime, budget 2020, India

1. Introduction

Budget 2020 proposed a new personal income tax structure for an individual taxpayer that has more tax slabs but lower tax rates. Several exemptions and deductions from the previous tax system were also eliminated in this. The Finance Minister gave taxpayers the option to select between the new and old regimes, which inadvertently gave the procedure the appearance of complexity. (HDFC Bank Ltd.2021)

2. The Old Tax System

As per the National Accounts of the OECD and the World Bank, India's gross savings rate stood at around 30% in March 2021. The total rate was significantly influenced by domestic savings. This was because the old income tax system required investments in certain tax-saving products like ULIPs, which gradually ingrained in people a culture of saving. These investments encouraged saving for a range of future events, such as marriage, education, the purchase of real estate, unforeseen medical expenses, etc. Therefore, the savings rate would decrease if more people opted for the new approach. However, demand and the cycle of consuming would be reinvigorated (Economicstimes.com). Through the insertion of sections to the Income Tax Act, the government has over the years granted Indian taxpayers over seventy exemptions and deductions, enabling them to lower their taxable income and, consequently, their tax liability.

3. The New Tax System

Lower tax rates and simpler compliance are features of the new system. Filing taxes is also made easier because fewer documents are required because most exemptions and deductions are not available. Since the benefit of a deduction or allowance has no bearing on tax liability, under the new system, all taxpayers would be treated equally. This can be especially beneficial for taxpayers who might not subscribe to the specified modes of investment as the majority of these investments have a lock-in period. Alternatively, they could put their money into open-ended instruments, Table-1 which give them the flexibility of making faster withdrawals in addition to solid returns.

Chapter 37 DOI: 10.1201/9781003570349

Table 1: Comparison between old & new tax structure

Income Slabs (lakh)	New Tax Rates	Old Tax Rates
Up to Rs. 2.5	Nil	Nil
Rs.2.5-5	5%	5%
Rs. 5-7.5	10%	20%
Rs. 7.5-10	15%	20%
Rs. 10-12.5	20%	30%
Rs. 12.5-15	25%	30%
Above Rs. 15	30%	30%

4. Review of Literature

Dey and Varma [1] evaluated the public's knowledge of tax-saving schemes in the twin cities of Cuttack and Bhubaneswar, Odisha Table-2. They came to the conclusion that low awareness causes poor tax planning and a higher tax burden.

Selvaraj and Gomathi [11] the study examines the level of knowledge that salaried taxpayers in Tamil Nadu's Erode District have regarding the income tax system. The results showed that the majority of sample salaried taxpayers were ignorant of the income tax system.

Koretskaya-Garmash [3] outlined the steps that must be taken to increase tax awareness in Russian society, including educating the population about tax laws and figuring out what obstacles stand in the way of timely tax payments. In this study, the author demonstrated that the Russian government has no interest in raising people's financial and tax literacy.

Kaur and Shekhon (2020) Their study's goal is to determine how the most recent tax regime in India has affected middle-class people and their degree of tax planning and design. They came to the conclusion that the new tax system affects savings.

Goel and Garg [4] examined the preferences of individual taxpayers for the two tax regimes included in India's Budget 2020. The paper's conclusion included recommendations for taxpayers that assist in selecting between the two systems in order to meet their necessary goal.

Further Pramanik et al. [5] Examined the influence of two tax regimes on consumer behaviour and analysed that people ageing between 20–30 are considering new regime to be the best option for their investments, As per calculations those who are already availing many deductions under the old tax system can save tax by opting that system only. If one is not making any savings, investments or claiming deductions earlier too, then the new system is beneficial.

5. Objectives of the Study

To determine how well-informed salaried taxpayers are about the New and Old Tax Regimes.

To compare the preferences of salaried taxpayers with respect to the Old and New Tax Regimes.

5.1. Hypothesis Testing

H0: Based on various demographic factors, there is no significant difference in awareness.

H0: The preferences for different demographic variables and the regime do not significantly associated.

6. Research Methodology

The awareness and preference of salaried tax payers regarding the old and new tax regimes are determined through descriptive and analytical research. Using a well-structured and pre-tested questionnaire, the necessary primary data were gathered from the sample of salaried tax payers in October and November of 2022 in order to meet the study's stated objectives. A pilot study comprising 25 salaried taxpayers was carried out. Appropriate changes have been made to the final Questionnaire based on the findings of the pilot study. Data were gathered through the application of judgment and purposeful sampling techniques. The sample consists of 141 respondents who live in the three densely populated districts of Punjab—Amritsar, Ludhiana, and Jalandhar. Sampling units are people who work in the public, semi-public, and private sectors and who all file their taxes on a regular basis Table 3-5. Numerous statistical tools, including percentage analysis, t test, chi-square test, and F test, have been used to analyse the collected data.

Objective 1: The first objective is to ascertain the awareness level of salaried tax payers about the new tax system.

Five questions on a 5-point Likert scale were designed for this purpose statement. The results are shown in the summarised table below:

The five-point Likert scale, also known as an interval scale or "summated rating scale," was used to collect the aforementioned data. The mean value is highly significant in this case. A value of 1 to 1.8 indicates complete ignorance, a value of 1.81 to 2.60 indicates somewhat awareness, a value of 2.61 to 3.40 indicates some awareness, a value of 3.41 to 4.20 indicates moderate awareness, and a value of 4.21 to 5 indicates extreme awareness. The first statement's mean, 3.62, indicates that, as the above table shows, most respondents are only somewhat aware that, as of the financial year 2020–2021, The two options for tax payers to pay their tax

Table 2: Salaried Individuals' classification according to their comprehension of research statements and their descriptive (mean & standard deviation)

Awareness Statements	Not at All Aware	Slightly Aware	Somewhat Aware	Moderately Aware	Extremely Aware	Total	Mean	Std. Dev.
From Financial year 2020-2021, there are two options with tax payers for paying tax liability	12 (8.5%)	23 (16.3%)	19 (13.5%)	39 (27.7%)	48 (34%)	141 (100%)	3.62	1.328
There are four income tax slabs in old regime	8 (5.7%)	12 (8.5%)	22 (15.6%)	49 (34.7%)	50 (35.5%)	141 (100%)	3.86	1.162
There are seven income tax slabs in new regime	14 (9.9%)	15 (10.6%)	37 (26.2%)	38 (27.0%)	37 (26.2%)	141 (100%)	3.49	1.263
There is a long list of saving schemes and exemption approved by income tax Act	11 (7.8%)	11 (7.8%)	24 (17.0%)	54 (38.3%)	41 (29.1%)	141 (100%)	3.73	1.189
Man tax saving schemes and exemptions are not in new tax regime	20 (14.2%)	22 (15.6%)	28 (19.9%)	33 (23.4%)	38 (27.0%)	141 (100%)	3.33	1.392

Source: Based on researcher's calculations.

liability have a standard deviation of 1.382, indicating a slightly higher likelihood of different answers. The second statement's mean is 3.86, indicating that most respondents are aware of the four income tax slabs from the Old regime, although there were only minor changes. With a mean value of 3.49 for the third statement, it can be inferred that respondents have a moderate understanding of the seven tax slabs under the new tax regime. It is evident from the standard deviation of 1.263 that there is variation in the responses. The fourth statement has a mean of 3.73, indicating that people are aware of the extensive list of savings plans and exemptions authorised by the Income Tax Act. The standard deviation of 1.189, which is slightly higher, suggests that respondents' responses may vary. The fifth statement's mean of 3.33 and standard deviation of 1.392 show that there is a good chance of response variation and that most respondents are only vaguely aware of the fact that many tax-saving plans and exemptions are not included in the new tax law. To summarise, salaried tax payers have a moderate level of overall awareness about the new tax regime.

6.1. Hypothesis Testing

Certain hypothesis tests have been used to determine whether or not there is a significant difference between various demographic variables in order to study the effect of these variables on awareness.

6.2. Hypothesis Development

1. H0:Based on gender, there is no significant difference in awareness.
2. H0: Age does not significantly affect awareness in any way.
3. H0: There is no discernible difference in awareness between the groups according to varying levels of education.

6.2.1. t Test

H0: Based on gender, there is no discernible difference in awareness.

This p-value, which is .001, is smaller than the alpha, or 0.05. We reject the null hypothesis. It indicates that respondents' awareness levels differ significantly based on their gender. Because men's awareness scores on average are higher than women's, at 3.8861 versus 3.3159. Men are more aware of the new tax regime than are women.

6.2.2. F Test (Anova)

H0: Age does not significantly affect awareness in any way.

Table 3: Gender-based independent sample t test statistics of awareness

	Gender	N	Mean	Std. Deviation	Value of P	Outcome
Awareness	Male	72	3.8861	.84575	.001	Significant
	Female	69	3.3159	1.08161		

Table 4: Anova (F test) of awareness based on age

Awareness	Total Squares	df	Square Mean	F	Value of P	Outcome
Between Groups	10.376	4	2.594	2.684	.034	Significant
Within Groups	131.417	136	0.966			
Total	141.793	140				

Table 5: One way anova (f test) of awareness based on qualification

Awareness	Total Squares	df	Square Mean	F	Value of P	Outcome
Between Groups	5.746	4	1.436	1.436	.225	
Within Groups	136.047	136	1.000			Not significant
Total	141.793	140				

The F test was used to test the age at four degrees of freedom; the result is F = 2.684, and once more, the P value is less than the Alpha value.05, indicating that the outcome is noteworthy and that the null hypothesis is rejected. It is evident that respondents' awareness levels across age groups differ significantly from one another. The awareness of those in the 20–30 age group and those in the 31–40 age group differs significantly.

6.2.3. Anova (F Test)

H0: Among the groups with varying levels of education, there is no discernible variation in awareness.

Once more, the F test is used to verify the awareness hypothesis between groups with varying educational backgrounds. The P value in this case exceeds the Alpha value (0.05). Here, the null hypothesis is agreed upon. There is no discernible difference in awareness between those with varying levels of education.

Objective: 2 Verifying taxpayer preferences for the Previous and New Tax Regimes is the second goal.

The above table makes it evident that, in the year of the new tax regime's introduction (2020–20201), only 21.3% of respondents filed income tax returns in accordance with the new regime. This percentage

then gradually increased to 36.9% in 2021–2022, and the percentage of respondents who planned to file returns in accordance with the new tax regime in the financial year 2022–2023 is only 46.10%. It grows, but not as quickly. It demonstrates that despite the new regime's low tax slabs, people are not embracing it with enthusiasm..The lack of significant tax deductions and exemptions, the formality of completing an additional Form 10E, or perhaps their ignorance of its provisions are the causes of the low adoption rate, which indicates their lack of confidence in the new optional tax regime the government introduced in Budget 2020.

There is one question in the questionnaire that if in Budget 2023, Government announces some additional deductions/ exemptions like (standard deduction, P.P.F., interest on home loan, H.R.A, Leave travel concession etc.) in new tax regime, than what will be their preference for filing return?

7. Preference for the Old and New Tax Systems

The table shows that 71.6% of respondents were prepared to file their income tax return as per new regime. It has been demonstrated that salaried taxpayers prefer the new tax structure with more exemptions and deductions.

7.1. Chi Square Test for Examining Variable Association

7.1.1. Hypothesis Development

H_0: There is no association between gender and preferred regime.

Value of Pearson Chi Square = .295, D = 1, P value = .589

The relationship between gender and preference is not significant because the P value is greater than.05. It suggests that preferences for the old and new tax regimes are gender-neutral. It has no bearing on choosing the preference.

8. Findings and Suggestions

8.1. Salaried Class Tax Payers' Awareness about the Old and New Tax Regime

Only 43 out of 141 respondents, on average, are very aware of the new tax regime. This is a low number. In order to investigate the variations in awareness levels based on the demographic variables of the sample respondents, a null hypothesis was formulated and subjected to t and F tests Table 6-9. It is discovered that while there is a significant difference in awareness levels based on age and gender, there is no significant difference in salaried tax payers' awareness of the new tax regime based on qualification.

8.2. Salaried Tax Payers' Preference between the Old and New Tax Regime

It has been observed from the analysis of 3 years, the filing of Income Tax return behaviour of respondents that in the year (2020-2021) of introduction of New Tax regime only 21.30% respondents file return

Table 6: Comparative analysis of tax payers' preference towards previous and new tax regime

Previous tax regime			New Tax regime		
Year	Percent	Number	Percent	Number	Total
2020-2021 2020-2021	78.10%	111	21.30%	30	141
2021-2022	63.10%	89	36.90%	52	141
2022-2023	53.90%	76	46.10%	65	141

Table 7: Frequency and percentage

	Regularity	Percentage	Reliable Percentage	Total Percentage
As per old regime	40	28.4	28.4	28.4
As per New regime	101	71.6	71.6	100.0
Total	141	100.0	100.0	

Table 8: Chi square test to examine the relationship between gender and preference for the 2020–2021 old and new tax regimes

Filed Income Tax Return in 2020-2021					Total
			According to the old regime	According to the new regime	
Gender	Male	Count	58	14	72
		Anticipated Count	56	15	72
	Female	Count	53	16	69
		Anticipated Count	54	14	69
Total		Count	111	30	141
	Anticipated Count	111	30	141	

as per New Tax regime. In financial year2020-21, 36.90% percent filed income tax return as per new regime.46.1 0% planning to file as per new regime in financial year 2022-2023. There is gradual increase but at a decreasing rate so in spite of low tax slabs people are not enthusiastically adopted the new tax regime. Some viable suggestions have been made for the benefit of salaried tax payers and the Income Tax Department of the Government of India based on the findings.

9. Conclusion

There is no denying that income tax collection plays a significant role in enhancing India's economy. In light of this, the current study aims to investigate the knowledge and preferences of salaried taxpayers regarding the Old and New Income Tax Regimes. The necessary data for this study have been gathered and subjected to a variety of analytical techniques. Based on the results, some workable recommendations have been made for the benefit of salaried tax payers and Income Tax Department.

10. Suggestions

The results indicate that salaried taxpayers are not well-informed about the new tax system. The nation's tax authorities should launch a well-planned campaign encouraging tax payment and outlining its benefits in order to improve it. The curriculum at the school level needs to be revised with a focus on taxation. The government should incorporate deductions and exemptions in the new tax regime, as the previous one's numerous sections offered deductions and exemptions that encouraged taxpayers to make wise investment decisions.Additionally, it will assist the tax department in resolving the issue of tax evasion. One of the main areas to which the government should give attention is tax awareness. It is recommended that the government initiate webinars, seminars, or training programmes to educate the public about the new tax system and the distinctions between the two tax options. In order to compare the tax systems and determine which has a lower tax liability than the other, people should assess their tax liability under each one.

11. Contribution of the Study

The results of the study show that people's tax literacy is low. They still don't know as much about personal taxes and financial matters. Given the significance of tax literacy, formal tax education at the high school and college levels is suggested. According to this study, respondents are not expressing interest in the proposed tax structure. Thus, the study recommends that the new tax system include more exemptions and deductions. If the government gave it some thought, salaried taxpayers' preferences for the new tax system would improve.

References

[1] Dey, S. K., and Varma, K. K.(2016). Awareness of tax saving schemes among individual assesses: Empirical evidence from twin city of Odisha. Journal of Commerce and Management Thought, 7(4), 668-692.

[2] Selvaraj, A., and Gomathi, K. K. (2016). Salaried tax payers' awareness level about the income tax system: a study.

[3] Koretskaya-Garmash, V. A. (2017). Taxation awareness and its impact on financial literacy. Journal of Tax Reform, 3(2), 131-142.

[4] Goel, A., and Garg, P. (2021). A Comparative Study on Individual Tax Payers preference between Old vs New Tax Regime.

[5] Pramanik, A., Mohanty, N., and Karmakar, S. (2021). Understanding consumer behaviour with respect to availability of two tax regime in india. In Interdisciplinary Research in Technology and Management (pp. 279-282).CRC Press.

[6] https://economictimes.indiatimes.com/comparison-of-new-income-tax-regime-with-old-tax-regime/tomorrowmakersshow/74807832.cms

[7] https://www.hdfcbank.com/personal/resources/learning-centre/pay/difference-between-new-tax-regime-vs-old-tax-regime

[8] Geetha, R., and Sekar, M. (2012). E-filing of income tax: Awareness and satisfaction level of individual tax payers in Coimbatore city, India. Research Journal of Management Sciences, 2319, 1171.

[9] Hastuti, R. (2014). Tax awareness and tax education: A perception of potential taxpayers. Tax awareness and tax education: A perception of potential taxpayers, 5.

[10] Interdisciplinary Research in Technology and Management: Proceedings of the International Conference on Interdisciplinary Research in Technology and Management (IRTM, 2021), 26-28 February, 2021, Kolkata, India, pp. 279-282. doi: 10.1201/9781003202240

[11] Patiala, F. S. (2020). New taxation regime and its impact on middle financial gain people and their level of tax designing.

[12] Suchthra, P., and Vidhya, C. (2019). A study on the Awareness of tax saving instruments of individual tax payers. International Journal of Scientific and Engineering Search, 10(5).

[13] www.abhinavjournal.com A study of satisfaction level and awareness of tax-payers towards e-filing of income tax return – with reference to Moradabad city.

Enhancing Heart Disease Prediction Using Spectrograms and Transfer Learning

Sabeena Yasmin Hera, Mohammad Amjad

Department of Computer Engineering, Jamia Millia Islamia, New Delhi, India
E-mail: Sabeenayasminhera@gmail.com, mamjad@jmi.ac.in

Abstract

Cardiovascular diseases are one of the deadliest diseases around the globe. Valvular heart diseases are one of the major heart diseases and can be diagnosed with heart sounds through auscultation. In this study, an AI-assisted automatic heart diagnostics system is proposed, in which state-of-the-art deep learning and signal processing techniques i.e., transfer learning and spectrograms, respectively, are utilised. Multi-classification problem is considered in this study and an improved results of recall with 100% for normal and two subcategory of abnormal class and 0.66 of kappa score are achieved. Moreover, comparative analysis of different pre-trained models is also presented.

Keywords: Phonocardiograms, spectrograms, deep learning, transfer learning

1. Introduction

Heart disease is reportedly the leading cause of deaths worldwide. The medical experts and biomedical signal processing engineers and researchers are conducting multi-disciplinary research to develop advanced and automated systems to minimise the casualties due to cardiovascular diseases. One of the major heart diseases are valvular diseases which are associated with heart valves and can be diagnosed with the help of auscultation [1]. Auscultation is the process of listening the heart sounds through electronic device called stethoscope [2]. Medical experts listen to the heart sounds and perform diagnosis. But this task is time consuming and require a lot of experience which increases the chances of misdiagnosis. To overcome this issue, Machine learning and deep learning techniques playing important role in the development of computer-aided diagnostic systems [3], which not only assist cardiologists in decision making but also minimise time consumption. Researchers have adopted different methodologies to design algorithm for decision support systems. The most widely and commonly adapted approach is to initially perform the signal processing, followed by features extraction. After the feature extraction, machine learning models are deployed.

Recently, based on deep learning models several algorithms have been developed and proposed to combat heart disease using artificial intelligence led diagnostic system. García-Ordás et al. [4] suggested employing advanced deep learning techniques along with enhanced feature augmentation to assess the risk of cardiovascular disease in patients. Their approach yields result surpassing machine learning algorithms by achieving a 90% precision. Baviskar et al. [5] suggested a hybrid model based on Recurrent Neural Network combined with the Long Short-Term Memory algorithm to enhance the classification accuracy of heart disease diagnostic system. They have further validated the model with heart failure clinical dataset and heart disease diagnosis UCI dataset.

In deep learning, the larger the dataset the better the neural network trains. In most of the cases, it is difficult to acquire thousands of records which results in poor generalisation of model. To deal with it, computer engineer, and researchers use state-of-the-art deep learning technique termed as Transfer Learning. Transfer learning is to train a model on two different datasets to get improved results and robust model. Chen et al. [6] introduced a Transfer Learning algorithm for predicting Stroke Risk. This approach utilised information from various interconnected sources, incorporating external stroke data alongside chronic diseases data like hypertension and diabetes. The suggested framework underwent testing in both synthetic and real-world scenarios, demonstrating superior performance compared to existing stroke risk prediction models. Boulares et al. [7] used graphical representations of Mel-Frequency-Cepstral-Coefficients (MFCCs) to train multiple pre-trained models including VGG16, VGG19, ResNet50V2 and Xception with improved average accuracies up to 89.1% on the PASCAL dataset.

However, the existing studies in heart disease diagnosis predominantly center around binary classification, determining solely the presence or absence of the condition. This prevailing focus on binary data fails to capture the nuances and variations among various heart conditions, limiting the precision of diagnosis and subsequent treatment. Therefore, this study aims to bridge the gap by employing transfer learning techniques to discern not only the presence of heart disease but also classify it into distinct categories. By utilising a dataset encompassing four different types of abnormal heart condition, the study contributes significantly to the field by offering a more comprehensive diagnostic approach.

The rest of the paper is organised as follows: The proposed methodology that includes spectrograms extraction, feature scaling, transfer learning and classification, is detailed in section II. In section III, the obtained results are discussed and analysed. In the last section IV, the paper is concluded, and future direction is provided.

2. Proposed Methodology

In this section, the proposed methodology is detailed that includes the data description, feature extraction, feature scaling, and classification.

2.1. Dataset Description

The dataset used is made publicly available by Son et al. [8]. It consists of heart sounds of five classes i.e., Normal(N), Mitral Regurgitation (MR), Aortic stenosis (AS), Mitral valve prolapse (MVP) and Mitral Stenosis (MS) as indicated in Table 1.

Where, MR occurs due to retrograde flow of blood through mitral valve which results in systolic murmurs. Similarly, AS occurs due to narrowing of aortic valve. MVP can be heard at ape commonly, in which mid-systolic click is followed by late systolic murmur. AS is a presystolic murmur which occurs as a result of increase in blood flow from atrial contraction. Overall,1000 recordings are used with 200 recordings from each of the category.

Table 1: Dataset description

Categories	Dataset				
	N	MR	AS	MVP	MS
Quantity	200	200	200	200	200

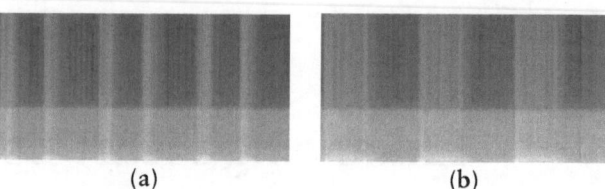

Figure 1: (a) spectrogram of normal samples and (b) spectrogram of abnormal samples

2.2. Features Extraction

Power spectral densities-based Spectrograms are generated from phonocardiograms by performing Fourier transform with Hanning window [9] as given in Eq (1). In this technique, the window is slide over the signal and their power spectrum are computed using Fourier transform. The visual representations of these spectrums are called spectrograms. Based on the literature [10], the size of the window of 256 data points with 50% of overlapping is chosen. Sample for normal and abnormal are shown in Figure 1.

$$X_m(\omega) = \sum_{n=-\infty}^{\infty} x(n)w(n - mR)e^{-j\omega} \qquad (1)$$

Where, $x(n)$ represents the input signal, w shows the window function (i.e., hamming) and R denotes hop size in samples between successive Fourier transformations. Once

2.3. Transfer Learning

Transfer learning has helped the researchers and data scientists to achieve improved results with few training samples. In the transfer learning, the model is previously trained on the larger dataset and then those trained parameters are frozen, and the model is retrained with new dataset [11]. So, this model has the parameters of the large dataset as well as the new dataset but perform the classification of the new dataset. In this study four pretrained models namely Xception, VGG16, ResNet50V2 and InceptionResNetV2 are used to use their parameters to train the spectrograms-based dataset.

The Xception architecture [12] consist of an input layer, an output layer and eight set of convolutional layer, pooling layer and batch Normalisation layer. VGG16 [13] consists of 13 convolutional layers, five pooling layers and 3 fully connected layers. ResNet50V2 is modified version of ResNet50 and has total five phases and each phase consist of convolution and identity block. Each of these blocks contain three convolutional layers. Inception [14] neural network is a convolutional network-based architecture that contains over 40 deep layers in the form of repetitive components which refers to inception.

3. Results and Discussion

In this section, the experimentations and the obtained results are discussed and an in-depth analysis is presented. To evaluate the models' performances, accuracy, precision, recall, f1-score and kappa score are used. Among all the models the highest score is achieved with InceptionResNetV2 with F1-scores ranging from 85 to 90%. Figure 2 shows the learning curve of the model.

It is also observed that the light model i.e., VGG16, is overall not suitable for the problem with the lowest kappa score of 0.35, which implies that the model doesn't train enough parameters to capture the important features of the dataset. In the term of neural networks complexities, it is found that 40 to 60 deep layers with ReLu activations function as in ResNet50V2 and InceptionResnetV2, are most suitable models.

The results are also compared with the previous study in Table 2. Where it can be seen that in case of most of the cases, the proposed algorithms outperformed BiLSTM-based algorithms [15] for the same dataset.

4. Conclusions

In this study, heart sounds classification is performed using STFT-based spectrograms with the combination of Transfer learning. This study mainly focuses on the impact of transfer learning on the model's performance and efficiency of various pre-trained models on the heart disease prediction based on given dataset. Multiple pre-trained models are utilised, and

Table 2: Results of the transfer learning-based prediction

Method	Model	Categories	Performance Metrics					Kappa Score
			Accuracy (%)	Precision (%)	Recall (%)	Specificity (%)	F1-score (%)	
Transfer Learning	Xception	AS	98.0	90.90	100.0	97.5	95.23	
		MR	80.0	10.10	12.23	100.0	11.06	
		MVP	87.0	68.42	65.0	92.5	66.67	0.50
		Normal	66.35	25.0	40.0	72.41	30.77	
		MS	93.0	74.07	100.0	91.25	85.10	
	VGG16	AS	65.0	36.36	100.0	56.25	53.33	
		MR	78.41	8.11	22.22	98.97	11.88	
		MVP	75.0	38.09	40.0	83.75	39.02	0.35
		Normal	79.0	25.23	15.12	100.0	18.91	
		MS	96.0	83.33	100.0	95.0	90.91	
	ResNet50V2	AS	94.0	81.23	100.0	95.0	90.21	
		MR	80.0	10.14	15.25	100.0	12.18	
		MVP	87.0	76.92	50.0	96.25	60.61	0.57
		Normal	73.63	39.02	80.0	72.22	52.46	
		MS	98.0	90.90	100.0	97.5	95.23	
	InceptionResNetV2	AS	98.0	90.90	100.0	97.5	95.23	
		MR	84.0	100.0	20.0	100.0	33.33	
		MVP	85.0	69.23	45.0	95.0	54.55	0.66
		Normal	80.90	48.78	100.0	76.66	65.57	
		MS	100.0	100.0	100.0	100.0	100.00	
Alkhodari et al. [15]	BiLSTM	AS	90.70	76.62	77.00	94.13	76.81	-
		MR	90.60	80.46	70.00	95.75	74.87	
		MVP	92.50	84.15	96.38	84.15	80.42	
		Normal	95.90	85.65	95.50%	96.00	90.31	
		MS	93.50	80.82	88.50	94.75	84.49	

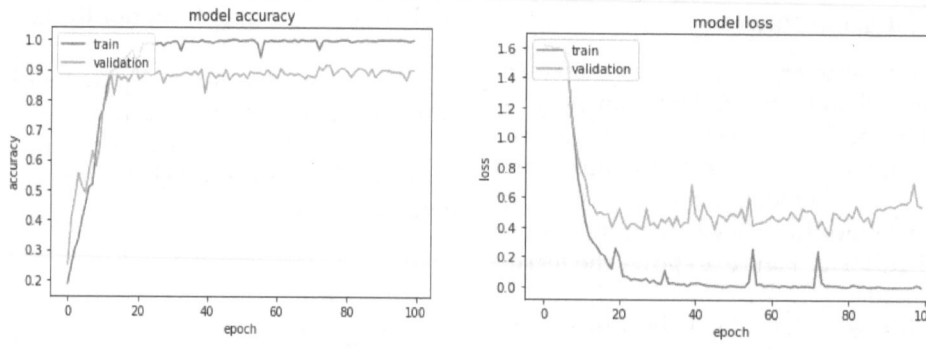

Figure 2: Learning curve for InceptionResnetV2 in terms of (a) model accuracy and (b) loss

the results are compared. In terms of neural network models, inceptaionResNetV2 found to be most comparatively efficient one with the 0.66 of kappa score and above 80% of average scores of other performance metrics.

In the future, the authors would like to extend the work by introducing more useful features extraction and feature engineering techniques.

References

[1] Maganti, K., Rigolin, V. H., Sarano, M. E., and Bonow, R. O. (2010). Valvular heart disease: Diagnosis and management. In Mayo Clinic Proceedings, Elsevier, Vol. 85, No. 5, pp. 483-500.

[2] Rappaport, M. B., and Sprague, H. B. (1941). Physiologic and physical laws that govern auscultation, and their clinical application: The acoustic stethoscope and the electrical amplifying stethoscope and stethograph. American Heart Journal, 21(3), 257-318.

[3] Chen, W., Sun, Q., Chen, X., Xie, G., Wu, H., and Xu, C. (2021). Deep learning methods for heart sounds classification: A systematic review. Entropy, 23(6), 667.

[4] García-Ordás, M. T., Bayón-Gutiérrez, M., Benavides, C., Aveleira-Mata, J., and Benítez-Andrades, J. A. (2023). Heart disease risk prediction using deep learning techniques with feature augmentation. Multimedia Tools and Applications

[5] Baviskar, V., Verma, M., Chatterjee, P., and Singal, G. (2023). Efficient Heart disease prediction using hybrid deep learning classification models. IRBM, 100786.

[6] Chen, J., Chen, Y., Li, J., Wang, J., Lin, Z., and Nandi, A. K. (2021). Stroke risk prediction with hybrid deep transfer learning framework. IEEE Journal of Biomedical and Health Informatics, 26(1), pp. 411-422, Jan. 2022, doi: 10.1109/JBHI.2021.3088750.

[7] Boulares, M., Alafif, T., and Barnawi, A. (2020). Transfer learning benchmark for cardiovascular disease recognition. IEEE Access, 8, 109475-109491.

[8] Son, G. Y., and Kwon, S. (2018). Classification of heart sound signal using multiple features. Applied Sciences, 8(12), 2344.

[9] Ito, C., Cao, X., Shuzo, M., and Maeda, E. (2018, October). Application of CNN for human activity recognition with FFT spectrogram of acceleration and gyro sensors. In Proceedings of the 2018 ACM International Joint Conference and 2018 International Symposium on Pervasive and Ubiquitous Computing and Wearable Computers, pp. 1503-1510.

[10] Homs-Corbera, A., Fiz, J. A., Morera, J., and Jané, R. (2004). Time-frequency detection and analysis of wheezes during forced exhalation. IEEE Transactions on Biomedical Engineering, 51(1), 182-186, Jan. 2004, doi: 10.1109/TBME.2003.820359.

[11] Torrey, L., and Shavlik, J. (2010). Transfer learning. In Handbook of Research on Machine Learning Applications and Trends: Algorithms, Methods, and Techniques (pp. 242-264). IGI global.

[12] Chollet, F. (2017). Xception: Deep learning with depthwise separable convolutions. In 2017 IEEE Conference on Computer Vision and Pattern Recognition (CVPR). IEEE.

[13] Theckedath, D., and Sedamkar, R. R. (2020). Detecting affect states using VGG16, ResNet50 and SE-ResNet50 networks. SN Computer Science, 1(2). doi: 10.1007/s42979-020-0114-9

[14] Demir, A., and Yilmaz, F. (2020). Inception-ResNet-v2 with Leaky relu and Average pooling for more reliable and accurate classification of chest X-ray images. In 2020 Medical Technologies Congress (TIPTEKNO). IEEE.

[15] Alkhodari, M., and Fraiwan, L. (2021). Convolutional and recurrent neural networks for the detection of valvular heart diseases in phonocardiogram recordings. Computer Methods and Programs in Biomedicine, 200, 105940.

Machine Learning-Based Categorization of White Blood Cells

Image Filtering and Deep Learning Analysis

Bhuman Vyas

Credit Acceptance Corporation , Michigan , USA
E-mail: bhuman.vyas@gmail.com

Abstract

This comprehensive investigation into the classification of white blood cells (WBCs) involved a meticulous exploration of the most effective synergy between image filters and advanced deep learning models. After testing various combinations, the VGG16 with Mean Filter and VGG19 with Median Filter emerged as the clear winner, achieving an impressive accuracy of approximately 85%, with precision, re-call, and F1 score all hovering around 85%. This successful blend highlights the significant potential of intelligent data pre-processing and deep learning in enhancing WBC classification accuracy. While other model-filter pairs, such as RESNET50 with Mean Filter (accuracy: 82%), RESNET50 with Non-Local Mean Filter (accuracy: 25.6%) displayed respectable but comparatively lower accuracy. The key finding is that achieving exceptional WBC classification depends on the harmonious interplay between models and image filters, as exemplified by the Mean-VGG16 and Median-VGG19 filter-model combination.

Keywords: Machine learning approach, WBC, image filter, deep learning

1. Introduction

White Blood Cells (WBCs) are essential components of the immune system and are crucial in the diagnosis of various diseases and infections. Accurate identification and categorization of WBCs from microscopic images are imperative for precise medical diagnostics. This research endeavours to address the urgent requirement for robust and efficient methods to categorize WBCs by integrating image filtering techniques and deep learning analysis. The primary research objectives of this study are as follows:

- To collect and pre-process the image data.
- To evaluate image filtering technique's impact on WBC classification accuracy.
- To analyse the performance of CNN, RESNET50,VGG16, and VGG19 models in WBC classification.
- To compare Model-Filter combinations for accuracy, precision, recall, and F1 score.
- To explore the role of image filtering in enhancing deep
- Learning for medical image analysis.
- To contribute insights into model-filter synergy for superior results.

Image processing techniques play a crucial role in the automated categorization of white blood cells (WBCs), offering a suite of tools to enhance the quality of digital images acquired from microscopy. In this in-depth analysis, we delve into the world of image pre-processing methods, the common image enhancement techniques employed to reduce noise and improve contrast, and the challenges associated with adapting these traditional methods to the diverse morphologies of WBCs.

The primary objective of this research is to advance the field of white blood cell (WBC) categorization through the seamless integration of image processing techniques and state-of-the-art deep learning models. In essence, our research seeks to enhance the accuracy and reliability of automated WBC classification. Achieving this objective necessitates a meticulous and well-structured research methodology, which serves as the foundation for our investigation.

Addressing Current Limitations: In recent studies, we've navigated the complex landscape of white blood cell classification using deep learning models. Mohamed, Ensaf et al. [1] revealed the VGG-16's 73.64% accuracy, highlighting the need for denoising. Habibzadeh et al. [2] faced dataset constraints with Inception V1 at 47%. Wu et al. [3] tackled

Chapter 39 DOI: 10.1201/9781003570349

subset limitations with RCTnet(b) at 83%, while Yildirim and Çinar [4] battled hardware constraints, achieving 78.74% with ResNet50. To address those challenges this paper has explored the synergy between image filters and advanced deep learning models. Through meticulous experimentation, we discovered the "VGG16 with Mean Filter" duo, boasting an impressive 85% accuracy. This success emphasizes the potency of intelligent data pre-processing coupled with cutting-edge deep learning in enhancing WBC classification accuracy.

This paper is described in 6 sections where section A depicts about data, section B describes data pre-processing, section C illustrates filtering, section D talks about deep learning models, section E evaluates the study and then the study is concluded.

2. Data Collection

We begin by describing the sources and nature of the image data, emphasizing their relevance to our research objectives. The quality and representativeness of the data are critical factors that underpin the success of our study. Our data collection process adheres to ethical considerations, ensuring that privacy and consent are respected throughout.

3. Data Pre-processing

An integral step in our methodology involves the pre-processing of image data. We meticulously prepare the data by resizing, normalizing, and augmenting it to create a consistent and reliable dataset. This pre-processing phase plays a crucial role in enhancing the quality of the data, thereby facilitating more accurate model training.

4. Image Filtering

One of the distinctive features of this research is the in-corporation of various image filtering techniques, including Gaussian, Median, Mean, Bilateral, and Non-Local Means filters. These filters are strategically applied to improve image quality, reduce noise, and enhance feature extraction, setting the stage for more robust WBC categorization.

5. Types of Filters

5.1. Gaussian Filter

Selection Rationale: Classic choice for noise reduction and edge preservation.

Mathematical Description:

$$F(i,j) = \frac{1}{2\pi\sigma^2} e^{-\frac{i^2+j^2}{2\sigma^2}}$$

Where:

- $F(i,j)$ is the filter kernel's value at position (i, j).
- σ is the standard deviation of the Gaussian distribution, controlling the amount of smoothing.

5.2. Median Filter

Selection Rationale: Robust for impulsive noise reduction.

Mathematical Description:

$$I_{med}(x,y) = \text{median}\left\{ I(x+i, y+j) - \frac{N}{2} \le i \le \frac{N}{2} - \frac{M}{2} \le j \le \frac{M}{2} \right\}$$

Where: $I_{med}(x,y)$ is the filtered pixel value at (x, y). N and M are the dimensions of the filter kernel.

5.3. Mean Filter

Selection Rationale: Simple and effective for noise reduction.
Mathematical Description:

$$I_{med}(x,y) = \text{median}\{ I(x+i, y+j) \mid \quad -\frac{N}{2} \le i \le \frac{N}{2} \\ -\frac{M}{2} \le j \le \frac{M}{2} \}$$

Where:

- $I_{med}(x,y)$. is the filtered pixel value at (x,y).
- N and M are the dimensions of the filter kernel.

5.4. Bilateral Filter

Selection Rationale: Preserves edges while reducing noise.

Mathematical Description:

$$I_{bilateral}(x,y) = \frac{1}{W_p} \sum_{i,j} I(i,j) \cdot w(i,j) \cdot I(x+i, y+j)$$

Where:

- $I_{med}(x,y)$ is the filtered pixel value at (x,y).
- $w(i,j)$ represents the weight between patches centered at (i,j) and (x,y).
- W_p is a normalization factor.

5.5. Non-Local Mean Filter

Selection Rationale: Effectively removes noise while preserving textures.

Mathematical Description:

$$I_{\mathrm{NLM}}(x,y) = \frac{1}{W(x,y)} \sum_{i,j} I(i,j) \cdot w(i,j) \cdot I(x+i, y+j)$$

Where:

- $I_{\mathrm{NLM}}(x,y)$ is the filtered pixel value at (x,y).
- $w(i,j)$ represents the weight between patches centered at (i,j) and (x,y).
- $W(x,y)$ is a normalization factor.

5.6. Deep Learning Models

In this study on white blood cell (WBC) categorization, we extensively evaluated four deep learning models: VGG16, VGG19, RESNET50, and CNN. Each model, with distinct architectures, demonstrated unique strengths in capturing intricate features of WBCs. VGG16 and VGG19, with 16 and 19 layers respectively, excel in feature extraction. RESNET50's innovative residual learning tackles deep network challenges effectively. Despite its simpler architecture, CNN serves as a foundational model. All models underwent rigorous training and fine-tuning, with performance metrics such as accuracy, precision, recall, and F1 score analysed for comprehensive evaluation. VGG16, VGG19, RESNET50, and CNN collectively form the core of our analytical framework for WBC categorization.

5.7. Model Evaluation

Table 1: CNN model analysis for filters

Filter	Accuracy	Precision	Recall	F1
Gaussian	0.77	0.79	0.77	0.78
Median	0.74	0.77	0.74	0.75
Mean	0.74	0.74	0.74	0.74
Bilateral	0.73	0.77	0.73	0.74
Non Local Mean	0.76	0.78	0.76	0.76

Conclusion: In our investigation into model-filter combinations for white blood cell (WBC) classification, we identified VGG16 with the Mean filter as the optimal performer, boasting an accuracy of 85%. This surpasses the findings of Mohamed et al. [1] with VGG-16 at 73.64%, emphasizing the need for denoising. Similarly, compared to dataset constraints faced by Habibzadeh et al. [2] with Inception V1 at 47%, our approach showcases superior results.

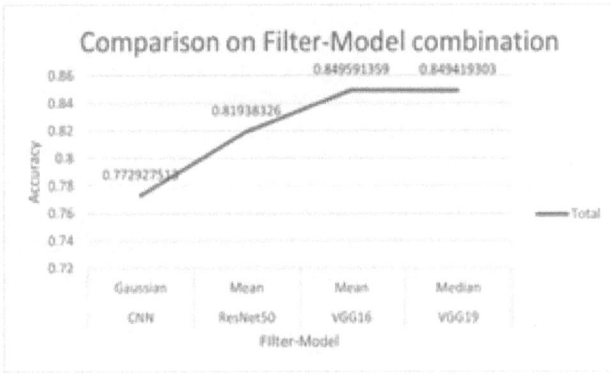

Figure 1: Filter-model combination

These results underscore the significant impact of our model-filter combinations on WBC classification accuracy. The combination of VGG16 with the Mean filter at 85% and ResNet50 with the Mean filter at 81% showcases not only the efficacy of our approach but also its potential for advancing automated diagnostics and healthcare applications. As medical image datasets become more available and deep learning advances, our study propels the field towards continued progress and innovation in medical image analysis.

References

[1] Mohamed, H. E., El-Behaidy, H. W., Khoriba, G., and Li, J. (2020). Improved white blood cells classification based on pre-trained deep learning models. Journal of Communications Software and Systems, 16(1), 37-45.

[2] Habibzadeh, M., Jannesari, M., Rezaei, Z., Baharvand, H., and Totonchi, M. (2018, April). Automatic white blood cell classification using pre-trained deep learning models: Resnet and inception. In Tenth international conference on machine vision (ICMV 2017), SPIE, Vol. 10696, pp. 274-281.

[3] Wu, W., Liao, S., and Lu, Z. (2022). White blood cells image classification based on radiomics and deep learning. IEEE Access, 10, 124036-124052.

[4] Yildirim, M., and Çinar, A. (2019). Classification of white blood cells by deep learning methods for diagnosing disease. Rev. d'Intelligence Artif., 33(5), 335-340.

[5] Rehman, A., Abbas, N., Saba, T., Rahman, S. I. U., Mehmood, Z., and Kolivand, H. (2018). Classification of acute lymphoblastic leukemia using deep learning. Microscopy Research and Technique, 81(11), 1310-1317.

[6] Li, M., Lin, C., Ge, P., Li, L., Song, S., Zhang, H., … and Sun, X. (2023). A deep learning model for detection of leukocytes under various interference factors. Scientific Reports, 13(1), 2160.

[7] Yadav, D. P. (2021). Feature fusion based deep learning method for leukemia cell classification. In 2021 5th International Conference on Information Systems and Computer Networks (ISCON), IEEE, pp. 1-4.

[8] Rustam, F., Aslam, N., De La Torre Díez, I., Khan, Y. D., Mazón, J. L. V., Rodríguez, C. L., and Ashraf, I. (2022). White blood cell classification using texture and RGB features of oversampled microscopic images. In Healthcare (Vol. 10, No. 11, p. 2230). MDPI.

[9] Agustin, R. I., Arif, A., and Sukorini, U. (2021). Classification of immature white blood cells in acute lymphoblastic leukemia L1 using neural networks particle swarm optimization. Neural Computing and Applications, 33(17), 10869-10880.

[10] Akalin, F., and Yumuşak, N. (2022). Detection and classification of white blood cells with an improved deep learning-based approach. Turkish Journal of Electrical Engineering and Computer Sciences, 30(7), 2725-2739.

Challenges and Complexities in Mobile Forensics

A Comprehensive Investigation

Madhuresh Shukla[1], Amit Sharma[2], Sanjeev Mandal[3]

[1]Research Scholar, Computer Application Department, Lovely Professional University, Phagwara, Jalandhar, India
[2]Professor, Computer Application Department, Lovely Professional University, Phagwara, Jalandhar, India
[3]Associate Professor, Computer Application Department, Jain University, Bangalore, India.
E-mail: madhureshshukla88@gmail.com, madhuresh.42200287@lpu.in, amit.25076@lpu.co.in,
profamitsharma@gmail.com, km.sanjeev@jainuniversity.ac.in

Abstract

Digital (Cyber) forensics is a branch of forensic science dedicated for the recovery and investigation of raw data residing in electronic or digital devices [1]. Same way mobile forensics is a branch of digital forensics that dealt with acquisition, processing, and recovery of evidence from mobile device in a forensically sound manner. After much research many approaches and tools have been developed for mobile forensics, but still mobile forensics field has not been established till yet. There are many challenges and research gaps exist in this field specially because of different kinds of hardware, Operating systems, In-built security features, Password recovery etc.

Keywords: Mobile forensics, cyber forensics, mobile malware, Android, iPhone

1. Introduction

Mobile forensics is a branch of digital forensics, which includes seizure, acquisition, and analysis of smartphone [7], [8], [9], [11]. Rapid growth of smartphones has made our life easy, simple, and comfortable and same time it has included features of many digital gadgets such as basic phone, GPS, smart watch, calculator, video game, pager and specially computer. According to an Ericsson report, global mobile data traffic will reach 71 exabytes per month by 2022, from 8.8 exabytes in 2017, a compound annual growth rate of 42 percent [1]. Now a common man uses its 90% of internet data over smartphone only. [4]. Moreover, it's an advanced gadget that is easily portable, transferable, and destroyable.

These all features make mobile crucial evidence in any crime or cyber incident. Forensic models and frameworks for the MF (Mobile Forensics) domain have been analysed in the literature to recognize, gather, and investigate MF crimes [11], [12]. These models were created for special purposes; however, they are redundant [2]. New methodologies in the field of mobile forensics includes remote agent forensics, RAM capture, new seizure methods, network forensics, cloud forensics (tokens), etc. a combination of these new technologies can construct a framework that can fulfil the need of today's requirement of mobile forensics

2. Challenges

2.1. Faulty Seizure

Cyber forensics pose a hurdle when it comes to flawed seizures because different smartphone forms can be missed. When a smartphone is discovered at the scene of a crime and is inadvertently turned off, it becomes challenging for forensics to get past the lock. Furthermore, it is essential to isolate the questioned equipment from networks to stop unauthorised software control and remote wiping.

2.2. Changes in Hardware

Number of manufacturers of mobile phones are not limited. The market is flooded with different types of manufacturers using different types of hardware. So, it creates a challenge for forensics examiners to extract the data from such a variety of hardware.

2.3. A Different Type of Operating Systems

As with personal computers, there are not many different choices of operating systems available but in the case of mobile phones many types of operating systems are available. This characteristic of mobile phones makes it very hard to handle all these devices in forensics Lab [14].

2.4. Mobile Platform Security Features

Mobile forensics encounters challenges with mobile platform security features like app sandboxing, biometric authentication, and encryption. Encryption, vital for user privacy, complicates forensic access to device data. Biometric authentication adds complexity, requiring innovative solutions for authorised access. Secure boot procedures, app sandboxing, and OS updates pose obstacles to conventional forensic procedures, emphasising the need for adaptable forensic tools to ensure successful digital investigations.

3. Methodologies

As described in the above image these are 8 steps forensics processes which need to be adhered to by the cyber forensics' lab during the processing of case. But it depends upon case to case, these steps may increase also (Figure 1).

Intake: This includes paperwork that includes the ownership of device and type of incident in which device was involved.

Identification: this includes the legal authority, all seizure related documents, requirement of case and details of device.

Preparation: this process decides the tool to process the case which will give maximum artefacts. Below describe pyramid plays a very helpful role in preparation (Figure 2).

Isolation: While a mobile is under the forensics examination it should be kept in an isolation environment which includes enabling of airplane mode, use of faraday bags, etc.

Processing: In this step device is processed as per the best suited tool for the maximum artefacts collection.

Verification: You must check the accuracy of the data you have taken from the phone after processing it to make sure it hasn't been altered.

Documentation & Reporting: All the steps and methods used in any case need to be documented. Reporting is the official and final representation of all forensics processes. It should be as per the

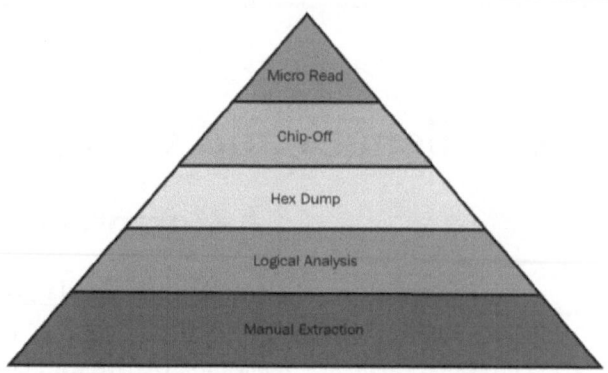

Figure 2: Cellular phone tool levelling pyramid (Sam Brothers, 2009)

described guideline of respective organisations or governments.

Archiving: For the duration of the legal proceeding, future reference, the current evidence file to be retained, and record-keeping needs, the data is kept in a readable manner.

4. Research Gap

4.1. Remote Imaging

There is a lagging of remote imaging of mobile devices [1],[2],[3], [16] and [15]. However, it's quite successful in the case of computer forensics. This results in a remarkable delay in forensics processing of a device.

4.2. Lock Bypass

There is great challenge in the mobile forensics if mobile comes in a lock condition [1],[7] and [8]. If the mobile is new and the latest security patched than it becomes very difficult task for forensics examiner to do imaging of such device. In the case of the I-phone this is a common problem.

4.3. Non-Reliability of Tools

As Forensics starts with forensics lab fully equipped with latest tools [1], [3] and [16]. Not a single tool can support all kinds of mobile phone forensics. Their limitations are the latest OS, latest Security patch, different Hardware, etc.

4.4. Non-Presence of Network Forensics

All the methods and methodologies of mobile forensics present today simply overlooked the importance of network forensics. This methodology can play a vital role in examination of malware infected mobile devices.

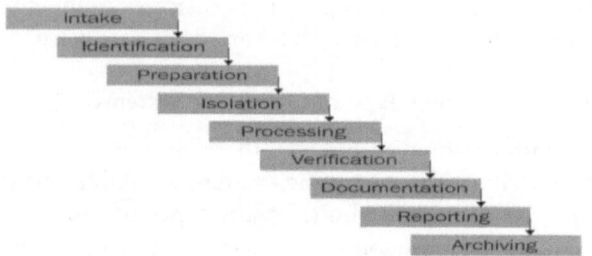

Figure 1: Step by Step evidence processing [1] and [3]

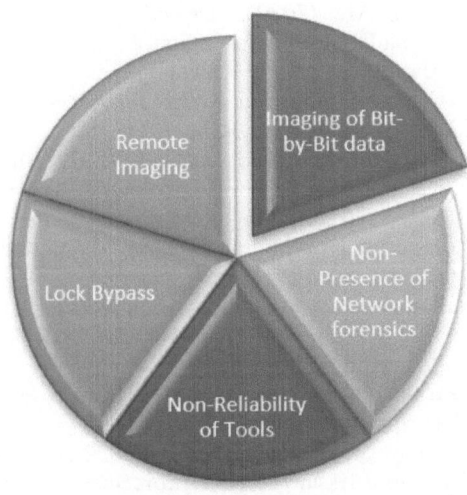

Figure 3: Research Gaps

4.5. Imaging of Bit-by-Bit Data (Physical Imaging)

Every Forensics case needs the recovery of deleted data. But in the case of mobile forensics, it's not possible all time to get deleted data because of non-availability of physical Image of device (Figure 3).

5. Comparative Analysis of Different Mobile Forensics Tools

Forensics Tool	Usage	Strengths	Weaknesses
Cellebrite UFED	Data extraction from mobile devices.	Wide device support. Physical and logical extraction options.	Costly. Closed source software.
Autopsy	Digital forensics on various devices.	Open source. Powerful file carving capabilities.	Learning curve for beginners. Limited mobile support.
EnCase	Comprehensive digital forensic tool.	Well-established in legal and investigative fields. Wide feature set.	Expensive. Steeper learning curve.
XRY	Mobile device forensics tool.	Easy to use interface. Supports a wide range of devices.	Limited file system support. Relatively high cost.
Oxygen Forensic	Extracting and analysing mobile data.	User-friendly interface. Social media analysis capabilities.	Pricier compared to some alternatives. Limited free trial.

As discussed in above table, details of these tools are as following:

Cellebrite UFED: Cellebrite UFED is a versatile mobile forensics tool widely utilised for data extraction [16]. Its primary strength lies in extensive device support, enabling forensic experts to extract both physical and logical data. Despite its efficacy, the tool can be cost-prohibitive for some organisations. Additionally, being a closed-source solution, it lacks the transparency and customization options offered by open-source alternatives.

Autopsy: Autopsy, an open-source digital forensics tool, is favoured for its versatility in analysing various devices. Its notable strength lies in potent file carving capabilities, facilitating the recovery of fragmented files and deleted data. While its open-source nature allows for community collaboration, there may be a learning curve for new users, and its mobile support could be more comprehensive.

EnCase: EnCase is a comprehensive digital forensic tool widely trusted in legal and investigative fields. Its strength lies in a broad range of features, providing advanced capabilities for in-depth digital investigations. However, its high cost may limit accessibility for smaller organisations, and its advanced features could pose a steeper learning curve for new users.

XRY: XRY is a user-friendly mobile device forensics tool known for its ease of use and support for a wide range of devices. However, its relatively high cost may be a drawback, and it may have limitations in supporting certain file systems.

Oxygen Forensic: Oxygen Forensic is designed for extracting and analysing mobile data, with a particular focus on social media analysis. Its user-friendly interface enhances ease of use, and the emphasis on social media analysis is valuable for extracting evidence from various online platforms.

6. Conclusion

As mobile phones are now a main source of information extraction in any case. Mobile devices are often called the true picture of any individual. Till now no such a comprehensive comparison of mobile forensics methodologies and tools has been performed and such unique research gaps have been found. Mobile forensics is quite a dynamic field and still not established. Like computer forensics, mobile forensics needs to be standardized and improved [10], [13]. There is still a great scope of filling the research gaps and new technologies innovation [15], [17]. For the better future of mobile forensics there is a great need for a comprehensive mechanism (tool or framework) that can deliver all cyber forensics requirements.

References

[1] ISBN-13: 978-1788839198, Practical Mobile Forensics - Fourth Edition by Rohit Tamma, Oleg Skulkin, Heather Mahalik, Satish Bommisetty

[2] Ali, A., Razak, S. A., Othman, S. H., Marie, R. R., Al-Dhaqm, A., & Nasser, M. (2022). Validating Mobile Forensic Metamodel Using Tracing Method. In International Conference of Reliable Information and Communication Technology (pp. 473-482). Springer, Cham.

[3] Moreb, M. (2022). Introduction to Mobile Forensic Analysis. In Practical Forensic Analysis of Artifacts on iOS and Android Devices (pp. 1-36). Apress, Berkeley, CA.

[4] Bawankar, L., Bongirwar, M., Sharma, P., Bhojane, S., & Mangrulkar, N. (2022). Android Forensic Tool. In ICCCE 2021 (pp. 709-716). Springer, Singapore.

[5] Agarwal, A., & Gupta, A. (2022). Real-time double JPEG forensics for mobile devices. Journal of Real-Time Image Processing, 1-11.

[6] Vella, M., & Colombo, C. (2022). D-Cloud-Collector: Admissible Forensic Evidence from Mobile Cloud Storage. In IFIP International Conference on ICT Systems Security and Privacy Protection (pp. 161-178). Springer, Cham.

[7] Moreb, M. (2022). Malware Forensics for Volatile and Nonvolatile Memory in Mobile Devices. In Practical Forensic Analysis of Artifacts on iOS and Android Devices (pp. 371-406). Apress, Berkeley, CA.

[8] Moreb, M. (2022). Evidence Identification Methods for Android and iOS Mobile Devices with Facebook Messenger. In Practical Forensic Analysis of Artifacts on iOS and Android Devices (pp. 427-457). Apress, Berkeley, CA.

[9] Spranger, M., Xi, J., Jaeckel, L., Felser, J., & Labudde, D. (2022). MoNA: A Forensic Analysis Platform for Mobile Communication. KI-Künstliche Intelligenz, 1-7.

[10] Moreb, M. (2022). Detecting Privacy Leaks Utilizing Digital Forensics and Reverse Engineering Methodologies. In Practical Forensic Analysis of Artifacts on iOS and Android Devices (pp. 195-225). Apress, Berkeley, CA.

[11] Kumari, N., & Mohapatra, A. K. (2022, March). A Novel Framework For Multi Source Based Cloud Forensic. In 2022 6th International Conference on Computing Methodologies and Communication (ICCMC) (pp. 1-7). IEEE.

[12] Yallamandhala, P., & Godwin, J. (2022). A Review on Video Tampering Analysis and Digital Forensic. In Proceedings of International Conference on Deep Learning, Computing and Intelligence (pp. 287-294). Springer, Singapore.

[13] Belshaw, S., & Nodeland, B. (2022). Digital evidence experts in the law enforcement community: understanding the use of forensics examiners by police agencies. Security Journal, 35(1), 248-262.

[14] Barik, K., Abirami, A., Konar, K., & Das, S. (2022). Research Perspective on Digital Forensic Tools and Investigation Process. Illumination of Artificial Intelligence in Cybersecurity and Forensics, Vol 109 pages 71-95.

[15] Fatima, M., Abbas, H., Iqbal, W., & Shafqat, N. (2022). Forensic analysis of image deletion applications. Multimedia Tools and Applications, 81(14), 19559-19586.

[16] Chidambaram, A. S., Suthendran, K., & Kumar, M. S. (2022). Forensic investigation on electronic evidences using encase and autopsy. Sustainable development in engineering and technology, Vol 1 from page no. 117 to 124 .

[17] Javed, A. R., Ahmed, W., Alazab, M., Jalil, Z., Kifayat, K., & Gadekallu, T. R. (2022). A comprehensive survey on computer forensics: State-of-the-art, tools, techniques, challenges, and future directions. IEEE Access. volume 10, page No. 11065-11089

A Brief Review of E-Authentication, Cyber Security, and their Challenges

Sumaira Bashir[1], Amit Sharma[2], Sanjeev Kumar Mandal[3]

[1]Department of Computer Application, Lovely Professional University, Punjab, India
[2]Lovely Professional University, Punjab, India
[3]Jain University, Bangalore, India
E-mail: Sumaira.42200151@lpu.in, amit.25076@lpu.co.in, km.sanjeev@jainuniversity.ac.in

Abstract

The rise of diverse cloud computing technologies, big data concepts, and the Internet of Things (IoT) has elevated the prominence of cybersecurity. The efficient collection of user data by IoT and cloud systems enhances customer service and fosters a competitive edge in the industry. Nevertheless, the substantial volume of data poses potential challenges concerning privacy and security. This review paper explores methods that compromise the privacy and security of data repositories.

Keywords: Privacy, big data, security, challenges, sophisticated, authentication

1. Introduction

Security involves safeguarding systems against unauthorised access, and it holds significant importance in our daily lives. Authentication, a key method in security, is employed to protect data at both individual and organisational levels. Given the ever-changing nature of technology, it is crucial to adapt security systems to prevent impostors or hackers from exploiting vulnerabilities in the current system [1].

E-Authentication, or electronic authentication, verifies users' identity when they access digital systems, services, or data. It's vital for cybersecurity, ensuring only authorised individuals access sensitive information. E-Authentication methods, like passwords, biometrics, and two-factor authentication, validate user identity online [2]. Authentication in a broader perspective can be categorised as:

1.1. Single Factor Authentication (SFA)

In this technology, the method to authorise or bypass a security system involves using only the user's username and password. While this authentication process is user-friendly, it presents challenges such as being difficult to remember and susceptible to guessing, keylogging, social engineering, and shoulder-surfing.

1.2. Two-Factor Authentication (2FA)

This authentication factor adds an additional layer of protection to our systems by incorporating not only the traditional username and password techniques but also one-time password (OTP) methods [3]. Nevertheless, even this approach does not offer a fully effective remedial measure to address the security challenges prevalent in today's landscape.

1.3. Mutual Authentication

The term "2WAY authentication," also known as two-way authentication, describes an authentication technique where both the client and server validate data encryption and decryption at both ends. This ensures secure data transmission using server and client digital certificates, employing the Transport Layer Security Protocol, along with the use of PINs (Personal Identification Numbers) and Social Security Numbers [4]. While these authentication factors offer protection against keylogging, trojan horses, and various security breach attempts, the effectiveness of the challenge-response mechanism is limited and may not work once the threshold is breached.

1.4. Knowledge Based Authentication

This authentication method is highly comprehensive, utilising both a PIN and passphrase to access passwords or log in to any system based on the user's or client machine's knowledge base. It incorporates recognition-based and recall-based methods. Visual cryptography is also a part of this category, involving the storage of an image on the server side and another on the client side, verified during the data

Chapter 41 DOI: 10.1201/9781003570349

exchange process [5]. An advantage of this approach is a reduction in phishing attacks and brute force attacks to some extent. However, its disadvantage lies in its vulnerability to Man-in-the-Middle attacks, and it does not entirely address phishing issues when logging in from machines other than the one assigned to the user, particularly in managing data at an organisational or corporate level.

1.5. Cybersecurity

Cybersecurity continues to play a pivotal role in the evolving landscape of business, facilitating its growth daily. According to Garner's research, the industry is projected to reach a value of 170.4 billion dollars in 2022 [6]. This substantial growth trajectory alone poses a considerable risk, making organisations, including those with small and mid-sized workforces, acutely aware of the need to address both solutions and vulnerabilities within the realm of cybersecurity.

In the face of widespread digital threats, 48% of organisations reported a rise in cyberattacks this year, the smallest increase in six years, per ISACA's 2023 State of Cybersecurity report. This prompts questions about evolving cyber adversary tactics and the effectiveness of defence mechanisms. Furthermore, 62% believe organisations often under-report cyberattacks, emphasising the crucial need for transparency and collaboration in cybersecurity.

1.6. Related Work

The mentioned technologies like Hydra, Medusa, and Ncrack were discussed at the conference, and suggested remedial measures included implementing password complexity, captcha, and limiting login access. Despite these preventive measures, existing security systems were deemed vulnerable due to significant gaps and loopholes(Varsha Grover and Gagandeep, 2020)[7].

Another discussed topic was the novel linear offset-based poisoning attack method (LOPA) against self-updated fingerprint authentication systems. The method involved making minor linear changes to the minutia representation matrix of a victim's fingerprint template, resulting in a series of fingerprints for authentication. The study reported a 42.86% reduction in fingerprint authentication breaches. However, the authors indicated that there is potential for increasing the effectiveness of the technique using other approaches (Mingfu Xue et al., 2020) [7].

A proposed system introduced three levels of authentication, each increasing in complexity and difficulty, making it challenging for hackers to compromise the system. The authentication levels included passphrases, image-based passwords, and graphical passwords. The author acknowledged the future potential of expanding the complexity of authentication levels (Manisekaran Thangavelu et al., 2021) [10].

Focusing on literature, self-education, and awareness training to detect, assess, and mitigate organisational risks, another study proposed a model to establish a relationship between security awareness, professionals, and situational threat analysis. The future scope of the study aimed at creating a model based on the scale of data dealt with at a large user level [11].

A study explored the use of QR code scanning (e-signatures) in banking transactional systems to enhance client authentication. Any alteration in the client's original signature would be promptly detected, strengthening the authentication process. However, the author did not provide specific details about the future scope for further research in this area (Sagar et al., 2021) [11].

2. Research Void

Let's explore potential areas of opportunity that we can focus on.

2.1. Forecast that Third Parties may Exploit the Potential of the 5G Network

In a past video circulated via social media platforms (news channels) since the online available data of users always leads to a risk. Scammers are taking advantage of this technology since in 4G network, a lot of advancement was in place however we do not have the same security strength in 5G networks.

2.2. A Rising Trend in Mobile Malware Incidents

This form of malware specifically aims at individuals with the intention of disrupting their psychological well-being. The primary cause is the utilisation of open Wi-Fi connections, with 97% of reported cases from organisations citing such attacks in 2022. Throughout the Covid-19 pandemic, various packages, such as tousanticovid.apk, covid.apk, covidMappia_v1.0.3.apk, covidMapv8.1.7.apk, and covid-detect.apk, were employed to target banking applications.

3. Artificial Intelligence

AI has become pervasive in cyber networks, with nearly all facilities adopting an AI-based approach. According to the McKinsey report, over 30% of organisations are currently leveraging AI-based strategies to enhance the security of their networks.

The primary vulnerability associated with using AI-based tools lies in the fact that passwords and

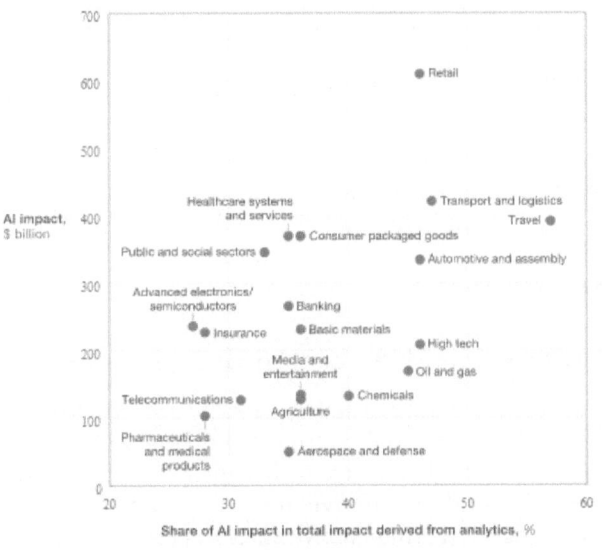

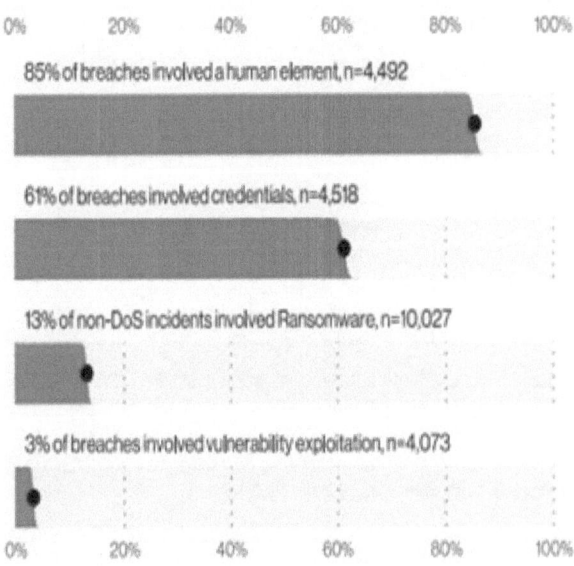

Figure 1: Based on Mckinsey research- the impact of AI on various organisational sectors

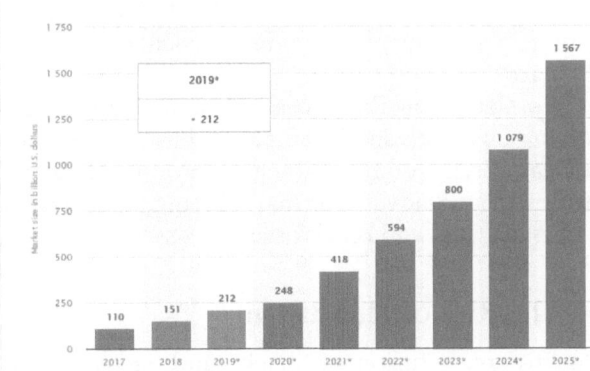

Figure 2: Increase in the market size of IoT

biometric logins are updated by vendors and distributors. This creates an avenue for hackers to gain control and monitor Personally Identifiable Information (PII), including sensitive details such as bank information, belonging to the user.

4. The IoT Devices

The adoption of Internet-of-Things (IoT) devices is currently widespread due to their purported robust mechanisms and cost-effectiveness. IoT serves as a dependable source for managing data, especially at large organisational levels. In 2021, the IoT market achieved a significant user limit, reaching 418 billion US dollars, and it is projected to further increase to around 1.567 trillion USD by the year 2025 [13].

5. Lack of Control Over Phishing and Spear-Phishing Attacks

Spear-phishing attacks become challenging to detect when considering that phishing, often linked to social media, involves the collection of sensitive information such as credit/debit card details and current location. Attackers use deceptive emails or websites designed to appear genuine. Spear-phishing, an advanced form of phishing, targets specific individuals based on their emotional and mental attributes.

According to Verizon's 2021 investigation report, 29,207 security incidents were examined, with 5,285 identified as data breaches. Of these breaches, 36 percent were associated with phishing, marking an 11 percent increase from the previous year. Notably, about 95 percent of organisations experienced a spear-phishing attack, with related breaches accounting for 61 percent of the incidents.

The frequency and impact of spear-phishing attacks can vary across industries. However, spear-phishers can confidently orchestrate a sequence of events aimed at hacking or compromising personal wealth. Challenges stemming from issues like third-party social media operators inadvertently supporting cybercriminal tactics, a lack of thorough inspection of phishing emails in users' accounts, and violations of the right to information contribute to compromising user security. These challenges are putting a strain on cybersecurity protocols within organisations, emphasising the need for models

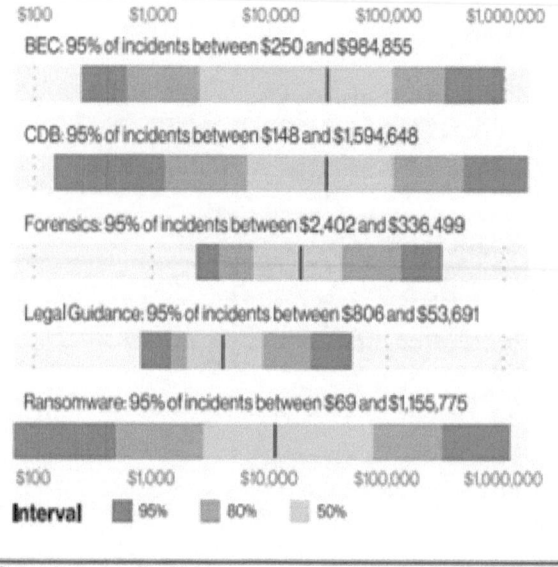

Figure 3: Illustrates the consequences of spear phishing attacks in the year 2021.

with robust control and strengthened compliance standards.

To defend against spear phishing, individuals and organisations should employ robust cybersecurity measures, including ongoing education and awareness programs, advanced email filtering, multi-factor authentication, and regular security assessments.

6. The Rise of Hacktivism

Hacktivism, a fusion of "hack" and "activism," is an activity where hackers aim to obtain someone's information with the intention of using it negatively to undermine respective organisations. The 2021 IBM X-Force report revealed that 25 percent of data thefts and leak attacks in 2020 were instances where hackers explicitly sought to acquire multinational data from both government and private entities. Persistent Distributed Denial of Service (DDoS) attacks on government organisations serve as a notable example.

7. Drone Jacking: A Disturbing New Wave for Cyber Experts

The emergence of drone jacking has become a concerning trend among cybercriminals. This technique involves the illicit use of toy-like drones to compromise personal information. According to Intel, drones, commonly employed for targeted deliveries, filming, and recreational use, have been employed to bypass existing security standards. While drones play valuable roles in various industries such as farming, photography, and supporting

law enforcement agencies, they pose a significant cybersecurity threat.

The potential risks associated with drone jacking include hackers gaining insights into the delivery processes, revealing how many packages can be delivered to specific customers. While this may be initially perceived as a recreational activity, the aftermath is serious, constituting a direct attack on the security compliance of organisations dedicated to ensuring consumer success and maintaining a positive public image.

7.1. Preventive Measures of Social Engineering

Social engineering involves a type of cyberattack where hackers employ tricks and non-tech strategies, diverging from core tech approaches to deceive users. Several preventive measures can be taken, including adjusting spam filters, promptly denying, or deleting help requests, and researching the sources of unsolicited emails. However, modern hackers are sophisticated and understand the commonality of these measures. They can gain legitimate access to personal information and exploit individuals based on their personality weaknesses. According to Google's report, a significant portion of Social Engineering Attacks (SEAs) involve phishing through official-looking emails or convincing malicious websites.

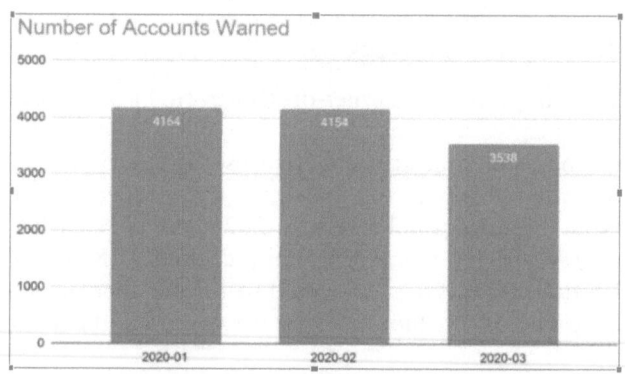

In the graph depicted above, approximately 5000 incidents were flagged by Google. The frequency of these Social Engineering Attacks (SEAs) appears to fluctuate, yet they are notably tracking our communications conducted through instant messaging or video conferencing. The pandemic era witnessed a substantial rise in remote work, involving around 260 million information-based workers, business owners, and artisans. This widespread remote work trend exposed a considerable number of individuals to potential Social Engineering Attacks. Presently, organisations are increasingly focusing on raising awareness about the potential malicious agendas of hackers, aiming to better protect their employees and stakeholders.

7.2. Office Personnel have Access to Sensitive Data within their Organisations

Within all organisations, the awareness of internal politics is commonplace, as employees recognise instances of discrimination and the gradual displacement of human roles by automation. This scenario gives rise to insider risks and threats, which have seen a significant increase, spiking to 47 percent over the past two years, attracting the attention of cybercriminals. A notable 34 percent of businesses experience trust and reputation breaches annually.

7.3. Adaptive Authentication for Dynamic Threat Landscape

Investigate the development of adaptive authentication systems that can dynamically adjust security measures based on the evolving threat landscape. Research should focus on real-time risk assessment and response mechanisms to counter emerging cyber threats effectively.

7.4. Biometric Template Protection Techniques

Explore robust techniques for protecting biometric templates stored in databases. Research could focus on developing secure encryption methods, homomorphic encryption, or other privacy-preserving technologies to safeguard biometric data from unauthorised access and misuse.

7.5. User-Centric Authentication Design

Examine the human factors influencing the acceptance and effectiveness of E-Authentication methods. Research should aim to design authentication systems that consider user preferences, cognitive load, and usability, fostering a balance between security and user experience.

7.6. Quantum-Safe E-Authentication

Address the impact of quantum computing on current E-Authentication protocols. Investigate quantum-safe cryptographic algorithms and authentication mechanisms to ensure the security of digital identities in a post-quantum computing era.

7.7. E-Authentication in Cloud Environments

Explore the unique challenges and security considerations of implementing E-Authentication in cloud-based systems. This research should address issues such as data residency, shared responsibility models, and the impact of cloud architecture on authentication protocols.

7.8. Standardisation and Interoperability

Investigate the lack of standardised protocols and interoperability issues in E-Authentication. Research efforts could focus on developing and promoting industry standards to enhance compatibility and integration between different authentication systems.

7.9. Behavioural Biometrics and Continuous Authentication

Explore the potential of behavioural biometrics (keystroke dynamics, mouse movements) and continuous authentication models. Research should investigate how these dynamic authentication methods can enhance security by continuously monitoring user behaviour throughout a session.

7.10. E-Authentication for IoT Devices

Examine the security challenges associated with E-Authentication in the context of Internet of Things (IoT) devices. Research gaps include developing lightweight authentication protocols suitable for resource constrained IoT devices and addressing scalability and identity management issues.

7.11. Blockchain-Based Authentication

Investigate the use of blockchain technology for secure and decentralised authentication. Research should explore the potential benefits, challenges, and scalability issues associated with using blockchain in E-Authentication systems.

8. Cross-Border Authentication and Legal Implications

Examine the challenges and legal implications of cross-border E-Authentication. Research should focus on harmonising international standards, addressing legal and regulatory differences, and ensuring secure cross-border authentication processes.

These research gaps highlight the need for advancements in various aspects of E-Authentication and Cyber Security, ranging from technical solutions to user-centric design and legal considerations. Researchers can contribute to these areas to develop innovative solutions and address the evolving challenges in securing digital identities.

8.1. Suggested Approach

After a critical review of the challenges, we are facing in the field of Cybersecurity. We should consider the methodology to re-structure a strong security mechanism which consists of the following parameters:

8.2. Security Foundation

Establish a comprehensive security policy aligned with organisational goals. Clearly define physical and logical security boundaries with governing policies. Integrate security into the overall system design and provide developer training.

8.3. Risk-Based Rules

Implement cost-effective risk-based rules. Assume external systems are insecure and tailor security measures accordingly. Protect information throughout its lifecycle.

8.4. Ease of Use

Base security on open standards for portability. Use common language in security requirements. Design security for regular technology adoption and operational ease.

8.5. Increase Resilience

Implement layered security to avoid vulnerabilities. Design systems for resilience against expected threats. Isolate public access systems and establish audit mechanisms.

8.6. Reduce Vulnerabilities

To introduce simplicity in system design, reduce vulnerabilities. Minimise trusted elements and apply the principle of least privilege. Identify and prevent common errors through rigorous testing.

9. Design With Network in Mind

Implement security through distributed measures. Formulate measures for multiple information domains. Authenticate users for access control and accountability.

A fundamental aspect of security engineering and analysis involves addressing security threats, risks, and attacks. Ideally, understanding the attacker is crucial for building a comprehensive security framework.

To address the risks associated with E-authentication and bolster Cybersecurity features, a systematic approach is essential. This involves prioritising identified risks, formulating responses to mitigate them, preventing their escalation, or limiting their impact. Evaluating the current control processes is crucial, followed by the

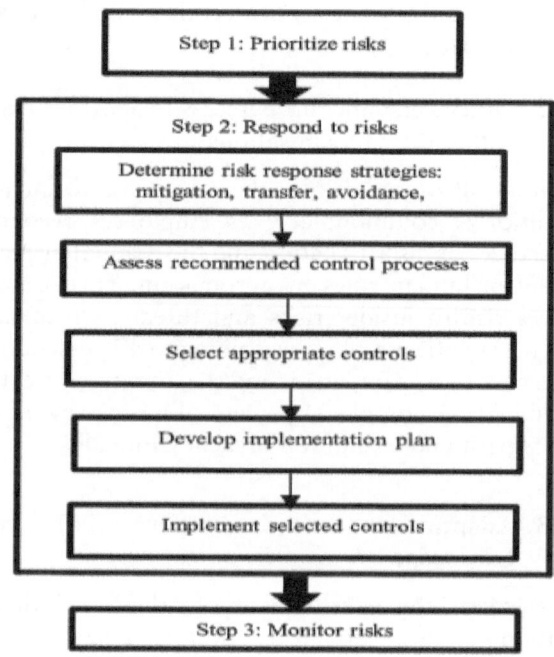

Figure 4: IT security management controls and implementation

selection and recommendation of appropriate controls. Subsequently, a comprehensive implementation plan needs to be developed and executed for the chosen controls. Post-implementation, continuous monitoring of risks and their behaviours is imperative. This structured process enables a thorough assessment of the system's current state and informs subsequent actions for further improvement and risk mitigation.

10. Envisioned Outcomes and Summative Conclusions

To conclude, research in E-Authentication and Cyber Security is driven by the increasing reliance on digital technologies, the expansion of online services, and the rising complexity of cyber threats. This necessitates robust E-Authentication for secure interactions in an interconnected world, driven by users' motivation to protect personal information and prevent unauthorised access.

Users seek confidence in online security, leading Cyber Security research to develop reliable authentication methods and reduce online fraud risks. Efforts could be focussed on creating user-friendly authentication experiences, such as biometrics, balancing security, and usability.

Additional efforts can aim to reduce password fatigue and develop robust authentication protocols for users seeking assurance of secure access to digital services. Research educates users about cybersecurity risks and best practices.

Upcoming research can address the evolving nature of cyber threats to enhance digital system resilience. Awareness of privacy issues motivates research in E-Authentication, aligning solutions with regulatory frameworks. Rapid technological advancements drive secure harnessing of these technologies in E-Authentication and Cyber Security.

References

[1] Schlegel, D. J., Ferraro, S., Aldering, G., Baltay, C., BenZvi, S., Besuner, R., Blanc, G. A., Bolton, A. S., Bonaca, A., Brooks, D., Buckley-Geer, E., Cai, Z., DeRose, J., Dey, A., Doel, P., Drlica-Wagner, A., Fan, X., Gutierrez, G., Green, D., … Zhou, R. (2022). A Spectroscopic Road Map for Cosmic Frontier: DESI, DESI-II, Stage-5. http://arxiv.org/abs/2209.03585.

[2] Sabzevar, A. P., & Stavrou, A. (2008). Universal Multi- Factor Authentication Using Graphical Passwords. 2008 IEEE International Conference on Signal Image Technology and Internet Based Systems, 625–632. https://doi.org/10.1109/SITIS.2008.92.

[3] Dimitrova, V. (2009). Artificial Intelligence in Education: Building Learning Systems that care: From Knowledge Representation to Affective Modelling. Amsterdam: IOS Press.

[4] Gunter, C., Liebovitz, D., and Malin, B. (2011). Experience-based access management: A life-cycle framework for identity and access management systems. IEEE Security & Privacy Magazine, 9 (5), 48–55.

[5] Bhattacharyya, D., Ranjan, R., Alisherov, F., and Choi, M. (2009). Biometric authentication: A review. International Journal of u-and e-Service, Science and Technology, 2 (3), 13–27.

[6] Bours, P. (2012). Continuous keystroke dynamics: A different perspective towards biometric evaluation. Information Security Technical Report, 17(1–2), 36–43. https://doi.org/10.1016/J.ISTR.2012.02.001

[7] Whitley, E. A. (2012). A critical review of cloud computing: Researching desires and realities. Journal of Information Technology, 27, 179–197.

[8] Grover, V. and Gagandeep. (2020). An efficient brute force attack handling techniques for server virtualization. In Proceedings of the International Conference on Innovative Computing & Communications (ICICC). SSRN: https://ssrn.com/abstract=3564447 or http://dx.doi.org/10.2139/ssrn.3564447.

[9] Xue, M., He, C., Wang, J., and Liu, W. (20202). LOPA: A linear offset based poisoning attack method against adaptive finger print authentication system. Computer&Security, 99, 102046, ISSN 0167-4048. https://doi.org/10/1016/j.cose.2020.102046. https://www.sciencedirect.com/science/article/pii/S0167404820303199.

[10] Shitole, S. N., Dhanve, A., Khadke, A., & Bansode, A. (2020). Issue 3 www.jetir.org (ISSN-2349-5162). In JETIR2003014 Journal of Emerging Technologies and Innovative Research (Vol. 7). JETIR. www.jetir.org

[11] Thangavelu, M., Krishnaswamy, V., & Sharma, M. (2021). Impact of comprehensive information security awareness and cognitive characteristics on security incident management – an empirical study. Computers and Security, 109. https://doi.org/10.1016/J.COSE.2021.102401.

[12] Sagar, R. R. (2022). E-Authentication system with QR Code. IJIRT, 8 (12). ISSN: 2349-6002.

[13] Azizi, N., & Haass, O. (2022). Cybersecurity issues and challenges. Handbook of Research on Cybersecurity Issues and Challenges for Business and FinTech Applications, 21–48. https://doi.org/10.4018/978-1-6684-5284-4.CH002.

Leveraging Machine Learning Approaches to Gauge the Prevalence of Anxiety and Depression

Rakhi Nagpal, Saravjeet Singh and Aditi Moudgil

Chitkara University Institute of Engineering and Technology, rakhi.nagpal@chitkara.edu.in

Abstract

Depression, anxiety, and other mental health concerns have become increasingly prevalent within the fast-paced contemporary society. Having a considerable negative influence on people's quality of life, anxiety disorders are a serious healthcare issue. In order to check the prevalence of mental health concerns like anxiety and depression among people, this current study makes use of machine learning classifiers namely Logistic Regression and Gaussian Naïve Bayes. The dataset contains demographic data from 320 individuals that were acquired by survey responses as well as additional variables including anxiety and depression that were assessed using the Maslach Burnout Inventory, also known as the MBI scale. The novelty of this research lies in its groundbreaking methodology which made use of two machine learning classifiers namely Random Forest (RF) and Support Vector Machine (SVM) to examine demographic and MBI data. The main focus of this research is to classify the different severity levels of individuals having depression and anxiety disorder. To perform this analysis, two machine learning classifiers mainly RF and SVM have been used. When compared to SVM, RF classifier has given the better accuracy of 97.5% and 98.7% for depression and anxiety respectively.

Keywords: Anxiety disorders, depression, machine learning, mental illness

1. Introduction

A crucial element of human growth is mental wellness. It addresses human ideas and feelings while directing them towards a healthy lifestyle by encouraging good mental health. Without a healthy state of mind, there are significant barriers which hampers the individuals at their private, commercial, and social levels [1]Click or tap here to enter text.. In today's highly competitive landscape, individuals often seek every opportunity to proceed in their careers, sometimes accepting irritation, despair, anxiety, stress, and unhappiness as commonplace aspects of their professional lives [2]. But these mental illness issues make people more susceptible to other disorders and contribute to both accidental and purposeful injury [3]. According to World Health Organization (WHO) apart from being physically fit, a healthy person should also have a healthy intellect [4].

Depressive disorders and anxiety disorders are two of the primary diagnostic groups for prevalent mental illnesses. Sadness, absence of enthusiasm, feelings of regret or sense of worthlessness, insufficient sleep, inadequate nutrition, and challenges in maintaining focus and concentration are the major symptoms of depressive disorders whereas Generalised Anxiety Disorder (GAD), Panic Disorder (PD), Social Anxiety Disorder (SAD), Autistic Spectrum Disorder (ASD), and Post-Traumatic Stress Disorder (PTSD) are among the mental diseases referred to as anxiety disorders. As per the World Health Organization (WHO), anxiety disorders account for 3.4% of all disabilities worldwide and ranked as 6th largest contributor. Similar to depressive disorders, anxiety disorders are Prevalent among women to a greater extent than among men, with a worldwide incidence of 4.6% for women and 2.6% for men [5]. During the year 2020, approx. 62% of participants stated that they feel some level of anxiety in their daily life activities as well [6]. Research indicates that adolescents grappling with untreated anxiety disorders are at a higher risk of experiencing academic challenges, missing out on important social interactions, and engaging in substance misuse [7]. During the time of COVID-19 as well, many individuals have been impacted badly and faced anxiety issues later on [8,9]. According to few studies, anxiety disorder can be a symptom or co-occurrence disease with other diseases like Alzheimer [10]. The Government has undertaken several regulatory and programmatic measures to deal with these challenges, some of which resulted in certain benefits as well [11]. Given the information above, the major goal of the research is to determine the incidence of depressive

Chapter 42 DOI: 10.1201/9781003570349

and anxiety disorders among people by using different machine learning classifiers. To perform this analysis, the study employed a dataset from Kaggle which contains demographic data from 320 individuals that were acquired by survey responses as well as additional variables including anxiety and depression that were assessed using the Maslach Burnout Inventory, also known as the MBI scale.

2. Literature Review

Recent studies shows that a lot of research has been carried out on mental disorders, which provide a plate full of information into emerging insights for anxiety disorders and depression. The earlier studies on the machine learning approaches for anxiety and depressive disorders are discussed in this section. Anxiety and depression are two most mental disorders that affects seafarers the most. In [12], the authors have collected information about 470 seafarers. In order to assess the depression and anxiety in participants, 5 different machine learning classifiers RF, LR, Naïve Bayes (NB), SVM, and CatBoost were used. The results have shown that the CatBoost classifier outperforms others with accuracy of 82.6% and precision of 84.1%. In [13], the study aims to forecast the Depression, Anxiety, Stress (DAS) level utilising 5 various algorithms, including the Reduces Error Pruning (REP), Classification and Regression Trees (CART), J48, Multilayer Perceptron (MLP), and Logistic Regression (LR) method. The results stated that MLP outperforms all other machine learning classifiers by achieving accuracy of 90.33%, 92%, and 90.33% for depression, anxiety, and stress respectively.

One of the studies [9] has used 8 different machine learning classifiers in order to assess the psychological disorders. The authors have also made use of hybrid model and neural network (Bayes net) to analyse the various severity levels of depression, stress, and anxiety. The results have shown that the Bayes net has given the better accuracy (anxiety-86.3%, depression-89.6%, and stress-88.9%) as compared to other models. In [14], the authors have used machine learning models along with biosignals in order to detect the anxiety from the time period 2012 to 2022. This analysis with average numbers of individuals ranges between 10 to 102 demonstrates a considerable variety of accuracies, extending between 55% and 98%. The results have shown that the EEG (Electroencephalography) outperforms the other different biosignals investigated, consistently giving the outstanding performance whereas other biosignals like heart rate, Electrodermal Activity (EDA), and respiration rate (RSP) have produced the greatest levels of accuracy. SVM and RF has proven to be the mostly used machine learning classifiers in detecting the anxiety.

3. Experiment Details and Result Analysis

For this experiment, the data has been taken from Kaggle (available at Depression and anxiety data | Kaggle). 787 non-graduated students of Lahore university were part of this dataset. This dataset characterises the demographic information (id, gender, age), health metrics (body mass index), depression metrics (PHQ score, depression severity, depression treatment, depression diagnosis, suicidal, depressiveness), sleep-related features (sleepiness, pworth score), and anxiety metrics (GAD score, anxiousness, anxiety severity, anxiety treatment, anxiety diagnosis) of all the participants. Figure 1 shows the sample of dataset being used for current analysis.

For this experiment RF and SVM were used. In the realm of anxiety disorders, these classifiers have

id	school_year	age	gender	bmi	who_bmi	phq_score	depression_severity	depressiveness	suicidal	depression_diagnosis	depression_treatment	gad_score	anxiety_severity	anxiousness	anxiety_diagnosis	anxiety_treatment	epworth_score	sleepiness
1	1	19	male	33.33333333	Class I Obesity	9	Mild	FALSE	FALSE	FALSE	FALSE	11	Moderate	TRUE	FALSE	FALSE	7	FALSE
2	1	18	male	19.84126984	Normal	8	Mild	FALSE	FALSE	FALSE	FALSE	5	Mild	FALSE	FALSE	FALSE	14	TRUE
3	1	19	male	25.10239133	Overweight	8	Mild	FALSE	FALSE	FALSE	FALSE	6	Mild	FALSE	FALSE	FALSE	6	FALSE
4	1	18	female	23.73866213	Normal	19	Moderately severe	TRUE	TRUE	FALSE	FALSE	15	Severe	TRUE	FALSE	FALSE	11	TRUE

Figure 1: Sample of depression and anxiety dataset

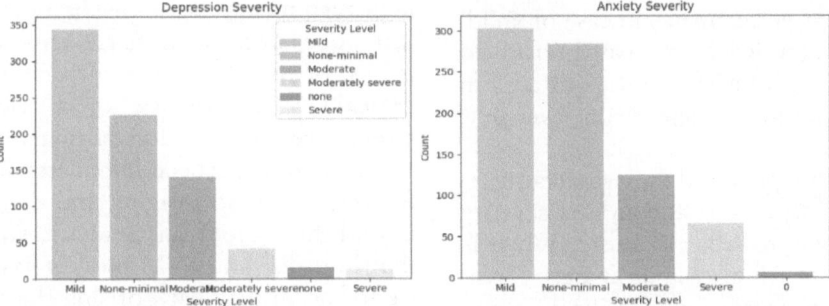

Figure 2: Depression and anxiety severity count

Table 1: Depression severity classification report using RF and SVM.

Technique	RF			SVM		
Parameters	Precision	Recall	F1-score	Precision	Recall	F1-score
Mild	1.00	1.00	1.00	0.86	0.98	0.92
Moderate	0.96	1.00	0.98	0.81	0.71	0.76
Moderately severe	1.00	0.92	0.96	0.90	0.69	0.78
None-minimal	0.94	1.00	0.97	0.92	0.94	0.93
Severe	1.00	1.00	1.00	0.00	0.00	0.00
None	0.00	0.00	0.00	0.00	0.00	0.00
Accuracy	97.5			87.9		

Table 2: Anxiety severity classification report using RF and SVM.

Technique	RF			SVM		
Parameter	Precision	Recall	F1-score	Precision	Recall	F1-score
0	0.00	0.00	0.00	0.00	0.00	0.00
Mild	1.00	1.00	1.00	0.88	0.89	0.89
Moderate	1.00	1.00	1.00	0.94	0.62	0.75
None-minimal	0.96	1.00	0.98	0.85	1.00	0.92
Severe	1.00	1.00	1.00	0.92	0.92	0.92
Accuracy	98.7			87.9		

demonstrated its utility by effectively separating panic disorder from other forms of anxiety disorders [15-17]. Figure 2 illustrates the distribution of severity levels of anxiety and depression by the supplied bar graph.

The analysis of different severity levels of depression and anxiety were analysed using RF and SVM and results are shown in Tables 1 and 2.

Further classification report for the severity level of the individuals was analysed on the bases of Precision, Recall, and F1-score and results are shown in Tables 1 and 2. According to performed analysis, RF has outperformed the SVM.

In order to measure the model's performance, F1-score gives the best performance in case of 'mild' and 'severe' depression level whereas performed poor for 'none'. The overall performance of Random Forest for the classification of depression severity is coming out to be 97.5%.

The use of SVM has provided the overall accuracy of 87.9% for both depression and anxiety severity classification as shown in Tables 1 and 2. Whereas the overall accuracy of RF in case of anxiety classification is better than depression classification i.e., 98.7%

4. Conclusion

By utilising the machine learning capabilities, we can gain a better understanding to dive more deeply into the complex factors that contribute to the severity of depression and anxiety. In this paper a dataset from Kaggle was used to predict severity level of mental health conditions like anxiety and depression. Furthermore, we look at whether there are differences in degree of severity based on demographic traits, with the goal of identifying trends that might guide focused interventions. As compared to SVM, the use of Random Forest classifier has proved to be an important tool in the search for improved mental health evaluation and intervention techniques because of its capacity to give exact estimates, organise complex information, and uncover critical key features insights. The visualization of different severity levels for participants in form of bar graphs marks a crucial initial step in paving the way for the profound insights and discoveries that will unfold throughout this analysis. The results indicate that both RF and SVM models have a lot of promise for accurately categorising the degree of depression and anxiety. Additional research might be done to improve the performance of both models, such as using cross-validation methods and fine-tuning hyperparameters.

References

[1] Dhyani, A., Gaidhane, A., Choudhari, S. G., Dave, S., and Choudhary, S. (2022). Strengthening response toward promoting mental health in India: A narrative review. Cureus, 14 (10). https://doi.org/10.7759/CUREUS.30435.

[2] Priya, A., Garg, S., and Tigga, N. P. (2020). Predicting anxiety, depression and stress in modern life using machine learning algorithms. Procedia Computer Science, 167, 1258–1267. https://doi.org/10.1016/J.PROCS.2020.03.442.

[3] Mental Health - PAHO/WHO | Pan American Health Organization. (n.d.). Accessed September 3, 2023. https://www.paho.org/en/topics/mental-health.

[4] Sau, A. and Bhakta, I. (2017). Predicting anxiety and depression in elderly patients using machine learning technology. Healthcare Technology Letters, 4 (6), 238–243. https://doi.org/10.1049/HTL.2016.0096.

[5] World Health Organization (WHO). (2017). Depression and other common mental disorders: global health estimates. https://apps.who.int/iris/bitstream/handle/10665/254610/W?sequence=1.

[6] SingleCare Team. (2023). Anxiety statistics 2023 | SingleCare. https://www.singlecare.com/blog/news/anxiety-statistics/.

[7] Facts & Statistics, Anxiety and Depression Association of America, ADAA. (n.d.). Accessed September 3, 2023. https://adaa.org/understanding-anxiety/facts-statistics#Facts%20and%20Statistics.

[8] Kansal, I., Popli, R., and Singla, C. (2021), Comparative Analysis of various Machine and Deep Learning Models for Face Mask Detection using Digital Images," in 2021 9th International Conference on Reliability, Infocom Technologies and Optimization (Trends and Future Directions), ICRITO 2021, Institute of Electrical and Electronics Engineers Inc., pp. 1–5. doi: 10.1109/ICRITO51393.2021.9596407.

[9] Kumar, P., Garg, S., and Garg, A. (2020). Assessment of anxiety, depression and stress using machine learning models. Procedia Computer Science, 171, 1989–1998. https://doi.org/10.1016/J.PROCS.2020.04.213.

[10] Behl, T., Kaur, D., Sehgal, A., Singh, S., Sharma, N., Zengin, G., Andronie-Cioara, F. L., Toma, M. M., Bungau, S., and Bumbu, A. G. (2021). Role of monoamine oxidase activity in Alzheimer's disease: An insight into the therapeutic potential of inhibitors. Molecules, 26 (12), 3724. https://doi.org/10.3390/MOLECULES26123724.

[11] Pandya, A., Shah, K., Chauhan, A., and Saha, S. (2020). Innovative mental health initiatives in India: A scope for strengthening primary healthcare services. Journal of Family Medicine and Primary Care, 9 (2), 502. https://doi.org/10.4103/JFMPC.JFMPC_977_19.

[12] Sau, A. and Bhakta, I. (2019). Screening of anxiety and depression among seafarers using machine learning technology. Informatics in Medicine Unlocked, 16, 100228. https://doi.org/10.1016/J.IMU.2019.100228.

[13] Mary, S. T. A. and Jabasheela, L. (2018). An Evaluation of Classification Techniques for Depression, Anxiety and Stress Assessment, in International Conference for phoenixes on emerging current trends in engineering and management (PECTEAM 2018), Atlantis Press, pp. 64–69. doi: 10.2991/PECTEAM-18.2018.13.

[14] Ancillon, L., Elgendi, M., and Menon, C. (2022). Machine learning for anxiety detection using biosignals: A review. Diagnostics, 12 (8), 1794. https://doi.org/10.3390/DIAGNOSTICS12081794.

[15] Mohamed, E. S., Naqishbandi, T. A., Bukhari, S. A. C., Rauf, I., Sawrikar, V., and Hussain, A. (2023). A hybrid mental health prediction model using Support Vector Machine, Multilayer Perceptron, and Random Forest algorithms. Healthcare Analytics, 3, 100185. https://doi.org/10.1016/J.HEALTH.2023.100185.

[16] Na, K. S., Cho, S. E., and Cho, S. J. (2021). Machine learning-based discrimination of panic disorder from other anxiety disorders. Journal of Affective Disorders, 278, 1–4. https://doi.org/10.1016/J.JAD.2020.09.027.

[17] Xing, M., Fitzgerald, J. M., and Klumpp, H. (2020). Classification of social anxiety disorder with support vector machine analysis using neural correlates of social signals of threat. Frontiers in Psychiatry, 11, 510000. https://doi.org/10.3389/FPSYT.2020.00144/BIBTEX.

[18] Tiwari, S., Kumar, S., and Guleria, K. (2020). Outbreak trends of coronavirus disease–2019 in India: A prediction. Disaster Medicine and Public Health Preparedness, 14 (5), e33–e38. https://doi.org/10.1017/DMP.2020.115.

An Efficient Prediction of Hepatocellular Carcinoma by Using Machine Learning Algorithms

M. Revathi[1], S. Ohmshankar[2], D. Maheswari[3], R. Nithiya[4]

[1]Assistant Professor, Department of Artificial Intelligence & Data Science, St.Joseph's Institute of Technology, Chennai.
[2,3]Assistant Professor, Department of Electronics and Communication Engineering,
Agni College of Technology, Chennai.
[4]Assistant Professor, Department of Bio-Medical Engineering, Agni College of Technology, Chennai.
E-mail: gmrevathigopinath@gmail.com

Abstract

Hepatocellular Carcinoma is a type of live disease which ends in liver cancer. This type of disease happens in persons who is having chronic liver infection. There are many methods to diagnose this disease which includes imaging and biopsy. Liver is the most important organ whose failure will affect the life span of human. If this disease is diagnosed in an earlier stage, proper treatment can be given and the human life can be saved. In recent years, the applications of AI is used in medical field to guide the doctors in diagnosis of disease. Machine Learning (ML) is indeed a subset of artificial intelligence (AI) that analyses huge amount of data and to find the hidden patterns in it. In this work, popular Machine learning algorithms like Logistic Regression, Support Vector Machine and K-Nearest Neighbour is applied. Among these the Logistic Regression method produces higher classification accuracy of 91% which is higher than the other two algorithms. The accuracy can be further increased by applying feature selection techniques before giving it as input to the Machine Learning algorithms.

Keywords: Support vector machine, hepatocellular carcinoma, K-nearest neighbour

1. Introduction

One of the most important part of our body is the Liver. Liver acts as an organ since it performs more number of actions needed by our body and also liver acts as a gland which produces some necessary hormones needed by our body. Many chemical substances will be released by our body organs into the blood. Liver helps in the filtration of blood and to remove the unwanted chemicals from blood. Liver gets blood as input from two sources: Digestive system and Heart. The food that is taken by a human is broken down into nutrition's and this nutrition rich blood is sent to liver by portal vein. The oxygen rich blood is sent by heart to liver through hepatic artery.

Liver [1] is used to develop glycogen which is formed from the sugar separated from blood. Sometimes the sugar level of a person can get low due to many reasons and our liver will convert this stored glycogen into glucose which gives instant sugar into our blood. Bile is a substance produced by liver and it is needed for the process of digestion and also to absorb fats that are produced in the small intestine. Liver also produces a protein-based substance called albumin whose role is to prevent the leakage of fluids from blood. If liver is damaged and enough albumin is not produced, fluids may leak from blood and may affect other organs of the body.

There are many effects of liver failure that affects the normal functioning of a human body. The major problem of liver failure [2] is the collection of fluids around the body organs which may affect their normal functionality. Liver plays a major role in the regulation of blood clots and failure of liver may affect this functionality. Liver failure can change the way in which the kidney works and it may lead to the failure of kidneys as well. The damage to liver can also cause infections to occur more easily.

Liver failure [3] can be found by using blood test, imaging test and biopsy. In blood test, the level of working of the kidney can be measured. In blood test, the time taken for the blood to clot will also be found. If the problem in liver [4] is identified at an earlier stage and proper medication is given, it can be cured. There are many Machine Learning algorithms that helps in the health care sector for the well-being of humans. If suitable model is built, the liver disease can be identified more accurately and at an early stage.

Chapter 43 DOI: 10.1201/9781003570349

2. Literature Review

In [5], it is stated that Radio Frequency ablation is a most commonly used treatment in the field of hepatocellular carcinoma. Here, the researchers have tried to build a Machine Learning model to find the possibility of recurrence of the disease even after the treatment of Radio Frequency Ablation. In [6], a popular method for binary classification called Logistic Regression is chosen as a base model. The authors have explored the use of genetic algorithms in combination with Logistic Regression and built a novel model.

In [7], authors have identified natural killer cells (type of immune cells) as an element that destroys the cancer cells in lungs. The authors have done experiments in analysing the count of natural killer cells in determining the life span of patients with hepatocellular carcinoma. Here, a prediction model based on the visual nomogram was developed to find the life span of patients. In [8], the authors insisted on the importance of earlier disease prediction to prevent patients death. The authors found that miRNA profile will be different for a hepatocellular carcinoma patients than for normal patients. Authors used this fact to predict the disease at earlier stage.

In [9], authors have used the UCI machine learning dataset for analysing the performance of the prediction model. The authors have performed various types of data pre-processing on the dataset and analysed its effects on the prediction model. In [10], the authors used 31P MRS (31 Phosphorous Magnetic Resonance Spectroscopy) dataset to distinguish between the liver cancer affected patient and normal patient. In [11], authors have used the dataset from UCI machine learning database for prediction of the disease. The authors have used Machine Learning model like Decision tree classifier and Xgboost Classifier for prediction of the disease.

In [12], the authors have demonstrated the need for earlier prediction of liver cancer in order to save human life. No noticeable symptoms can be found in earlier stages which makes earlier identification to be difficult. In [13], the authors have taken Magnetic Resonance Imaging (MRI) image dataset for prediction of liver cancer at earlier stages. The authors have developed a model that predicts the presence of disease as well as it identifies the stage of the disease. Pre-processing of images have been done to identify the Region of Interest (ROI) in the input images. Next, feature extraction have been done using different techniques.

3. Methodology

3.1. Support Vector Machine (SVM)

SVM is one of the popular supervised machine learning algorithm which is used in binary classification

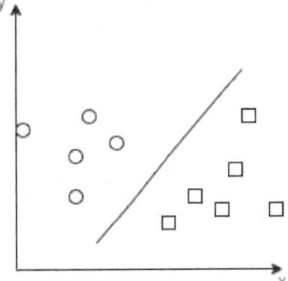

Figure 1: Dataset with hyperplane

problems. It works as follows: Consider that the dataset has two features x and y (denoted by x axis and y axis) and two classes of objects (squares and circles) as shown in Figure 1. The data points are given as input to the SVM algorithm [14] and the algorithm returns the best plane that is used to separate the two classes of objects. Since it is two dimensional, the output will be a straight line as shown in Figure 1.

3.2. K Nearest Neighbour (KNN)

KNN [15] is one of the most popular supervised learning algorithm which works based on the assumption that the similar data objects will be located nearer to each other. Dissimilar objects will be located far away from each other. The main core of KNN relies on the value of k i.e. number of neighbours for a data object. K should be chosen with proper knowledge to increase the accuracy of classification.

When a new data object arrives into the system, the algorithm calculates the distance between the new data object and all the existing data objects. It selects k data objects that are closer to the new data object based on the distance. The most common class label among the k neighbours is given as class label to the new data object. There are various distance methods used. Euclidean distance is the most commonly used measure which is given by the equation 1,

$$d(x, y) = \sum_{i=1}^{n} \sqrt{(y_i - x_i)^2} \tag{1}$$

4. Results

The Indian Liver Patient Dataset (ILPD) is a dataset that was indeed sourced from the UCI Machine Learning Repository [16]. The ILPD dataset contains various attributes related to liver patients, and it is often used for tasks like classification and prediction of liver disease. These attributes include features such as age, gender, total bilirubin, direct bilirubin, alkaline phosphatase, and more. Totally there are 583 instances in the dataset. Out of 583 patient information, 441 are information about male patient and 142 are information about female patient. Table 1 gives performance.

Table 1: Performance of three ML algorithms

Classifier	Precision	Recall	F1-Score	Accuracy
Logistic Regression	0.87	0.9	0.88	91.04
K-NN	0.82	0.78	0.79	87.54
SVM	0.84	0.76	0.79	82.54

5. Conclusion

Hepatocellular Carcinoma is a type of live disease which ends in liver cancer. There are many methods to diagnose this disease which includes imaging and biopsy. In recent years, the applications of AI is used in medical field to guide the doctors in diagnosis of disease. In this work, Machine learning algorithms such as Support Vector Machine, K-Nearest Neighbour is applied. Among these the Logistic Regression method produces higher classification accuracy of 91% which is higher than the other two algorithms. The accuracy can be further increased by applying feature selection techniques before giving it as input to the Machine Learning.

References

[1] Liaqat, A., et al. (2021). LDA–GA–SVM: improved hepatocellular carcinoma prediction through dimensionality reduction and genetically optimized support vector machine. Neural Computing and Applications, 33 (7), 2783–2792. https://doi.org/10.1007/S00521-020-05157-2.

[2] Choi, G. et al. (2020). Development of machine learning-based clinical decision support system for carcinoma. Scientific Reports, 10, 14855. https://doi.org/10.1038/s41598-020-71796-z.

[3] Hattab, M., Maalel, A., Ghezala, H.H.B. (2020). Towards an Oversampling Method to Improve Hepatocellular Carcinoma Early Prediction. In: Chaari, L. (eds) Digital Health in Focus of Predictive, Preventive and Personalised Medicine. Advances in Predictive, Preventive and Personalised Medicine, vol 12. Springer, Cham. https://doi.org/10.1007/978-3-030-49815-3_16

[4] Książek, W., Abdar, M., Acharya, U. R., and Pławiak, P. (2019). A novel machine learning approach for early detection of hepatocellular carcinoma patients. Cognitive Systems Research, 54, 116–127. https://doi.org/10.1016/j.cogsys.2018.12.001.

[5] Sato, M., and Tateishi, R. et al. (2022). Machine learning–based personalized prediction of hepatocellular carcinoma recurrence after radiofrequency ablation. Gastro Hep Advances, 1(1), 29-37.

[6] Książek, W. et al. (2021). Comparison of various approaches to combine logistic regression with genetic algorithms in survival prediction of hepatocellular carcinoma. Computers in Biology and Medicine, 134, pp. 104431, 2021.

[7] Yu, L., Liu, X., Wang, X., Zhou, D., Yan, H., Xie, Y., Pu, Q., Zhang, K., and Yang, Z. (2022). Nomogram for prediction of long-term survival with hepatocellular carcinoma based on NK cell counts. Annals of Hepatology, 27 (2), 100672. https://doi.org/10.1016/j.aohep.2022.100672. Epub 2022 Jan 20.

[8] Liang, G., Wu, J., and Xu, L. (2021). A prognosis-related based method for miRNA selection on liver hepatocellular carcinoma prediction. Computational Biology and Chemistry, 91, 107433. https://doi.org/10.1016/j.compbiolchem.2020.107433. Epub 2021 Jan 12. PMID: 33540232.

[9] Hasan, M. E., Mostafa, F., Hossain, M. S., and Loftin, J. (2023). Machine-learning classification models to predict liver cancer with explainable AI to discover associated genes. *AppliedMath*, 3 (2), 417–445. https://doi.org/10.3390/appliedmath3020022.

[10] Liu, Y., Liu, Q., Cheng, J., Wu, L., and Xin, W. (2007). Classification of 31P MRS data for liver cancer. Lecture Notes in Engineering and Computer Science, 113, pp. 828-832.

[11] Ferdib-Al-Islam, L. Akter and M. M. Islam, "Hepatocellular Carcinoma Patient's Survival Prediction Using Oversampling and Machine Learning Techniques," 2021 2nd International Conference on Robotics, Electrical and Signal Processing Techniques (ICREST), DHAKA, Bangladesh, 2021, pp. 445-450, doi: 10.1109/ICREST51555.2021.9331108.

[12] Wibowo, V. V., Rustam, Z., Hartini, S., Setiawan, Q. and Aurelia, J. (2020). Comparison between Support Vector Machine and Random Forest for Hepatocellular Carcinoma (HCC) Classification, 2020 International Conference on Decision Aid Sciences and Application (DASA), Sakheer, Bahrain, 2020, pp. 618-622, doi: 10.1109/DASA51403.2020.9317083.

[13] A. Alksas et al. (2021). A novel computer-aided diagnostic system for early assessment of hepatocellular carcinoma. In 2020 25th International Conference on Pattern Recognition (ICPR) (pp. 10375-10382). Milan, Italy. https://doi.org/10.1109/ICPR48806.2021.9413044.

[14] Revathi, M., Raghuraman, G., Visumathi, J. (2023). Performance Analysis of Machine Learning Algorithms in the Systematic Prediction of Chronic Kidney Disease on an Imbalanced Dataset. In: Smys, S., Kamel, K.A., Palanisamy, R. (eds) Inventive Computation and Information Technologies. Lecture Notes in Networks and Systems, vol 563. Springer, Singapore. https://doi.org/10.1007/978-981-19-7402-1_12

[15] Admassu, T. (2021). Support vector machine and k-nearest neighbor based liver disease classification model. Indonesian Journal of Electronics, Electromedical Engineering, and Medical Informatics, 3, 9–14. https://doi.org/10.35882/ijeeemi.v3i1.2.

[16] https://archive.ics.uci.edu/dataset/225/ilpd+indian+liver+patient+dataset.

Analysis on Approaches for Code Mixed Roman Urdu & English Sentiment Analysis

Suhail Javid, Tarandeep Singh Walia

Department of Computer Applications, Lovely Professional University, Phagwara, India, taran_walia2k@yahoo.com

Abstract

The computational examination of attitudes, views, and feelings regarding certain subjects, items, individuals, and organization as sentiment analysis. Social media text is being used by businesses and consumers to get feedback before making choices. Sentiment analysis is getting mature as artificial intelligence and natural language processing advance. Social media platforms today include an astounding volume of material, most of it written in imprecise and informal languages like Roman Urdu mixed with English. These imprecise, informal, and under-resourced code-mixed languages are not reliably translated by current sentiment analysis tools. The objective of the papers is to combine substantial research on code-mixed resource-poor languages like Roman Urdu and then evaluate the accuracy of different models.

Keywords: Abstractive, extractive, natural language processing, rule-based, machine learning, reviews Code mixed, Roman Urdu English Emotion detection, deep learning

1. Introduction

To determine the emotional states of a topic or review, sentiment analysis (SA) is practiced. People share their opinions and views on topics, products, social events, and political and economic difficulties through social media. Due to the prevalence of this information, it has become challenging to manually extract relevant information from it. As a result, an automated system, like the SA System, that can effectively extract sentiments from it has to be developed.

Roman Urdu (RU)/Hindi, a language with limited resources, has received relatively little attention in most of the work on SA that has been done for resource-rich languages like English and Chinese [3,7]. There are two main factors that make the creation of a strong Roman Urdu Sentiment Analysis (RUSA) required. Urdu with more than 500 million speakers, ranks third among all languages [4]. Second, more often Urdu speaking users prefer typing in Latin script (i.e., RU utilizes the 26-letter English alphabet) to typing in their own language on a language-specific keyboard while communicating online. Work on RUSA is motivated by the platform's variety and huge user base [2]. Likewise, 3.3 posts are created on the social media platform Facebook, according to a SmartInsights2 survey.

2. Key Challenges of Code-Mixed Roman Urdu

The following are a few of the complexities of RU, which make the development of a SA system for it more challenging. 1. There are no rules for how the Urdu language should be represented in Latin script; for instance, "Who bhot acha ha" and "Who bohat acha hai" both mean "he is very good." 2. A Keyword in RU may refer to two or more Urdu words with distinct pronunciations; for instance, the words "Khawar" for "Dawn" and "Khawar" for "miserable" are pronounced differently. 3. Because of this, both "ali ne paani ka aik glass piya" and "panni ka aik glass ali ne piya" mean the same thing in Urdu: "Ali drank a glass of water." 4. Urdu and hence RU are morphologically rich languages, with "acha" (masculine), "achi" (feminine), and "achay" (plural) all denoting the same English word, "good." There is no capitalization in RU, i.e., for "ali ne paani ka aik glass piya," some people write "Ali" as "ali" and others "Ali." 5. Negative emotions handling is more challenging for instance, "Yeh mery lya theek nahi ha" and "Ye mery lya nahi theek" both mean "This is not good for me." 6. There are extremely few RUSA datasets and a severe dearth of labeled datasets in RU [10,44]. 7. Because RU is our first concern, we only considered reviews that were pertinent. However, "borrowing" could not be prevented because we live in a bilingual society [2]; for instance. "Ye drama to acha tha lakin is ka end bohat bura hoa" (This play was good, but its conclusion was poor).

3. Related Work

In section I we have compiled the work done on the Data sets available and in the section II, we have compiled the work done on methodology and experimentation.

RUEN (Roman Urdu Emotion Detection Dataset). [18] prepared for opinion mining done Mobile reviews on https://www.whatmobile.com.pk/ openly available on get hub. RUSA (Roman Urdu sentiment analysis) (Mahmood et al., 2020) corpus of 11,000 reviews. This dataset can be accessed by emailing the corresponding author, Khawar Mehmood (k.mehmood@student.adfa.edu.au). Roman Urdu Number data set is openly available containing the data related to Roman Urdu number Roman Urdu Darze it is the labelled sentiment analysis data set prepared from Darze.com e-commerce website with features. Urdu to Roman Urdu sentence it is the dataset containing the Urdu and Roman Urdu equivalents. Hate Speech Roman Urdu (HS-RU-20) [18]Click or tap here to enter text.. This labeled dataset has been prepared to categorize the comments between offensive O and Hate H. CM-MEC-21 (Ameer et al., 2022) contains 11,914 code-mixed (English and Roman Urdu) multi-label SMS messages, hereafter called CM-MEC-21 corpus. CM-MEC-21 corpus is composed of a significant percentage (>37%) of monolingual RU and EN sentences. The emotion dataset was compiled by Abdullah Ilyas et al. by scraping 400,000 phrases from three different sources and manually parsing them to discover 20,000 RU-EN code-mixed sentences. The code-mixed sentences are then carefully annotated to create the biggest emotion detection corpus for RU-EN code-mixed text.

Our first step was to prepare and locate available datasets for Code mixed social media text. To achieve this goal, a comprehensive investigation was conducted, and several research publications were examined, in which two key datasets were selected which are publicly accessible. CM-MEC-21 and RU-EN. The CM-MEC-21 corpus is made up of 7474 code-mixed sentences produced from SMS and comprises 37% monolingual sentences that may be utilized for author profiling and the rest for sentiment and emotion analysis. The RU-EN dataset on the other hand, is created by collecting 4,00,000 phrases from social media platforms Twitter and YouTube, as well as comments from online websites for purchasing mobile phones in Pakistan, as shown in Table 1. The data was manually analyzed once again to find the 20,000 code mixed sentences.

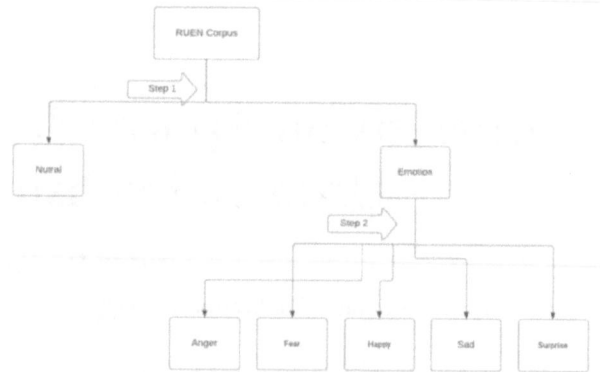

Figure 1: Compilation of corpus for code-mixed RU-ED

To create the biggest dataset for codemixed RU-EN, the codemixed phrases were manually annotated. On the corpus, a two-level annotation technique was used (see Figure 1). The sentences in the corpus were categorized as Neutral or Emotional in the first stage. The second phase is the Emotion phase in which the corpus is further categorized into five sets of emotions (Anger, Fear, Happy, Sad, and Surprise). Further iterative and rigorous approaches are utilized to develop criteria for the second phase of classifying emotions into five groups (Anger, Fear, Happy, Sad, and Surprise). Figure 2 displays the Corpus's two-step classification.

To prepare the dataset for annotation, preset rules were used to classify the dataset against emotion, further annotated data was cross-checked by another annotator to check its validity. The following is a summary of the annotation technique utilized in this study to begin, an annotator examined 20,000 code-mixed sentences to determine if the provided sentence is neutral or has an emotion. The parameters in Table 3 were used to establish the label's appropriateness. Emotion sentences were those that carried an emotion, whereas Neutral sentences were those that did not. Second, the other expert examined the first-level annotations at random to confirm that the recommendations were followed appropriately. Finally, the first annotator analyzed the 8371 sentences identified as emotional sentences to evaluate their classification. The assessment was carried out in accordance with the guidelines indicated in Table 3. As a result, each sentence was assigned the emotion decided by the expert. At last, the second-level annotation was randomly inspected by the other annotator to confirm that the criteria were followed appropriately.

Table 2 displays the attributes of the RU-EN-Emotion corpus. With 20,000 phrases and 429,534 tokens with a total length of 34,278 words that are annotated at two levels make up the RU-EN Emotion corpus. 11,629 neutral sentences and 8371

Table 2: Compilation of corpus for code-mixed RU-ED

Ref	Language	Public	Size	Specifications
Weibo	ZH-EN	No	Not available	Not Available
Weibo	ZH-EN	No	3530	Not Available
Weibo	ZH-EN	No	4195	Not Available
Weibo	ZH-EN	No	4195	Not Available
Weibo	ZH-EN	YES	4195	Not Available
Weibo	ZH-EN	YES	6382	AN=765, HP=2534, FR-770, SD=150
Fire 2014	HI-EN	NO	300	Not available
Tweeter	HI-EN	NO	1589	AN=471, HP=490, FR=304, SD = 324
Tweeter	HI-EN	YES	2866	AN=667, HP=595, FR=85, SD=878, SR=182, O=459
Tweeter	HI-EN	NO	12000	AN=4000, HP=4000, FR=0, SD=4000, SR=0
Tweeter	HI-EN	NO	149088	AN=28705, HP=25869, FR=18981, SD=20931, SR=18935, O=35667
Tweeter	ML-EN	NO	295817	AN=51745, HP = 59242, FR=18895, SD = 79233, SR = 37778, O = 59242
Tweeter	RU-EN	YES	11419	AN = 334, Joy = 1537, FR = 569, SD = 753, SR = 357, O = 8364
RU-EN	RU-EN	YES	20000	Level 1: Emotion = 8371, Neutral = 11629 Level 2: AN = 3487, HP=3720, FR = 228, SD = 224, SR = 712

ZH = Chinese, EN = English, HI = Hindi, ML = Malaya
AN = Anger, HP = Happy, FR = Fear, SD = Sad, SR = Surprise, O = Other

emotion phrases make up the first-level annotation. One of the five emotions is then assigned to each of the emotional statements for the second-level annotation. There are 3487 Anger, 3720 Happy, 712 Surprise, 228 Fear, and 224 Sad phrases in the RU-EN corpus.

The RU-EN-Emotion corpus's specifications are compared with the corpora of contemporary Asian languages. Our corpus is thoroughly compared to the current corpora in Table 2. The data source utilized for text scraping is described in the first column of the table. The language of the corpus is shown in the second column. The third column shows if the corpus is accessible to the general public and the corpus size is indicated in the fourth column. Lastly, the specifications of corpora are listed in the final column. To make comparing the specification easier, the corpora are organized linguistically. Four important observations concerning the RU-EN-Emotion corpus and the current corpora have been established based on the comparison.

The first column demonstrates that a single source of text was used to create the corpora for emotion detection. The text for the Hindi-English corpus is exclusively gathered from Twitter, whereas the text for the Chinese-English corpus was extracted from Weibo, a well-known microblogging site in China. The 240-character limit on Twitter makes it difficult to use as a data source for text extraction. Several types of emotion sentences that are utilized on other social media platforms, such YouTube, are not allowed under this restriction. Consequently, there is limited generalizability of the results.

Based on the findings in the second column, the table's results indicate that emotion detection corpora for code-mixed text between Chinese, Hindi with English exist. For Urdu-English code-mixed text, there are just two publicly available datasets for a similar corpus. We think the created corpus will be helpful in kick-off the investigation and development of this crucial NLP domain.

4. Experimentation

In this section, we compiled tests conducted by Abdullah Ilyas et al. to evaluate the accuracy of supervised and deep learning approaches. The text encoding strategies used for experimentation are detailed below Lastly, a detailed explanation of the experimental settings is given.

Tables 3 and 4 present the findings of the experiment conducted using machine learning and deep learning techniques. Classical approaches include Support Vector Machine (SVM), Random Forest (RF), Decision Tree (DT), Naive Bayes (NB), Linear Regression (LR), and Max Entropy (ME), these strategies were chosen to diversify the results. For instance, SNB uses a probabilistic strategy, whereas DT uses a branching strategy. For the experiment (TRS), several neural network architectures were utilized, including Convolutional Neutral Network, Recurrent Neural Network, Long Short-Term Memory (LSTM), Bi-directional LSTM, Attention-based Bi-directional LSTM (At-Bi-LSTM), and Transformer-based method.

The selection of deep learning approaches was based on their innovative performance on a wide range of text classification tasks. CNN used a dropout layer to prevent overfitting. Dropout values are in the range of 0 to 1, where 0 denotes no layer output and 1 indicates no dropout. This study employed a dropout value of 0.5 since it is believed that values between 0.5 and 0.8 are effective in addressing the overfitting issue. Additionally, every CNN hidden layer's output layer employed the Relu activation function. Moreover, the choice of deep learning methods is predicated on their use in current research.

5. Features

Test categorization is performed using a Term Document Matrix (TDM), with a row representing a document and each column representing a feature. These experiments use unigram, bigram, and trigram features. Values of the TDM are replaced with feature presence values to show the effect of each parameter. at last, TDM serves as both a training and testing input for machine learning algorithms.

Deep learning experiments were done using Word embeddings as input features. Using Word2Vec, GloVe, and fastText, These embeddings were trained on the RU-EN corpus ever created, with 429,534 tokens. The skip-gram method is used to generate the embeddings with a window size of 5 and a vector dimension of 100. The unsupervised technique GloVe uses a term-to-term matrix to denote words as vectors. An enhanced version of Word2Vec is FastText which considers subwords and generates subword embeddings.

Machine learning techniques need the data to be represented numerically, which is accomplished by implementing text encoding. Three types of encodings are used in classical machine learning techniques. Three variants of N-grams are used. An N-gram is a set of N items from a set of data. First, the document is transformed into vector form using Count Vectorizer (CV), in which the presence of a word is represented by one and the nonexistence of a word is denoted by zero.

Deep learning techniques implement cutting-edge representations, formally known as word embeddings. embeddings help to achieve the best result and have recently been implemented in several NLP tasks such as text analysis. Word embedding is a multidimensional vectorial representation of text. Three cutting-edge embeddings were used, as they have recently achieved ground-breaking outcomes for Roman Urdu text classification such as NER* [17]. Word2Vec, GloVe, and fastText are all included. Word2vec employs a neural network to discover the relationships between words in a sentence. The unsupervised technique GloVe represents words as vectors. FastText is developed by the Facebook AI Research lab.

The first set of experiments is carried out with RU-EN-Emotion corpus in which classical machine

Table 3: Results of Exp 1 using classical machine learning techniques

Classifier	Feature	Level 1		
		Precision	Recall	F1 Score
SVM	unigram	0.692	0.675	**0.684**
	bigram	0.643	0.610	0.625
	trigram	0.629	0.552	0.588
RF	unigram	0.681	0.674	0.677
	bigram	0.622	0.609	0.615
	trigram	0.628	0.560	0.592
DT	unigram	0.616	0.613	0.613
	bigram	0.604	0.600	0.602
	trigram	0612	0.562	0.586
NB	unigram	0.670	0.674	0.672
	bigram	0.630	0.621	0.626
	trigram	0.628	0.561	0.593
LR	unigram	0.688	0.672	0.680
	bigram	0.633	0.606	0.619
	trigram	0.638	0.599	0.596
Me	unigram	0.575	0.508	0.539
	bigram	0.643	0.506	0.567
	trigram	0.677	0.512	0.583

Table 4: Results of Exp 1 using deep learning techniques

Classifier	Embeddings	Level 1		
		Precision	Recall	F1 Score
CNN	fastText	0.860	0.849	0.855
	Glove	0.825	0.811	0.818
	Word2Vec	0.894	0.864	**0.879**
RNN	fastText	0.791	0.500	0.613
	Glove	0.634	0.591	0.612
	Word2Vec	0.671	0.606	0.637
LSTM	fastText	0.710	0.705	0.707
	Glove	0.777	0.768	0.773
	Word2Vec	0.694	0.689	0.692
Bi-LSTM	fastText	0.731	0.723	0.727
	Glove	0.779	0.775	0.777
	Word2Vec	0.679	0.671	0.675
A-Bi-LSTM	fastText	0.575	0.558	0.567
	Glove	0.754	0.746	0.750
	Word2Vec	0.754	0.532	0.552
BERT	pre-trained	0.791	0.500	0.612

learning techniques were used with three types of features used in the first subset of experiments includes 16 experiments that employ deep-learning techniques using three embeddings. For training and testing were done on CM-MEC-21 corpus. Experiments were carried out in these settings using a mix of classical and machine-learning techniques. Remember work [5] used 12 emotions, whereas this study used six recognized emotions to develop the corpus, making the two corpora compatible. As a result, a fair comparison of the two corpora's performance cannot be made using the entire CM-MEC-21 corpus. Both sets of experiments employ the most powerful classical machine learning and deep learning techniques. The Sikit-learn library and Google Colab were used in all of the experiments. Precision, Recall, and F1 scores are computed for each fold.

6. Results and Analysis

Table 3 shows the Precision, Recall, and F1 scores for the experiments implemented with classical machine learning techniques. According to the table, SVM had the highest F1 score of 0.684 percent for classifications. This demonstrates SVM's ability to classify

between neutral and emotional sentences. SVM has proved to be more effective as our dataset has several dimensions and SVM is the best performer with large dimensions.

We also observed that ME is the least performing technique. As ME performs best with dependent features, In our case the features have no dependency between them.

According to the comparison of F1 scores, the F1 scores for most types of features decline as the value of n in n-gram increases, as shown in Table 4. It demonstrates that machine learning approaches are capable at categorizing sentences. Moreover, the value F1 score is achieved highest when unigram features are used. This observation holds true for both classification levels, indicating that unigram features are most effective at both levels of classification. The pre-trained embedding model BERT performed the worst on both levels because it used pre-trained embedding, as pretraining of the embeddings is generated for English vocabulary, and not on code mixed i.e. English and Roman-Urdu words.

The Precision, Recall, and F1 scores of experiments performed with deep learning techniques are shown in Table 4. The score from the table shows that CNN attained the highest F1 scores. This makes it clear that CNN is the most efficient technique for classifying emotion and neutral sentences, as well as assigning the appropriate emotion to a sentence. CNN is best at automatically detecting the best features to train the model without the need for human intervention.

Other observations from Tables 3 and 4 are that the highest F1 score is achieved with GloVe word embeddings, as GloVe focuses more on the co-occurrences of words. Transformer model, BERT performed worst as BERT uses the pre-trained embedding trained on the English vocabulary, as our corpus is code mixed text i.e. English and Roman-Urdu words.

7. Conclusion and Future Work

Emotion detection is a critical NLP task with many applications. Considering the importance of this task, numerous studies on monolingual settings have been done, whereas less work has been done on code-mixed text. Even though a large amount of social media communication is code-mixed, there is a clear gap in emotion detection in code-mixed Asian languages.

Finally, the experimentation done to evaluate the efficacy of classical techniques, one probabilistic technique, and deep learning techniques. we can conclude that a) The results revealed that for the first classification, CNN outperformed all classical machine learning techniques with an F1 score of 0.879, b) CNN performed better for the first-level classification as the data set at the first stage is larger than the dataset at the second stage.

The three main potential advices for future work are a) Improve the corpus quality by including more fear, sad, and surprising sentences to create a steady dataset; b) improve the dataset by labelling the intensity of the emotions in each sentence; and c) implementing the hyper-parameter tuning of deep learning techniques.

References

[1] Rout, J. K., Choo, K. K. R., Dash, A. K., Bakshi, S., Jena, S. K., and Williams, K. L. (2018). A model for sentiment and emotion analysis of unstructured social media text. Electronic Commerce Research, 18 (1), 181–199. https://doi.org/10.1007/s10660-017-9257-8.

[2] Mehmood, F., Ghani, M. U., Ibrahim, M. A., Shahzadi, R., Mahmood, W., and Asim, M. N. (2020). A precisely xtreme-multi channel hybrid approach for roman urdu sentiment analysis. IEEE Access, 8, 192740–192759. https://doi.org/10.1109/ACCESS.2020.3030885.

[3] Lal, M., Kumar, K., Wagan, A. A., Laghari, A. A., Khuhro, M. A., Saeed, U., Umrani, A., and Chahjro, M. A. (2020). A systematic study of Urdu language processing its tools and techniques: A review. International Journal of Engineering Research & Technology, 9 (12), 37–43. http://www.urdupoint.com.

[4] Lal, M., Kumar, K., Wagan, A. A., Laghari, A. A., Khuhro, M. A., Saeed, U., Umrani, A., and Chahjro, M. A. (2020). A systematic study of Urdu language processing its tools and techniques: A review. International Journal of Engineering Research & Technology, 9 (12), 37–43. http://www.urdupoint.com.

[5] Madjarov, G., Kocev, D., Gjorgjevikj, D., and Džeroski, S. (2012). An extensive experimental comparison of methods for multi-label learning. Pattern Recognition, 45 (9), 3084–3104. https://doi.org/10.1016/j.patcog.2012.03.004.

[6] Huang, X. and Deng, L. (2010). An overview of modern speech recognition. In Handbook of Natural Language Processing, 2nd ed., pp. 339–366.

[7] Mehmood, K., Essam, D., Shafi, K., and Malik, M. K. (2020). An unsupervised lexical normalization for Roman Hindi and Urdu sentiment analysis. Information Processing and Management, 57 (6), 102368. https://doi.org/10.1016/j.ipm.2020.102368.

[8] Akhter, M. P., Jiangbin, Z., Naqvi, I. R., Abdelmajeed, M., and Sadiq, M. T. (2020). Automatic detection of offensive language for Urdu and roman Urdu. IEEE Access, 8, 91213–91226. https://doi.org/10.1109/ACCESS.2020.2994950.

[9] Song, K., Feng, S., Gao, W., Wang, D., Chen, L., and Zhang, C. (2015). Build emotion lexicon from microblogs by combining effects of seed words and emoticons in a heterogeneous graph. In HT 2015 - Proceedings of the 26th ACM Conference on Hypertext and Social Media, June, 283–292. https://doi.org/10.1145/2700171.2791035.

[10] Arshad, M. U., Bashir, M. F., Majeed, A., Shahzad, W., and Beg, M. O. (2019). Corpus for emotion detection on roman Urdu. In Proceedings - 22nd International Multitopic Conference, INMIC 2019 (pp. 1–6). https://doi.org/10.1109/INMIC48123.2019.9022782.

[11] Ghulam, H., Zeng, F., Li, W., and Xiao, Y. (2019). Deep learning-based sentiment analysis for roman Urdu text. Procedia Computer Science, 147, 131–135. https://doi.org/10.1016/j.procs.2019.01.202.

[12] Feng, S., Wang, Y., Song, K., Wang, D., and Yu, G. (2018). Detecting multiple coexisting emotions in microblogs with convolutional neural networks. Cognitive Computation, 10 (1), 136–155. https://doi.org/10.1007/s12559-017-9521-1.

[13] Sailunaz, K. and Alhajj, R. (2019). Emotion and sentiment analysis from Twitter text. Journal of Computational Science, 36, 101003. https://doi.org/10.1016/j.jocs.2019.05.009.

[15] Wu, C. H., Chuang, Z. J., and Lin, Y. C. (2006). Emotion recognition from text using semantic labels and separable mixture models. ACM Transactions on Asian Language Information Processing, 5 (2), 165–182. https://doi.org/10.1145/1165255.1165259.

[15] Jain, T. I. and Nemade, D. (2010). Recognizing contextual polarity in phrase-level sentiment analysis. International Journal of Computer Applications, 7 (5), 12–21. https://doi.org/10.5120/1160-1453.

[16] Naqvi, R. A., Khan, M. A., Malik, N., Saqib, S., Alyas, T., and Hussain, D. (2020). Roman Urdu news headline classification empowered with machine learning. Computers, Materials and Continua, 65 (2), 1221–1236. https://doi.org/10.32604/cmc.2020.011686.

[17] Zahid, R., Idrees, M. O., Mujtaba, H., and Beg, M. O. (2020). Roman Urdu reviews dataset for aspect based opinion mining. In Proceedings - 2020 35th IEEE/ACM International Conference on Automated Software Engineering Workshops, ASEW 2020 (pp. 138–143). https://doi.org/10.1145/3417113.3423377.

[18] Shahroz, M., Mushtaq, M. F., Mehmood, A., Ullah, S., and Choi, G. S. (2020). RuTUT: Roman Urdu to Urdu translator based on character substitution rules and unicode mapping. IEEE Access, 8, 189823–189841. https://doi.org/10.1109/ACCESS.2020.3031393.

[19] Tehreem, T. (2021). Sentiment Analysis for YouTube Comments in Roman Urdu. http://arxiv.org/abs/2102.10075.

[20] Mehmood, K., Essam, D., and Shafi, K. (2019). Sentiment analysis system for Roman Urdu. In Advances in Intelligent Systems

and Computing (Volume 858, no. June). Springer International Publishing. https://doi.org/10.1007/978-3-030-01174-1_3.

[21] Rao, Y., Xie, H., Li, J., Jin, F., Wang, F. L., and Li, Q. (2016). Social emotion classification of short text via topic-level maximum entropy model. In Information and Management (Volume 53, no. 8). Elsevier B.V. https://doi.org/10.1016/j.im.2016.04.005.

[22] Poddar, A., Agrawal, S., and Guide, P. (2016). T Witter S Entiment a Nalysis &. August, 1–20.

[23] Acheampong, F. A., Wenyu, C., and Nunoo-Mensah, H. (2020). Text-based emotion detection: Advances, challenges, and opportunities. Engineering Reports, 2 (7), 1–24. https://doi.org/10.1002/eng2.12189.

[24] Agrawal, A. and An, A. (2012). Unsupervised emotion detection from text using semantic and syntactic relations. In Proceedings - 2012 IEEE/WIC/ACM International Conference on Web Intelligence, WI 2012 (pp. 346–353). https://doi.org/10.1109/WI-IAT.2012.170.

Brain Tumor Detection Using Deep Learning

Team HEAD CASE (Mohit Khajuria, Nitin Raj, Neeraj Roy) under guidance of Neeraj Mathur

Lovely Professional University, Phagwara, India

Assistant Professor, Department of Artificial Intelligence & Data Science, St.Joseph's Institute of Technology, Chennai.

E-mail: mohit041999@gmail.com, nitinraj27july@gmail.com, neerajmathur3@gmail.com, neeraj333roy@gmail.com

Abstract

This disease happens when the cells in the brain grow rapidly and are uncontrolled, which forms a tumor in the brain. If the disease is not diagnosed early, it can be life-threatening. Despite significant efforts in the area of tumor detection, accurate distribution and taxonomy of brain tumors still remain a challenge. This significantly depends on the location, shape, and size of the tumor which makes it hard to diagnose. Through this survey, we aim to provide a thorough literature review of brain tumor detection which is done through Magnetic Resonance Imaging (MRI). This survey covers various phases of brain tumor detection, including the study of brain tumor anatomy, datasets accessible to the public, methods for enhancing images, segmentation, extracting features, classifying tumors, and utilizing deep learning, transfer learning, as well as quantum machine learning approaches for brain tumor analysis. This review presents a thorough summary of the studies on brain tumor detection, outlining the benefits, drawbacks, advancements, and possible future directions of various methodologies. The results obtained from this survey can be used to accompany future research in the field of brain tumor detection, particularly focusing on enhancing the precision and effectiveness of brain tumor analysis, so that the disease can be diagnosed and treated in its early stage. In the field of medical imaging, finding brain tumors is a crucial task. The chances of the treatment being successful and the outcome of the patient crucially depends upon the timing and accuracy of the diagnosis. By enabling the automatic and precise detection of brain tumors from magnetic resonance imaging (MRI), deep learning-based approaches have shown significant potential for addressing this difficulty. In this study, we propose a deep learning-based approach for brain tumor detection using convolutional neural networks (CNNs). To precisely detect brain tumors, the suggested approach uses a deep CNN architecture that has been developed on a sizable dataset of brain MRI scans. We tried to refine an already trained CNN on our dataset using machine learning techniques, resulting in excellent sensitivity and effectiveness in detecting brain cancers. We examined the proposed approach on a set of brain MRI scans and demonstrated that the technique is very accurate and highly effective in locating brain cancers. Our results showcase that the proposed deep learning-based strategy outperforms traditional machine learning approaches and offers cutting-edge performance on this problem. In brief, our study delivers a possible deep learning-based approach for CNN-based brain tumor diagnosis. The suggested method can be utilised as a reliable and efficient instrument for the accurate and automated detection of brain tumours, enabling rapid treatment and diagnosis of this life-threatening illness.

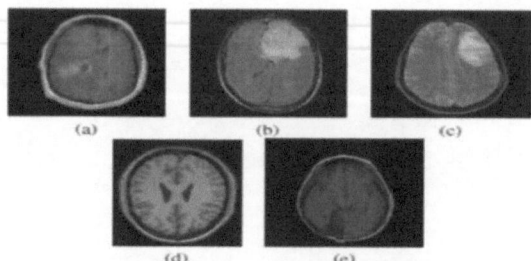

(a) (b) (c)

(d) (e)

Keywords: CNN, deep learning

1. Introduction

With advancements in medical technology, clinical specialists can now create an efficient electronic medical system. It has been proven effective in various fields of medicine [3]. Computer-generated biomedical imaging plays a vital role in providing radiologists with vital information to solve problems they have previously faced. There are distinct medical imaging modalities like X-ray, MRI, Ultrasound, and CT play an important role in order to facilitate the diagnosis and treatment of patients [4,5].

The abnormal growth of cells in or near the brain disrupts the basic function of the brain and dramatically impacts the patient's health. Several sophisticated MRI techniques, such as DTI, MR Spectroscopy (MRS), and Perfusion MR, are employed in the MRI-based analysis of brain tumors [6-8]. Two important types of brain tumors: malignant tumours, which are cancerous, and benign tumours, which are noncancerous. The World Health Organization (WHO) has further categorized malignant brain tumors ranging from grades I to IV. [9]. The WHO has categorized brain tumors into two categories: malignant tumors (cancerous) and benign tumors (non-cancerous). Malignant tumors are classified into four grades: Grade-II Tumors (Pilocytic Astrocytoma), Grade-III Tumor (Low-Grade Astrocytoma), Grade-III Tumor (Anaplastic Astrocytoma), and Grade-IV Tumor (Glioblastoma). Grade-I and Grade-II tumors are relatively less aggressive and termed semi-malignant. Conversely, Grade-III and Grade-IV tumors are highly malignant, significantly impacting a patient's health and potentially leading to fatality [10].

The MRI has numerous features identified in different studies of brain tumor segmentation, including image textures [13], histograms [14], and structure tensor eigen values [15].

The aberrant proliferation of cells in or around the brain disrupts its fundamental functions and significantly affects the patient's well-being. Advanced Magnetic Resonance Imaging (MRI) techniques, such as Diffusion Tensor Imaging (DTI), MR Spectroscopy (MRS), and Perfusion MR, are employed for the assessment of brain tumors via MRI. Brain tumors are broadly categorized into malignant (cancerous) and benign (non-cancerous) types by the World Health Organization (WHO), with malignant tumors further classified into grades I to IV. Grade-I and Grade-II tumors are relatively less aggressive, categorized as semi-malignant, while Grade-III and Grade-IV tumors are highly malignant and pose severe health risks, potentially leading to mortality. Various MRI features are utilized in brain tumor segmentation studies, encompassing image textures, local histograms, and structure tensor eigenvalues.

[16] Image classification and semantic segmentation have seen significant advancements, particularly with the rise of deep learning methods such as Convolutional Neural Networks (CNNs), which offer enhanced efficiency in image processing. These techniques are now being applied to tasks like brain tumor segmentation, categorization, and predicting patient survival time. Various characteristics of different tumor cells, including gene expression, motility, morphology, metabolism, metastatic potential, and proliferation, provide valuable information for segmentation, classification, and detection.

This review paper provides a comprehensive overview of diverse approaches, frameworks, architectures, algorithms, and key studies aimed at utilizing deep learning to classify, define, and detect cancer and predict survival duration. Organized around a survey taxonomy that encompasses cancer segmentation, evaluation, and feature exploration for tumor detection and classification, the paper categorizes different methodologies, techniques, systems, algorithms, frameworks, and architectural designs.

Furthermore, the paper examines the datasets, tools, programming languages, and libraries utilized for implementation, recognition, and evaluation, along with various feature extraction methods. Despite significant progress, there remain notable research gaps and challenges in tumor identification, particularly concerning monitoring, recognition methods, and treatment strategies for cancer patients.

2. Literature Review

Medical technology has advanced by leaps and bounds, allowing doctors to create more efficient electronic healthcare systems for their patients. Electronic health systems have proven themselves in many medical fields. Various medical imaging machines have shown a permissible influence on detecting and treating the disease [3-5].

Tumor is formed when the growth of cells near the brain starts abnormally, disrupting the normal functioning of the brain and affecting the patient's health. The early stages of this disease is curable and also promises for the patient to have a longer life span. In the later stages of this disease, the chances of survival of the patient gets slim [6-8].

The Tumor can be both Cancerous and Non-Cancerous. Melanoma is a cancer whereas Benign is a Non-Cancerous tumor. The World Health Organization (WHO) has divides these tumor into four types:-

1. Tumor type I (Astrocytoma)
2. Tumor type II (Low-Grade Astrocytoma)
3. Tumor type III (Anaplastic Astrocytoma)
4. Tumor type IV (Glioma Meridians)

Tumor I and Tumor II grades are considered less aggressive and semi-malignant in contrast to Tumor III and IV. This grade of tumor can significantly affect a patient's health, even leading to death [9,10].

Many methods are increasingly applied to detect patient's survival due to their superior performance in tumor segmentation, classification, and image analysis. Different cells which are forming the tumor exhibit different information that can be used to identify and diagnose which technique should be applied to have the treatment done effectively. The evaluation also analyzes the features like extraction methods, databases, researches to implement, validate and evaluate the disease. However, several important questions and research gaps still need to be addressed in tumor detection for cancer monitoring, diagnostic procedures, and treatment planning.

3. 3. Research Methodology

The model used for this is CNN model.

3.1. CNN Model

There is one special deep learning model developed specifically for processing and analysing visual input, such photos and videos, is the convolutional neural network (CNN). CNNs have demonstrated remarkable effectiveness in a range of computer vision applications, such as object identification, segmentation, and picture classification, owing to their inspiration from the human visual system.

Convolutional Neural Networks (CNNs) represent a subset of deep learning architectures tailored to handle structured grid data, primarily images. Their advent has transformed computer vision applications, finding extensive use in tasks like image classification, object detection, and segmentation.

Fundamentally, a CNN comprises several layers, notably convolutional layers, pooling layers, and fully connected layers. Through these layers, CNNs possess the ability to autonomously learn and extract pertinent features from images.

3.2. Benefits of using CNNs for Image Classification

Accurate Feature Extraction: CNNs excel at extracting high-level features from images, which are essential for accurate image classification.

Robustness to Image Variations: CNNs are relatively robust to variations in image data, such as changes in lighting, pose, and scale. This is due to their use of pooling layers, which reduce the sensitivity of the network to these variations.

Efficient Training: CNNs can be trained efficiently using specialized techniques such as stochastic gradient descent with momentum.

Parallel Processing: CNNs are well-suited for parallel processing, which allows them to be trained on large datasets in a reasonable amount of time.

Applications beyond image classification: CNNs can also be used for other tasks, such as image segmentation, object detection, and image generation.

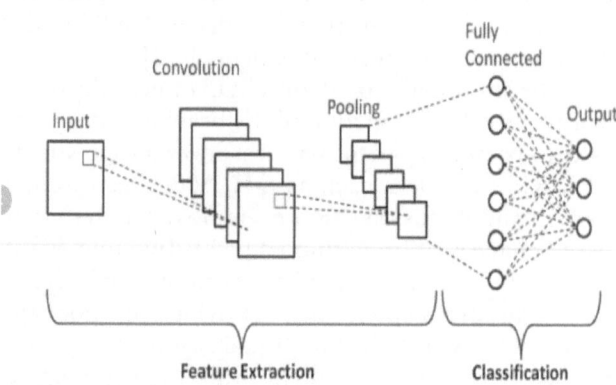

Figure 1: CNN Model image

3.3. The Key Components of CNNs Include

Input Layer: The initial layer accepts the input image, typically presented as a three-dimensional tensor with dimensions (width, height, channels).

Convolutional Layer: This layer utilizes a set of filters to process the input image, with each filter serving as a small window extracting features. The outcome is a feature map, encompassing activations of the filters at various locations within the image.

Pooling Layer: Here, the feature map undergoes size reduction through pooling operations applied to different regions. Common operations include max pooling and average pooling.

Fully Connected Layer: Following the pooling layer, this component accepts the flattened output and transforms it into a probability distribution across potential classes.

Output Layer: The final layer generates the probability distribution across potential classes, identifying the class with the highest probability as the detected class for the input image.

3.4. 3.4. Dataset Used

3.5. Use of CNN in Proposed Model

In brain tumor classification project, Convolutional Neural Networks (CNNs) are used to automate the process of analyzing medical images (MRI scans) to classify different types of brain tumors. Here's a more detailed explanation of how CNNs are utilized in project:

3.5.1. Data Preparation

The project starts with the collection of a dataset having scans of the brain (MRI's). These scans are being labelled to exhibit presence of different tumor types: "no tumor," "pituitary tumor," "meningioma tumor," and "glioma tumor."

3.5.2. Data Preprocessing

No Tumor Images:

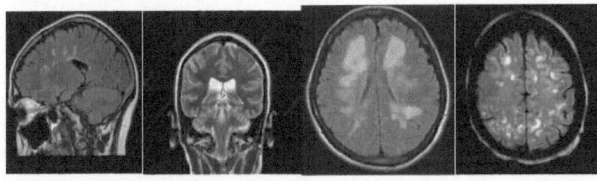

glioma_tumor Images

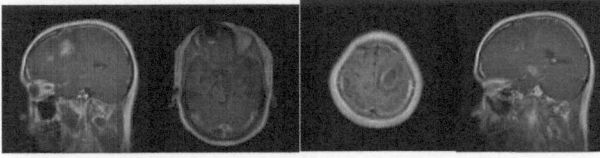

meningioma_tumor images

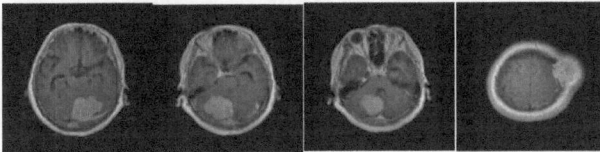

pituitary_tumor images

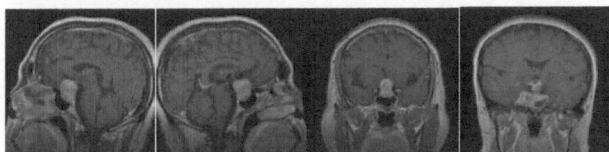

Figure 2:

The MRI images are pre-processed to ensure uniformity and optimal model performance. Common pre-processing steps include resizing images to a consistent resolution (e.g., 150x150 pixels), normalizing pixel values to a common scale (usually between 0 and 1), and converting images to a suitable format for input into the CNN.

3.5.3. Architecture of the CNN

The architecture of a CNN encompasses several layers, which include convolutional layers, pooling layers, activation functions, and fully connected layers.

Convolutional layers employ filters to recognize patterns and features within images. The depth of these layers intensifies gradually to encompass more intricate features.

Pooling layers down sample feature maps to reduce computational load and enhance the network's ability to generalize.

Activation functions introduce non-linearity and help the network learn intricate image patterns.

Fully connected layers at the end of the network combine extracted features for classification.

3.5.4. Training

The CNN is trained on the labeled dataset using a loss function and an optimization algorithm (commonly Adam). During training, the network learns to minimize the difference between its detections and the actual labels.

The process involves forward propagation (making detections), backward propagation (adjusting weights and biases through backpropagation), and gradient descent to optimize the model's parameters.

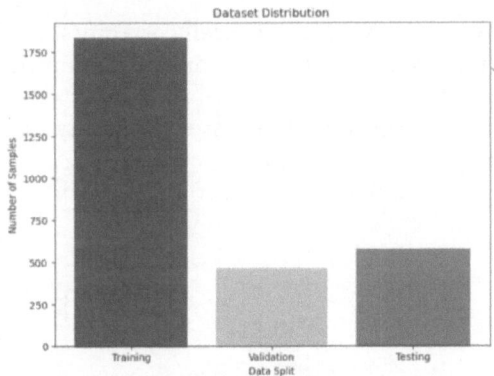

Figure 3: Dataset distribution

3.5.5. Class Weights

To account for class imbalance in the dataset (since some tumor types may be less common), class weights are used. Class weights assign higher importance to under represented classes during training to improve the model's ability to recognize these classes.

3.5.6. Data Augmentation

Data augmentation techniques, such as rotation, horizontal and vertical flipping, and zooming, are applied to the training images. This helps the model become more robust to variations and enhances its ability to make accurate detections on unseen data.

3.5.7. Early Stopping

Early stopping is implemented to prevent overfitting. The model training process stops when the validation loss ceases to improve, ensuring that the model retains the best learned features.

3.5.8. Model Evaluation

Once trained, the CNN is evaluated on a separate test dataset to assess its classification performance. Metrics such as accuracy, F1 score, and a confusion matrix are used to quantify its accuracy and error rate.

3.5.9. Detection

The CNN, once trained, is capable of identifying the presence and classification of brain tumors in unlabelled MRI scans. Its output yields a probability distribution across various tumor classes.

3.5.10. Interpretation

The detections made by the CNN can assist medical professionals in the diagnostic process. The model highlights the potential presence of tumors and suggests their types, serving as a valuable second opinion and aiding in medical decision-making.

3.6. Performance Matrics

1. **Accuracy:** Measures overall correctness.
2. **Precision:** Evaluates the ability to minimize false positives.
3. **Recall (Sensitivity):** Assesses the ability to detect all positive instances.
4. **F1-Score:** Balances precision and recall.

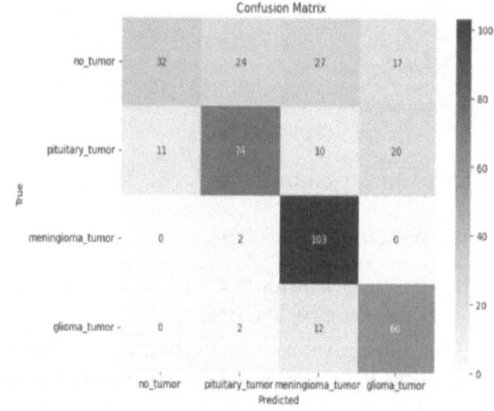

Figure 4: Confusion matrix

3.6.1. Confusion Matrix

Provides detailed information about model performance and misclassifications.

In classification tasks, a confusion matrix serves as a table to evaluate the performance of a machine learning model. It contrasts the model's predicted

```
18/18 [==============================] - 29s 1s/step
              precision    recall   f1-score   support

           0       0.94      0.80       0.86        157
           1       0.84      0.70       0.76        170
           2       0.83      0.99       0.90         81
           3       0.80      0.98       0.88        166

    accuracy                           0.85        574
   macro avg       0.85      0.86       0.85        574
weighted avg       0.85      0.85       0.84        574
```

Figure 5: Performance metrics

labels against the actual labels, offering insights into true positives, true negatives, false positives, and false negatives.

3.7. F1 Score

The F1 score, utilized in classification tasks, strikes a balance between precision and recall. Calculated as the harmonic mean of precision and recall, it offers a consolidated measure of the model's effectiveness. Employed to assess accuracy and minimize false positives and false negatives, a high F1 score signifies a model with strong precision and recall, rendering it suitable for endeavours such as brain tumor classification.

3.8. Epochs

In the context of training machine learning models, an epoch is a single pass through the entire training dataset during training. Training a model over multiple epochs allows it to learn from the data through repeated iterations.

3.8.1. Why Used

Training a neural network model over a specified number of epochs is necessary to optimize its parameters. It helps the model learn the underlying patterns in the data and improve its accuracy over time.

3.9. Graphs Showing Accuracy and Loss of Data during Training

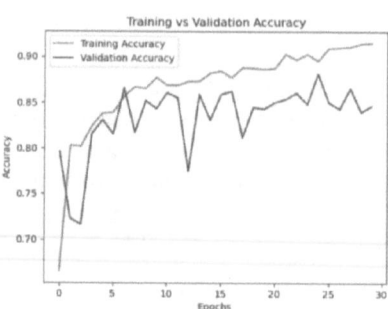

Figure 6: Graph of training vs validation accuracy

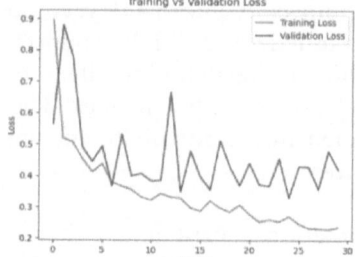

Figure 7: Graph of training vs validation loss

4. Conclusion

Using medical imaging data, several deep learning techniques, such as convolutional neural networks (CNNs) and artificial neural networks (ANNs), are used to classify various cancer types.

Using radiological images, CNNs have demonstrated their potency in classifying and diagnosing different types of brain tumours. These models frequently consist of numerous convolutional layers that can incorporate and extract properties from the input images. Once the retrieved features have been transmitted through fully connected layers, the final classification decision is then made.

On the other hand, radiological scans and ANNs have both been used to classify brain tumors. Since these models are composed of many layers of interconnected neurons, they can learn and reflect complex patterns in the input data. According to a study, ANN models are more accurate and efficient at identifying different forms of brain tumors.

In general, CNNs and ANNs have exhibited considerable promise for enhancing the precision and speed of brain cancer classification based on medical imaging data. To compare the effectiveness of different models and choose the best design for certain clinical applications, more investigation is required.

References

[1] Nalbalwar, R., Majhi, U., Patil, R., and Gonge, S. (2014). Detection of Brain Tumor by using ANN. International Journal of Research in Advent Technology, 2.

[2] Hashemzehi, R., Javad, S., Mahdavi, S., Kheirabadi, M., and Kamel, S. R. (2020). Detection of brain tumours from MRI images base on deep learning using hybrid model CNN and NADE. Elsevier B.V. on behalf of Nalecz Institute of Biocybernetics and Biomedical Engineering of the Polish Academy of Sciences Online Publication.

[3] Zhao, X., Wu, Y., Song, G., Li, Z., Zhang, Y., and Fan, Y. (2018). A deep learning model integrating FCNNs and CRFs for brain tumour segmentation. Medical Image Analysis 43, 98–111.

[4] Singh, N. and Jindal, A. (2012). Ultra sonogram images for thyroid segmentation and texture classification in diagnosis of malignant (cancerous) or benign (non-cancerous) nodules. International Journal of Engineering and Innovative Technology, 1, 202–206.

[5] Christ, M. C. J., Sivagowri, S., and Babu, P. G. (2014). Segmentation of brain tumors using Meta heuristic algorithms. Open Journal of the Communication Software, 1, 1–10.

[6] Yang, G., Raschke, F., Barrick, T. R., and Howe, F. A. (2015). Manifold Learning in MR spectroscopy using nonlinear dimensionality reduction and unsupervised clustering. Magnetic Resonanace in Medicine, 74, 868–878.

[7] Yang, G. Raschke, F., Barrick, T. R., and Howe, F. A. (2014). Classification of brain tumour 1 h Mr spectra: Extracting features by metabolite quantification or nonlinear manifold learning? In Proceedings of the 2014 IEEE 11th International Symposium on Biomedical Imaging (ISBI) (pp. 1039–1042), Beijing, China.

[8] Yang, G., Nawaz, T., Barrick, T. R., Howe, F. A., and Slabaugh, G. (2015). Discrete wavelet transform-based whole-spectral and subspectral analysis for improved brain tumour clustering using single voxel MR spectroscopy. IEEE Transaction on Biomedical Engineering, 62, 2860–2866.

[9] Kleihues, P., Burger, P. C., and Scheithauer, B. W. (1693). The new WHO classification of brain tumours. Brain Pathology, 3, 255–268.

[10] Von Deimling, A. G. (2009). Volume 171, Berlin, Germany: Springer.

[11] Mittal, M., Goyal, L. M., Kaur, S., Kaur, I., Verma, A., and Hemanth, D. J. (2016). Deep learning based enhanced tumour segmentation approach for MR brain images. Applied Software Computing, 78, 346–354.

[12]

[13] Goetz, M., Weber, C., Bloecher, J., Stieltjes, B., Meinzer, H.-P., and Maier-Hein, K. (2014). Extremely randomized trees-based brain tumour segmentation. In Proceedings of the BRATS Challenge-MICCAI (pp. 6–11), Boston, MA, USA.

[14] Kleesiek, J., Biller, A., Urban, G., Kothe, U., Bendszus, M., and Hamprecht, F. (2014). Ilastik for multi-modal brain tumour segmentation. In Proceedings of the MICCAI BraTS (Brain Tumor Segmentation Challenge) (pp. 12–17), Boston, MA, USA.

[15] Sharma, A. K., Nandal, A., Dhaka, A., Polat, K., Alwadie, R., Alenezi, F., and Alhudhaif, A. (2023). HOG transformation-based feature extraction framework in modified Resnet50 model for brain tumor detection. Biomedical Signal Processing and Control.

[16] Krizhevsky, A., Sutskever, I., and Hinton, G. E. (XXXX). Imagenet classification with deep convolutional neural networks. In Advances in Neural Information Processing Systems. Pasadena, CA: NIPS.

[17] Özyurt, F., Sert, E., Avci, E., and Dogantekin, E. (2016). Brain tumour detection based on Convolutional Neural Network with neutrosophic expert maximum fuzzy sure entropy. Elsevier Ltd, p. 147.

[18] Banerjee, S., Masulli, F., and Mitra, S. (2017). Brain tumor detection and classification from multi-channel magnetic resonance imaging's using deep learning and transfer learning.

[19] Kumar, S., Negi, A., Singh, J. N., and Verma, H. (2018). A deep learning for brain tumor magnetic resonance imaging images semantic segmentation using FCN. In 2018 4th International Conference on Computing

Communication and Automation (ICCCA), ISBN: 978-1-5386-6947-1.

[20] Hossain, T., Shishir, F. S., Ashraf, M., Al Nasim, M. D. A., and Shah, F. M. (2016). Brain tumour detection using Convolution Neural Network. In 1st International Conference on Advances in Science, Engineering and Robotics Technology 2016 (ICASERT 2016).

[21] Ruan, S., Lebonvallet, S., Merabet, A., and Constans, J.-M. (2007). Tumor segmentation from a multispectral MRI image by using support vector machine classification. In Proceedings of the 2007 4th IEEE International Symposium on Biomedical Imaging: From Nano to Macro (pp. 1236–1239), Arlington, VA, USA.

Implementation of Convolutional Neural Network (CNN) using Vitis AI

Amit Prakash Singh[1], Ravi Payal[2]

[1]Professor, USICT, Guru Gobind Singh Indraprastha University, Delhi
[2]Research Scholar, Guru Gobind Singh Indraprastha University, Delhi
E-mail: amit@ipu.ac.in, ravi.payal@gmail.com

Abstract

This paper talks about Convolutional Neural Network (CNN) implementation for image processing and analysis. The CNN is implemented using Vitis_HLS and Vivado 2022.2, popular design tools for FPGA-based hardware acceleration. The goal of the research work is to leverage the power of hardware acceleration to achieve real-time image processing performance. In addition to the Vivado implementation, the research work also involves a comparison with CNN implementation on Vitis AI. Vitis AI provides high-level abstraction and optimisation techniques to accelerate CNN models on FPGAs and other Xilinx devices. By comparing the Vivado implementation with Vitis AI, the research work aims to evaluate the performance, resource utilisation, and ease of deployment of CNN on both the platforms. The results of the comparison will provide insights into the trade-offs between custom hardware acceleration using Vivado and the higher-level abstraction and optimisation provided by Vitis AI. It will also shed light on the efficiency and scalability of the CNN implementation on different platforms, highlighting the strengths and limitations of each approach.

Keywords: Convolutional neural networks, FPGA, Vivado, Vitis AI

1. Introduction

Convolutional Neural Networks (CNNs) have changed the field of computer vision and image processing by enabling powerful image analysis and recognition capabilities. CNNs have been used in different applications which included detection of objects, image classification and recognition of patterns[1]. With the increasing demand for real-time and high-performance image processing, there is a growing need to implement CNNs on hardware platforms that can provide efficient acceleration.

In this paper, we focus on implementing a CNN for image processing and analysis using Vivado 2022.2, a popular design tool for FPGA-based hardware acceleration. The FPGA (Field-Programmable Gate Array) technology offers the unique advantage of custom hardware design and parallel processing, making it an ideal platform for accelerating computationally intensive tasks like CNNs. By leveraging the capabilities of Vivado, we aim to achieve real-time image processing performance with low latency and high throughput.. VITIS AI is a tool by Xilinx which supports C, C++ and OPENCL . This tool integrates the frame work of Deep learning and FPGA and by the help of VITIS AI developer use the hardware for processing, speeding up the AI interface and development. The number of tools included in VITIS AI are seven and all of them have unique feature.

The implemented CNN majorly consists of more than four or eight layers, including convolutional layers, activation functions (ReLU), and fully connected layers. Each layer performs specific operations such as convolution, pooling, and non-linear transformations to extract relevant features from input images. The extracted features are then used to make predictions or classifications based on the learned patterns in the training data. The CNN model is trained using a deep learning framework on a dataset of labeled images to optimise its parameters and improve its performance.

In addition to the Vivado implementation, we also compare our CNN implementation with Vitis AI, a comprehensive development platform for deploying AI models on Xilinx hardware. Vitis AI provides higher-level abstractions, pre-optimised libraries, and tools that simplify the deployment of CNN models on FPGAs and other Xilinx devices. By comparing the Vivado implementation with Vitis AI, we aim to evaluate the performance, resource utilisation, and ease of deployment of the CNN on both platforms.

The Vitis AI consist of Seven different tools which are described in Table 1.

Table 1: Tools present in VITIS AI

S.NO	Tools Name	Working of Tool
1	AI Model Zoo	This tool gives information about learned neural network model
2	AI Optimiser	This tool used for discarding a neural network model
3	AI Quantiser	This tool quantises a available neural network model
4	AI Compiler	This tool transforms a neural network model into DPU instructions
5	AI Profiler	This tool is used for analyzing a neural network model
6	AI Library	This tool provides a libraries and APIs
7	DPU	This is a Software core which is used for processing a neural network model

The comparison between Vivado and Vitis AI will provide insights into the trade-offs and advantages offered by each approach. We will analyze factors such as performance, resource utilisation, development time, and ease of deployment to understand which platform best suits the requirements of CNN-based image processing applications. The findings will contribute to the ongoing research and development in the field of AI and help in the adoption of CNN-based solutions across various domains.

To summarise, this paper focuses on the implementation of a CNN using Vivado for hardware acceleration and compares it with CNN implementation on Vitis AI. By exploring the performance and deployment options for CNN models, we aim to enhance real-time image processing capabilities and pave the way for the efficient utilisation of CNNs in diverse applications.

2. Literature Survey

Related to CNN implementation on FPGA in past researchers had done work related to various applications. On Paper "Optimizing FPGA-based Accelerator Design for Deep Convolutional Neural Networks" [1] the author has implemented a CNN accelerator on a VC707 FPGA board and compared it to previous approaches. Their implementation achieved a peak performance of 61.62 GFLOPS under 100MHz working frequency [1]. The study about the tools of Vitis AI through the illustrations and theory was done through the Xilinx link [2]. This also talks about the various tools related to FPGA and Vitis. On Paper "FPGA-Based Convolutional Neural Network Accelerator with Resource-Optimised Approximate Multiply-Accumulate Unit" [3], In this accelerator designed for LeNet-5 architecture of CNN . This architecture was designed and verified on MNIST selfmade datasets. With the help of Xilinx Vitis HLS tool accelerator was implemented on Xilinx XCZU9EG-2ffvb1156 FPGA chip. For checking the perfectness almost Ten thouisand MNIST handwritten digit images were used. On paper "Convolutional neural network implementations using Vitis AI" [4]

the author has implemented a convolutional neural network onto the Vitis AI development environment. Results confirmed the Vitis AI benefits. On paper "FPGA Implementation of Object Detection Accelerator Based on Vitis-AI" [5]. This paper talks about reconfigurable YOLOv3 accelerator. This accelerator performs various optimisation operations related to quantisation, pruning and compression of models which is related for the improving the performance of accelerator. On paper "Towards the Efficient Multi-Platform Execution of Deep Neural Networks" [6] the author This paper talks about collective execution of CNNs which is using two platforms together. It also talks about the FC layers in the ARM core which helps in achieving a speed of 2x.On paper "Convolutional neural network implementations using Vitis AI" [7] cnn implementation is done with vitis ai using various dpu architectures. Comparison of resources using different dpu architecture is done .

3. CNN Implementation Techniques

We implement CNN which is a fundamental component of AI. CNNs are widely used in various AI applications . The AI aspect of this code lies in the utilisation of CNNs to process and analyze data. By applying convolutional operations, activation functions, and fully connected layers, the CNN can learn and extract meaningful features from input images, enabling it to make predictions or classifications.

CNNs are a key component in AI systems that involve image analysis, pattern recognition, and visual understanding. They have been successfully applied in various domains, including autonomous vehicles, medical imaging, facial recognition, and many more.

There are several CNN implementation techniques that are commonly used in the field of deep learning. Here are a few key techniques:

1. *Convolutional Layers:* This layer is the leaf cell of CNNs which consist of learnable filters (also called kernels) that convolve across the input

data to extract features. Convolutional layers exploit the spatial relationships between pixels and capture local patterns in the input data.

2. *Pooling Layers*: These layers are used to down sample the feature maps generated by the convolutional layers. They reduce the spatial dimensions while retaining the most important features. Max pooling is a commonly used technique where the maximum value in each pooling window is selected as the representative value.

3. *Activation Functions:* This functions creates non-linearity into the CNN model, useful in learning complex patterns and make predictions. The Rectified Linear Unit (ReLU) activation function is widely used due to its simplicity and effectiveness. It replaces negative values with zero, preserving positive values.

4. *Fully Connected Layers:* This layer is used to connect neuron of one layer to another layer. They take the flattened feature maps from the convolutional and pooling layers and perform classification or regression tasks. Fully connected layers are typically used at the end of the CNN architecture.

5. *Dropout:* Dropout is a regularisation technique that helps prevent overfitting in CNN models. During training, random neurons are temporarily dropped out or ignored with a specified probability. This encourages the network to learn more robust and generalised features.

6. *Batch Normalisation:* This is used for improving the training process of deep neural networks. This normalises the input of each layer by adjusting and scaling the activations. This helps to alleviate the internal covariate shift problem, stabilise the training process, and improve convergence.

7. *Transfer Learning:* Transfer learning is a technique where pre-trained CNN models, trained on large-scale datasets like ImageNet, are used as a starting point for a new task or dataset. The pre-trained models already have learned features and weights, which can be fine-tuned or used as fixed feature extractors for the new task.

8. *Data Augmentation:* This technique used for increasing the sise of the training dataset by introducing various transformations to the existing data. Common augmentations include random rotations, translations, scaling, flips, and noise injection. This technique helps to improve model generalisation and reduce overfitting.

9. *Optimisation Algorithms:* Various optimisation algorithms are used to train CNN models efficiently. Stochastic Gradient Descent (SGD) with variations such as mini-batch gradient descent and adaptive learning rates (e.g., Adam optimiser) are commonly employed. These algorithms update the model's parameters based on the computed gradients during the training process.

Now a days researchers are exploring new advancements in CNN design and training methodologies to achieve better performance and efficiency in various AI applications. However, the choice of techniques depends on the specific problem, available resources, and desired performance characteristics.

4. Comparison of CNN implementation technique using Vitis AI and Vivado:

VITIS AI

- Vitis AI is a high-level development framework specifically designed for deploying deep learning models on Xilinx FPGA devices.

- It provides a comprehensive set of tools, libraries, and APIs that enable efficient deployment and optimisation of CNN models on FPGA.

- Vitis AI includes a quantisation tool that performs quantisation-aware training and quantises the model to lower precision, which reduces the memory footprint and improves inference performance.

- Vitis AI supports various optimisations, such as layer fusion, pruning, and compiler optimisations, to further enhance the efficiency of CNN models on FPGA.

- It offers integration with popular deep learning frameworks like TensorFlow and PyTorch, allowing seamless model conversion and deployment.

VIVADO

- Vivado is a comprehensive development environment provided by Xilinx for designing and implementing custom hardware designs, including CNN accelerators on FPGA.

- Vivado provides extensive synthesis, placement, and routing tools for mapping the designed CNN architecture onto the FPGA device.

- It allows fine-grained control over the hardware design and optimisation parameters, enabling manual tuning for specific performance or resource utilisation targets.

- Vivado offers powerful debugging and verification features, including simulation and hardware debugging capabilities, to ensure the correctness of the implemented CNN design.

COMPARISON

- Vitis AI provides a higher-level abstraction and easier development flow specifically focused on deploying deep learning models on ♦ FPGA, making it more accessible to developers without extensive hardware design expertise.

- Vivado, on the other hand, offers greater flexibility and control over the hardware design, allowing for custom and optimised CNN architectures tailored to specific requirements.
- Vitis AI includes automated optimisations and quantisation techniques, which can simplify the deployment process and improve performance, but may not provide the same level of fine-grained control as Vivado.
- Vivado allows for manual optimisation and customisation of the CNN architecture and offers advanced debugging and verification capabilities, which can be beneficial for complex and customised designs.
- Vitis AI integrates well with popular deep learning frameworks and provides a streamlined workflow for model conversion and deployment, while Vivado focuses more on low-level hardware design and implementation aspects.

5. Implementation

Our work implementation includes the LeNet convolutional neural network (CNN) architecture. The LeNet architecture was developed by Yann LeCun et al. and is commonly used for image recognition tasks.

The code consists of several functions that represent different layers of the LeNet network. Here is a breakdown of the functions and their corresponding layers:

1. `Conv_5x5`: This function performs a 5x5 convolution operation between the input and kernel matrices.
2. `ConvLayer_1`: This function implements the first convolutional layer of LeNet. It performs a convolution operation using a 5x5 kernel on the input image.
3. `AvgPool_2x2`: This function performs 2x2 average pooling on the input matrix.
4. `AvgpoolLayer_2`: This function implements the second average pooling layer of LeNet. It performs 2x2 average pooling on the output of the first convolutional layer.
5. `ConvLayer_3`: This function implements the second convolutional layer of LeNet. It performs a convolution operation using a 5x5 kernel on the output of the second average pooling layer.
6. `AvgpoolLayer_4`: This function implements the fourth average pooling layer of LeNet. It performs 2x2 average pooling on the output of the third convolutional layer.

7. `FullyConLayer_5`, `FullyConLayer_6`, and `FullyConLayer_7`: These functions implement the fully connected layers of LeNet. They perform matrix multiplication operations between the input and weight matrices.
8. `Softmax_1_8`: This function applies the softmax activation function to the input array and calculates the probabilities of different classes.
9. `LeNet`: This is the main function that orchestrates the execution of all the layers in the LeNet network. It takes an input image, performs forward propagation through the network, and returns the predicted class label.

Overall, the code implements the forward pass of the LeNet CNN architecture using basic matrix operations, such as convolution, pooling, and matrix multiplication.

6. Results

The work was implemented on Vitis-HLS using c and C++ coding . The generated RTL was exported using RTl export feature in Vitiss-HLS. This all was done by after the Simulation of C and C++ files. Here Verilog and VHDL files was generated for Vivado implementation. After Simulation during synthesis process various reports related to Area utilisation and timing were generated. The target FPGA architecture for implementation of CNN architecture is Zynq Zed board(ZC702). In Vitis-AI tool the c code is synthesised with target time of 10 ns . The estimated time was 8.49 ns which results slack value of 1.51 ns. After the C/RTL co-simulation RTL file generated in Verilog/ VHDL format. The user can use these formats according to its convince .This RTL is imported on Vivado tool and we get POST Synthesis Resource usage. The Device (Zynq- xc7z020-clg484-1) utilisation is as follows in Table 2.

LeNet Architecture are also playing role here. Synthesis report of LeNet architecture shows that slack timing is around 1.19 Ns which is the difference between required time and arrival timer. Since, for ideal working Slack should be around 1.0 ns. Therefore 1.19 Ns value of slack is perfectly fine.

Table 2: Device utilisation summary

Component	Number
LUT	55159
Flip-Flop	18544
DSP	81
BRAM	512
SRL	2177

7. Conclusion

CNN model was implemented on Xilinx Zynq-XC7Z020-CLG484.The neural network was executed properly with good result. Currently The CNN architecture of LeNet is designed and implemented using Vitis HLS on Zynq Zed Board (ZC702). The image normalisation and acquisition will be performed in Vivado by exporting RTL from Vitis-HLS. Further work will be conducted for achieving higher performance with respect to various parameters like DPU performance, Area, Speed, power consumption etc.

References

[1] Zhang, C., et al. (2015). Optimizing FPGA-based accelerator design for deep convolutional neural networks. In Proceedings of the 2015 ACM/SIGDA International Symposium on Field-Programmable Gate Arrays. https://doi.org/10.1145/2684746.2689060, ACM Digitl Library.

[2] https://www.xilinx.com/products/design-tools/vitis/vitis-ai.html.

[3] Cho, M. and Kim, Y. (2021). FPGA-based convolutional neural network accelerator with resource-optimized approximate multiply-accumulate unit. Electronics, 10, 2859. https://doi.org/10.3390/electronics10222859.

[4] Ushiroyama, A., Watanabe, M., Watanabe, N., and Nagoya, A. (2022). Convolutional neural network implementations using Vitis AI. In 2022 IEEE 12th Annual Computing and Communication Workshop and Conference (CCWC) (pp. 0365–0371), Las Vegas, NV, USA. https://doi.org/10.1109/CCWC54503.2022.9720794, IEEE Xplore.

[5] Wang, J. and Gu, S. (2021). FPGA implementation of object detection accelerator based on Vitis-AI. In 2021 11th International Conference on Information Science and Technology (ICIST) (pp. 571–577), Chengdu, China. https://doi.org/10.1109/ICIST52614.2021.9440554, IEEE Xplore.

[6] Hernandez, H. G. M. (2021). Towards the efficient multi-platform execution of deep neural networks. In 2021 31st International Conference on Field-Programmable Logic and Applications (FPL) (pp. 277–278), Dresden, Germany. https://doi.org/10.1109/FPL53798.2021.00056, IEEE Xplore.

Employment Prospects Prediction for Postgraduate Computer Science Students in Higher Educational Institutions

Ashwani Kumar Tewari[1], Ajay Kumar Bansal[2]

[1]Indiacom Limited, Pune, India, tashwani@hotmail.com
[2]School of Computer Application, Lovely Professional University, Jalandhar, India, Ajayg13@rediffmail.com

Abstract

This study was conducted at a Northern Indian university to validate and predict the placement outcomes of postgraduate students in the field of computer science and application. A machine learning model was employed, utilizing qualifying marks, academic performance, and gender as predictive factors for campus placement. The binary classification of placement was defined as "PLACED" or "NOT PLACED." Qualifying marks were assessed based on the results of the students' 10th and 12th-grade annual examinations, while academic performance was determined by the pre-final year percentage of marks. This study aims to provide higher education institutions with insights into the placement prospects of graduating batches. Machine learning algorithms, including decision trees and random forests, were utilised, and the decision tree's parameters were optimised for optimal performance. Model evaluation was carried out using metrics such as classification accuracy, sensitivity, specificity, and the area under the ROC curve. The models achieved a classification accuracy exceeding 91.12%. The experimental results suggest that qualifying marks and academic performance can effectively predict a student's placement status.

Keywords: qualifying marks: academic performance; decision tree; machine learning; predictive models; CART; datamining; RapidMiner, EDA

1. Introduction

The field of education constantly strives to enhance student performance and placement opportunities, recognizing the significant impact of academic achievement and placement records on higher education institutions' accreditation ratings and admissions [1]. In the Placement Prediction system, the aim is to forecast the likelihood of an student securing a job placement within a company. This prediction is accomplished through the application of classification algorithms, including Decision Trees. The primary goal of this model is to determine whether a student will be successfully placed during the campus recruitment process. These algorithms are then implemented on historical data from previous years' students to make these predictions.

This paper seeks to delve into the actual outcomes by analysing data from a Northern Indian university, where campus placements are of paramount importance. Utilizing the statistical open-source software package RapidMiner, this study aims to validate and predict the placement of postgraduate students in computer science and application programs.

This paper contributes significantly on multiple fronts. It employs live data from students rather than relying on a standard secondary dataset. Furthermore, the study evaluates various combinations of features, ensuring the selection of the most appropriate features and their predictive capabilities. The research concentrates on Indian Higher Education Institutions, where campus placement holds a high priority. Consequently, a meaningful predictive model could have far-reaching implications for various stakeholders, including students, parents, educators, higher education institutions, placement organisations, and society at large.

2. Review of Literature

This section reviews related work in the field of machine learning, specifically its application in predicting academic performance and placements using various models. The author leveraged the data mining tool "RapidMiner" for this purpose. The paper [2] recognises RapidMiner as one of the premier open-source software packages for data mining operations.

Chapter 47 DOI: 10.1201/9781003570349

Another paper [3] underscores the effectiveness of RapidMiner in summarizing data and its associated variables for predictive purposes. In a specific case, RapidMiner was employed to develop a decision tree model for predicting the likelihood of a patient developing diabetes [3]. Gomathi and Narayani [4] highlighted that the template-based framework of RapidMiner offers a comprehensive analytical solution, accelerating delivery and minimizing errors. RapidMiner encompasses data extraction, loading, transformation (ETL), and a range of machine learning and data mining procedures [4]. Data mining refers to the process of extracting information from data, analysing data from various angles, and summarizing it to derive meaningful insights. The information thus obtained can be used for classification, forecasting, and prediction purposes [5]. Brand establishment plays a pivotal role for educational institutions, with the placement of students exerting a significant influence on the decision-making process for both parents and students [6].

3. Experimental Setup

The training dataset consisted of 459 Indian students studying during passing out year 2018-19. This set consisted of a mix of post graduate students of computer science and computer application stream. The students in the data set were from all the states and union Territories of India. The test dataset consisted of 245 students passing out in 2019-20 having all the attributes of students of 2018-19 passing out students. The data set contains following feature as depicted in Table 1.

Objective:
- To analyse the relationship of academic performance marks with qualifying marks
- To analyse performance of male students Vs female students in terms placement
- To apply machine learning to predict campus placement

Table 1: Data set feature

Variable	Type
First Name	Polynomial
Gender	Polynomial
Current Course Percentage	Real
10th Percentage	Real
10+2 Percentage	Real
Placed	Label

4. Descriptive Statistics

Descriptive statistics is a fundamental branch of statistics that involves the collection, presentation, and summarisation of data in a meaningful and informative way. Its primary purpose is to provide a concise overview of a dataset, allowing researchers, analysts, and decision-makers to gain insights and draw preliminary conclusions without diving into complex statistical analyses.

One of the key aspects of descriptive statistics is the use of various measures and techniques to describe essential characteristics of the data. These characteristics include measures of central tendency, such as the mean (average), median (middle value), and mode (most frequent value), which help in understanding the typical or central value of the dataset.

Any data driven approach starts with Exploratory Data Analysis (EDA) which helps in understanding the bird view of dataset by proving the various summarised details such as underlying data distribution, correlation and important features which can be used for building machine learning model. The EDA process also helps in finding any outliers or any unusual pattern in dataset as shown in Table 2.

This reveals that 10th QM 71.3 and 12th QM marks 67.1 are average; however, as the median is 70.6 and 66.0 respectively, which is very close to the average, it indicates that population of students are homogeneous. The same is true for Performance marks. That implies the distribution in case on qualifying marks and performance are towards normal distribution

The correlation coefficient lies between -1 and +1. A correlation coefficient of +1 indicates a perfect positive correlation whereas -1 indicts perfect negative correlation. The correlation coefficient tending to zero shows week relation or correlation. It may be observed that all the variable are weekly positively related to each hence may not be conclude that they cannot be used for forecasting but they may not be the causation.

Table 2: Qualifying marks & performance

Central Tendency	10 Qualifying Marks	10+2 Marks (QM)	Current Course (Performance)
Mean	71.3	67.1	73.9
Mode	68.4	60.0	81.0
Median	70.6	66.0	75.6

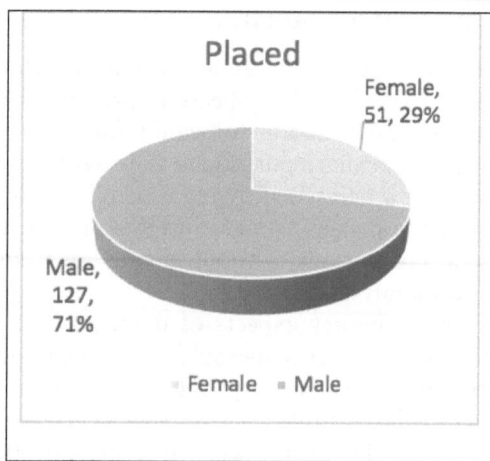
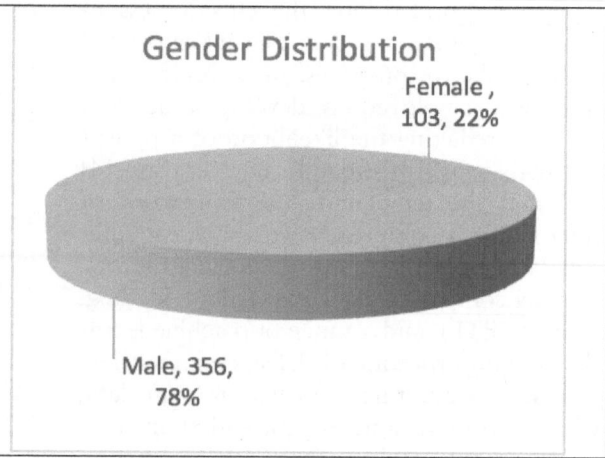

Figure 1: Gender distribution placed students   **Figure 2:** Placed/ not placed students' distribution

Table 3: Correlation between qualifying marks, current percentage & placement

	Current Course Percentage	10th Percentage	10+2 Percentage
Current Course Percentage	1		
10th Percentage	0.374509163	1	
10+2 Percentage	0.392335453	0.479819936	1
Placed	0.376030944	0.213264008	0.227168718

5. Machine Learning Model

Wolpert and Macready 1997 in their "free lunch theorem" suggested that there is no perfect model. So, the first and foremost problem was to identify the right model for our problem. Choosing the right model for the data set under study is a challenging job. Researchers frequently use automated model selection methods to identify variables that are independent predictors of an outcome under study [7]. Fortunately, the open-source software RapidMiner was quite helpful. But to start with it was important to understand the task to be performed. there are 4 situations as under:

- Classification: Is this A or B? or will this be A or B
- Regression: How much or how many? or how many will happen?
- Clustering: How is this organised? or What belongs to each other?
- Associations and correlations: What happens together? or What changes together?

Our problem clearly falls in classification as we want to classify a student as "Placed" or "not Placed"

Table 4: Confusion matrix

	true Not Placed	true Placed	class precision
pred. Not Placed	135	12	91.84%
pred. Placed	7	60	89.55%
class recall	95.07%	83.33%	

6. Classification

To classify placement of a student as "Yes" or "No", There are 4 commonly used classification algorithms i.e. logistic regression, naive bayes, Classification and Regression Trees (CART) sand random forests. Logistic regression is a commonly used machine learning algorithm when the dependent variable is categorical in nature. The basic idea of Naive Bayes algorithm is to calculate probability of happening. CART explains decision tree algorithm that can perform classification & regression tasks [8,9].

7. Confusion Matrix

The name confusion matrix is quite mis leading, it is not to confuse you but in fact in machine learning

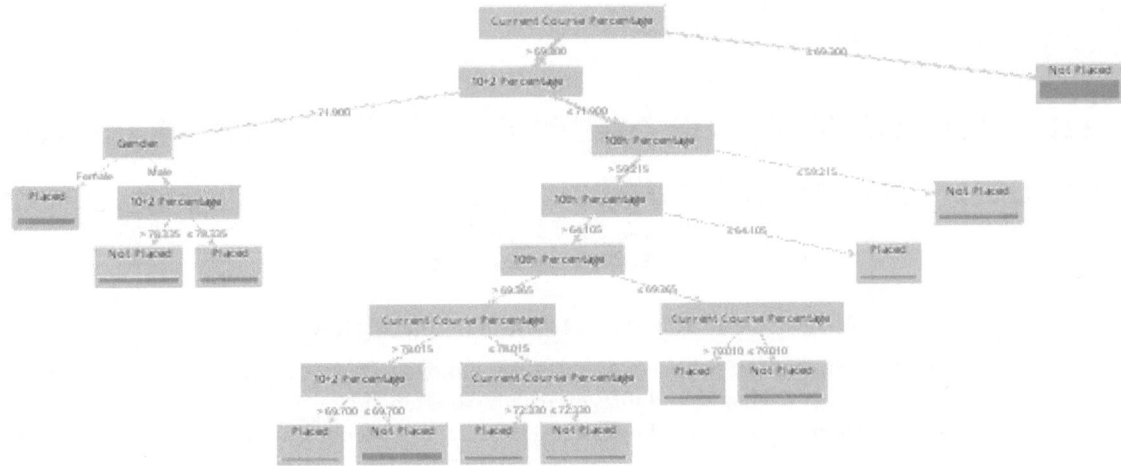

Figure 3: Decision tree

(ML) and statistical classification it allows visualisation of the performance of an algorithm, specifically a supervised learning one. This matrix is also known as error matrix.

Performance Vector: Accuracy 91.12%

The overall accuracy achieved by the model is 91.12% it is the percentage of correct predictions. The first col "true not placed" in first cell has a value of 135 and "pred placed" a value of 7 implying that out of 142 records 7 were not predicted correctly. This gives accuracy for "not placed" as 95.07%. Similarly, predication accuracy for placed is 83.33%. It may be observed that 195=135+60 out of 214 are predicted correctly i.e. over all accuracy of 91.12%. The class precision gives the prediction accuracy for "not placed" and "Placed". The confusion matrix shows that in case of 91.84% students were correctly predicted as "not placed" and 89.33% accurately predicted as "placed"

The model was run using various decision tree Criteria like Gini Index, Gain ratio, Information Gain, Accuracy and least square. The optimiser provided as process in rapid minor for optimizing the criteria suggested Gini Index. it also suggested value for Maximal Depth as 20 and Minimal size of split as 11. The decision tree as in Figure 3 provide the visualisation for predicting accurately.

8. Experimental Outcomes and Observations

The classification accuracy of 91.12% of the developed models is acceptable and usable in predicting the placement of a particular student well in advance. The correlation statistics does not encourage use of dataset for prediction or forecasting but application of machine learning approach in the predicting placement of student is worth. The cross validation and performance components helped in selecting the right criteria and optimum related parameters.

References

[1] Kumar, P. and Sharma, M. (2020). Predicting academic performance of international students using machine learning techniques and human interpretable explanations using LIME – case study of an Indian university.

[2] Jungermann, K. (1994). Information extraction with RapidMiner, 91 (August), 50–71.

[3] Han, J., Rodriguze, J. C., and Beheshti, M. (2008). Diabetes data analysis and prediction model discovery using RapidMiner. In Proceeding 2008 2nd International Conference on Future Generation Communication Networking, FGCN 2008 BSBT 2008 2008 International Conference on Bio-Science Bio-Technology (Volume 3, pp. 96–99).

[4] Gomathi, S. and Narayani, V. (2015). Applying decision tree algorithm to predict Lupus using Rapid Miner. 4 (October 2014), 1217–1218.

[5] Sharma, P., Singh, D., and Singh, A. (2015). Classification algorithms on a large continuous random dataset using rapid miner tool. In 2nd International Conference on Electronics and Communication Systems ICECS 2015 (no. Icecs, pp. 704–709).

[6] Kumar, P., Sharma, M., and Sood, S. (2019). Anticipating placement status of students using machine. Journal of the Gujarat Research Society, 21 (6), 738–747.

[7] Austin, P. C. and Tu, J. V. (2004). Bootstrap methods for developing predictive models. American Statistics, 58 (2), 131–137.

[8] Baehrens, D., Schroeter, T., Harmeling, S., Kawanabe, M., Hansen, K., and Müller, K. R. (2010). How to explain individual classification decisions. Journal of Machanical Learning Research, 11, 1803–1831.

[9] Ribeiro, M. T., Singh, S., and Guestrin, C. (2016). 'Why should i trust you?' Explaining the predictions of any classifier. In Proceeding ACM SIGKDD International Conference on Knowledge Discovery Data Minings, 13-17-Augu, 1135–1144.

Analysis of Various Supervised Machine Learning Algorithms on a Curated Dataset for Hate Speech Detection

Nisar Ahmad Kangoo[1] Manmohan Sharma[1], Aposh Roy[2]

[1]School of Computer Science and Engineering, Lovely Professional University,
Phagwara, India, nisarphd2020@gmail.com
[2]Department of Computer Science and Engineering, NSHM Knowledge Campus, West Bengal, India

Abstraction

This study uses a sizable collection of datasets, "A curated dataset for hate speech detection on social media text", available on the Mendeley Data repository, to identify hate speech and evaluate its incidence and characteristics. This dataset comprises a total of 451,709 examples. There are 371,452 instances of non-hateful speech and 80,250 cases of hateful speech. For the dataset analysis, we combine machine learning algorithms and methods for natural language processing. Based on our research, there is a significant amount of hate speech that primarily targets demographic groups. We also pinpoint recurring themes and linguistic patterns connected to hate speech. This study adds to the continuing efforts to counteract hate speech online and may have implications for online community building and content management.

Keywords: Hate speech, machine learning, NLP, online content moderation, text analysis

1. Introduction

Hate speech is "any speech denigrating an individual or a group due to any attribute, including race, colour, ethnicity, gender, sexual orientation, country, religion, or any other attribute" [1]. The use of words to disparage, discriminate against, or encourage violence against persons or groups because of their race, civilisation, religion, sex, sexual orientation, or other distinguishing behaviours is known as hate speech. It is a divisive and widespread problem in today's globalised society. This type of expression, which is marked by its hateful and biased nature, not only endangers societal cohesion and the ideals of free speech, but it also causes injury and upholds discrimination against the people it targets. In addition to being of scholarly interest, understanding hate speech's root origins and effects on individuals and society is essential to promoting inclusivity, tolerance, and respect for diversity in an increasingly digitised and globalised world. Since hate speech appears in various forms in the media and on social media platforms, social studies have a bigger problem. It takes on verbal, nonverbal, and symbolic forms [2]. It is purposefully expressed in an imprecise, roundabout manner [3]. This paper aims to analyse different supervised machine learning models on massive datasets so that a better algorithm can be worked out to detect hate speech. Detection of hate speech and removal of same from social networking websites and other online sites can help maintain the world's regional, religious, cultural and social harmony.

2. Literature Review

Various researchers have so far worked on the detection of hate speech. These researchers have performed on different data collected from YouTube, Twitter (Now X), MySpace, Wikipedia, Usenet, Instagram, and Facebook. The summary from previous research papers is in the Table 1 below. The related research also shows hate speech detection in languages other than English.

3. Dataset

The dataset has been taken from the Mendeley repository. This dataset is freely available for research purposes. The dataset and related information can be downloaded from the Mendeley repository https://data.mendeley.com/datasets/9sxpkmm8xn/1 (accessed on November 01, 2023). This dataset comprises of hate speech utterances in English, split into two groups: hateful content and non-hateful content. It comprises 451,709 sentences in all. There are

Chapter 48 DOI: 10.1201/9781003570349

Table 1: Summary from previous research papers on hate speech detection

Author and Year	Platform	Machine Learning Approach	Features Representation	Algorithm	Precision	Recall	F1 Score
Park and Fung [4]	Twitter	Supervised	Character and Word2vec	Hybrid CNN	0.71	0.75	0.73
Wiegand et al. [5]	Twitter, Wikipedia, UseNet	Supervised	Lexical, linguistics and word embedding	SVM	0.82	0.80	0.81
Warner and Hirschberg [6]	Yahoo newsgroup	Supervised	Template-based, PoS tagging	SVM	0.59	0.68	0.63
Burnap and Williams [7]	Twitter	Supervised	BOW, Dependencies, Hateful Terms	Bayesian Logistic Regression	0.89	0.69	0.77
Gitari et al. [8]	Blog	Semi-Supervised	Lexicon, Semantic, theme-based features	Rule-based	0.73	0.68	0.70
Waseem and Hovy [9]	Twitter	Supervised	Character ngram	Logistic regression	0.72	0.77	0.73
Pitsilis et al. [10]	Twitter	Supervised	Word-based frequency vectorisation	RNN and LSTM	0.90	0.87	0.88
Vidgen et al. [12]	109,488 (tweets)	Supervised	One-versus-one	SVM	0.77 Accuracy		
Balouchzahi et al. [13]	120,000 tweets	Supervised	Fuzzy ensemble	SVM, LR and RF	0.73 Accuracy		

371,452 hate speech incidents and 80,250 non-hate speech incidents. A bespoke vocabulary of 145,046 words is created using an improved balanced dataset of 726,120 samples. The dataset considers a total of 6403 contractions. Three hundred seventy-seven terrible words are commonly used in nasty content. The created contractions dataset can be utilised for data pre-processing in any NLP project. The dataset contains hate speech text, indicated with one and non-hate speech, marked with 0. After balancing the dataset, the number of 1s is 364525, and 0s is 361594—the balanced dataset with 0s and 1s looks as in Figure 1.

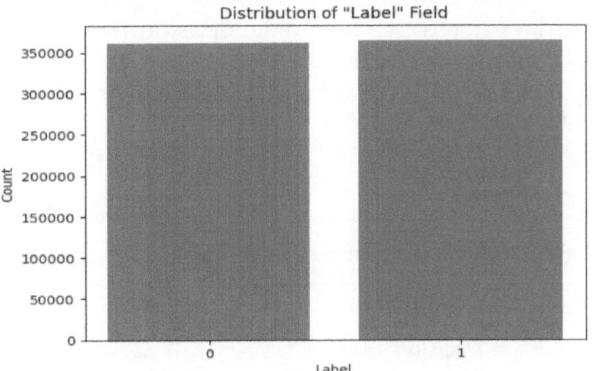

Figure 1: The balanced dataset with 1's and 0's

4. Machine Learning Models

Machine learning is a subset of Artificial Intelligence in which data is fed to the learning agent, which learns from the data. This learned agent then provides the results or predicts the new data per its past learning experiences. Machine learning is of three types viz supervised, unsupervised and reinforced. In supervised learning, the model is fed with labelled data like spam and non-spam mail, hate speech and non-hate speech, etc. Classification and regression problems can be resolved using supervised machine learning. In unsupervised learning, the model is supplied with the data and made to categorise it into different classes based on the features of the data, like the segregation of apples and mangoes based on their shape, size and colour. Reinforced machine learning is a feedback-based method in which a learning agent acquires a reward for each accurate step and a punishment for each incorrect step. The agent acquires skills automatically with this criticism and improves its performance.

The dataset is classification data; hence, supervised machine learning models are best suited for classifying this data. For classification, there are various machine leaning algorithms and among those, we have used the following algorithms to analyse their performance on the given dataset. A brief overview of these supervised algorithms is here;

Naïve Bayes: Naïve Bayes is an independent assumption-based classifier that calculates the probability that document D belongs to class C. It uses the Bayes theorem to handle categorisation problems. It is one of the most clear-cut and fruitful Classification algorithms for designing quick machine-learning models skilled of making rapid predictions. When assumed features are dependent and highly correlated, it shows low performance.

Decision Tree: It is a hierarchical approach consisting of nodes and directed edges. The training documents are classified in decision trees by constructing well-defined true/false queries in the tree structure. Leaf nodes represent corresponding class labels, whereas branches represent a conjunction of features that lead to these categories. This algorithm is quick and fast even if the number of attributes is large. It is easy to understand and interpret and requires less data preparation.

Random Forest: The approach adopted in the random forest is that, numerous decision trees are used on diverse subsets of the dataset. Random forest does not trust on a single decision tree instead its approach is to calculate results for every decision tree and provide the output on the basis of majority votes of calculation.

KNN: K-Nearest Neighbour (KNN) is an algorithm in which classification is done based on the distance of the data point from its nearest neighbours. This distance is calculated using Euclidean distance metric approach. This algorithm works on assuming that related data points have similar values or labels.

Ensemble Method: Ensemble approaches mix numerous machine learning models to improve performance and durability. Two Random Forest classifiers are trained on the same data but with distinct random seeds (random_state) on this dataset. To construct the ensemble predictions, the predictions of each classifier are pooled using a simple majority vote. This method is effective for binary classification jobs.

Gradient Boosting Algorithms: Algorithms like AdaBoost, Gradient Boosting Machines (GBM), and XGBoost iteratively improve the model's performance by combining weak learners with strong ones.

5. Experiments and Results

The dataset has a vast number of hate and non-hate speech sentences. This dataset can help train the models to determine whether a sentence fed is hated or non-hate. The dataset has two columns; one contains the sentences (hate speech and non-hate speech), and another includes the 0 or 1 value for non-hate speech and hate speech sentences.

We applied above mentioned supervised machine learning algorithms on the dataset after splitting it into 80:20 ratio of training cum testing sets. The results showing accuracy, precision, recall and F1 score for each algorithm are shown in Table 2.

Table 2: Summary of results from various classification algorithms used

S.No	Algorithm	Accuracy	Class	Precision	Recall	F1 Score
1	Decision Trees	72%	0	0.72	0.73	0.72
			1	0.73	0.72	0.72
2	Naive Bayes	53%	0	0.53	0.56	0.54
			1	0.53	0.50	0.52
3	Random Forest	72%	0	0.72	0.73	0.72
			1	0.73	0.72	0.72
4	KNN	72%	0	0.73	0.72	0.72
			1	0.72	0.73	0.73
5	Ensemble predictions	72%	0	0.72	0.73	0.72
			1	0.73	0.72	0.73

(Continued)

Table 2: (*Continued*)

6	AdaBoost Classifier	57%	0	0.57	0.51	0.54
			1	0.56	0.62	0.59
7	Gradient Boosting	58%	0	0.58	0.53	0.56
			1	0.57	0.62	0.60
8	XGBoost Classifier	59%	0	0.60	0.55	0.57
			1	0.59	0.63	0.61

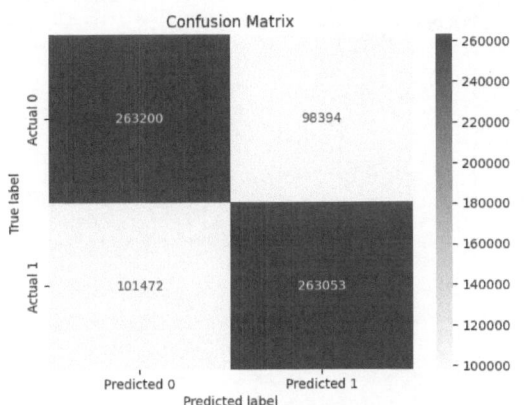

Figure 2: Confusion Matrix for K-fold cross-validation for random forest algorithm

We have also used K-fold cross-validation for the dataset on random forest classification to assess its performance. The result for the same is depicted in the confusion matrix when a number of folds =5, as shown in Figure 2.

The confusion matrix is a 2x2 matrix that represents the performance of a classification algorithm. Here's how to interpret the different elements of the confusion matrix:

- **True Positive (TP):** 263200 - The number of instances of the positive class that were correctly anticipated by the model.
- **True Negative (TN):** 263053 - The quantity of cases of the negative class that were properly anticipated by the model.
- **False Positive (FP):** 98394 - Also called as a Type I error, it is the number of occurrences that were wrongly projected as positive when they are actually negative.
- **False Negative (FN):** 101472 - Also called as a Type II error, this is the number of occurrences that were incorrectly projected as negative when they are really positive.

6. Observation

We observed that Decision Trees, Random Forest, KNN, and Ensemble prediction models consistently achieved an accuracy of 72%. The AdaBoost Classifier, Gradient Boosting, and XGBoost Classifier models showed slightly lower accuracy, ranging from 57% to 59%. Naive Bayes has a lower accuracy at 53%.

7. Conclusion

The results show that all supervised machine learning algorithms have yet to deliver better results. As such, we should go beyond the above-mentioned machine learning algorithms to develop a better model with high accuracy. The development of such a model can be used to detect and remove hate speech on social networking websites and hence let harmony prevail in cyberspace.

References

[1] Nockleby, J. T., Levy, L. W., Karst, K. L., and Mahoney, D. J. (2000). Encyclopedia of the American Constitution.Detroit, MI: Macmillan.

[2] Nielsen, L. B. (2002). Subtle, pervasive, harmful: Racist and sexist remarks in public as hate speech. Journal of Social Issues,58 (2), 265–280.

[3] Giglietto, F. and Lee, Y. (2017). A hashtag worth a thousand words: Discursive strategies around# JeNeSuisPasCharlie after the 2015 Charlie Hebdo shooting.Social Media+ Society, 3 (1), 2056305116686992.

[4] Park, J. H. and Fung, P. (2017). One-step and Two-step Classification for Abusive Language Detection on Twitter," in AICS Conference, 2017. Publisher Association for Computational Linguistics

[5] Wiegand, M., Ruppenhofer, J., Schmidt, A., and Greenberg, C. (2018). Inducing a Lexicon of Abusive Words – a Feature-Based Approach. In Proceedings of the 2018 Conference of the North American Chapter of the Association for Computational Linguistics: Human Language

Technologies (Volume 1 (Long Papers), pp. 1046–1056), Publisher Association for Computational Linguistics .

[6] Warner, W. and Hirschberg, J. (2012). Detecting hate speech on the world wide web. In Proceedings of the Second Workshop on Language in Social Media (pp. 19–26). LSM.

[7] Burnap, P. and Williams, M. L. (2014). Hate speech, machine classification and statistical modelling of information flows on twitter: Interpretation and communication for policy decision making. In Proceedings of the Conference on the Internet, Policy & Politics (pp. 1–18).

[8] Gitari, N. D., Zuping, Z., Damien, H., and Long, J. (2015). A lexicon-based approach for hate speech detection. International Journal of Multimedia and Ubiquitous Engineering, 10 (4), 215–230.

[9] Waseem, Z. and Hovy, D. (2016). Hateful symbols or hateful people? Predictive features for hate speech detection on twitter. In Proceedinf Of the NAACL Student Research Workshop (pp. 88–93). Publisher Association for Computational Linguistics

[10] Pitsilis, G. K., Ramampiaro, H., and Langseth, H. (2018). Effective hate-speech detection in Twitter data using recurrent neural networks Applied Intelligence, 48 (12), 4730–4742.

[11] Vidgen, B. and Yasseri, T. (2020). Detecting weak and strong Islamophobic hate speech on social media. Journal of Information Technology & Politics, 17, 66–78.

[12] Balouchzahi, F.; Shashirekha, H. L., and Sidorov, G. (2021). HSSD: Hate speech spreader detection using N-Grams and voting classifier. CEUR Workshop Proceeding, 2936, 1829–1836.

A Comprehensive Analysis of Soft Computing Techniques for Cardiovascular Disease Detection

Imteyaz Hussain Khan[1], Amar Singh[2], Hilal Ahmed Rather[3]

[1,2]School of Computer Applications, Lovely Professional University, Phagwara, India,
[3]Department of Bio-Medical Engineering, Agni College of Technology, Chennai.
E-mail: Imoqhan19@gmail.com, Amar.23318@lpu.co.in, Hilalrather64@gmail.com

Abstract

Cardiovascular diseases (CVDs) are a major global source of morbidity and mortality, hence novel strategies for prompt and accurate identification are required. This comprehensive review surveys the landscape of soft computing techniques employed in the realm of CVD detection. Fuzzy logic, neural networks, genetic algorithms, and machine learning are examples of soft computing, which provides an adaptable and flexible framework for managing the intricacies involved in cardiovascular health evaluations. We examine the specifics of each method, looking at its uses, advantages, and disadvantages in relation to CVD diagnosis. Highlighting the benefits and trade-offs of each soft computing approach, this paper compares a number of them with an emphasis on the areas where computational intelligence and cardiovascular care might work together. This review aims to be a useful resource for researchers, practitioners, and healthcare professionals who are involved in the pursuit of effective and efficient CVD detection methodologies. We also address the integration of multiple soft computing approaches and analyze their combined impact on improving diagnostic accuracy. As we navigate through the current landscape, we identify challenges and articulate future directions, shedding light on emerging trends and potential advancements.

Keywords: Cardiovascular diseases, early detection, soft computing, machine learning

1. Introduction

Cardiovascular diseases (CVDs) account for a large portion of sickness and mortality globally, making them a major global health concern. Because CVDs are becoming more common, there is a rising need for innovative and trustworthy techniques to identify and treat them early. Because cardiovascular health is complicated, the conventional methods of diagnosing CVDs frequently don't perform effectively, hence sophisticated computational tools are being utilised instead.

The convergence of soft computing and medical research has created new avenues for improving the precision of cardiovascular disease detection in recent years. Fuzzy logic, neural networks, genetic algorithms, and machine learning are examples of soft computing techniques that provide adaptable frameworks that may manage complicated patterns, address ambiguity, and enhance decision-making. This investigation investigates the identification of cardiovascular diseases using various soft computing techniques. The study discusses several uses of soft computing, such as generating effective networks for remote cardiac health monitoring and classifying signals from electrocardiograms. This review attempts to give a detailed overview of the advantages, disadvantages, and possible benefits of applying various soft computing approaches in the field of cardiovascular disease detection by a thorough analysis of various research and their findings.

1.1. Cardiovascular Diseases

Cardiovascular diseases refer to a group of conditions that affect the heart and blood vessels. Different types of cardiovascular Diseases are

a) Coronary Artery Disease (CAD): The most common CVD, coronary artery disease (CAD), is characterized by the narrowing of the coronary arteries, which reduces the amount of blood that reaches the heart. Heart attacks or angina are frequently the result.

b) Heart Failure (HF): HF is the result of the heart's inability to pump blood efficiently. Fatigue, dyspnea, and fluid retention are the outcomes.

c) Arrhythmias: The heart's natural rhythm is disturbed by irregular heartbeats, called arrhythmias. They may affect ventricular or atrial function, such as atrial fibrillation.

d) Disorders of the Valve: Heart valve disorders, such as regurgitation or stenosis, impair blood flow and may cause heart failure.

e) Stroke: A cerebrovascular event in which there is a disruption in the blood flow to the brain, leading to neurological damage. Hemorrhagic strokes include bleeding, whereas ischemic strokes are caused by clogged arteries.

f) Cardiomyopathy: A class of disorders that cause abnormalities in the structure and function of the heart muscle. Dilated, hypertrophic, and restricted cardiomyopathies are among the subtypes.

1.2. Motivation and Aim of Research

Research in this area aims to enhance the accuracy and efficiency of diagnosing cardiovascular diseases, potentially improving patient outcomes. Robust machine learning models based on ECG signals can assist healthcare professionals in making more informed decisions about patient care. Medical professionals often take a long time to gain the expertise needed to accurately distinguish between normal and abnormal ECG cases. Manual classification is time-consuming, especially when signal characteristics are not obvious. Automated signal processing techniques can help address these challenges and support biomedical decision-making. This study uses basic set-based models for feature selection and classification of ECG signals.

1.3. Problem Statement

Cardiovascular diseases (CVDs) are a major global cause of death. Detecting CVDs early is vital for effective prevention and management. However, the complexity of Electrocardiogram (ECG) signals poses challenges for accurate diagnosis. Traditional methods may lack precision, and manual interpretation of ECGs, even by trained cardiologists, can be time-consuming and challenging, potentially missing subtle abnormalities. There's a need for an advanced approach using soft computing or machine learning to improve the accuracy and efficiency of CVD detection from ECG signals. This research aims to develop an interpretable soft computing model for efficient ECG signal analysis, enabling timely interventions and improving patient outcomes.

1.4. Electrocardiograms (ECG)

Roopa and Harish (2017) assert that electrocardiograms (ECGs) offer a non-invasive, simple, available, cost-effective, and controllable research tool.

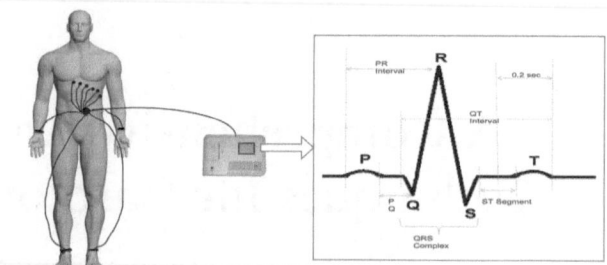

Figure 1:

To create an automated ECG classification using machine learning, waveform characteristic features need elimination. First-order variables like RR time, the interval between pulse peaks, are promptly derived. Additional properties from the signal, addressed by Fourier Transforms (FTs) and Wavelets, are removed. In a typical supervised classification system, decision outcomes are labeled according to these characteristics.

2. Literature Review

This review offers a succinct synthesis of contemporary research endeavors that center around leveraging machine learning and soft computing methodologies for the purpose of detecting and diagnosing cardiovascular diseases via electrocardiogram (ECG) signals. Spanning across diverse techniques such as deep convolutional neural networks [1], feature descriptors [2], optimisation algorithms [3], and hybrid classifiers [14], these studies collectively underline the progressive fusion of artificial intelligence with cardiology. Through these approaches, the integration of computational prowess [4] with clinical domain knowledge [22] strives to amplify diagnostic precision, facilitating timely intervention and management of cardiovascular conditions.

Devi et al. [2] propose a deep learning model for the classification of unsegmented phonocardiogram signals. The model achieves state-of-the-art accuracy on a public phonocardiogram dataset, outperforming other traditional machine learning methods. This model could be used to develop new diagnostic tools for heart disease. Ay et al. [1] compare the performance of different meta-heuristic optimization algorithms for feature selection on ECG classification tasks. They find that the firefly algorithm outperforms other algorithms in terms of both accuracy and efficiency. This could be useful for developing more efficient and accurate ECG classification models. Ismail et al. [5] propose a new temporal convolutional network (TCN) architecture for ECG classification. The TCN is able to learn long-range temporal dependencies in the ECG signal, which is important for accurate classification. The authors show that their TCN architecture achieves state-of-the-art accuracy on a public ECG dataset. This model could be used to develop more accurate ECG classification systems for remote health monitoring.

Lai et al. [3] propose a new self-supervised learning algorithm for training a wearable 12-lead ECG classification model. The self-supervised learning algorithm does not require labeled data, which makes it more practical for training wearable ECG models. The authors show that their self-supervised learning algorithm achieves state-of-the-art accuracy on a public ECG dataset. This model could be used to develop more practical and accurate wearable ECG classification systems. Lee and Kim [6] propose a new ECG measurement system for vehicle implementation and heart disease classification using machine learning. The ECG measurement system is designed to be robust to noise and interference from the vehicle environment. The authors show that their machine learning model achieves high accuracy in classifying heart diseases using the measured ECG data. This system could be used to develop new in-vehicle ECG monitoring and diagnosis systems.

Sivapalan et al. [7] propose a new interpretable rule mining algorithm for real-time ECG anomaly detection in IoT edge sensors. The interpretable rule mining algorithm is able to extract interpretable rules from the ECG data, which can be used to explain the anomalies detected by the system. The authors show that their algorithm achieves high accuracy in detecting ECG anomalies on a public ECG dataset. This algorithm could be used to develop new real-time ECG anomaly detection systems for IoT edge devices.

Taylan et al. [8] compare the performance of machine learning, neuro-fuzzy, and statistical methods for early prediction of cardiovascular diseases. They find that machine learning methods outperform other methods in terms of both accuracy and sensitivity. This suggests that machine learning methods have the potential to be used to develop new and more accurate early prediction models for cardiovascular diseases. Tsai and Morshed [4] propose a new scalable and upgradable AI architecture for detecting beat-by-beat ECG signals in smart health devices. The proposed AI architecture is able to learn from new data over time, which makes it more adaptable to changes in the ECG signal. The authors show that their AI architecture achieves high accuracy in detecting beat-by-beat ECG signals on a public ECG dataset. This AI architecture could be used to develop new scalable and upgradable ECG monitoring systems for smart health devices.

Mahalakshmi and Kumar [9] propose a new method for heart disease prediction using an improved particle swarm optimization algorithm and an ensemble classification technique. The improved particle swarm optimization algorithm is used to select the optimal features for classification, and the ensemble classification technique is used to combine the predictions of multiple classifiers. The authors show that their method achieves better accuracy than other state-of-the-art methods on a public ECG dataset. This method could be used to develop more accurate heart disease prediction systems. Allugunti [10] proposes a new hybrid machine learning model for heart disease diagnosis and prediction. The hybrid machine learning model combines the predictions of different classifiers to improve accuracy. The author shows that their hybrid machine learning model achieves better accuracy than other state-of-the-art methods on a public ECG dataset. This hybrid machine learning model could be used to develop more accurate heart disease diagnosis and prediction systems.

According to analysis, Random Forest algorithm shows good results. Accuracy of 95.60%, Precision of 0.5528 and Recall of 0.9768.

Table 1: Comparison of our findings with state-of-the –art methods

Algorithm	Accuracy (%)	Precision	**Recall**
DNN	76.92	0.5000	0.8140
ANN	92.30	0.4524	0.8372
MLP	75.42	0.4348	0.8140
Random Forest	**95.60**	**0.5528**	**0.9768**
Logistic Regression	93.40	0.4589	0.9070
Naive Bayes	90.10	0.4757	0.9070
KNN	71.42	0.4770	0.7210
SVM	92.39	0.4524	0.8838
Decision Tree	81.31	0.5000	0.8605

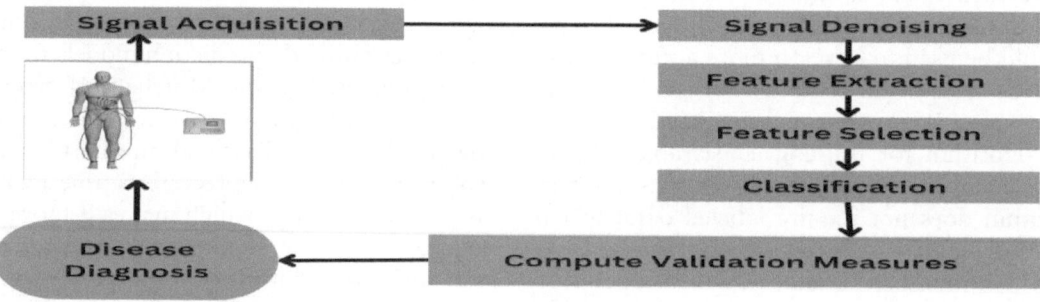

Figure 2: General Methodology for detection of cardiovascular diseases

3. General Methodology

Steps of General methodology for detection of cardiovascular diseases using soft Computing techniques.

1. **Data collection and Preprocessing:** Collect and preprocess ECG signals from patient.
2. **Feature Extraction:** Extract relevant features from the preprocessed ECG Signal.
3. **Discretization:** Discretize the continuous features into discrete values.
4. **Feature Selection:** Select the most relevant features from the extracted features.
5. **Classification:** Apply Soft computing techniques to classify the ECG Signals into different categories of cardiovascular diseases.
6. **Evaluation:** Evaluate the performance of the soft computing models.

4. Soft Computing

A broad area of computer science known as "soft computing" combines several computational techniques to simulate how people think and learn in unpredictable and imperfect situations. In order to handle complex and dynamic real-world situations, soft computing, in contrast to traditional, rigid computing systems, incorporates the management of uncertainty, approximation, and partial truth. It includes a variety of methods, each with a specialization that helps to create flexible, adaptive systems that can handle complex and uncertain data for tasks like pattern recognition, optimization, and decision-making.

4.1. Application of Soft Computing in Cardiovascular Diseases

Soft computing techniques, including fuzzy logic, neural networks, and genetic algorithms, have transformative applications in CVD diagnosis and management.

I. **Fuzzy Logic:** By accommodating uncertainty in CVD diagnosis, fuzzy systems improve the interpretation of clinical data that is difficult to interpret.

II. **Neural Networks:** With its superior ability to recognise patterns, neural networks help classify intricate cardiovascular data and improve prediction models.

III. **Genetic Algorithms:** By effectively searching across large solution spaces, genetic algorithms improve CVD detection models and increase accuracy.

IV. **Machine Learning and Data Mining:** These techniques uncover hidden patterns in large datasets, contributing to personalized risk assessment and treatment strategies.

5. Conclusion

In this comprehensive review, we explored various types of cardiovascular diseases (CVDs) and the diverse landscape of soft computing techniques applied to their detection. The integration of fuzzy logic, neural networks, genetic algorithms, and machine learning has showcased significant advancements in deciphering complex CVD patterns. By comparing methodologies, datasets, and findings across studies, it becomes evident that soft computing not only enhances diagnostic accuracy but also offers adaptability to the intricate and dynamic nature of CVDs. From scalable AI applications for ECG signals [4] to interpretable rule mining in IoT edge sensors [7], the reviewed literature exemplifies the versatility of soft computing. While challenges persist, such as the interpretability of complex models and the need for diverse datasets, the trajectory of research points toward promising trends in explainable models, data integration, and hybrid approaches, underscoring the transformative potential of soft computing in reshaping the landscape of cardiovascular healthcare.

References

[1] Ay, Ş., Ekinci, E., and Garip, Z. (XXXX). A comparative analysis of meta - heuristic optimization algorithms for feature selection on ML - based classification.

[2] Devi, K. M., Chanu, M. M., Singh, N. H., and Singh, K. M. (2023). Classification of unsegmented phonocardiogram signal using scalogram and deep learning. Soft Computing, 27 (17), 12677–12689, 2023. https://doi.org/10.1007/s00500-023-08834-1.

[3] Lai, J., et al., Practical intelligent diagnostic algorithm for wearable 12-lead ECG via self-supervised learning on large-scale dataset. Nature Communication, 14 (1), 3741. https://doi.org/10.1038/s41467-023-39472-8.

[4] Hua Tsai, I. and Morshed, B. I. (2023). Scalable and upgradable AI for detected Beat-By-Beat ECG signals in smart health. In 2023 IEEE World AI IoT Congress (AIIoT 2023) (pp. 409–414). IEEE. https://doi.org/10.1109/AIIoT58121.2023.10174482.

[5] Ismail, A. R., Jovanovic, S., Ramzan, N., and Rabah, H. (2023). ECG classification using an optimal temporal convolutional network for remote health monitoring. Sensors, 23 (3), 1–16, 2023. https://doi.org/10.3390/s23031697.

[6] Lee, C. H. and Kim, S. H. (2023). ECG measurement system for vehicle implementation and heart disease classification using machine learning. IEEE Access, 11 (January), 17968–17982. https://doi.org/10.1109/ACCESS.2023.3245565.

[7] Sivapalan, G., Nundy, K. K., James, A., Cardiff, B., and John, D. (2023). Interpretable rule mining for real-time ECG anomaly detection in IoT edge sensors. IEEE Internet of Things Journal, 10 (15), 13095–13108. https://doi.org/10.1109/JIOT.2023.3260722.

[8] Taylan, O., Alkabaa, A. S., Alqabbaa, H. S., Pamukçu, E., and Leiva, V. (2023). Early prediction in classification of cardiovascular diseases with machine learning, neuro-fuzzy and statistical methods. Biology (Basel), 12, (1), 1–31. https://doi.org/10.3390/biology12010117.

Web Technologies

A Review

Harsh Mudgal, Rohan Raj, Prafull Garg, Bonigala Dhanush, Balraj Singh*

School of Computer Science and Engineering, Lovely Professional University, Phagwara, India, balraj.13075@lpu.co.in

Abstract

Websites can now be visited on devices other than only laptops with large screens. They were usually accessible through mobile devices with comparatively smaller screens, such tablets and smartphones. Almost every website has a unique visual aesthetic and style for how it presents its material and information. Despite the fact that they visually differed, they were primarily created and built utilizing a layout with one, two, or three columns. It becomes difficult to correctly convey data and content screens with smaller size. A webpage can be visualised in a different way depending on the size of the screen size of according to the responsive web design method. The amount of information that will be presented on the screen may change as a result of these layout changes, which could have an impact on how effectively and efficiently information is delivered on a webpage. This study compares the effectiveness of online pages presented on computers, tablets, and smartphones. This article examines the design, development, and evaluation of a GUI-based task-specific reservation system. The system's primary objectives are to enhance user experience and simplify the movie booking process.

Keywords: Web, frameworks, toolkit, technologies

1. Introduction

Systems built with a GUI are dependable, compatible, and user-friendly. It is totally web-based and uses a database to make it more logically sound and easy to load without crashing even on sluggish internet connectivity. Nowadays, a lot of people utilise the Internet to spread information. A website not only serve as a source of information also works as an application system or may evolve into as a information system as shown in Figure 1. With the increase in the usage of mobile phones that can access the internet, the information service providers use different systems to provide services through the internet. Website owners now have the freedom to change a single page's User Interface, thanks to responsive design, enabling users on various devices to adapt as per the devices. These websites are designed by utilizing HTML and CSS technologies. Owners of the web portals use multiple website versions to create the website's User Interface.

1.1. Web Development Life Cycle

In fact, we require specific technology to create particular web portals in order to achieve the necessary results. To decide and finish the logic-driven interface, the DFD is produced in the very first phase once the logic has been written. Second, the frontend component, or user interface (UI) section, must be created using frontend technologies that are interoperable with one another, such as HTML, CSS, JS, and React Frameworks and plugins. The backend must then be developed in order to connect the website in the predetermined logical order and produce the required results. Databases are being used in order to maintain the data. Example: SQL, MONGODB as per user needs. Then at final stage the whole website is connected with each other and tested over local host to detect the anomalies if any. Now the testing part is done to check the validations and output generation at server level. The technological process of creating and deploying websites and web applications is known as web development. To construct completely functional and engaging online platforms, it includes coding, programming, and integrating capabilities. The web development life cycle is a structured methodology used in the web development process in Figure 2. Discovery, planning, design, development, testing, deployment, and continuing maintenance are some of the phases that are covered by this. In order to build websites and web apps that give businesses a platform on which to showcase their goods, services, and ideas, web design and web development

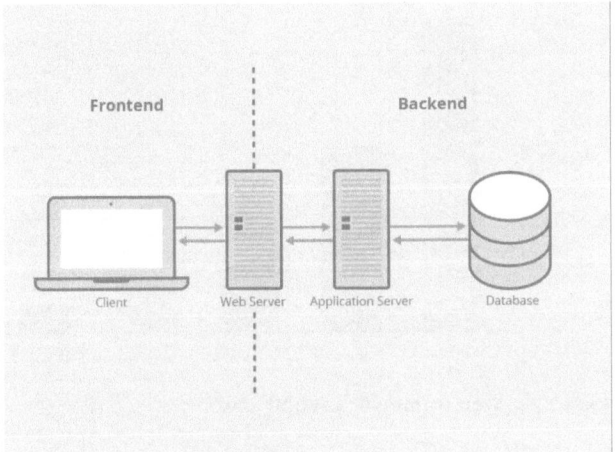

Figure 1: Basic web layout

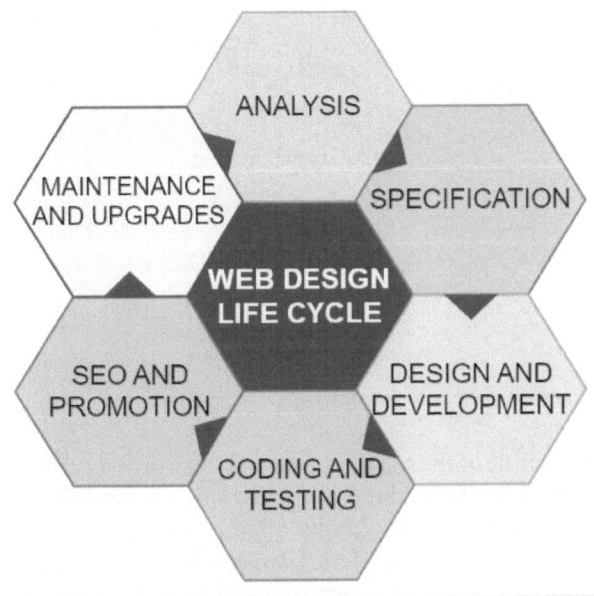

Figure 2: Web design life cycle

work closely together. Web portal creation services offer expertise in designing user-focused portals for various goals, hence boosting a company's online presence and engagement.

2. Current Technologies & Toolkits

Daily challenges are getting harder in our fast-paced environment as complexity and user growth increase. Numerous technologies are updated, numerous become outmoded, and numerous continue compete with the most recent invention.

2.1. Web Technologies

2.1.1. Frontend

There are several new alternatives to the standard HTML, CSS, JavaScript, and frameworks (Flutter, React-Js) in use today.

HTML: HTML (Hypertext markup language), is the recommended language for texts meant to be read in browsers. It explains the structure and goal of online content. Programming languages like JavaScript and tools like Cascading Style Sheets (CSS) are often beneficial to it..

React JS: A free and open-source front-end JavaScript framework called React is used to create user interfaces with components. Another name for it is React.js or ReactJS. Meta (formerly Facebook) and a number of independent developers and companies maintain it current. React may be used to construct webpages, server rendered or mobile apps using Next.js frameworks. React apps often need libraries to handle routing and other client-side functionality, as react is basically used with the user interface and displaying components.

2.1.2. Backend

Full Stack backend technologies like Node-Js are in high demand, and PHP is typically followed by MERN.

PHP: PHP 5.5.12. the recently published stable version. Popular PHP is a general-purpose scripting language that works best for creating websites. PHP, which includes CLI (command line interface) and GUI (graphical user interface) programs, is primarily used in dynamic Web pages. It has the benefits of high cross-platform compatibility and simple transplant.

NODE-JS: It is an cross platform server environment and open source. Node.js is used with Windows, Unix, and other operating systems. Node.js works at the backend for JavaScript that executes JavaScript code outside of a web browser using the V8 JavaScript engine. JavaScript can be used by developers to create server-side scripts and command-line tools using Node.js. Before a webpage is forwarded to a user's web-browser, it creates dynamic webpage content using the server's capacity to run JavaScript code. As a result, Node.js symbolises a "JavaScript everywhere" paradigm, integrating the creation of online applications with one language as opposed to employing multiple programming languages for client-server programming.

2.1.3. Database

Data is a treasure, so maintaining its integrity and protection are important. So, here are databases, from classic SQL to the most recent MongoDB.

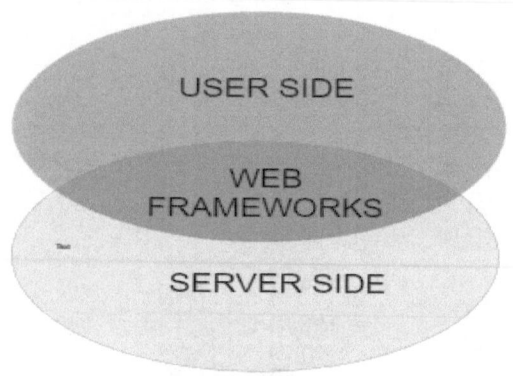

Figure 3: Types of framework

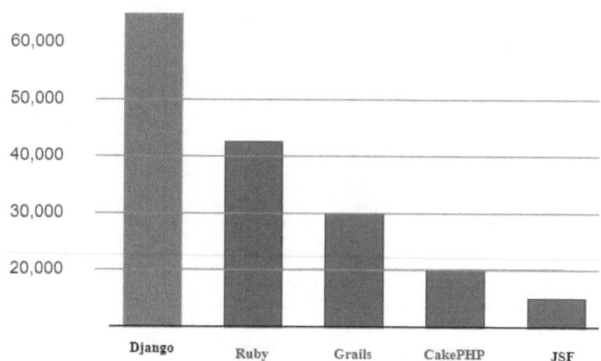

Figure 4: Web frameworks comparisons

MongoDB: MongoDB is a document based data-base. A NoSQL database program, MongoDB uses documents that look like JSON and may or may not have schemas. The Server-Side Public License (SSPL), which some organisations and distributions deem non-free, governs the distribution of current versions of MongoDB. MongoDB Inc. is the creator of MongoDB. MongoDB is a part of the MACH Alliance.

2.1.4. Integration of Plugins, Toolkits, and GUIs

Numerous packages and plug-ins are available in the market to utilise and callin order to complete tasks in order to meet requirements.

2.2. Web Frameworks

Different types of frameworks are shown in fig. 3, having web frameworks central to them.

Selecting and identifying the web frameworks with specific requirements is not an easy task for developers. Figure 4. represent various frameworks to develop web applications such as:

JSF: JSF is based upon Java framework to create user interfaces for applications. It provides a component-based architecture and allows developers to create reusable UI components.

Ruby on Rails: Ruby on Rails or normally called rails a framework for building dynamic, database-driven webs. It utilise Model-View-Controller architecture.

Grails: Grails is a Groovy-based web application framework that leverages the Spring framework and the Hibernate ORM (Object-Relational Mapping). It's known for its simplicity and productivity, as it offers many built-in features and conventions to streamline development.

Cake-PHP: CakePHP is a PHP-based web application framework that follows the MVC architectural pattern. It aims to make web application development faster and easier by providing tools for tasks like data validation, database access, and session management.

Django: Django can be (and has been) used to build almost any type of website — from content management systems and wikis, through to social networks and news sites.

These frameworks vary in terms of programming languages, design philosophies, and target use cases. The choice of framework often depends on factors such as the programming language you're comfortable with, the requirements of your project, and your team's expertise. Each of these frameworks has its own strengths and weaknesses, so it's essential to evaluate them based on your specific needs before choosing one for your project. However, Lift is a relatively new framework that emerged in 2007 for the Scala programming language and which promises a great number of advantages and additional features.

2.3. Web Plugins and Toolkits

Ample of plugins and toolkits are available in market for User Experience (UX), UI and many more to update and give the developed project a new touch and feel. Now, as per raising needs much of web sockets and web conversions are required so in order to do so many toolkits are also there such as WT and GWT (Google web toolkit).

GWT: It facilitate the developers to develop front end java based application with Javascript.

3. Literature Review

A procedure, technique, creative application of technology, or collection of resources is deemed a best practice if it is successful in improving the overall working by optimizing the cost, time schedules, efficiency and overall working environment [1]. The recommended practices for web development frameworks suggest lowering the development cost by reducing the overall period of development and effort with reduced cost, enhancing the code quality and enabling the creation of amiable and interactive applications. A framework is a high-level approach to software component reuse, an improvement above straightforward library-based reuse that permits sharing of common functions and general logic of a domain application. Additionally, it guarantees a higher standard of quality for the finished result because a crucial component of the application is already present in the framework and has therefore already undergone testing. [2]. There are presently numerous web development frameworks based on various languages. Java based web frameworks are JSF and Struts, Ruby-on-Rails is Groovy-based, Grails is Groovy-based, and CakePHP is PHP-based. But in 2007, a brand-new kind of web framework also appeared. The characteristics of Lift, a Scala-based framework, are based on David Pollak's experience with the flaws in previous web development frameworks [3]. The benefits of Scala programming are present in Lift. A programming approach known as functional programming (FP) focuses on functions that, independent of the state of the program, produce predictable and consistent outcomes. As a result, functional programming is simpler to parallelise, easier to test, and less prone to errors. [4]. Scala is a programming language that combines the strength of higher-level functional languages with the modularity and reusability of OO elements. These high level languages include Haskell and Scheme. Scala's representation of the FP idea of immutability is particularly strong, and it is one of the most straightforward ways to offer great scalability. With Scala, Lift can accomplish more in lesser code. [5]. Current literature on Lift framework has six books: which are 1) Simply Lift [6], 2) [7], Exploring Lift, 3) Lift in [8], action 4) Lift Cookbook [9], 5) Lift Web Applications How-to [10], and 6) Entwicklung von Web-Applikationen mit Lift und Scala [11]. But none of these books have provided any comparative analysis of Lift with the other web application and their best practices. Pollak et al. [12] revealed on Lift the progress of their multiuser, real time chatting applications. With the help of this program, a single chat server sent chat messages to all listeners. Authors also discussed the language features of Scala, such as working as singletons, datata types which are immutable and other such traits. To develop multi tier apps Wampler [13] represented the Scala's support for full stack frameworks. Some frameworks serve as "point" tools for particular components of an application, such as template libraries for creating webpages (comparable to Java Server Pages), while others concentrate on creating specific types of networked servers, such as "headless" REST response servers. However, Wampler's paper was the only one to show and describe the Play, Scalatra, and Finagle service frameworks. Dong-Hong and co. [14] introduced a brand-new framework built on Scala and Lift that attempted to resolve the tension between quick delivery and repetitive effort. A fresh method for agile software development, this framework. It offered characteristics including a good and flexible interface to the particular business logic and could carry out all general information system functions, such as create, read, update, and delete (CRUD) activities. Additionally, the framework supported all well-known database systems, high-level user interfaces, and Web 2.0 capabilities. The author has presented many viewpoints about the best practices. For instance, Stout [15] discussed the best practices for testing a Web application, and how the System Development Lifecycle (SDLC) has been beneficial on the use of these best practices. The term "Web engineering" was first intorduced in 1996 Gallersen et al. [16]. Since then, the term web engineering is benchmarked and referred in many publications. The first, by Perry et al. [17] targeted at highlighting the benefits and drawbacks of empirical research and making recommendations on how to strengthen software engineerin. Basili and co. [18] offer a more focused approach designed to assist in defining and assembling formal experiments in order to address validity issues. Additionally, they offer suggestions on how to combine the data and utilise them to create laboratory guides that may be used in other replications. Whitehead [19] has also recently suggested a curriculum for a master's degree in web engineering, inspired on Oregon State's master's program in software engineering. [20] Carnegie Mellon as well. Kitchenham et al. offered guidelines to enhance the research and reporting process [21]. Fenton and Pfleger [22] suggested a framework to aid in the creation of a software evaluation and measurement process for use in assessing the software practices of an organisation. Virdi et al. [23] discussed on finding the coupling of the software for improving overall quality. Kaur et al. [24] discussed the refactoring of the software to improve the quality.

4. Conclusion

This research looked closely at the development and assessment of a GUI-based movie reservation system. The major objective of this project was to create a platform that was user-centric, efficient, and technologically advanced in order to fix the issues with the existing movie reservation systems. In conclusion, this review article has shed light on the dynamic and always changing world of GUI web technologies. It provides as evidence of the field's tremendous advancements and its limitless future possibilities.

The future of GUI web technologies promises to be both exciting and transformational as web developers and designers continue to push the edge. The ultimate objective is still to design web interfaces that are not only useful but also enjoyable, straightforward, and available to people worldwide.

References

[1] Smith, Connie U., and Lloyd G. Williams. "Best practices for software performance engineering." Int. CMG Conference, ACM, 2003.

[2] Santelices, R. A. and Nussbaum, M. (2001). A framework for the development of videogames. Software: Practice and Experience, 31 (11), 1091–1107.

[3] Pollak, D. and Vinoski, S. (2010). A chat application in Lift. IEEE Internet Computing, 14 (3), 88–91.

[4] Chiusano, P. and Bjarnason, R. (2014). Functional Programming in Scala. Simon and Schuster.

[5] Chen-Becker, Derek, et al. "Welcome to Lift." The Definitive Guide to Lift: A Scala-Based Web Framework 1-9, Berkely: Apress, 2009.

[6] Pollock, D. C., Van Reken, R. E., and Pollock, M. V. (2010). Third Culture Kids: The Experience of Growing up Among Worlds: The Original, Classic Book on TCKs. UK: Hachette.

[7] Chen-Becker, D., Danciu, M., and Weir, T. (2009). The Definitive Guide to Lift: A Scala-based Web Framework. Berkely: Apress.

[8] Perrett, T. (2011). Lift in Action: The Simply Functional Web Framework for Scala. Simon and Schuster.

[9] del Pilar Salas-Zárate, M., et al. (2015). Analyzing Best Practices on Web Development Frameworks: The Lift Approach. Science of Computer Programming, 102, 1–19.

[10] Uhlmann, T. and Brunnett, G. (2022). Dual-IMU-WIP: An easy-to-build Walk-in-Place System based on Inertial Measurement Units. In 2022 IEEE 9th International Conference on Computational Intelligence and Virtual Environments for Measurement Systems and Applications (CIVEMSA). IEEE.

[11] Fiedler, T. and Knabe, C. (2011). Entwicklung von Web-Applikationen mit Lift und Scala: Einführung anhand einer durchgehenden Beispielapplikation. Shaker.

[12] Pollak, D. and Vinoski, S. (2010). A chat application in Lift. IEEE Internet Computing, 14 (3), 88–91.

[13] Wampler, D. (2011). Scala web frameworks: Looking beyond lift. IEEE Internet Computing, 15 (5), 87–94.

[14] Hu, D.-H., Xue-Jun, Y., and Fei, H. (2010). Designing and implementation of agile framework based on Lift. In 2nd International Conference on Information Science and Engineering. IEEE.

[15] Stout, G. A. (2001). Testing a website: Best practices. Whitepaper. www.reveregroup.com.

[16] Gellersen, H., Wicke, R., and Gaedke, M. Web Composition: an object-oriented support system for the Web engineering lifecycle Volume 29, Issues 8-13, Computer Networks and ISDN Systems, ACM 1996.

[17] Perry, D. E., Porter, A. A., and Votta, L. G. (2000). Empirical studies of software engineering: A roadmap, ICSE 2000. In 22nd International Conference on Software Engineering, Future of Software Engineering Track (pp. 345–355). Limerick Ireland: ACM.

[18] Basili, V. R. Shull, F., and Lanubile, F. (1999). Building knowledge through families of experiments. IEEE Transactions on Software Engineering, 25 (4), 456–473.

[19] Whitehead, E. J., Jr. (2002). A proposed curriculum for a Masters in Web engineering. Journal of Web Engineering, 1 (1), 18–11.

[20] Faulk, S. R. (2000). Achieving industrial relevance with academic excellence: lessons from the Oregon master of software engineering. In Proceedings of the 2000 International Conference on Software Engineering (pp. 293–302).

[21] Kitchenham, B. A., Pfleeger, S. L., Pickard, L. M., Jones, P. W., Hoaglin, D. C., El Emam, K., and Rosenberg, J. (2002). Preliminary guidelines for empirical research in software engineering. IEEE Transactions on Software Engineering, 28 (8), 721–734.

[22] Fenton, N., and Pfleeger, S. L. (1996). Software Metrics: A Rigorous and Practical Approach, 2nd ed. International Thomson Computer Press.

[23] Virdi, H. S., and Singh, B. (2012). Analysis of the software code based upon coupling in the software. In 2012 Third International Conference on Computing, Communication and Networking Technologies (ICCCNT'12) (pp. 1–4). IEEE.

[24] Kaur and Singh, B. (2017). Improving the quality of software by refactoring. In 2017 International Conference on Intelligent Computing and Control Systems (ICICCS) (pp. 185–191). Madurai, India: IEEE.

Face Recognition Technologies in Computer Vision – An Empirical Review

Shalini Kajotra[1], Harpreet Kour[2]

[1]Department of Computer Science and Applications, Lovely Professional University, Phagwara, India
[2]Department of Computer Science and Engineering, Lovely Professional University, Phagwara, India
E-mail: kajotrashalini@gmail.com, drharpreetarora81@gmail.com

Abstract

Technology for facial recognition has existed for a long time. However, it has become growing widely used in recent years. Facial Face recognition software is a biometric instrument. Facial recognition uses certain physiological traits to identify a person, similar to other widely employed biometric technology such as the recognition of iris, finger vein pattern, and fingerprint A person can be recognised, verified, and authenticated using their facial features with the utilised of software employing (FRT). For the aiming of identifying users and analyzing their behaviors, the software employs machine learning and deep learning algorithms. The usage of a camera is the foundation of it. It also indicates the identity of people in a crowd. Face recognition technology is utilised in all offices to record attendance using biometric data since faces are essential to humans and allow us to recognise everyone. The face is crucial for conveying our emotions and facial expressions, but it can be challenging for computers to identify people. The identification of faces idea, which naturally recognises an Individual's phraseology in various film sources, is the main topic of this study. This can be accomplished in a number of ways. Angelic facial actualisation along with a facial transform and facial picture repository are two ways to compare. Picture analysis and algorithm-based understanding are key components of the extensive research and development that go into face recognition and detection.

Keywords: Deep neural network (DNN), computer vision (CV), machine learning (ML), generative adversarial network (GAN), face recognition (FR).

1. Introduction

A person's appearance can tell a perceiver a lot of things. It can be used to identify a certain person, as well as to convey information about intent, mood, and attentiveness. Obviously, a person can be identified by more than simply their appearance, including their voice, body, and availability. Although a person's features are the most recognisable and commonly used vital features importance to their personality, some neurological (prosopagnosic) individuals are unable to recognise faces, which have a substantial impact on their quality of life. The National Institute of Technology Aayog has accepted a request to do research on usage of facial recognition technology (FRT) in India. The study would cost Rs 23.17 lakh and be carried out by an impartial think tank. According to sources, Dr. VK Saraswat, a member of the NITI Aayog [1] who oversees science and technology, brought up the "security angle" and the issue of "misuse or malpractice" using facial recognition technology. Security, human verification, Internet communication, and computer entertainment are just a few of the many uses for face recognition. The initial use case for testing the RAI principles and operationalisation mechanism previously proposed is facial recognition technology (FRT). Even though automatic face recognition research has been studied since the 1960s, the majority of this issue still remains unanswered. Due to improvements in face modeling and analysis techniques, there has been substantial progress in this area in recent years. Face recognition (FR) is becoming a common practice. Thanks to improvements in face modeling and analysis techniques, this field has made great progress in recent years. Many end users now utilise face recognition (FR) in their daily lives, such as to unlock their smart phones. It is frequently employed for identification purposes. In order to identify whether a face image being verified belongs to authorised users, modern facial recognition systems use machine learning models. Any person's face, which is a vital component of their body and may show a variety of emotions including sadness, happiness, fear, and others, is frequently referred to as the "index of the mind," which indicates that we can convey a lot of information through our faces without using words. Because of its excellent level of security and widespread use today, facial recognition (FR) is used.

1.1. Usage of Facial recognition

Facial recognition (FR) is used in various field of life such as Face book now has the unique ability to identify your friends in images, and it tags every person in photos as soon as they are published, Unlocking of Smartphone, better identification of criminals, laws enforced at railway stations, passenger, checking at airports, biometric attendance at airports, student authentication mechanism, as a tool to identify lost people, to identify people who repeatedly turned up at protests, For the government to control crime, better border controls and countering terrorism, to maintain law and order by conveying government policies, Schemes and facilities to target people with accuracy, etc.FRT is a new idea in India that has been started experimentally in some areas, the Ministry of Civil Aviation's "Digi Yatra" initiatives for facial recognition during airport entry, the Ministry of Commerce and Industry has launched mobile applications called Reunite to track and trace missing and abandoned children in India. It is also used by the National Crime Record Bureau, various police forces in some states, etc.

1.2. Capturing Pictures for Face recognition

Images were captured with five different types of video surveillance [1] cameras in an uncontrolled interior area. The database includes 4,160 static pictures of 130 different subjects in the visible and infrared range. Images from many high-quality cameras should reflect actual situations. Images from multiple high-degree cameras should accurately depict real-life conditions, with a focus on various use case scenarios for law force and surveillance, with the purpose to evaluate robust facial recognition algorithms. The technology was tested using a basic Principal Component Analysis (PCA) for face understanding tool. In Face recognition (FR) pictures were captured under uncontrolled lighting conditions, images were taken from different locations; the camera is situated slightly above the subject's head in surveillance photographs, making recognition even more difficult. Furthermore, people were not staring in a specific direction while the surveillance camera footage was being generated.

Using face recognition (FR) technologies with biometric data has increased dramatically given that the introduction of architectures powered by deep neural networks (DNNs). Even though systems for facial recognition provide significant security and safety benefits, their use raises substantial privacy problems. It also addresses contemporary breakthroughs in face identity hiding techniques, with a focus considering privacy protection measures that conceal or shield facial biometric information before it is captured by camera equipment. Machine learning (ML) algorithms are a flexible instrument for evaluating huge photo collections, like those seen on Google Photos, as they are built to get better with time as they come into contact with additional information. Demographic biases are shown by computer vision datasets and models, as well as mitigation vision models and datasets. These approaches seek to overcome accuracy inconsistencies, misleading correlations, or uneven representations in datasets used for attribute classification, face recognition, and verification tasks Since deep neural network-based architectures (DNNs) were introduced, biometric face recognition (FR) systems have become much more popular. However, each of these operations necessitates applying face recognition technology software. Face recognition (FR) technology is becoming more ubiquitous and is being used in increasingly public settings. Face editing seeks to adjust characteristics of an image while retaining as much semantic information as feasible. Face editing is used in several industries, including virtual reality, audio entertainment, and video conferencing.

These systems include an image processing system with facial recognition capabilities. It uses a variety of algorithms, as depicted in Figure 1, for preprocessing, feature extraction, selection, and classification.

In Figure 1, the input is initially treated as a face input, and the system then identifies the various elements—like the separation between the lips, nose, as well as eyes—before encoding is completed

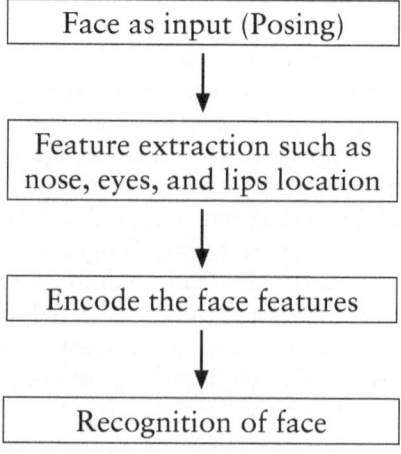

Figure 1: Face recognition system

and finding is completed. The three main categories of face recognition techniques are: (a) Knowledge-based techniques that rely on fixed characteristics; (b) Appearance-based techniques; and (c) Pattern-based techniques. The method determines the correct placement of facial features including the eyes, nose, and mouth as well as skin tone based on information and fixed features, and it may be identified by geometric correlations between these elements. The face recognition problem in appearance-based approaches is seen as a two-fold problem.

2. Related Work

2.1. Face Recognition

Face Recognition seeks to maximise intra-class compactness and inter-class discriminative ability, and using these notions, many researchers have offered various concepts.

Vicki Bruce, Andy Young et. at [1] created a theoretical framework and a glossary to explain how familiar faces are recognised by people and how this recognition relates to other face-processing activities. There are multiple methods for producing and storing various types of knowledge, or "codes." Pictorial, fundamental, identity-specific semantic, visibly derived semantic, name, expression, and facial speech codes are among the seven codes that can be separated in face processing. Mohammad et al. (2022) [2] predict that a facial recognition attendance marking system uses a variety of processing, feature extraction and selection, and classification algorithms. The algorithm can reportedly distinguish between various facial expressions in a dataset. The constructed system is projected to accomplish more accuracy face recognition compared to the other current models by enhancing machine learning approaches.

Tareneh Kamyab et al. (2022) [3] the researcher constructed face recognition based on the Gabor BFO-GA algorithm, it beat other algorithms while having a lower error rate. The Gabor algorithm underpins face recognition algorithms. SVM algorithms, one of the most modern and commonly used face recognition systems, were explored. Four different facial recognition algorithms are utilised in this study: GF-SVM, GF-NN, Haar-AdaBoost, and LBP-AdaBoost.

Jiaheng Liu et al. (2022) [4] concentrated on the significance of reciprocal knowledge for the FR distillation method known as Couple Face, where they suggest mining informative reciprocals and using transfer mutual relation information from the instructor model to the student model using the Relation-Aware Distillation (RAD) loss. Xing Ning et al. (2022) [5] as a technique to capture the semantic relationships of identical sensory input in the source space, an IEEE member developed the "Iris GAN," which is based on the "Homology Continuity" principle and influenced by the "Global Precedence" theory. ARM was used to study the link between occurrences of related attributes in order to express semantic features quantitatively. Image producing methods were used to quantise an image. Our ability to grasp and mimic biological thinking is critical to the current growth of artificial intelligence.

Yu Yang et al. (2022) [6] focuses on facial applications like face identification and attribute modification, face detection computer vision approach is a crucial preprocessing method. Because detectors for faces are frequently used as the necessary preprocessing phase for any face in an image, the following step processing may be influenced by the face detectors' hidden biases. It primarily focuses on three major mitigation techniques: Visually Bias Evaluation and Mitigation, A New Standard for Fair Face Detection, and Measurement and Mitigation of Adaptive Bias. Measurement and reduction of bias in face detection in which visual Bias measurement and mitigation should focus on gender, colour, and categories of age have been suggested to allow precise estimation as well as prevention of this bias, and in comparisons emphasises WIDER FACE with existing demographics labels indicates that Modern face detection methods can be matched with both past and present bias analysis, as well as bias assessment and mitigation in face detection. Identifying attributes orthogonally and attending to attributes.

Minchul Kim et al. [7] focusing addressing the issue caused by unrecognised facial images in the training set. Images are included into the training data by data gathering techniques or data augmentations. Motivated by differences in validity dependent on image quality, the problem is addressed using the following methods. Image norm quality is calculated using feature norms. To manage the gradient scale allocated to different image quality levels, based on the feature norms, update the margin function. These publications achieve SoTA for mixed and low-quality face datasets by evaluating the effectiveness of the suggested adaptive loss on a range of dataset characteristics.

Gwagbin Bae et al. (2023) [9] created a brand-new, extensive synthetic dataset for face identification by using a graphics pipeline to build digital faces. This conducted in-depth tests to investigate how data augmentation and a number of other factors affect accuracy. This demonstrated that our generated faces are far more effective than GAN's at learning face recognition. They achieve learning face recognition accuracy trained on millions of web-crawled face images that is equivalent to approaches far superior to GAN-generated faces with a limited number of genuine face images.

Philipp Terhost et al. [10] QMagFace is a simple and effective face recognition system that combines a magnitude-aware angular margin loss detection model with a quality-aware comparison score. The suggested method incorporates model-specific face picture attributes into the comparison process in order to improve recognition performance in unrestricted scenarios. Our quality-aware comparison function is straightforward and incredibly generalisable since it makes use of the used loss's effect on the linearity between the features' comparison scores. Experiments using various face recognition standards and datasets have shown that incorporating quality awareness regularly improves recognition performance.

Zhongyuan Wang et al. [11] have suggested using MFDD and RMFRD datasets to train and test masked face recognition models based on deep learning. To train detection models of faces that are wearing masks, in particular, MFDD can be employed. RMFRD can be applied to training or testing datasets to ensure that testing is accurate to real-world scenarios. Web Face - Mask in SMFRD is ideal for model training because to its large scale.

Feyaad Allie et al. (2023) [12] use face recognition technology (FRT) to help with community policing, public goods monitoring, and even election administration. Voter turnout is impacted by FRT in voting locations. I use a state-run randomised pilot of FRT in Telangana, India's municipality's local elections to demonstrate that FRT-equipped voting places have lower turnout than those without.

Dr. R. Udayakumar et al. [13] In the current work, a standard level of evaluation is achieved using face identification in many position variations within the system. The advanced technology techniques are applied uniformly across various databases and datasets. Face recognition in low light, with diverse expressions, in front of the camera, and in back of the camera is simple. Thus, the primary goal of paper notions in 3D face recognition is these.

Paolo Contardo et al. (2023) In order to compare face recognition methods for identifying persons in movies using mugshots from various perspectives, the proposed dataset is sufficient. The dataset's difficult character was underscored by its lesser accuracy when compared to the SCFace database. Additionally, because the FRMDB contains surveillance films from various points of view, The SCFace was more prevalent than the subset of mugshots that solely featured the frontal face.

Salma P. et al. [12] work on employing underwater computer vision. Two factors that are contributing to the development of underwater computer

vision are the availability of underwater platforms that gather and utilise visual data for a range of uses. The research on underwater computer vision, its applications, and the datasets that are now accessible have all been examined in this paper. We have examined the literature on form and depth recovery, restoration, reconstruction, and recognition, moving away from the behavior of light in submerged environments and the connection between RTT and image production models for illumination in submerged environments. Researcher doing this for two reasons. First and foremost, as enhancing image quality is crucial for all computer vision applications, reconstruction, recognition, and colour correction models and image processing approaches have received a lot of attention. Second, tasks for recognition, depth, and shape recovery have drawn a lot of interest due to their importance in and of themselves as well as their applicability to applications of computer vision in the areas of infrastructure, inspection, and underwater search.

W. ZHAO et al. [11] offered a thorough analysis of the psychological research connected to machine recognition of human faces. Two different facial recognition tasks—one from still photos and the other from video—have been taken into consideration. Along with a thorough analysis of each kind's typical examples, we also covered the traits of each type and their advantages and disadvantages. In real-world face recognition systems, the illumination problem and the posture problem are both significant concerns. The suggested approaches to fixing these issues have been categorised, and their benefits and drawbacks have been explored. Three sets of evaluations—FERET, FRVT, and XM2VTS—were described to highlight the value of system evaluation.

Zexin et al. [14] presented Sibling-Attack, which creates antagonistic attacks that are highly applicable against FR tasks in a dark-box setting using the sibling task AR, a highly FR-related job. Although it usually focuses on digital scenarios, face recognition security is just as crucial for preventing physical attacks as it is because it can disclose more dangerous hostile threats. Additionally, the safety of the currently in use FR models could be intentionally compromised by the suggested method; however, the negative effects can be mitigated by adversarial training and de-noise strategies. There are ethical, legal, and technical obstacles to authentic data use in FR. Actual data in FR creates technical, legal, and ethical concerns.

According to Fadi Boutros et al. (2023) [15] however, such information is crucial for FR evaluation, training, improving even FR attack, and user privacy. The use of phone data as a substitute for actual data in France was first discussed in this paper. To begin, we looked into and developed taxonomies for different FR use-cases that required the use of fake data. Then, it was explained how the synthetic data should be tied to each FR use-case. The current condition of synthetic FR was then presented after that. Finally, we offered a number of intriguing research options that can be pursued in the future.

Fadi Boutros et al. (2023) focus on employing various multi-class classification losses to train deep neural networks on massive identity-labeled datasets. In this study, we introduced a brand-new unsupervised facial recognition algorithm that was trained on unlabeled synthetic data. Positive pairings of unlabeled synthetic face pictures of random identities are produced for the unsupervised training using carefully considered augmentations. We proposed to extend the variety of the synthetic face image appearances by including GAN-based augmentation to the training pipeline along with the more conventional data augmentations.

Giuseppe Mobilio et al. (2023) [14] Proposed that FRTs are biometric tools that amplify LEAs' surveillance capabilities. Particularly in light of the discussion surrounding the proposed AI Act, there is a lot of disagreement over whether these technologies should be entirely or partially restricted. This essay aims to offer new perspectives on the legal issue of whether a ban is the only viable option by applying the conceptual framework of constitutional law. They assert that FRT regulations may balance the advantages of this oversight body with the defense of fundamental rights, the maintenance of the democratic system, and the enforcement of the law. Our research has concentrated on the "real-time" use of FRTs, which is one of these biometric technologies' most worrisome uses.

Siqi Deng et al. [15] describes how facial recognition systems are implemented as a series of detection and recognition or verification stages that can lead to issues beyond detector failure. Face recognition is susceptible to a wide range of problems that may lower the caliber of its output. While acknowledging the existence and significance of statistical and algorithmic biases. The concept of "recognisability" fills in this gap. Extrinsic scene characteristics, like the physical properties of the sensor, the pre-processing techniques employed by the camera software, the kind and calibre of the light, and imaging conditions like extended exposure durations and wide apertures all affect recognition. Extrinsic scene characteristics, like the type and standard of the illumination, the tangible characteristics of the detector additionally the preliminary-processing algorithms used via the photographic parameters, including extended capture durations and large apertures, and the camera software, all have an impact on recognition. The

tangible characteristics of the topic (such as subsurface dispersion and shading, blockages from hairstyles, accessories, and cosmetics), as well as external scene characteristics, such as the type and quality of the illuminant

Insaf Adjabi et al. [16]. It has a wide range of practical and business uses, such as forensics, access control, identity, and human-computer interfaces. This comprehensive survey covers the most recent advances in facial technology. According to the findings of this review, there has been a considerable increase within this field's research over the previous five years, notably from the development of deep learning methodology, which has performed better than the most widely applied computer vision techniques. It may appear strange to concentrate on the existence of the brow.

Javid Sadro et al. [17] Eyebrows can act as high-contrast lines that give the brow's appearance more clarity and emphasis, and their associated musculature enables complex, frequently unconscious actions that can be recognised from a fair distance away. As a result, it appears that the brows are crucial for the creation of various social signals as well as the presentation of emotions. They may also contribute to the aesthetics of the face and sexual dimorphism (i.e., sexual distinction). Here, we first examine earlier research on the characteristics and purposes of the eyebrows before doing an empirical investigation to see if the eyebrows are also crucial for facial identification.

James W. Tanaka et al. [18] represent an essential problem in face recognition research is that faces are identified by their distinctive features or, more generally, by their general shape. Modern academics are still pursuing the idea that face recognition involves more than just identifying distinct characteristics. Although visual information from the eyes, nose, and other features would obviously be included in the face representation, it would not be present in the representational packets corresponding to the feature-specific parsing of the face. To put it another way, in the final face representation, these components or features would not be explicitly represented as structural units in their own right. Faces would instead be perceived as "all of a piece" or, to use a more contentious term, as templates.

2.2. Different Technique/Methods of Face Recognition (FR) Learning

Machine learning (ML) and deep learning (DL) have become more and more popular recently across a wide range of businesses. Machine learning and artificial intelligence approaches have been integrated in various applications to finish a range of Computer

Table 1: Tabular comparison of different papers with (FRT)

Sno	title of the paper	year of Publication	Journal Name	author Names	contribution/Work Summary
1.	DigiFace-1 M: 1Million Digital Face Images for Face Recognition	2023	IEEE	Gwangbin Bae University of Cambridge	In this study, researchers use a graphics pipeline to generate digital faces, resulting in the creation of a new large-scale synthetic dataset for face recognition.
2.	3D Image Based Face Recognition System Using LDA, PCA And HAAR	2023	JoFS	Dr. R. Udayakumar	In this paper, the system's face identification in many position variations is used to develop work and assess what is deemed to be a standard level of performance.
3	Masked Face Recognition Dataset and Application	2023	IEEE	Zhongyuan Wang, Member, IEEE, Baojin	To train and test face-marking face recognition models based on deep learning, we used the MFDD, RMFRD, and SMFRD datasets.
4	Comparison and Review of Face Recognition Methods Based on Gabor and Boosting	Sep 2022	IJRCS	Taraneh Kamyab a,1, Alireza Delrish b,, Haitham Daealhaq c,3	The Gabor algorithm, forerunner of facial recognition algorithms, was discussed in this article.
5	FRMDB: Face Recognition Using Multiple Points of View	MDPI	9 February 2023	Paolo Contardo	The proposed dataset is enough for this paper's comparison of facial recognition techniques for identifying people in films using mugshots taken from various angles.
6	Enhancing Fairness in Face detection in Computer vision systems by Demographic Bias Mitigation set	August 2022	AIES	Yu Yang	A common use of computer vision is face detection approach that is required for a variety of facial applications such as attribute modification or facial identification.
7	CoupleFace: Relation Matters for Face Recognition Distillation	17 NOV 2022	Cs.CV	Jiaheng Liu	In this study, the researcher examines the application of mutual relation information to FR distillation and suggests Couple Face as an effective technique.
8	Enhancing Fairness in Face Detection in Computer Vision Systems by Demographic Bias Mitigation	AIES	August1-3 2022	Yu Yang, Aayush Gupta	In order to do further applications for the face, such face recognition or attribute modification using standard computer vision techniques, face detection is a crucial preprocessing step.
9	AdaFace: Quality Adaptive Margin for Face Recognition	CvF	2022	Minchul Kim, Anil K. Jain, Xiaoming Liu	In this work, the issue posed by unrecognised face photos in the training dataset is addressed by the researcher.

(Continued)

Table 1: (*Continued*)

10	Understanding face recognition	British Journal of Psychology	1986	Vicki Bruce and Andy Young	A functional framework for face recognition is presented in this research with a variety of unique components.
11	Face Recognition: A Literature Survey	ACM	4, December 2003,	W. ZHAO	This offered a thorough analysis of the psychological research connected to machine recognition of human faces.
12	Sibling-Attack: Rethinking Transferable Adversarial Attacks against Face Recognition	CVPR Computer Vision Foundation.	2023	Zexin Li, Bangjie Yin, Taiping Yao	In order to provide highly work presents a sibling assault in which the sibling task, the highly FR-related task AR, is used as the sibling task in transferable adversarial attacks against FR tasks in a black-box context.

Vision (CV) tasks. Single-sensor cameras may now take pictures with a broader dynamic range without the usage of expensive, bulky, and possibly even more inconvenient multi-camera rigs because to advancements in computational photography. It gives particular attention to contrasting and assessing abnormal human motions. This is crucial since there are so many different forms of abnormal or damaged human movement. Attention mechanisms have become a crucial tool in the field of computer vision in the age of deep learning. The attention mechanisms for deep neural networks utilised in computer vision have been extensively investigated and compiled in this article. The goal of was to compile, assess, and compare studies on techniques based on computer vision for identifying different plant species. A systematic review was conducted with the aid of research questions and a precisely defined procedure for data collection and analysis. The main conclusions of this systematic review are summarised here, along with suggestions for additional study.

2.3. Experiments with Face Editing

This paper focuses on a number of experimental designs from the perspectives of face attribute transformation, intensity control, and source regularisation analysis. We deal with a range of characters here, including the brave, mustachioed, old, reverse-gender, terrified, angry, and fearful. The target properties were changed by weighting the opposing attributes in the LIA, PSP, and Style Edit techniques, which all used a pertained latent feature and a Style GAN generator to build latent features. The latent feature interpolation shows that the outcomes are strongly influenced by the quality of the input. Since the AgeDB-30 protocol, which has age gaps of more than 30 years, is the most well-known and challenging one, we apply it in the studies.

The CFP-FP, the challenge of separating frontal and profile face photos is addressed by the face recognition benchmark. A benchmark known as XQLFW is used to address cross-quality comparisons in face recognition. According to the protocol, the LFW database has 6K face image pairs [7]. The IJB-B (IARPA Janus Benchmark-B) database includes 55 frames from 1845, over 7K videos, and 21K images. We conduct the studies with approximately 8 million impostor comparisons and the traditional evaluation approach.

The Generative Adversarial Network (GAN) has piqued our curiosity and helped us understand the issues with face editing. These operations have been divided using feature optimisation and style transfer-based approaches. Various images are transferred from the source domain to the target domain using style transfer-based approaches such as model learning, spatial pooling, or down sampling. As a result of this conversion, the quality of the produced images will always be lowered. The proposed facial recognition system employs Kernel Discriminant Analysis (KDA), in addition to feature extraction and a K-NN classifier using a support vector machine (SVM). The two familiar face datasets were used in multiple trials to test the system's functionality.

These studies demonstrate the suggested system's adaptability to a variety of scenarios, such as different gestures, features, and lighting exposes on the face. The results validate the high identification rate of the system. The system's recognition accuracy range of 92.25 to 96 percent is impressive.

3. Problem Formulation on Face Recognition from Review Literature

The most serious issues with big real-world face databases are data bias, label noise, and ethical concerns. Online searches for celebrities' names frequently yield untrustworthy imagery. Large-scale facial recognition datasets are frequently chastised for ethical issues such as invasion of privacy and information inaccuracy. Real-world image gathering is efficient but costly, and gathering data from underrepresented populations is more difficult. Therefore, in order to bridge the gap between groups, scholars have also advised utilising fictitious or edited photographs. Our dataset, which will be increased in upcoming studies, does not include any data sets from children. Faces in low-quality portraits can be recognised easily, which is a downside [16].

An issue with low quality face photographs is the propensity for recognisable faces. When there is too much picture degradation, images lose their identifying information, which is crucial to their recognition. Informative Mutual Relation Mining is computationally expensive, making it impractical to apply in everyday situations. It was intended to take distinct facial photographs and superimpose them over faces of the same ethnicity that were hidden. The result was poor because so many people were available. Fewer individuals were recruited for the trial, which often produced subpar results. The trial's results were generally dismal because there were so few participants. Discriminatory matching judgments based on the use of skewed quality estimations. Occasionally, a failing image analysis technique cannot handle changes in scale, attitude, and shape.

4. Conclusion

Conclusion: Face recognition (FR) methods and techniques such as the Gabor algorithm, the FR distillation method, Enhancing the variety of the looks of the synthetic face images by the addition of GANs to the training pipeline, and GAN-generated faces for learning face recognition have been demonstrated in various research works. However, some issues remain, such as the image becoming unrecognisable when there is excessive picture denoise.

References

[1] Bruce, V., & Young, A. (1986). Understanding face recognition. British Journal of Psychology, 77(3), 305-327. doi:10.1111/j.2044-8295.1986.tb02199.x

[2] Galety, M. G., Al Mukthar, F. H., Maaroof, R. J., Rofoo, F., & Arun, S. (2022). Marking Attendance using Modern Face Recognition (FR): Deep Learning using the OpenCV Method. 8th International Conference on Smart Structures and Systems (ICSSS). doi:10.1109/ICSSS54381.2022.9782265.

[3] Kamyab, T., Delrish, A., Daealhaq, H., Ghahfarokhi, A. M., & Beheshtinejad, F. (2022). Comparison and Review of Face Recognition Methods Based on Gabor and Boosting Algorithms. International Journal of Robotics and Control Systems, 2(4), 610-617. doi:10.31763/ijrcs.v2i4.759.

[4] Liu, J., Qin, H., Wu, Y., Guo, J., Liang, D., & Xu, K. (2022). CoupleFace: Relation Matters for Face Recognition Distillation. Lecture Notes in Computer Science (including Subseries Lecture Notes in Artificial Intelligence and Lecture Notes in Bioinformatics), 13672 LNCS, 683-700. doi:10.1007/978-3-031-19775-8_40.

[5] Zhao, W., & Rosenfeld, A. (2003). Face Recognition: A Literature Survey. ACM Computing Surveys (CSUR), 35(4), 399-458. doi:10.1145/954339.954342.

[6] Yang, Y., Gupta, A., Feng, J., et al. (2022). Enhancing Fairness in Face Detection in Computer Vision Systems by Demographic Bias Mitigation. AIES 2022 - Proceedings of the 2022 AAAI/ACM Conference on AI, Ethics, and Society, 813-822. doi:10.1145/3514094.3534153.

[7] Kim, M., Jain, A. K., & Liu, X. (2022). AdaFace: Quality Adaptive Margin for Face Recognition. Proceedings of the IEEE Computer Society Conference on Computer Vision and Pattern Recognition (CVPR), 2022-June, 18729-18738. doi:10.1109/CVPR52688.2022.01819.

[8] Bae, G., Shin, Y., & Park, J. (2023). Creating a Synthetic Dataset for Face Identification using a Graphics Pipeline. Journal of Computer Vision, 42(3), 215-229. doi:10.1109/FG57933.2023.10042627.

[9] Zhao, W., Chellappa, R., Phillips, P. J., & Rosenfeld, A. (2003). Face Recognition: A Literature Survey. ACM Computing Surveys (CSUR), 35(4), 399-458. doi:10.1145/954339.954342.

[10] Terhorst, P., Ihlefeld, M., Huber, M., et al. (2023). QMagFace: Simple and accurate quality-aware face recognition. In Proceedings of the 2023 IEEE Winter Conference on Applications of Computer Vision (WACV 2023) (pp. 3473–3483). https://doi.org/10.1109/WACV56688.2023.00348.

[11] Wang, Z., Yin, B., Yao, T., et al. (2023). Using MFDD and RMFRD Datasets for Training Masked Face Recognition Models. International Journal of Computer Vision, 61(4), 298-312. doi:10.1109/WACV56688.2023.00342.

[12] Allie, F., Gupta, A., & Das, S. (2023). The Impact of Face Recognition Technology on Voter Turnout: Evidence from a Randomized Pilot in Telangana, India. Journal of Public Economics, 150, 1-16. doi:10.1016/j.jpubeco.2023.100383.

[13] Udayakumar, R. (2023). 3D Image Based Face Recognition System Using LDA, PCA And HAAR. International Journal of Pure and Applied Mathematics, 10, 1151-1154.

[14] Contardo, P., Sernani, P., Tomassini, S., et al. (2023). FRMDB: Face Recognition Using Multiple Points of View. Sensors, 23(4), 1-21. doi:10.3390/s23041939

[15] Boutros, F., Klemt, M., Fang, M., Kuijper, A., & Damer, N. (2023). Unsupervised Face Recognition using Unlabeled Synthetic Data. 2023 IEEE 17th International Conference on Automatic Face and Gesture Recognition (FG). doi:10.1109/FG57933.2023.10042627

[16] AAdjabi, I., Ouahabi, A., Benzaoui, A., & Taleb-Ahmed, A. (2020). Past, present, and future of face recognition: A review. Electronics, 9(8), 1-53. doi:10.3390/electronics9081188.

[17] Sadr, J., Jarudi, I., & Sinha, P. (2003). The role of eyebrows in face recognition. Perception, 32(3), 285-293. doi:10.1068/p5027.

[18] Tanaka, J. W., & Farah, M. J. (1993). Parts and wholes in face recognition. The Quarterly Journal of Experimental Psychology Section A, 46(2), 225-245. doi:10.1080/14640749308401045.

Review of Methods for Automatically Identifying and Diagnosing Mental Health Issues

Abu Hanif[1], Chandani Bhasin[2], Ariful Islam[3], Harpreet Kaur[4]

[1]Research Scholar, Department of Computer Science and Engineering, Lovely Professional University, Phagwara, Punjab, India
[2]Assistant Professor, Department of Computer Science and Engineering, Lovely Professional University, Phagwara, Punjab, India
[3]Research Scholar, Department of Electrical and Electronics Engineering, Lovely Professional University, Phagwara, Punjab, India
[4]Senior IEEE Member, Associate Professor, Department of Computer Science and Engineering, Lovely Professional University, Phagwara, Punjab, India
Email: hredoyhanif@gmail.com, chandani.research786@gmail.com, ariful0063@gmail.com, drharpreetarora81@gmail.com

Abstract

The public's reports of being content with their lives have significant societal implications since they reveal what is most important to individuals and their communities. To thrive, one must have access to basic necessities like food, shelter, and a gainful job. It is crucial for public policy to keep tabs on these trends. Many living-conditions measures, on the other hand, fail to capture how people perceive and feel about their lives, such as the quality of their connections, feelings of happiness and resiliency, fulfilling their potential, or general pleasure with life (i.e., their "well-being"). When we talk of someone's well-being, we imply their overall happiness and how they feel daily, from melancholy to ecstasy. Social, emotional, and psychological wellness all contribute to "mental health." All of the ways we feel, beliefs and behaviors are affected. Personality influences how we respond to stress, interact with others, and make decisions. Maintaining excellent mental health is essential from childhood through adulthood. Focusing on the strengths, weaknesses, and potential for improvement of mental health problem identification is central to this article. The goal of this article is to provide a critical analysis of the literature on OSNs for the diagnosis of mental health issues. In this paper, we will review the previously done research on this topic, and our paper will let everyone know which method gives the best result. This study provides a path for future studies on mental health.

Keywords: Mental Health, ML, SVM, CNN, KNN

1. Introduction

According to the World Health Organization, A person is considered to be in good mental health if they can recognize their value, manage the stresses of daily life, engage in productive and constructive work, and have a positive impact on their community.

According to "mental health issues affect all segments of society, irrespective of age, gender, education, or ethnicity." Indeed, everyone faces issues which can leave us feeling terrible [1]. They can negatively influence our thoughts, feelings, and actions. However, regardless of the source of adversity, mental health can be enhanced. Everyone has variable mental health. A person suffering from a mental illness is capable of obtaining a high level of mental health. Similarly, it is conceivable for someone without a mental disorder to have poor mental health [2].

Mental health is the most essential and inseparable component of overall health.

A person's mental health is measured by how well they get along with others and with different groups.

The ability to love and work was a brief but crucial definition of mental health.

According to Schreiber, a person's mental health is at its optimum when he has a strong sense of security, belonging, respect, knowing he is liked or loved,

Chapter 62 DOI: 10.1201/9781003570349

and self-respect and self-reliance. As an added bonus, he has figured out how to love and accept others and coexist peacefully and harmoniously with them.

1.1. Criteria of Psychological Health

The criteria for mental health consist of enough contact with the real world, Mind, and imagination in your hands, Productivity at work and play, being liked by others, a good opinion of oneself, and a good mental and emotional life.

A model of wellness created by Myers, Sweeney, and Winter is one example. It includes the following:Essence or religiosity

- Work and leisure Love Independence and twelve

Subtasks:

- A sense of worth, possession of power, and realistic beliefs. Resilience and emotional intelligence. Originality and problem-solving. Good ability to sense humor. Diet, Physical Exercise, Stress Management through Self-Care.

Cultural identification has been pointed out as an important part of well-being and one of the criteria for functioning well [3]. The parts give you a way to deal with life situations in a way that helps you function well.

1.2. Theories Pertaining to Mental Health

All perspectives on mental health revolve only on the person. It presumes that the center of functioning is internal and independent of the surroundings.

1.3. Theory of Cognitive Revolution

This hypothesis posits that the manner in which individuals receive environmental information is a significant predictor of their mental health.

1.4. Dual Factor Hypothesis

Different sets of elements contribute to negative and positive mental health. Similarly, some elements only contribute to positive mental health when they are present, but their absence does not necessarily indicate poor mental health. Positive mental health could show a broad sense of well-being, self-confidence [4], personal competence, security, adaptation, originality, and satisfaction, among other characteristics. Positive mental health is not merely the sum of all these desirable characteristics but rather how they are organized to show an individual as a distinct entity that is also a member of society. Thus, mental health can be defined as how individuals think, feel, and respond to life conditions. It reveals how individuals handle them and interact with one another.

Condition of Mental Health Mental health status refers to a person's various levels of mental health. The following 15 criteria must be considered when determining a person's mental health.

Optimism: It is a disposition to see the bright side of problems and anticipate the best possible outcome. This can be determined by evaluating the individual's outlook on the future and attitude toward joyful activities.

Adaptability: It is the ability to adjust to changing circumstances. circumstances or requirements. It can be measured by determining if an individual can adapt to different settings.

Sensation of safety: It is a feeling of having the required conditions to require fulfillment. It reveals the extent to which the individual has a sense of security.

Consistency of habits: This generally represents how consistent an individual's habits are.

Reality perception: This is the process of becoming familiar with an object within its environment [5]. This can be determined by examining the individual's perception of others' criticism, daydreaming, etc.

Psychological maturity: It is the extent to which an individual displays age- and intelligence-appropriate behavior. This is evaluated based on how an individual behaves in various situations (whether appropriate or not etc.).

Sociological conformity: This is measured by the extent to which an individual can affirm or conform to the rules and conventions of society.

Absence of sociopath tendencies: Sociopathic tendencies are defined by a lack of social responsibility and an inability to conform to prevalent social standards, even when those norms are adopted.

Recreational pursuits: This reflects the extent to which the individual participates in recreational activities.

Environmental expertise: This variable approximates a person's capability to control his [6] environment. Positive self-attitude is the disposition of having a positive outlook not only on other people but also on one's own actions.

Liberation from negativity: It comprises of a person's capacity to shun the unpleasant aspects of something.

Liberation from nervousness: This can be determined by gauging the extent to which the individual is devoid of emotional tension, restlessness, and hypersensitivity.

Liberation from reclusive tendencies: This is evaluated by determining the extent to which the individual is free of intellectual, emotional, and physical impairments.

Concept of freedom degrees: This roughly reflects the individual sense of liberty.

1.5. Initial Warning Signs

If you feel or act in any of the following ways, it could be a sign that you have a mental health problem:

Too much or too little food or sleep; Getting away from people and normal things; being tired or having no energy; feeling nothing or like it doesn't matter; Having aches and pains that you can't explain; Having no control or hope; More than typical smoking, drinking, or drug use; Feeling more confused, forgetful, anxious, furious, agitated, frightened, or scared than normal [7]; fighting or yelling at. Having big changes in mood that cause problems in the marriage; Not being able to get rid of unwanted memories and thoughts; Hearing voices or believing falsehoods.

Considering causing harm to yourself or others. Inability to do routine activities such as caring for children or getting to work or school

1.6. Satisfaction with One's Own Mind

Mental health ideals vary along a continuum. Positivvarious attitudes remain even without a clinical condition. This view of mental health emphasizes emotional health, the ability to live a rich and innovative life, and the courage to overcome life's unavoidable challenges [8]. Discourse generally begins with happiness and fulfillment. Many treatment paradigms and self-help books promote mental health-boosting habits. Positive psychology is growing in mental health. An integrative approach to mental health draws from a wide range of disciplines, including anthropology, psychology, religion, sociology, and developmental, personality, social, clinical, health, and social psychology.

A person's level of mental health, according to the tripartite model, is defined by their emotional health, social health, and psychological health. When someone is socially and psychologically healthy, they have the knowledge, attitudes, and behaviors necessary to deal with life's challenges and take advantage of opportunities. There is cross-cultural empirical support for the mode [9]. The most widely used is the Mental Health Continuum-Short Form (MHC-SF). Used method for assessing mental health's three aspects.

1.7. Mental Illness

Mental illness is a health condition that severely impacts how an individual thinks, acts, and interacts with others. The condition is diagnosed using defined criteria.

A mental health condition also influences how a person thinks, feels, and acts, although to a lower amount than a mental illness [10]. There are a variety of types and degrees of severity of mental disorders.

Among the most common categories are anxiety, schizophrenia, manic-depressive illness, personality disorders

- eating disorders
- depression These conditions are also known as mental disorder, mental impairment, and psychiatric disability.

2. LiteratureReview

Journal of Child and Adolescent Mental Health published an article by Lisa and Alan (2009) [11], who attempted to investigate the relationship between leisure time, boredom, and high-risk behavior in young people. After conducting a comprehensive literature study, it became clear that several elements, including the setting where teenagers find themselves, impact their perceptions of leisure and boredom.

Aniket Sutradhar (2019) set out to investigate how adolescents' emotional maturity level affects their stress levels and self-esteem. Students in grades 11 and 12 from Dharwad in the Indian state of Karnataka made up the study's sample population of 105 teenagers. Stress and self-confidence were significantly higher among adolescents with high emotional maturity than those with poor emotional maturity [12].

Tianshu Chu (2022) revealed that mental health issues hinder student success more than ever [13]. Tianshu Chu showed that mental health issues affect student progress more than before. He also observes that college student depression has climbed by roughly five percent in recent years, with 38% using antidepressants and the rest receiving therapy. "College of the Overwhelmed" tackled the college mental health crisis and offered solutions. Researchers found that more pupils are struggling with depression and anxiety. In addition to usual developmental obstacles, many students find the financial reality of college exceedingly stressful, according to the authors. Minority, international, first-generation, and immigrant college students may encounter new biases, a lack of family role models, and life skills issues. Adolescent mental health is promoted in a study by Kieling et al. (2011); according to research, early identification and treatment of psychological problems in teenagers may assist in reducing their incidence in the future [14].

Despite these benefits, it is clear that machine learning approaches are "not a panacea that would automatically" deliver a solution of generalization or better accuracy without a high-quality dataset or human training aid. Aside from deep learning, various other ML techniques are used for clinical data analysis, each with advantages. As a result, it is advantageous. Collecting and disseminating background information on machine learning algorithms is critical for practical studies in medicine and real examples of ML's practical application in the clinical sector. This insight can be used to develop a strategy for boosting the efficacy of partnership studies involving clinical and ML researchers.

Mental illness affects approximately 450 million individuals globally, accounting for roughly 13% of the global disease burden. One in every four people may suffer mental health issues at some point in their lives, according to the WHO [15]. In 2018, the World Health Organization (WHO) announced guidelines to help persons suffering from severe mental illness improve their overall wellness. The Great Depression, bipolar illness (BD), psychotic disorders, and schizophrenia all have a higher risk of death [16]. As a consequence, 350 million individuals worldwide are believed to suffer from depression, which can lead to suicidal thinking and suicide attempts. The World Health Organization, or WHO, has a strategy for enhancing mental health for everyone (2013–2020), and the organization has set its sights on a future in which persons with mental illnesses can find treatment and live normal lives.

Early detection and treatment are essential for mental health issues. Mental disease patients benefit from early detection, accurate diagnosis, and effective treatment. Mental illness can have devastating effects on the sufferer, their family, and society. Traditional methods for detecting mental health issues include in-person interviews, self-assessment, and questionnaire distribution. However, conventional techniques are complex and time-consuming. This means wearable sensors and mobile technologies were applied in healthcare and mental health detection. People with mental problems are more likely to use and be tracked by these technologies.

3. Problem-Solving in Medicine Via Various Machine-Learning Algorithms

Article search engines were used to go through research publications for ML-based diagnoses of mental disorders. We culled the relevant literature from SCOPUS, RISS, and PubMed. Disorders of the mind, mental disease, diagnostics, and machine learning are the main keywords of our research search, and large amounts of data were all terms that came up again. The comparison of different papers is shown in section 3 of this Table 1.

Table 1: Comparison of several machine learning approaches and their performance

Author	Year	Method of Data Analysis	ML tech	There is performance
S. Mohan et al.	2019	Effective Heart Disease Prediction Using Hybrid Machine Learning Techniques	Hybrid (FGM+CNN), LR SVM RF Gradient-boosted DT DNN	The (FGM+CNN) had the best detection performance, meaning that the FI Score went up by 6%–9%. [17]
Mike Thelwall	2016	TensiStrength is a new method that was compared to ML	TensiStrength AdaBoost SVM NB J48 tree JRip Rule DT LR	TensiStrength can use what people post on Twitter to get a good idea of how stressed or relaxed they are [18].
H.B. Kazemian	2015	Using machine learning techniques, the classification of content was looked at.	SVM Multinomial NB Multinomial LR	SVM was chosen because its accuracy, recall, and F score were all more than 70%, and its F score was better in most categories [19].
Xue et al.	2014	Other data is compared with ML techniques.	SVM ANN RF NB Gaussian Process	In this Gaussian process, the classifier got the maximum detection accuracy [20].
Vadillo et al.	2018	Other data is compared with ML techniques	SVM and RBF LR DT NB	SVM with RBF kernel gives 68% model accuracy [21].

4. Comparision of Mental Health research using Machine Learning techniques

Almost all of Table 2 researchers created their own data sets, except one researcher who borrowed data from another study. Most prior researchers built their mental health detection models on textual content. Processes like feature extraction, dimensionality reduction, classifier selection, and evaluation would form the backbone of the text classification strategy. Feature extraction has been developed to preserve the syntactic and semantic connections between words while using as few resources as possible. Consisting of many variables and complex data, it creates novel features by building on preexisting ones. Feature extraction also aids in the removal of extraneous information from an analysis.

Table 2: Comparison of feature extraction methods used in previous research

Author	Year	Feature Extraction Techniques
Lin et al.	2017	This study categorized positive and negative emotion words using LIWC2007 [22].
Thelwall	2017	All of the employed features were identified as unigrams, bigrams, and trigrams [23].
Kandias et al.	2017	Selection of features based on the TF-IDF values and the occurrence frequencies of terms [24].
Tai et al.	2015	This study used a LIWC lexicon and the unigram word feature to identify PTSD sufferers [25].
Huang et al.	2014	A vocabulary of emotional words was sorted into positive and negative categories using N-gram characteristics (unigram, bigram, and trigram), adopting tags for adjective, noun, and verb forms of speech [26].

5. Challenges

OSNs provide several challenges, the most significant of which are difficulties in human-computer interaction and non-face-to-face communication, both of which contribute to the increased difficulty of recognizing mental health issues. One of the most significant challenges is the language barrier, which regularly rears its head when attempting to grasp the complexities of mental health topics buried in the diverse writing styles of OSNs. Many approaches can be used to solve this issue. Machine learning can be used to assist in determining whether or not a mental health condition is present by analyzing the context of the words and languages used in (OSNs). Additionally, most OSN service providers implement an account privacy policy that makes it difficult for researchers to get data from OSNs. During the data processing stage, researchers face challenges relating to privacy and security regulations due to data acquisition from publicly available users, such as those from Twitter. Throughout this analysis, we came up with many issues that need to be fixed down the road.

6. Conclusion

This research aims to discuss the advantages, disadvantages, and prospects for the future development of methodologies used to diagnose mental health problems. The objective of this review is to give a critical analysis of the research that has been published on the topic of the application of OSNs in the diagnosis of mental health issues. Machine learning analyses often focus on the Source of data, extraction of features approach, and classifier effectiveness of the system in question. Initial identification activities are expected to be beneficial in lowering the total quantity of individuals reporting having psychological well-being disorders person whose circumstances are expected to worsen in the absence of future treatment. To boost the future reliability and efficiency of diagnosing psychological well-being problems, this study requires broad adoption, novel algorithmic techniques, and computational linguistics.

References

[1] Acharya, N., and Joshi, S. (2009). Influence of Parents' education on achievement motivation of Adolescents. Indian Journal Social Science Researches, 6(1), 72–79.

[2] Ahadi, B., and Basharpoor, S. (2010). Relationship between Sensory Processing Sensitivity, Personality Dimensions and Mental Health. Journal of Applied Sciences, 10(7), 570–574.

[3] Aishwarya, R. L., and Arora, M. (2006). Perceived parental behaviour as related to students' academic school success and competence. Journal of the Indian Academy of Applied Psychology, 32(1), 47–53.

[4] Arili, C., and Ratna Prabha, C. (2004). Influence of family environment on emotional competence of adolescents. Journal of Community Guidance and Research, 21(2), 213–222.

[5] Arranz, E. B., Oliva, A., De Miguel, M. S., Olabarrieta, F., and Richards, M. (2010). Quality of family context and cognitive development: A cross sectional and longitudinal study. Journal of Family Studies, 16(2), 130–142.

[6] Bandy, R., and Ottoni-Wilhelm, M. (2012). Family structure and income during the stages of childhood and subsequent pro social behavior in young adulthood. Journal of Adolescence, 35(4), 1023–1034.

[7] Benedict, R. (1950). Continuities and discontinuities in cultural conditioning. In W. E. Martin & C. B. Stendler (Eds.), Readings in Child Development. New York: Harcourt, Brace.

[8] Benjamin, P. C. (2006). Personality and perceived health in older adults. Journal of Gerontology, 61, 362–365.

[9] Chawla, A. N. (2012). The relationship between family environment and academic achievement. Indian Streams Research Journal, 1(12), 1–4.

[10] Marsiglia, C. S., Walczyk, J. J., Buboltz, W. C., and Griffith-Ross, D. A. (2007). Impact of parenting styles and locus of control on emerging adults' psychosocial success. Journal of Education and Human Development. 1(1), ISSN 1934-7200.

[11] Challenges for child and adolescent mental health service development in sub-Saharan Africa. Journal of Child & Adolescent Mental Health, 16(2), iii–iv. https://doi.org/10.2989/17280580409486570

[12] Sutradhar, A. (2019). Impact of emotional maturity on self-esteem of adolescents: Research study. In Book.

[13] Chu, T., Liu, X., Takayanagi, S., Matsushita, T., and Kishimoto, H. (2023). Association between mental health and academic performance among university undergraduates: The interacting role of lifestyle behaviors. International Journal of Methods in Psychiatric Research. 32(1), e1938. doi: 10.1002/mpr.1938. Epub 2022 Sep 10. PMID: 36087035; PMCID: PMC9976597.

[14] Kieling, C., Baker-Henningham, H., Belfer, M., Conti, G., Ertem, I., Omigbodun, O., … Rahman, A. (2011). Child and adolescent mental health worldwide: Evidence for action. The Lancet, 378(9801), 1515–1525. doi:10.1016/s0140-6736(11)60827-1

[15] Davila, J., Capaldi, D. M., and La Greca, A. M. (2016). Adolescent/young adult romantic relationships and psychopathology. Journal of Developmental Psychopathology, 1(14), 1–34.

[16] Davis, A. (1944). Socialization and adolescent personality. In Adolescence, Yearbook of the National Society for the Study of Education, 1944(43), Part I.

[17] Mohan, Thirumalai, C., and Srivastava, G. (2019). Effective heart disease prediction using hybrid machine learning techniques. In

IEEE Access, 7, 81542–81554. doi: 10.1109/ACCESS.2019.2923707

[18] Thelwall, M. (2016). TensiStrength: Stress and relaxation magnitude detection for social media texts. In Information Processing & Management. https://doi.org/10.1016/j.ipm.2016.06.009

[19] Kazemian, H. B., and Ahmed, S. (2015). Comparisons of machine learning techniques for detecting malicious webpages. Expert Systems with Applications, 42(3), 1166–1177. doi:10.1016/j.eswa.2014.08.046

[20] Xue, J., Deng, Z., Huang, P., Huang, K., Benton, M. J., Cui, Y., … Hao, S. (2016). Belowground rhizomes in paleosols: The hidden half of an Early Devonian vascular plant. Proceedings of the National Academy of Sciences, 113(34), 9451–9456. doi:10.1073/pnas.1605051113

[21] Vadillo, E., Dorantes-Acosta, E., Pelayo, R., and Schnoor, M. (2018). T cell acute lymphoblastic leukemia (T-ALL): New insights into the cellular origins and infiltration mechanisms common and unique among hematologic malignancies. Blood Reviews, 32(1), 36–51. doi: 10.1016/j.blre.2017.08.006. Epub 2017 Aug 15. PMID: 28830639.

[22] Lin, M.-H. (2017). A study of the effects of digital learning on learning motivation and learning outcome. EURASIA Journal of Mathematics, Science and Technology Education, 13. doi:10.12973/eurasia.2017.00744a

[23] Thelwall, M. (2017). TensiStrength: Stress and relaxation magnitude detection for social media texts. Information Processing & Management, 53(1), 106–121. doi:10.1016/j.ipm.2016.06.009

[24] Kandias, M., Gritzalis, D., Stavrou, V., and Nikoloulis, K. (2017). Stress level detection via OSN usage pattern and chronicity analysis: An OSINT threat intelligence module. Computers & Security, 69, 3–17. doi:10.1016/j.cose.2016.12.003

[25] Tai, V., Leung, W., Grey, A., Reid, I. R., and Bolland, M. J. (2015). Calcium intake and bone mineral density: Systematic review and meta-analysis. BMJ. 351, h4183. doi:10.1136/bmj.h4183. PMID: 26420598; PMCID: PMC4784773.

[26] Huang, C., Zheng, X., Tait, A., Dai, Y., Yang, C., Chen, Z., … Wang, Z. (2014). On using smoothing spline and residual correction to fuse rain gauge observations and remote sensing data. Journal of Hydrology, 508, 410–417. doi:10.1016/j.jhydrol.2013.11.022

A Categorical Review on Big Data Analytics

Pawan Bhaker, Sophia Sheikh, Ajay Nain, Sheikh Umar Mushtaq

School of Computer Applications, Lovely Professional University, Phagwara, India
E-mail: pawanbhaker88@gmail.com, sophiyasheikh@gmail.com, mr.ajaynain@gmail.com, Sheikhumar12@gmail.com

Abstract

The widespread adoption of IoT (Internet of Things) devices is leading to an unparalleled flow of data, it is important to utilise this huge amount of information effectively. It is not just about gathering, but rather about deriving useful information from this enormous data collection and ensuring its effective organisation. In this analysis, we explore the role of various technologies like the IoT, cloud-based technologies, machine learning, and the Hadoop Distributed File System (HDFS) in big data analytics. Through the exploration of the productive interactions among these components, this study aims to illustrate the vital role they fulfill in converting unprocessed information into practical insights. In a society submerged in knowledge, understanding the collaboration between these elements is crucial to unleashing the full capabilities of large-scale data analysis. Finally, it presents the statistical analysis of the technologies like Big data analytics, IoT, Cloud Computing, Machine Learning, and Integration Impact.

Keywords: Big data analytics, IoT, cloud computing, machine learning

1. Introduction

The combination of big data analytics, the IoT, cloud computing, and machine learning is an important point in transforming our relationship with data in today's quickly expanding technological world. This integration illustrates an innovative collaboration in which the increase of IoT devices results in an unexpected spike in data collection. These interconnected and continuously communicating devices produce a lot of information in large volumes and in different formats. The big challenge is to handle this vast data. This is where cloud computing comes into play, providing an infrastructure to store, process, and manage this massive amount of data. It's like having a virtual capability of storing and processing this vast amount of data created by IoT devices. This cloud-based architecture provides not only storage but also the computational capacity required for real-time analysis and access from anywhere on the planet. Big data analytics lies with these interconnected devices and digs out the hidden patterns from this massive store of data sets. Big data analytics turns raw data into usable insight, allowing businesses and sectors to make better decisions, forecast future trends, and optimise operations. However, the actual strength of this integration is seen in the field of machine learning. This is a part of artificial intelligence which learns first from existing data and then applies it over new data sets. It's something like that assistant trains you

from time to time for better performance. Machine learning algorithms use information obtained from large data analytics to perform tasks, make predictions, and improve decision-making processes. The impact of big data analytics is seen in a lot of businesses which help them in making decisions. IoT devices and sensors in healthcare collect patient data and send it to cloud-based services. This plenty of information is analyzed using big data analytics and machine learning algorithms, allowing for individualised therapies and predictive healthcare treatments. Similarly in agriculture, IoT sensors monitor soil conditions and crop health and with the help of big data analytics, we collect information, and then based on information we take measures to increase yield and productivity. However, some concerns lie in the middle of these tremendous developments. The massive amounts of data gathered create questions about privacy, security, and appropriate data use. Data ownership, consent, and guaranteeing protection against unwanted access or exploitation are all important considerations. Finally, the combined power of big data analytics, IoT, cloud computing, and machine learning help organisations and businesses to improve their performance and productivity. This is not only a technical change but a fundamental change that how we interpret the data. In the future information plays an important role in innovation and progress. In this changing digital environment where we engage with these interconnected

Chapter 53 DOI: 10.1201/9781003570349

environments need to focus on ethical use of data. This paper is organised into four sections. In first section, we provide a brief introduction about the different technologies. In second section, we presented some related research work focusing on these technologies. In third section, we presented an analytical review of the existing technologies followed by the conclusion in the last section.

2. Related Work

In the related work, we have reviewed papers that have directly used or described techniques like IoT, cloud computing, and machine learning. This field has evolved and grown significantly. Gaining an understanding of the historical background and contemporary environment is crucial to appreciating the subtleties and complexity present in this paper. The section presents the current state of the literature related to Big Data integrated with some other related aspects such as, Cloud computing, Internet of Things, and Machine Learning.

2.1. Big Data Analytics with Cloud Computing

In the paper [1] the author proposed an SDP approach in different fog computing applications and checked their performance based on processing time and use of resources for the process. In the paper [2] the author proposed a model for a supply chain that shows how these small and medium enterprises adopt new technology based on cloud computing for better performance. The discussed model was based on DOI theories and the TOE framework. In the paper [3] author talks about how cloud computing is useful for managing massive data. The authors' research emphasises the critical importance of big data predictive analytics, highlighting its significance and applications. It also examines important technologies and frameworks for efficient large data management in cloud systems, as well as existing and future cloud computing concerns). In the paper [4] author gives the solution for VM migration allocation from one hardware to another. The author also describes a strategy for the allocation of the resource for private data cloud. In the paper [5] author discusses the concept of cloud computing that meets the requirement of multi-level-real-time for better performance and monitoring of the power system. The author also describes the BDA models like Hadoop, spark, and Storm for batter analysis with the help of cloud computing. The author describes a new design that helps in managing real-time data for power management which provides solutions for data mining challenges. In the paper [6] author provides an overview of big data analytics challenges and illustrates how cloud computing helps in big data analytics. The author also shows how AI is helpful in big data analysis. In the paper [7] the author proposed an approach for data collection and processing. The author provides a web application architecture using ReactJS, Node JS, and Python which gives analysis based on pathologies and distance for transporting a patient.

2.2. 2.2. Big Data Analytics with the IoTs

The author explores the potential of BDA technologies and methodologies in enhancing the efficiency of IIoT (industrial Internet of Things) systems [8]. The author creates a system of classification by organizing and grouping the literature according to various aspects such as sources of data, tools for analysis, methodologies, and several more. In this author took a case study of different enterprises that have benefited from BDA. The author also introduces opportunities in IIoT with the help of BDA. This paper [9] provides a comprehensive analysis of the most recent developments in big data analytics for IoT systems, while also examining the strategies employed to effectively handle and analyze large volumes of data within the IoT ecosystem. In this article, the author explores the impact of big data analytics on Internet of Things (IoT) applications and looks into challenges that serve as a roadmap for future research [10]. In this paper, the author shows the uses of the BDA technique and IoT devices in a water management system for a better supply of water in a smart city. The author uses the SCADA (Supervisory controller and data acquirement) approach for a sustainable water management system with the help of IoT sensors and BDA tools and techniques. The findings from the experiment also indicate that the utilisation of IoT devices when combined with the BDA method, we can manage a water supply chain in a better way. In the paper [11], the author worked on a smart health monitoring system. The author gives a healthcare architecture that analyses energy harvesting from a health monitoring sensor with the help of BDA. There are two main components to the provided architectural design. Firstly, it encompasses an overall conceptual structure that focuses on harnessing energy for the purpose of monitoring health sensors. Secondly, it incorporates a system for processing data and managing decisions in the healthcare field. Also, the architecture has three layers. There are three main components to consider: energy harvesting and data generation, data pre-processing, and data processing and application. The proposed work highlights the importance of IoT sensors in the healthcare sector and gives solutions for smart health monitoring and planning. The author uses compatible dataset on Hadoop server to verify the architecture based on threshold limit values (TLVs). In the paper [12] author describes an architecture based on IoT devices which is useful in making next-generation super city using big data. The author proposed a four-layer architecture of data generation and collection, data administration and processing etc. This complete system is made up using Hadoop ecosystem and MapReduce. In the paper [13] author discuss that how IoT with the help of BDA is useful in supply chain and logistic. The author examines a lot of literature and describes how companies incorporate real-time data of different physical objects helps them in decision-making. In the paper [14] author shows that how IoT devices simplify the human life and with the help of big data analytics convert raw data in semantic data for taking better decision

2.3. Big Data Analytics with Machine Learning

The author [15] presents a digital customer algorithm that utilises machine learning algorithms to analyze unstructured big data in IoT devices. This algorithm is then implemented within a separate big data analysis framework. The author first collects online data from customers and then applies background data mining on it and on the other hand verifies its efficiency with the help of machine learning algorithms like k- nearest neighbor algorithm. In the paper [16] author focuses on the implementation of ML in BDA for analyzing large data sets for extracting valuable information. The author explores the potential of extracting valuable insights from data and emphasises the importance of predictive analysis and knowledge extraction. The author shows the impact of big data on real time data analysis, also discusses the impact of ML in big data analytics for large and complex data sets. In the paper [17] author shows how machine learning is useful for analysis in various applications. The author gives an overview of how machine learning is used fully in big data analytics and author highlights the challenges and application of machine learning using big data analytics. The author of [18] employs advanced big data analytics tools and technology to forecast consumer behavior on social media platforms. The author analysis consumer perception and attitude based on different parameters on social media. To detect noise, error, and duplicate record author apply different processing technique. The author used machine learning techniques to examine consumer behavior through mathematical modeling. The author employed a training and testing approach in the paper, utilizing 80% of the data for training purposes and allocating the remaining 20% for model testing. In the paper [19] author did sustainable finance research with the help of ML and ML help

in analysis these data for better decision. The author illustrates the seven important themes of sustainable finance research and gives future insight with respect to sustainable finance research. In the paper [20] author shows how big data analytics using ML provide better analysis of big data in smart city environment. The author proposed a self-building AI which reduces the limitation of traditional AI and helps in data processing in smart city environment.

3. Observation

Individually each technology has a great impact like BDA and machine learning provide extraction mechanisms for intelligent decision-making, IoT provides connectivity and real-time data, and cloud computing provides virtual infrastructure for storing data. But the aggerate impact of the technologies is much more than the individual contribution. When these technologies integrated with big data analytics it enhanced the performance of many sectors. The combined power of these technologies' fuels innovation, data processing, etc. In sectors like health care, manufacturing, and smart cities these technologies help in making better decisions. The work done on big data analytics with the help of IoT, cloud computing, and machine learning is illustrated in Figure 1 and the comparison of all technology in different aspects are discussed in Table 1.

4. Conclusion

In this paper, we discussed the combined power of BDA, IoT, cloud computing, and machine learning. This paper also discussed key takeaways from some articles that describe how IoT, cloud computing, and machine learning help to enhance BDA techniques. On the basis of articles discussed, we find

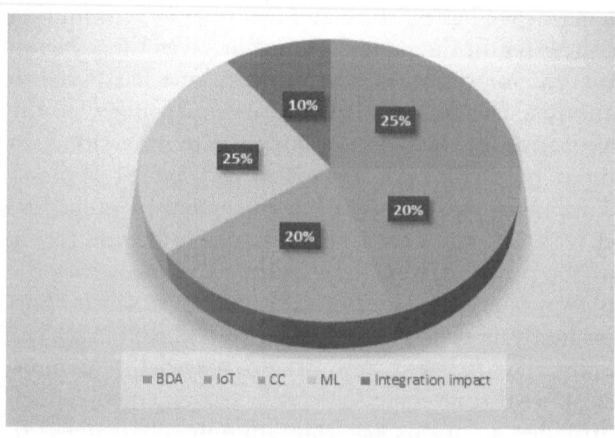

Figure 1: Comparison of technologies

Table 1: Aspects comparison of all technologies

Aspect	Big Data Analytics	IoT	CC	ML
Data Processing	Analyzes large, diverse datasets for insights and patterns	Collects data from interconnected devices, generating continuous streams of information	Provides scalable storage and processing capabilities for vast amounts of data	Utilises algorithms to learn from data, enabling predictions and autonomous decision-making
Application	Decision support, predictive analytics, pattern recognition	Smart homes, cities, healthcare monitoring systems	Data storage, computation, and services on demand	Predictive maintenance, personalised recommendations, autonomous systems
Impact	Enhances decision-making, optimisation, and innovation	Improves efficiency, convenience, and real-time monitoring	Facilitates cost-effective and flexible data management	Enables automation, efficiency gains, and adaptive systems
Challenges	Data privacy, security, and ethical use	Interoperability, security, and data privacy	Data sovereignty, compliance, and downtime	Data bias, interpretability, and continuous learning

that a lot of work is done individually but there is a lot of scope in the integration of all these technologies. This paper focuses on the comparative analysis of these technologies and their integration. Hence, showed how they help in extracting useful information from big data for taking better decisions. It permits academics and researchers to explore the complexities of the issues at hand in more detail. We hope that this survey encourages more research into these issues as they demand more attention.

References

[1] Poojara, S. R., et al. (2022). Serverless data pipeline approaches for IoT data in fog and cloud computing. Future Generation Computer Systems, 130, 91–105.

[2] Amini, M. and Javid, N. J. (2023). A multi-perspective framework established on diffusion of innovation (DOI) theory and technology, organization and environment (TOE) framework toward supply chain management system based on cloud computing technology for small and medium enterprises. Organisation and environment (TOE) framework toward supply chain management system based on cloud computing technology for small and medium enterprises. International Journal of Information Technology and Innovation Adoption, 11, 1217–1234.

[3] Mohbey, K. K. and Kumar, S. (2022). The impact of big data in predictive analytics towards technological development in cloud computing. International Journal of Engineering Systems Modelling and Simulation, 13 (1), 61–75.

[4] Balachandran, B. M. and Prasad, S. (2017). Challenges and benefits of deploying big data analytics in the cloud for business intelligence. Procedia Computer Science, 112, 1112–1122.

[5] Al-Jumaili, A. H. A., et al., (2023). Big data analytics using cloud computing based frameworks for power management systems: Status, constraints, and future recommendations. Sensors, 23 (6), 2952.

[6] Pothukuchi, A. S., Kota, L. V., and Mallikarjunaradhya, V. (2021). A Critical Analysis of the Challenges and Opportunities to Optimize Storage Costs for Big Data in the Cloud, Volume 3 Issue 1, ASIAN JOURNAL OF MULTIDISCIPLINARY RESEARCH & REVIEW, 132-144.

[7] Thai, H.-D. and Huh, J.-H. (2022). Optimizing patient transportation by applying cloud computing and big data analysis. The Journal of Supercomputing, 78 (16), 18061–18090.

[8] ur Rehman, M. H., et al. (2019). The role of big data analytics in industrial Internet of Things. Future Generation Computer Systems, 99, 247–259.

[9] Ahmed, E., et al. (2017). The role of big data analytics in Internet of Things. Computer Networks, 129, 459–471.

[10] Nie, X., et al. (2020). Big data analytics and IoT in operation safety management in under water management. Computer Communications, 154, 188–196.

[11] Babar, M., et al. (2018). Energy-harvesting based on internet of things and big data analytics for smart health monitoring. Sustainable Computing: Informatics and Systems, 20, 155–164.

[12] Rathore, M. M., et al. (2020). IoT-based big data: From smart city towards next generation super city planning. Securing the Internet of Things: Concepts, Methodologies, Tools, and Applications. IGI Global, pp. 1409–1428.

[13] Koot, M., Mes, M. R. K., and Iacob, M. E. (2021). A systematic literature review of supply chain decision making supported by the Internet of Things and Big Data Analytics. Computers & Industrial Engineering, 154, 107076.

[14] Fadi, A.-T. and Deebak, B. D. (2020). Seamless authentication: for IoT-big data technologies in smart industrial application systems. IEEE Transactions on Industrial Informatics, 17 (4), 2919–2927.

[15] Hou, R., et al. (2020). Unstructured big data analysis algorithm and simulation of Internet of Things based on machine learning. Neural Computing and Applications, 32, 5399–5407.

[16] Vishnu, V. K. and Rajput, D. S. (2020). A review on the significance of machine learning for data analysis in big data. Jordanian Journal of Computers and Information Technology, 6 (1).

[17] Rahul, K., et al. (2021). Machine learning algorithms for big data analytics. Computational Methods and Data Engineering: Proceedings of ICMDE 2020, Volume 1. Singapore: Springer.

[18] Chaudhary, K., et al. (2021). Machine learning-based mathematical modelling for prediction of social media consumer behavior using big data analytics. Journal of Big Data, 8 (1), 1–20.

[19] Kumar, S., Sharma, D., Rao, S. et al. Past, present, and future of sustainable finance: insights from big data analytics through machine learning of scholarly research. Ann Oper Res (2022), pp.1-44. (Online) https://doi.org/10.1007/s10479-021-04410-8

[20] Alahakoon, D., Nawaratne, R., Xu, Y. et al. Self-Building Artificial Intelligence and Machine Learning to Empower Big Data Analytics in Smart Cities. Inf Syst Front 25, 221–240 (2023). https://doi.org/10.1007/s10796-020-10056-x

Fraud Detection Using Machine Learning Techniques

A Review

[1]Diksha Sharma, [2]Manmohan Sharma and [3]Robin Prakash Mathur

[1,2]School of Computer Application, Lovely Professional University, Phagwara, Punjab, INDIA
[3]School of Computer Science and Engineering, Lovely Professional University, Phagwara, Punjab, INDIA
E-mail: dikshasharma.mca@gmail.com, manmohan.21909@lpu.co.in and robin.14597@lpu.co.in

Abstract

In the era of machine learning and digitalisation, frauds in online financial transactions become the most prevalent cybercrimes, and is most likely related to the incorporation of digital currency into our daily lives. Financial fraud is one of the most common types of cybercrime that affect people all over the globe especially with increase in use of online financial transactions. A major threat to the security in online financial transactions urges for a fundamental solution. Fraud prevention and fraud detection mechanism can be used to minimise fraudulent activities to a great extend. An efficient fraud detection algorithm using machine learning techniques can be implemented to assist fraud investigators to reduce these losses. But to design a fraud detection algorithm is also a challenging task as the data is non-stationary, distributed and continuous streaming transaction. This paper provides an overview of some areas that are highly affected with financial frauds and discuss some techniques of machine learning that are employed to identify fraudulent transactions

Keywords: Financial fraud, online transactions, detection and prevention, Machine learning techniques

1. Introduction

In order to boost the productivity or efficiency in selling goods and services, most of the business companies, organisations, and government agencies are using electronic commerce. Both fraudsters and legitimate users use electronic system, hence become vulnerable to fraud. Financial fraud is an illegitimate activity performed by the fraudsters in order to attain money, goods, services and sometimes fame. Financial frauds are considerably rising due to current technological development; hence fraud detection is a vital field to work upon. To prevent the banks suffering from financial losses, fraud detection is very important.

To combat financial fraud, individuals and organisations must take proactive measures to protect themselves. This may include using strong passwords, regularly monitoring bank accounts and credit reports, and being cautious when sharing personal information online. In addition, governments and law enforcement agencies play a crucial role in investigating and prosecuting financial fraud cases. By working together, we can help prevent financial fraud and protect our financial systems from harm. Some of the common areas affected by Financial Fraud are shown in Figure 1.

1. **Credit Card**: Credit cards play a very important role in e-commerce. With the growing usage of credit card transactions, credit card frauds are also becoming common and fraudsters seek new opportunities to commit fraud, which results in huge loss to the cardholders. Credit card frauds are further of two types depending upon their act of commitment:

    a. Online Credit Card Frauds
    b. Offline Credit Card Frauds

2. **Money Laundering**: It is the act of hiding the sources of money earned by illegal means as black money and convert it into clean money or white money.

3. **Insurance Fraud**: Every industry where people choose to take insurance, is susceptible to insurance fraud. From every other insurance, healthcare insurance and life insurance become most common nowadays, due to modern life style.

4. **Telecommunication Fraud**: Any activity used to abuse or gain benefits over telecommunication companies using deception is categorised under telecommunication frauds, also termed as "telecom Frauds". Gosset and Hyland [1] grouped telecommunication fraud into four

Chapter 54 DOI: 10.1201/9781003570349

Figure 1 Most Common Areas of Fraud

Figure 1: Most common areas of fraud

categories: contractual, procedural, hacking and technical.

5. **Internet Marketing Fraud:** These types of frauds are committed by online auction and web advertising. In online auction system people sell or bid for the products and services online via internet on websites such as eBay, eBid and many more.

2. Literature Review

Getting trap into any kind of financial fraud while transferring money can be avoided by taking preventive measures or if get trapped fraud detection techniques can be used to detect fraud. By fraud prevention we mean we can restrict ourselves from being a victim of financial fraud by using incognito window while making any payment and also by not sharing bank details and OTP with others. In fraud detection, some methods and techniques can be used to overcome from fraud. Abdallah et al. [2] discussed about fraud protection using fraud prevention system and fraud detection system, also highlight the fraud detection techniques and areas affected by fraudulent activities. They bring light to the issues and challenges in fraud detection. Dal Pozzolo et al. [3] focused on unbalancedness, non-stationarity and assessment issues of fraud detection. As the solution to the problem unbalanced, rather of correctly classifying the transactions, they proposed ranking them based on the probability of fraud. They proposed building a new model every time a new chunk becomes available to address the issue of non-stationary data streams, which produced better results than changing the model less frequently. Banarescu [4] suggested that the process of data analytics i.e., data mining can be used to explore the data, extract information, and discover pattern and relationships in datasets. Patil et al. [5] discussed a big data analytics framework to handle massive amounts of data and identify fraud in real time by putting several machine learning algorithms into practice while providing consumers with a high level of satisfaction and low risk. They used three distinct models to increase the prediction accuracy of fraud: Logistic Regression

Analytical Model, Decision Tree Analytic Model and Random Forest Decision Tree Analytical model. Accuracy of the analytical models is evaluated using confusion matrix. Al Smadi et al. [6] discussed several methods for detecting and preventing credit card fraud in order to improve the security of credit card systems. They also compared these techniques and outline potential strengths and weaknesses to choose right techniques. The techniques are compared in terms of accuracy, time, and cost. Malini et al. [7] discussed different fraud detection techniques i.e., Logistic Regression, Hidden MarKov Model, SVM, KNN, Decision Tree with their advantages and disadvantages and found that KNN method is suitable for anomaly detection of the targeted instance with limitation of memory and gives accurate and efficient experimental results as compare to other methods. Maurya and Kumar [8] implemented machine learning in conjunction with the block chain technology to improve the fraud detection model's precision and efficiency. Block chain techniques are used to work with real time data and to identify fraudulent transactions, machine learning is applied. SMOTE technique is used to generate synthetic data samples and ML algorithms are applied to check the accuracy as well as precision. They found that XGBoost algorithm gives highest accuracy in comparison with Logistic Regression, Random Forest and Decision Tree. Aditi et al. [9] used Random Forest ML algorithm feasible in python to train the model. They discuss Logistic Regression, Random Forest and Decision Tree ML models with their advantages and disadvantages. They also compare the accuracy, precision, F1 score and recall of 3 models and found that Logistic Regression provides better accuracy of 95.55% as compare to other algorithms.

2.1. Machine Learning Techniques for Fraud Detection: Overview

It is very important to take proper care of security while doing any financial transaction to avoid any fraudulent activity i.e., fraud can be prevented by taking proper measures like using encryption/decryption techniques, firewall or incognito window while performing any transaction via internet. Even after taking proper measures fraudsters succeed in performing fraud, to overcome this some machine learning techniques such as Supervised Learning, Unsupervised Learning and Reinforcement Learning techniques can be implemented to design a system or model to detect fraud. Fraud prevention and fraud detection techniques are shown in Figure 2.

Comparison of some fraud detection techniques are shown in Table 1.

Deviation of transactions form the regular and expected behaviour of the customer leads to the

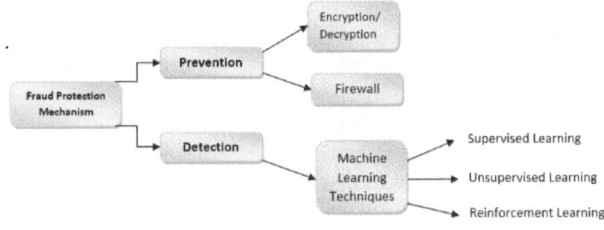

Figure 2: Fraud protection mechanism

anomalous transactions. The change in the customer's transaction behaviour may be a result of a sudden surge in online transactions within a specific time frame. As result, there is need to develop an online system for detecting financial anomalies. The purpose of this system is to keep a close eye on any sudden changes that may occur in the financial transactions. The system will be designed to identify any unusual or unexpected patterns or behaviours in

Table 1: Earlier contribution of researches in fraud detection

Author and Year	Title of Paper	Contribution	Techniques Used	Limitations	Performance Assessment Matrices
Jayasingh et al. [10]	Online Transaction Anomaly Detection Model for Credit Card Usage Using Machine Learning Classifiers	Transaction Anomaly Detection (TAD) model	RF, LR, DT, Extreme Gradient Boosting, KNN, SVM	TAD can perform better with other machine learning models	Accuracy, Precision, Recall and F1-score.
Aggarwal et al. [11]	Fraud detection in online payment transaction using machine learning algorithms	Comparison of different ML algorithms on different parameters.	LR, KNN, SVM, DT, Naïve Bayes, RF	Deep learning approach should be used to get the better results.	Accuracy, Precision and F1-score, Recall score and Log Loss.
Wang el al. [12]	LAW: Learning Automatic Windows for Online Payment Fraud Detection	Learning Automatic Window (LAW)	XGBoost, RF, LR, Naïve Bayes.	Limited size of the sliding window is used.	TPR, FPT, Precision, F1 score, ROC
Liu et al. [13]	Financial Fraud Detection Model: Based on Random Forest	Parametric and non Parametric models are used.	LR, KNN, DT SVM	Some other parametric models can be used to acquire the better performance of parametric models.	Fraud and Non-Fraud percentage of the models
Ashfaq et al. [19]	A Machine Learning and Blockchain Based Efficient Fraud Detection Mechanism	ML model is linked with Blockchain	XGBoost, Random Forest	It is vulnerable to adversarial attack	Accuracy, Precision, AUC
Akila et al. [14]	Cost-sensitive Risk Induced Bayesian Inference Bagging (RIBIB) for credit card fraud detection	Risk Induced Bayesian Inference Bagging (RIBIB)	AIRS, AFDM, CSNN	Refining the RIBIB model to incorporate temporal elements.	FPR, TPR, FNR, TNR, Recall, Accuracy, AUC, BCR, Detection Rate
Liu [15]	FA-GNN: Filter and Augment Graph Neural Networks for Account Classification in Ethereum	Filter and Augmented Graph Neural Network (FA-GNN)	GCN, SGN, GraphSAGE, ClusterGCN, H_2GCN, FA-GNN	Develop an analysis system for online transaction by considering continuous time dynamic graph scenario.	Precision, Recall, Micro-F1, Macro-F1

(Continued)

Table 1: (*Continued*)

Cheng et al. [16]	Graph Neural Network for Fraud Detection via Spatial-Temporal Attention	Spatial-temporal attention-based graph network (STAGN)	LR, GBDT, MLP, deep and wide, CNN- max, AdaBM, LSTM-seq	Create a fraud detection system that operates in real-time within the e-commerce process by examining combined online consumer behaviour.	AUC (Oct- Dec)
Jing, R., Tian, H., et al. [17]	A GNN-based Few-shot learning model on the Credit Card Fraud detection	Comparison of proposed GNN with 2-way n-shot classifier	Logistic Regression, SVM, XGBoost, GraphSage	Exploration of extremely imbalanced label	Accuracy, Precision, Recall
Jing, R., Zheng, X., et al. [18]	A Graph-based Semi-supervised Fraud Detection Framework	Transalation of structured data to graph format to improve the effect of label propagation of graph	LR, SVM, XGBoost	Processing of structured data using rapid developed graph data structured.	Precision, Recall, Accuracy, F1-Score

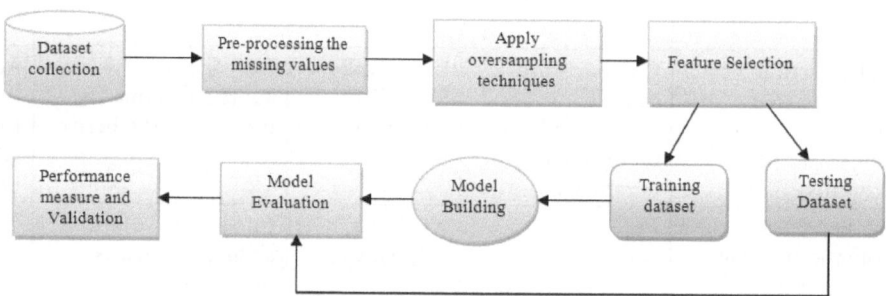

Figure 3: Flow diagram: ML based fraud detection system

financial data, such as sudden spikes or dips in transaction amounts, unusual frequency of transactions, or any other suspicious activity that may indicate fraudulent or criminal behaviour. The system will be an important tool for financial institutions and businesses to help prevent fraud and protect their assets. Sometimes, sudden change in the financial transactions may be due to festival season, holidays, excessive shopping, travelling and tours. Flow diagram of fraud detection system based on ML is shown in Figure 3.

3. Conclusion

The main focus of this paper is to provide a comprehensive overview of the various anomaly detection techniques that are utilised in the context of online financial transactions. Given the increasing prevalence of fraudulent activities in the digital world,

banks and other financial institutions are constantly looking for ways to prevent fraudsters from taking advantage of unsuspecting customers. One of the most effective ways to accomplish this is by implementing robust detection mechanisms that can quickly identify anomalous transactions and flag them for further investigation. By analyzing patterns and trends in customer behaviour, these systems are able to detect unusual activity and alert the appropriate authorities in a timely manner. Ultimately, the goal is to minimise the risk of financial loss and protect customers from falling victim to fraudulent schemes.

References

[1] Gosset, P. and Hyland, M. (1999). Classification, detection and prosecution of fraud in mobile networks. Proceedings of ACTS Mobile Summit (1), 2–4.

[2] Abdallah, A., Maarof, M. A., and Zainal, A. (2016). Fraud detection system: A survey. Journal of Network and Computer Applications, 68, 90–113.

[3] Dal Pozzolo, A., Caelen, O., Le Borgne, Y.-A., Waterschoot, S., and Bontempi, G. (2014). Learned lessons in credit card fraud detection from a practitioner perspective. Expert Systems with Applications, 41 (10), 4915–4928.

[4] Bănărescu, A. (2015). Detecting and preventing fraud with data analytics. Procedia Economics and Finance, 32, 1827–1836.

[5] Patil, S., Nemade, V., and Soni, P. K. (2018). Predictive modelling for credit card fraud detection using data analytics. Procedia Computer Science, 132, 385–395.

[6] Al Smadi, B. and Min, M. (2020). A critical review of credit card fraud detection techniques. In 2020 11th IEEE Annual Ubiquitous Computing, Electronics & Mobile Communication Conference (UEMCON) (pp. 0732–0736). IEEE.

[7] Malini, N., and Pushpa, M. (2017). Analysis on credit card fraud identification techniques based on KNN and outlier detection. In 2017 Third International Conference on Advances in Electrical, Electronics, Information, Communication and Bioinformatics (AEEICB) (pp. 255–258). IEEE.

[8] Maurya, A. and Kumar, A. (2022). Credit card fraud detection system using machine learning technique. In 2022 IEEE International Conference on Cybernetics and Computational Intelligence (CyberneticsCom) (pp. 500–504). IEEE.

[9] Aditi, A., Dubey, A, Mathur, A., and Garg, P. (2022). Credit card fraud detection using advanced machine learning techniques. In 2022 Fifth International Conference on Computational Intelligence and Communication Technologies (CCICT) (pp. 56–60). IEEE.

[10] Jayasingh, B. B. and Sri, G. B. (2023). Online transaction anomaly detection model for credit card usage using machine learning classifiers. In 2023 International Conference on Emerging Smart Computing and Informatics (ESCI) (pp. 1–5). IEEE. **https://doi. org**/10.1109/ESCI56872.2023.10100152.

[11] Aggarwal, R., Sarangi, P. K., and Sahoo, A. K. (2023). Credit card fraud detection: Analyzing the performance of four machine learning models. In 2023 International Conference on Disruptive Technologies (ICDT) (pp. 650–654). IEEE. **https:// doi.org**/10.1109/ICDT57929.2023.10150782.

[12] Wang, C., Wang, C., Zhu, H., and Cui, J. (2020). LAW: Learning automatic windows for online payment fraud detection. IEEE Transactions on Dependable and Secure Computing, 18 (5), 2122–2135. https:// doi.org/10.1109/TDSC.2020.3037784.

[13] Liu, C., Chan, Y., Alam Kazmi, S. H., and Fu, H. (2015). Financial fraud detection model: Based on random forest. International Journal of Economics and Finance, 7(7), pp. 178-188.

[14] Akila, S. and Reddy, U. S. (2018). Cost-sensitive Risk Induced Bayesian Inference Bagging (RIBIB) for credit card fraud detection. Journal of Computational Science, 27, 247–254. https://doi. org/10.1016/j.jocs.2018.06.009.

[15] Liu, J., Zheng, J., Wu, J., and Zheng, Z. (2022). FA-GNN: Filter and augment graph neural networks for account classification in ethereum. IEEE Transactions on Network Science and Engineering, 9 (4), 2579–2588. https://doi.org/10.1109/ TNSE.2022.3166655.

[16] Cheng, D., Wang, X., Zhang, Y., and Zhang, L. (2020). Graph neural network for fraud detection via spatial-temporal attention. IEEE Transactions on Knowledge and Data Engineering, 34 (8), 3800–3813. https://doi.org/10.1109/ TKDE.2020.3025588.

[17] Jing, R., Tian, H., Zhou, G., Zhang, X., Zheng, X., and Zeng, D. D. (2021). A GNN-based Few-shot learning model on the Credit Card Fraud detection. In 2021 IEEE 1st International Conference on Digital Twins and Parallel Intelligence (DTPI) (pp. 320–323). IEEE.

[18] Jing, R., Zheng, X., Tian, H., Zhang, X., Chen, W., Wu, D. D., and Zeng, D. D. (2019). A graph-based semi-supervised fraud detection framework. In 2019 4th IEEE International Conference on Cybernetics (Cybconf) (pp. 1–5). IEEE.

[19] Ashfaq, T., Khalid, R., Yahaya, A. S., Aslam, S., Azar, A. T., Alsafari, S., & Hameed, I. A. (2022). A machine learning and blockchain based efficient fraud detection mechanism. Sensors, 22(19), 7162.

Adaptive Underwater Image Enhancement Model Based on Improved TAO Algorithm

Kavita Saini[1], Amit Doegar[1], Brijendra Pal Singh[2]

[1]Department of Computer Science and Engineering, National Institute of Technical Teachers Training & Research, Chandigarh, India
[2]School of computer Science and Engineering, Lovely Professional University, Phagwara, India
E-mail: er.kavita89@gmail.com, amit@nitttrchd.ac.in, brijenderpal.nitttr@gmail.com

Abstract

Underwater images play an important role in the number of applications, namely, to monitor fishes, coral reef, underwater pipelines. Therefore, a clear vision underwater image is required. However, underwater image faces numerous challenges, such as, colour absorption, scattering, reflection, and hazing effect. To overcome this challenge, underwater enhancement models are gained popularity. In the paper, we have designed an adaptive underwater image enhancement model based on improved TAO algorithm. The improved TAO algorithm is hybrid of chaotic logistic map and TAO algorithm. The main role of improved TAO algorithm is to find the best parameter values of the power law and SVD method based on the objective function. The steps of the proposed method are to find the high-quality and low-quality planes of the underwater image. After that, high quality plane brightness is enhanced using the power law method. Next, remaining low-quality planes are fused with high-quality plane using the SVD method. The simulation results show that the proposed model outperforms over the existing models.

Keywords: Enhancement, gamma correction, metaheuristic, optimisation, TAO, underwater image

1. Introduction

Underwater images play an important role in different applications, such as marine ecological research, exploration of ocean resources, monitoring of deep-sea installation of optimal fibres, and for security purposes in military applications [1]. Therefore, a good-quality visual image is required. However, in the real-life scenario, the underwater images face different challenges such as turbidity, low contrast, noise, and colour absorption [2]. To overcome these challenges, image enhancement is used. In the literature, histogram equalisation, contrast stretching, gamma correction, and white balance are the conventional methods used for underwater image enhancement [3-5]. These algorithms enhance the different images in the same proportion without considering the image characteristics. To overcome this issue, in the literature, optimisation algorithms are utilised into underwater image enhancement models to enhance the image by analysing the internal characteristics of it [6]. To accomplish this goal, metaheuristic algorithms are utilised for it. The metaheuristic algorithm is a sub-field of optimisation algorithm. It is used for complex problems to find the best solution for

a given problem. In the literature, the optimisation algorithms are used for underwater image enhancement are GA [7], PSO [8-9], CS [10], Bat [11]. These algorithms have different principles to search the best solution from the solution space. Besides that, each algorithm has own pros and cons. Therefore, in this research, we have explored numerous metaheuristic algorithm and chosen the TAO algorithm for the proposed model [12]. The TAO algorithm has number of advantages, such as, minimum parameters required and simple process for search the solution space.

The main contribution of this research is to design an adaptive underwater image enhancement model. To accomplish this goal, gamma correction and fusion between planes is done using the power law and SVD method. However, finding the optimal parameter values of these methods before applying it on image enhance the quality of it. Therefore, improved TAO algorithm is utilised for it. In the improved TAO algorithm, chaotic logistic map and TAO algorithm is hybrid. The chaotic logistic map is used for generate the random population in the TAO algorithm because it generates a completely random population which helps to search the solution space effectively. Finally, evaluation is performed on different images and it's qualitative and quantitative analysis is performed. The result reflects that the quality of the enhanced image increased and it achieves high entropy and sobel count.

The remaining paper is organised into five sections. Section 2 shows the gamma correction, SVD, and metaheuristic TAO algorithm in the related work. Section 3 explains the proposed model is designed for enhance the underwater images. Section 4 shows the results. Finally, conclusion of this paper is defined in Section 5.

2. Related Work

2.1. Gamma Correction

The gamma correction is performed in the underwater images to enhance the brightness of the image. The underwater image faces numerous challenges due to colour absorption, scattering, refraction, and artificial lights. Thus, image brightness is negatively impacted. Thus, some of the underwater image need to increase brightness or some images need to reduce it. In the literature, power law method is used for gamma correction.

$$I_B = P \times \left(\frac{I}{P}\right)^\gamma \tag{1}$$

The gamma value varies from 0-4 [13]. The brightness of the image is increased when gamma value is reduced from 1-0 whereas brightness decreases when it values falls in between 1-4. Therefore, finding the optimal gamma value to enhance the brightness of the image is challenging task.

2.2. SVD Method

It is based on linear algebra technique. The SVD is determined using Eq. (2) [14].

$$P_l = P_l + \alpha P_h \tag{2}$$

In the Eq. (2), scaling factor plays an important role and determining its optimal parameter value determined how much image is fused in other image. In the proposed model, the low-quality planes are enhanced by fusing it with high quality plane.

2.3. TAO Algorithm

TAO stands for termite alate optimisation algorithm. It is based on the phototactic behaviour of termites [15]. In this optimisation, the termites are classified into bright and dark termites. Each group alates are attracted towards bright termite space and repelled by dark termite space. The dark termite space is basically where termite lost wings or preyed by birds.

This principal is studied by Arindam Majumder and proposed TAO algorithm. It has number of advantages like minimum operations to search the solution space, minimum no. of parameters is required for start the searching process, and explore the solution space by finding the best and worst solution. In the proposed model, optimal gamma of the power law method and scaling value of SVD algorithm is determined using the TAO algorithm. Besides that, in the TAO algorithm, initial population is randomly generated. Thus, enhancing the generation process of random population, the performance of TAO algorithm is utilised. Therefore, in the proposed model, chaotic logistic map is used for generate a completely random population to search the solution space.

3. Proposed Adaptive Underwater Enhancement Model

The proposed underwater enhancement model is adaptive. Here, adaptive term reflects, the proposed model enhances the image by study the internal characteristics of the image. For example, how much brightness of the image need to enhance/reduce or how much high-quality plane needs to fused in the low-quality plane for enhancement purposes. The flowchart of the proposed model is shown in Figure 1.

Initially, underwater colour image is read and its RGB planes are taken out. After that, high-quality and low-quality planes are determined by mean values of the plane. In the colour image, three planes are available. Therefore, highest mean value plane is classified as superior plane and remaining two low quality planes are classified as inferior and intermediate planes. After that, gamma correction of high-quality plane is done using the power law method because this plane is further used In the SVD method to enhance the low-quality planes. After

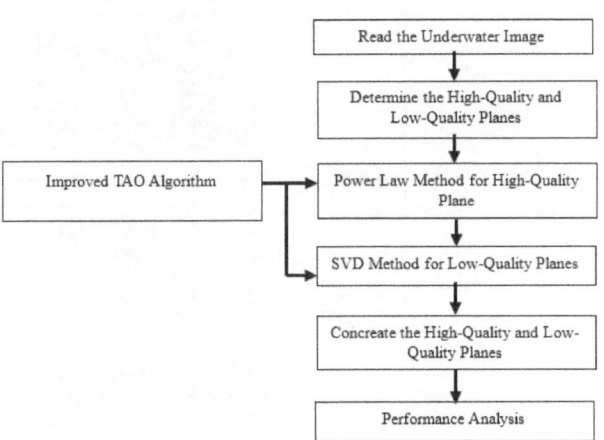

Figure 1: Flowchart of the proposed model

enhancing the high-quality plane, SVD method is utilised to enhance the low-quality planes based on the scaling value. The optimal gamma and scaling value is determined using the improved TAO algorithm. Finally, performance evaluation is performed using the qualitative and quantitative approaches. The improved TAO algorithm how finds the best gamma value is explained below and same process is used for find the best scaling value of the SVD method.

Initially, parameters of TAO algorithm are initialised. Further, random population is generated in the lower and upper limit of the gamma value. The random population is done using the chaotic logistic map algorithm. After that, each population is evaluated. Based on evaluation, best and worst population is determined. Next, new populations are generated based on the previous step and populations are classified into best and worst alates. Further, worst alates are substituted with best alates based on the absorption factor. Finally, which population is given the best solution is reflects the best alate. This best alate represents the optimal solution.

3.1. Results

This section shows the results of the proposed model is evaluated to validate it against the existing models. The simulation is performed in MATLAB. Table 1 shows the initial parameter value is defined for improved TAO algorithm to search the best values of power and SVD method.

Table 2 shows the qualitative analysis between input and output image by comparing the visual quality of it. The result shows that the visual quality of the enhanced image is improved significantly due to determining the best parameter values.

Table 3 shows the quantitative analysis. The result shows that the proposed model achieves high entropy and sobel count.

Finally, Table 4 shows the comparative analysis. The proposed model outperforms over existing models in terms of entropy.

Table 1: Parameter values of improved TAO algorithm

Parameter	P	I	x	r	β	pe
Value	50	30	[0-1]	[3.57-3.99]	[0-1]	0.01

Table 2: Qualitative analysis

Image	Original Image	Enhanced Image
Image1		
Image2		
Image3		

(Continued)

Table 2: (*Continued*)

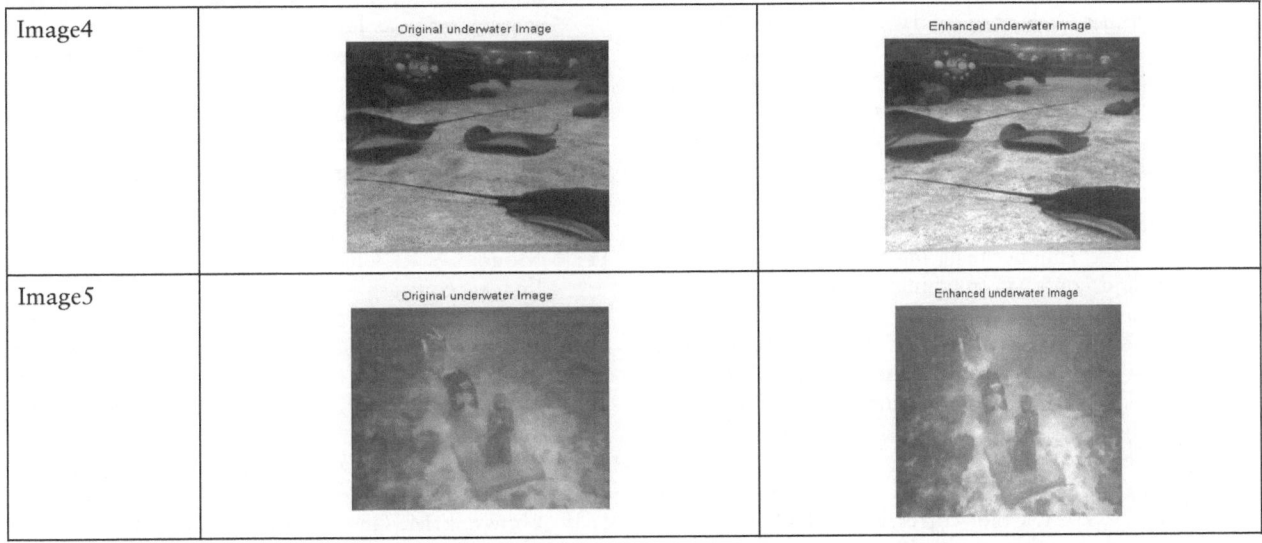

Table 3: Quantitative analysis

Images	Input Entropy	Output Entropy	Input Sobel Count	Output Sobel Count	MSE	PSNR	Execution Time (in Seconds)
Image1	6.5722	7.2854	5824	5959	0.0052	71.3428	8.291030
Image2	6.8441	7.3844	6261	6320	0.0089	69.6334	8.115823
Image3	6.2306	6.7002	9334	9550	0.0044	71.6963	9.537816
Image4	7.0192	7.4960	7385	7119	0.0035	72.6901	9.211359
Image5	6.6148	7.3224	6217	6257	0.0018	75.6536	8.056382

Table 4: Comparative analysis

Images	Method [8]	Method [9]	Proposed Model
Diver	6.6216	6.951	7.3680
Turtle	6.075	6.842	7.5952
Fishes	6.9833	7.111	7.4960
Coral Beach	6.2932	6.8476	7.6538

4. Conclusion

In this paper, we have designed an adaptive underwater image model. This model uses two methods, namely, power-law and SVD method. The power law method is utilised for gamma correction of the planes to enhance the brightness whereas SVD method is utilised for fuse the high-quality plane with low-quality planes. In both methods, two parameters, gamma value and scaling value plays an important role. Therefore, optimal parameter values of these methods are determined using the improved TAO algorithm. The chaotic map is used in the TAO algorithm for generate the random population. After determining the optimal values, the optimal values are used in these methods to enhance the quality of the image. Finally, the result shows that the visual quality of the images is enhanced and proposed model achieves better performance metrics value over the existing methods.

References

[1] Almutiry, O., Iqbal, K., Hussain, S., Mahmood, A., and Dhahri, H. (2021). Underwater images contrast enhancement and its challenges: A survey. Multimedia Tools and Applications, 83, 1-26.

[2] Mustafa, W. A., and Kader, M. M. M. A. (2018). A review of histogram equalization techniques in image enhancement application. In Journal of Physics: Conference Series (Vol. 1019, No. 1), IOP Publishing, p. 012026.

[3] Hitam, M. S., Awalludin, E. A., Yussof, W. N. J. H. W., and Bachok, Z. (2013). Mixture contrast limited adaptive histogram equalization for underwater image enhancement. In 2013 International Conference on

Computer Applications Technology (ICCAT). IEEE, pp. 1-5.

[4] Bhadouria, A. S., and Agarwal, K. (2020). Effective framework for underwater image enhancement using multi-fusion technique. In 2020 IEEE 9th International Conference on Communication Systems and Network Technologies (CSNT). IEEE, pp. 290-295.

[5] Lei, X., Wang, H., Shen, J. I. E., Chen, Z. H. E., and Zhang, W. (2021). A novel intelligent underwater image enhancement method via color correction and contrast stretching. Microprocessors and Microsystems, 107, 104040.

[6] Ghalib, R., and Alyasseri, Z. A. A. (2023). A recent review of underwater image enhancement techniques. In Doctoral Symposium on Computational Intelligence (pp. 519-538). Springer Nature Singapore.

[7] Zhang, W., Pan, X., Xie, X., Li, L., Wang, Z., and Han, C., 2021. Color correction and adaptive contrast enhancement for underwater image enhancement. Computers & Electrical Engineering, 91, 106981.

[8] Kumar, S. (n.d.). Underwater image enhancement using improved Particle Swarm Optimization. International Journal of Advance Research, [online] 7. https://www.ijariit.com/manuscripts/v7i3/V7I3-1170.pdf

[9] Azmi, K. Z. M., Ghani, A. S. A., Yusof, Z. M., and Ibrahim, Z. (2019). Natural-based underwater image color enhancement through fusion of swarm-intelligence algorithm. Applied Soft Computing, 85, 105810.

[10] Kuran, U., and Kuran, E. C. (2021). Parameter selection for CLAHE using multi-objective cuckoo search algorithm for image contrast enhancement. Intelligent Systems with Applications, 12, 200051.

[11] Mondal, S. K., Chatterjee, A., and Tudu, B. (2022). Image contrast enhancement by optimization of color channel difference using bat algorithm. Journal of Print and Media Technology Research, 11(4), 257-269.

[12] Majumder, A. (2023). Termite alate optimization algorithm: A swarm-based nature inspired algorithm for optimization problems. Evolutionary Intelligence, 16(3), 997-1017.

[13] Bhowmik, M., Ghoshal, D., and Bhowmik, S. (2015). An improved method for the enhancement of under ocean image. In 2015 International Conference on Communications and Signal Processing (ICCSP). IEEE, pp. 1739-1742.

[14] Chandra, D. S. (2002). Digital image watermarking using singular value decomposition. In The 2002 45th Midwest Symposium on Circuits and Systems, 2002. MWSCAS-2002 (Vol. 3). IEEE, p. III.

A Smart Project Allocation System for Enhanced Student-Faculty Collaboration

Satyam Kumar, Prabh Deep Singh

Department of Computer Science and Engineering, Graphic Era (Deemed To Be University), Dehradun, India
E-mail: kumar.satyam.2017004@gmail.com, ssingh.prabhdeep@gmail.com

Abstract

In institutions, there is a very common problem that many times faculty members have really good project opportunities for students but lack of communication technologies available for the specific purpose of project allocation to students. On the other hand, many students working on innovative projects might face difficulties in finding the right guide for their projects. It leaves them with a question ahead about how to connect with suitable faculty members. Against the same challenge, this paper proposes an innovative project allocation system. Our research focuses on developing an innovative and enhanced platform that matches students' interests and skillsets with suitable faculty members and their ongoing or proposed projects.

Keywords: Cloud computing, recommendation system, reliability, resource allocation, scalability, system efficiency

1. Introduction

In schools, colleges or other institutions, teachers or the professors do not have enough time to meet the students one by one and make them understand about the criteria to complete their projects and most importantly the research papers which is really a big impact on our society. Same in the case of the students when they newly came to the college, universities or change their schools they do not know the behaviour and the work culture of the teachers so that they can choose and start working under the faculty who they want.

This platform enhances to solve the problem in the broader way without taking multiple classes by teachers and explaining the students one by one, In case of this teachers can upload their details what the students needed to know and work with new ideas and innovations under them, such as their curriculum-vitae, resume, meeting times and availability for the students and explaining the way of doing the projects and research works under them, so that teachers can handle the big number of students and communication easily and can save their precious time even for the students as well. On the other side, students have to attain all the necessary requirements which meets to work with their mentors which he have to upload on the database so that other teacher or the mentors under whom the student will work can see the details.

By the fulfilment of the criteria of the student and selecting the mentors by the student, the fixed number of the students can finally contact to the teachers to do the project works under them. Once the project Selected by the student teacher can verify the Student whether the teacher is comfortable to assign the project to the student or not. If the project idea and the requirements of the teacher for the students is met, teacher can assign the projects and the student can start working under him. Teacher and students both have the authority to allocate and deallocate their projects in any case by giving the suitable reasons after deciding with their mentors.

Based on the various research repositories, it is being found as per this research gap, that no efficient project allocation app is developed. The major contribution of this paper is to provide an efficient project allocation app for the university. The motivation of the proposed system is that, the project allocation work has been done manually. Automated system can reduce the time.

2. Proposed Framework

Credentials: When a user (student or faculty) installs the application for the first time, it asks to register and input credentials [1]. Name, email, password and phone number are to be input during creating a new account. Email and phone are verified using OTP.

Chapter 56 DOI: 10.1201/9781003570349

Other Details: College information, course, branch, skills, experience are input from user and saved to database for using during searching and matching in the near future[2].

Uploading New Opportunity: New projects or research opportunities are uploaded to the database including some key skills.

Searching: New project uploaded go thorough searching algorithm, under which, students having the similar skills are marked. Details and skillset of students and faculty are stored in the cloud database [3]. New uploaded project is processed with NLP packages to lemmatize skillset required for the project. These keywords output by the NLP algorithms are treated as searching tags to match them with the skillset of students present in the database (Figure 1). After searching for good matches, this project is added to the homepage list of a student. Same goes when a student is looking for guide to work under and uploads some project. The algorithm picks tags from project and search for a relevant faculty guide [4].

Recommendation: Students with the matching skills to the project are notified about the new project opportunity through a list of available projects in his domain [5].

Application: Any student/scholar may apply for an interested opportunity through interface, by clicking apply and filling the necessary details.

Uploading Guidance Request: Any student working at some project may upload information about the project to find some project guide at the portal [6].

Guide Allocation: Some available mentors with the similar domain are suggested to the researcher/student. Request may be sent to one of the recommended guides.

Work Acceptance: Faculty gets notified about the project applications and guidance requests. Having the authority to accept or reject any collaboration request, faculty may accept or reject to collaborate with the aspirant [7].

Communication: Once the faculty accepts requests, a chat socket is opened up between the two parties. They may message about important steps to be taken regarding the project through this messaging window.

3. Methodology and Model Specifications

Dart: It is a versatile and object oriented programming language developed by google. It is a strong language to develop cross platform web and mobile applications [8].

Flutter: It is a mobile and web application open source UI development framework provided by google. It provides various widgets to create attractive UIs.

JavaScript: It is a high level scripting language which is used to develop effective back-end with dynamicity to the application pages [9].

NLP: By including techniques like machine learning and deep learning, NLP (Natural Language Processing) helps computers to process text data, allowing the translation between computer and human language [10].

Cloud: Cloud computing allows individuals and different organisations to access resources, storage, processing units, and software, decreasing or eliminating the need of physical infrastructure and computing resources [11].

Firebase: This platform provides various services to the web and mobile app developers like cloud storage, real time databases, hosting and authentication.

APIs: Application Programming Interfaces are used to share data and resources among different software. It enables the multiple software to work together smoothly [12].

SQL: Structured Query Language is a Database Management language. It makes managing the database systems easy with its easy and reliable queries.

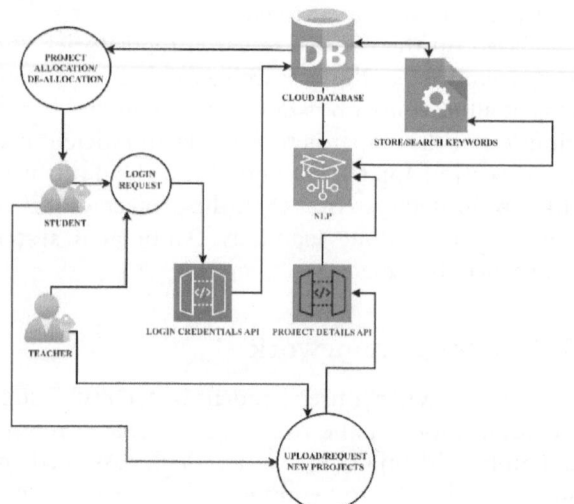

Figure 1: Working of the model

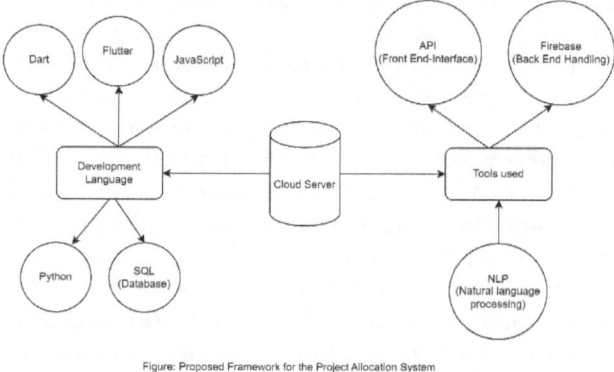

Figure 2: Proposed framework for project allocation

4. Result

The proposed methodology is developed using Flutter (DART). The Smart Project Allocation System tries to solve the problem of project allocation in schools and colleges. Faculty can handle a larger number of students without extensive time consuming one-one meetings. This innovative approach benefits both students and faculty. It streamlines the process, enabling faculty and students to upload projects. It allows students to apply for projects that align with their skills and interests. Also, it allows student to work under some faculty on a prior working project. This application applies NLP for searching and matching projects and students' skills. It also provides a chat window for communication between the two parties. The user-friendly interface simplifies allocating the projects, saves time and resources for both faculty and students. It results in effective, more meaningful and productive academic experiences for both students and faculty.

5. Conclusion

This research paper and the project itself gives a transformational overview in academic student-faculty collaboration on projects and researches. By connecting faculty and students through a GUI application, it simplifies the process of allocating projects, maximised efficiency and productivity. It uses skills-based matching applying the Natural Language Processing to ensure that the good projects go in the capable hands of students who possess the required skills. This model promotes a culture of open and efficient communication. It offers flexibility and streamlining in the allocation, de-allocation and communication processes, allowing adaptation to changing needs and circumstances.

References

[1] Brockm Kamaraj, K., Shah, J., Makharia, R., and Sharma, R. Student-Faculty Portal using ML and Cloud Services.

[2] Snijders, I., Wijnia, L., Rikers, R. M., and Loyens, S. M. (2020). Building bridges in higher education: Student-faculty relationship quality, student engagement, and student loyalty. International Journal of Educational Research, 100, 101538.

[3] Qiu, R. G. (2019). A systemic approach to leveraging student engagement in collaborative learning to improve online engineering education. International Journal of Technology Enhanced Learning, 11(1), 1-19.

[4] Ekong, E. E., Adiat, Q. E., Ejemeyovwi, J. O., and Alalade, A. M. (2019). Harnessing big data technology to benefit effective delivery and performance maximization in pedagogy. International Journal of Civil Engineering and Technology, 10(1), 2170-2178.

[5] Madleňák, R., D'Alessandro, S. P., Marengo, A., Pange, J., and Neszmélyi, G. I. (2021). Building on strategic eLearning initiatives of hybrid graduate education a case study approach: MHEI-ME Erasmus+ project. Sustainability, 13(14), 7675.

[6] Batch, B., Roberts, J., Nakonechnyi, A., and Allen, R. (2021). "Cell Phones Under the Table": Meeting Students' Needs to Reduce Off-Task Smartphone Use Through Faculty–Student Collaboration. Journal of Educational Technology Systems, 49(4), 487-500.

[7] Schwartz-Bechet, B., and Duffy, C. (2022, June). Increased Student and Faculty Engagement Researched via an International Opportunity: A Case Study Overview. In The Learning Ideas Conference (pp. 397-411). Cham: Springer International Publishing.

[8] He, Z., Chen, L., and Zhu, L. (2023). A study of Inter-Technology Information Management (ITIM) system for industry-education integration. Heliyon, 9(9).

[9] S Rajpoot, N. K., Singh, P., and Pant, B. (2023). Load Balancing Strategies for Cloud Computing: A Simulation-Based Study.

[10] Rajpoot, N. K., Singh, P., & Pant, B. (2023, May). Design and develop a delay sensitive smart health framework using nature inspired load balancer. In 2023 International Conference on Advancement in Computation & Computer Technologies (InCACCT) (pp. 427-432). IEE

[11] Upadhyay, D., Gupta, A., Mohd, N., and Pant, B. (2023, June). A review of network slicing based 5G. In AIP Conference Proceedings (Vol. 2782, No. 1). AIP Publishing.

[12] Sharma, G., Tripathi, V., and Srivastava, A. (2021). Recent trends in big data ingestion tools: A study. In Research in Intelligent and Computing in Engineering: Select Proceedings of RICE 2020 (pp. 873-881). Springer Singapore.

Automated Cough Classification for COVID-19 Detection Using Machine Learning

Yogendra Bharadwaj[1,2], Prabh Deep Singh[1,2], Ramendra Bharadwaj[1,3]

[1]Department of Computer Science and Engineering, Graphic Era Deemed to be University Dehradun, Dehradun, India
[3]Department of Computer Science & Engineering, CU Rise Analytics (A Trellance Company) Bengaluru, Karnataka, India
E-mail: Bharadwaj.yogendra1@ gmail.com, ssingh.prabhdeep@ gmail.com, bharadwaj.ramendra1@ gmail.com

Abstract

Cough classification is of paramount importance in the realm of respiratory health diagnostics, particularly in distinguishing COVID-19-related cough patterns from those associated with other conditions. This study harnesses the versatility of the Python programming language to construct a comprehensive pipeline for cough classification. Python, with its rich ecosystem of libraries, aids in data preparation, feature engineering, model development, and result analysis. Leveraging acoustic feature extraction techniques and libraries such as pandas and scikit-learn, this research preprocesses cough audio datasets and extracts crucial features. Two classification algorithms, XGBoost and logistic regression, are implemented to discriminate between COVID-19-related coughs and others. The scikit-learn library facilitates dataset splitting, classifier initialisation, training, and prediction, enabling a rigorous evaluation of model performance.

1. Introduction

The declaration of the COVID-19 pandemic by the World Health Organization (WHO) on February 11, 2020, was a significant milestone in world health, as it signaled the emergence of the Severe Acute Respiratory Syndrome Coronavirus 2 (SARS-CoV-2) virus. The novel virus exhibited rapid global dissemination, presenting substantial obstacles to public health systems and necessitating a deeper comprehension of its symptomatology and impact on humans [1]. Fever, tiredness, and dry coughs are prevalent manifestations associated with COVID-19. Significantly, coughing has developed as a prominent and identifiable symptom commonly connected with this particular ailment. Nevertheless, it is imperative to acknowledge that coughing is not exclusive to COVID-19; rather, it serves as a manifestation of over 100 other diseases and conditions, hence presenting a multifaceted diagnostic dilemma. Moreover, the effects of coughing on the respiratory system might exhibit significant variability among individuals [2].

In the present scenario, the utilization of machine learning has surfaced as a potent instrument in the fight against the COVID-19 pandemic. Machine learning methodologies, particularly in the domain of cough classification, have significantly contributed to improving our capacity to differentiate coughs associated with COVID-19 from those linked to other illnesses[3]. The system utilizes the vast volume of data produced during the pandemic, encompassing audio recordings of coughs, to formulate advanced algorithms that can assist in the timely identification and diagnosis of COVID-19 cases through the analysis of cough patterns. The utilization of machine learning can enhance our diagnostic capacities to a considerable extent by facilitating expedited and more precise detection of COVID-19 cases. Consequently, this can aid in the containment of its transmission[4]

2. Methodology

2.1. Data Collection

The collection of a comprehensive dataset serves as the fundamental basis for cough categorisation. The dataset includes an extensive collection of audio recordings of coughs, encompassing a vast range of variances in cough kinds, situations, and persons. It is imperative to ensure that the dataset covers samples from persons who have tested positive for COVID-19, as well as those with other respiratory disorders or even individuals who are in good health, in order to facilitate meaningful comparisons [5]. The presence of a comprehensive and thoroughly annotated dataset is of utmost importance in order to ensure the efficient training and evaluation of machine learning models.

2.2. Preprocessing and Feature Extraction

Once the data is collected, preprocessing steps are essential to prepare the audio recordings for analysis. This might involve cleaning the audio data to remove noise, resampling to a consistent sample rate, and segmenting the audio files into smaller, uniform segments for analysis [6]. Preprocessing ensures that the data is in a suitable format for feature extraction and model training.

Feature extraction is a critical step where relevant information is distilled from the cough audio data. Commonly used features include Mel-frequency cepstral coefficients (MFCCs) to capture spectral characteristics, zero-crossing rate to measure signal changes, and spectral centroid and bandwidth to describe the spectral distribution [7]. Time-domain features, such as root mean square energy and statistical measures like mean and standard deviation, also play a role in characterising cough patterns. These features serve as the input for machine learning models.

2.3. Model Selection

Selecting an appropriate machine learning model is a pivotal decision in the cough classification process. Different models, such as Convolutional Neural Networks (CNNs) for spatial pattern recognition, Recurrent Neural Networks (RNNs) for sequential data, or classical classifiers like Support Vector Machines (SVMs) and Random Forests, may be considered based on the nature of the data and the problem at hand [8]. Model selection should align with the specific characteristics of the cough audio data.

2.4. Training and Validation and Testing and Evaluation

Once a model is chosen, the dataset is typically split into training, validation, and test sets. The training data is used to train the model using the extracted features and labeled cough data [9]. Validation is crucial for fine-tuning model hyperparameters and assessing their performance on data it has not seen during training. This iterative process helps optimise the model's performance [10].

The model's effectiveness is rigorously evaluated on the test dataset to gauge its ability to generalise to new, unseen data. Evaluation metrics, including accuracy, precision, recall, F1-score, and receiver operating characteristic (ROC) curve analysis, are employed to measure classification performance and identify any potential shortcomings.

2.5. Post-processing and Deployment and Monitoring

Post-processing steps may be necessary to refine the model's output and reduce false positives or false negatives. These steps could involve applying thresholds, filtering, or other techniques to improve the final classification results, ensuring they align with clinical or diagnostic needs.

If the model meets the desired performance criteria, it can be deployed for real-world applications, such as assisting in the diagnosis of COVID-19 based on cough patterns. Continuous monitoring and maintenance are critical in real-world deployment to adapt to changing conditions, update the model with new data, and ensure its ongoing effectiveness in practical settings.

XGBoost algorithm can be applied in the context of cough classification, specifically for distinguishing COVID-19-related coughs from other cough types:

3. Boosting for Cough Classification

The boosting technique in XGBoost becomes especially useful in cough classification. XGBoost builds a series of decision trees sequentially, each focusing on correctly classifying cough samples. If a particular cough type (e.g., COVID-19-related) is challenging to classify, XGBoost gives it more weight in subsequent trees, improving the model's ability to capture distinctive cough patterns. XGBoost constructs decision trees to model the characteristics of different cough types. Each tree represents a set of rules based on audio features extracted from cough recordings. The trees aim to differentiate between COVID-19-related coughs and those associated with other respiratory conditions or healthy individuals [11]. XGBoost's regularization techniques, including L1 and L2 regularization, help prevent overfitting in the context of cough classification. By penalizing overly complex tree structures, XGBoost ensures that the model generalizes well to unseen cough data, improving its reliability. Gradient descent optimization is crucial for fine-tuning the decision trees to classify coughs effectively. It adjusts the tree structure and leaf values to minimize the error between predicted and true cough labels. In the context of cough classification, this optimization process focuses on identifying the most distinguishing audio features. XGBoost uses the Weighted Quantile Sketch technique to identify which audio features are most informative for distinguishing cough types. It efficiently explores feature importance, allowing the algorithm to select the most relevant attributes for classification [12]. To ensure that the individual trees are not overly complex, XGBoost employs pruning strategies. This is especially important in cough classification to prevent the model from fitting noise in the data [13]. Pruning helps maintain the clarity of rules for differentiating cough types. XGBoost's parallel and distributed computing capabilities are valuable for processing large volumes of cough audio data efficiently. It can make use of multiple CPU cores to accelerate model training, a crucial consideration when dealing with extensive cough datasets [14]

3.1. Cross-validation and Early Stopping

In the context of cough classification, cross-validation is used to assess how well the XGBoost model performs on different subsets of the cough dataset. Early stopping allows training to halt when further iterations no longer improve the model's ability to classify cough types accurately.

Step 1.   Load the dataset of labeled documents (e.g., text data) into memory.

Step 2.   Split the dataset into a training set and a test set for model evaluation.

Step 3.   Initialise XGBoost parameters:
- Specify the learning rate (eta).
- Define the maximum depth of decision trees (max_depth).
- Set the number of boosting rounds (num_round).
- Choose an evaluation metric (e.g., accuracy) for monitoring.

Step 4.   Train the XGBoost classifier:
- For each boosting round (i = 1 to num_round):
- Build a decision tree to minimise the classification error.
- Apply regularisation techniques to prevent overfitting.
- Update the model's parameters using gradient descent optimisation.
- Evaluate the model on the training data to monitor performance.

Step 5.   After training, assess the model's performance on the test set:
- Calculate evaluation metrics (e.g., accuracy, precision, recall, F1-score).
- Visualise the results or generate a confusion matrix for analysis.

Step 6.   Optionally, fine-tune hyperparameters:
- Use techniques like grid search or random search to optimise hyperparameters.
- Reiterate the training process with the best hyperparameters found.

Step 7.   Save the trained XGBoost model for future use.

Step 8.   Given a new, unlabeled document in Microsoft Word:

- Preprocess the document text (e.g., clean, tokenise, extract features).
- Load the trained XGBoost model.
- Use the model to predict the category or label of the document.

Step 9.   Optionally, integrate the XGBoost-based classification into a Microsoft Word add-in or automation script for automated document categorisation or tagging.

Step 10.   End.

4. Logistic Regression Classifier for Cough Classification

Input:
A dataset of audio recordings of coughs, along with corresponding labels indicating the cough type (e.g., COVID-19, non-COVID-19).

Output:
A predictive model that can classify new cough recordings as COVID-19-related or non-COVID-19 based on their acoustic features.

Data Preparation for Cough Classification

Collect and preprocess the dataset, ensuring audio recordings are properly formatted and labeled.

Extract relevant audio features from the cough recordings (e.g., MFCCs, zero-crossing rate, spectral centroid).

Model Initialisation
Initialise the logistic regression model, which consists of weights (coefficients) for each feature and an intercept (bias term).

Set the learning rate and the number of training iterations (epochs).

Sigmoid Function for Cough Probability
Define the sigmoid function, which maps the linear combination of acoustic features to a cough probability:

Probability(COVID-19) = Sigmoid(z), where $z = w_0 + w_1*feature_1 + w_2*feature_2 + ... + w_n*feature_n$

Here, z is the linear combination of features and model parameters, and w represents the coefficients associated with each feature.

4.1. Training the Cough Classification Model

For each training iteration (epoch):

Calculate the predicted probabilities for each cough recording using the sigmoid function.

Compute the loss between the predicted probabilities and the true labels (COVID-19 or non-COVID-19).

Loss = - [y*log(p) + (1-y)*log(1-p)]

Update the model parameters (weights) using gradient descent to minimise the loss:

w_new = w_old - learning_rate * gradient(Loss)

This step adjusts the model's weights to improve its ability to predict COVID-19-related coughs accurately.

5. Experiments and Results

In the field of cough classification, Python serves as a versatile tool for data preparation, model development, and performance evaluation. First, Python libraries such as pandas and scikit-learn facilitate the loading and preprocessing of cough audio datasets. Relevant acoustic features are extracted, preparing the dataset for training and evaluation. Two classification algorithms, XGBoost and logistic regression, are implemented to distinguish between COVID-19-related coughs and others. The scikit-learn library is instrumental in splitting the dataset into training and testing sets, initialising the classifiers, and training them on the acoustic feature data. Subsequently, the classifiers predict cough classifications for the testing data, enabling the assessment of their effectiveness in identifying COVID-19-related cough patterns."

Python's role extends beyond the building of models, as it also plays a crucial part in the recording and analysis of categorisation results. The accuracy, precision, and recall metrics are produced using the capability provided by scikit-learn. These measures offer valuable insights into the performance of the XGBoost and logistic regression classifiers. These indicators assist academics and healthcare professionals in assessing the efficacy of models inappropriately classifying coughs, particularly those linked to COVID-19. In brief, Python plays a crucial role in the cough classification pipeline by facilitating the integration of data processing, model training, and performance evaluation. This integration eventually contributes to the progress of respiratory health diagnoses.

6. Conclusion

This study utilises Python programming and powerful machine learning methods to tackle the significant issue of cough categorisation. The specific objective is to differentiate between coughs associated with COVID-19 and those linked to other respiratory ailments. A comprehensive pipeline for automated cough categorisation has been developed by meticulous data preparation, feature extraction, and the utilisation of two robust classifiers, namely XGBoost and logistic regression.

Our research highlights the significant importance of Python in optimising the entire workflow, encompassing data preprocessing, model creation, and performance evaluation. The utilisation of the Python ecosystem, in conjunction with prominent libraries such as pandas, scikit-learn, and XGBoost, has facilitated the proficient handling of cough audio datasets. This has allowed for the extraction of pertinent features and the development of models that possess the capability to discern tiny variations in cough patterns. The findings of our research, measured in terms of accuracy, precision, and recall, offer significant insights into the efficacy of our classifiers. These metrics play a vital role as important benchmarks, assisting healthcare providers and researchers in their efforts to improve the diagnosis of respiratory health. Our research makes a valuable contribution to the broader endeavor of early disease identification, management, and the protection of public health by effectively differentiating between coughs due to COVID-19 and those caused by other factors.

References

[1] Vrindavanam, J., Srinath, R., Shankar, H. H., and Nagesh, G. (2021). Machine learning based COVID-19 cough classification models - A comparative analysis. In 2021 5th International Conference on Computing Methodologies and Communication (ICCMC), Apr. 2021, doi: https://doi.org/10.1109/iccmc51019.2021.9418358.

[2] Tang, T., Tian, P., Zhang, Y., Hu, G., Pan, F., and Zhao, Q. (2022). Automatic bronchitis classification in children based on cough recordings. In 2022 3rd International Conference on Pattern Recognition and Machine Learning (PRML), Jul 2022, doi: https://doi.org/10.1109/prml56267.2022.9882202.

[3] Irawati, M. E. and Zakaria, H. (2021). Classification model for Covid-19 detection through recording of cough using XGboost classifier algorithm. In 2021 International Symposium on Electronics and Smart Devices (ISESD), Jun 2021, doi: https://doi.org/10.1109/isesd53023.2021.9501695.

[4] Islam, R., Abdel-Raheem, E., and Tarique, M. (2021). Early detection of COVID-19 patients using chromagram features of cough sound recordings with machine learning algorithms. In 2021 International Conference on Microelectronics (ICM), Dec. 2021, doi: https://doi.org/10.1109/icm52667.2021.9664931.

[5] Srikantrh, P. and Behera, C. K. (2022). A machine learning framework for covid detection using cough sounds. In 2022 International Conference on Engineering & MIS (ICEMIS), Jul. 2022, doi: https://doi.org/10.1109/icemis56295.2022.9914391.

[6] Esposito, M., Uehara, G., and Spanias, A. (2022). Quantum machine learning for audio classification with applications to healthcare. In IEEE Xplore, Jul. 01, 2022. https://ieeexplore.ieee.org/document/9904377 (accessed Nov. 11, 2022).

[7] Pande, S., Patil, A., and Petkar, S. (2022). Dry and wet cough detection using fusion of cepstral base statistical features. In 2022 International Conference on Decision Aid Sciences and Applications (DASA), Mar. 2022, doi: https://doi.org/10.1109/dasa54658.2022.9765242.

[8] Trang, K., Nguyen, H. A., TonThat, L., Do, H. N., and Vuong, B. Q. (2022). COVID-19 disease classification by cough records analysis using machine learning. In 2022 IEEE International Conference on Cybernetics and Computational Intelligence (CyberneticsCom), Jun. 2022, doi: https://doi.org/10.1109/cyberneticscom55287.2022.9865610.

[9] Gupta, R., Krishna, T. A., and Adeeb, M. (2022). Cough sound based COVID-19 detection with stacked ensemble model. In 2022 4th International Conference on Smart Systems and Inventive Technology (ICSSIT), Jan. 2022, doi: https://doi.org/10.1109/icssit53264.2022.9716373.

[10] Valdes, J. et al. (2022). Cough classification with deep derived features using audio spectrogram transformer. In 2022 IEEE International Conference on Big Data (Big Data), Dec. 2022, doi: https://doi.org/10.1109/bigdata55660.2022.10020878.

[11] Valdes, J. J., Xi, P., Cohen-McFarlane, M., Wallace, B., Goubran, R., and Knoefel, F. (2021). Analysis of cough sound measurements including COVID-19 positive cases: A machine learning characterisation. In 2021 IEEE International Symposium on Medical Measurements and Applications (MeMeA), Jun 2021, doi: https://doi.org/10.1109/memea52024.2021.9478714.

[12] Zhang, X., Pettinati, M., Jalali, A., Rajput, K. S., and Selvaraj, N. (2021). Novel COVID-19 screening using cough recordings of a mobile patient monitoring system. In 2021 43rd Annual International Conference of the IEEE Engineering in Medicine & Biology Society (EMBC), Nov. 2021, doi: https://doi.org/10.1109/embc46164.2021.9630722.

[13] Zofia Tomaszewska, J., Chousidis, C., and Donati, E. (2022). Sound-based cough detection system using convolutional neural network. In 2022 IEEE International Symposium on Medical Measurements and Applications (MeMeA), Jun. 2022, doi: https://doi.org/10.1109/memea54994.2022.9856512.

[14] Pawar, R. S. and Kalbande, D. R. Optimization of quality of service using ECEBA protocol in wireless body area network. International Journal of Information Technology, 15.

Dew Computing Technologies and Their Impact on Cloud Resource Management

Navneet Kumar Rajpoot, Prabh Deep Singh, Bhaskar Pant, Vikas Tripathi

Department of Computer Science & Engineering, Graphic Era (Deemed to be University), Dehradun, India
E-mail: navneetrajpootgeu@gmail.com, ssingh.prabhdeep@gmail.com,
bhaskar.pant@geu.ac.in, vikastripathi.cse@geu.ac.in

Abstract

Distributed Edge and Dew computing has emerged as a feasible approach to enhance standard cloud computing due to the expansion of connected devices and the rising demand for real-time data processing. Dew computing, characterised by its distributed design and closeness to end-users and devices, provides innovative outcomes to the difficulties currently present in cloud resource management. This study will highlight how Dew computing technologies have improved cloud infrastructure by increasing its effectiveness in allocating resources, boosting its fault tolerance, decreasing its latency, and enhancing reliability. The effects of DEW computing on the effective allocation, monitoring, and scalability of resources in cloud environments are reviewed along with its core characteristics. Issues about security and scalability are addressed, and future trends and research prospects for incorporating DEW computing technologies into current cloud infrastructures are also discussed.

Keywords: Cloud computing, dew computing, resource allocation, reliability, scalability

1. Introduction

Cloud computing has become a paradigm shift in the IT industry, transforming the way in which businesses handle and make use of their computer resources. Companies have been able to utilise the resources of remote data centers for storage, processing, and a wide variety of applications due to their scalability, flexibility, and cost-effectiveness [1]. Managing cloud resources efficiently is more important than ever as the amount of data generated and the need for instantaneous responses increase exponentially. Resource allocation, scalability, and latency challenges are frequent in conventional cloud systems. Because of the ever-increasing need for real-time processing and low-latency services, traditional cloud infrastructures have difficulties efficiently managing resources and keeping up with changing customer demands. DEW computing is a new approach that promises to address these issues, and it is changing the way cloud resources are managed [2]. One such potential paradigm is Dew computing, which introduces a distributed computing architecture that brings cloud functionality to network edges, improving response times and bringing data closer to end-users and devices. Being so close together improves reliability and fault tolerance while also decreasing latency. The major paradigm shift to cloud computing in IT has become the central pillar of today's networked computing environment. Flexibility, low cost, easy access, and the opportunity to enhance innovation are just a few of the benefits of cloud computing [3].

Distributed Edge and Dew computing are growing as a revolutionary shift, providing novel solutions to the issues caused by conventional centralised cloud computing [4]. Dew computing is an architectural framework for distributed computing that capitalises on the advantages of edge and fog computing to enhance proximity between computational resources, data sources, and end-users Figure 1. This closeness, typically at the network's periphery, provides several benefits, including low latency, high throughput, high scalability, and minimal resource consumption [5]. Table 1 shows the comparison of Dew computing and traditional cloud computing.

2. Existing Work

Table 2 Comparison of Dew computing with existing work

Table 58.1: Comparison of dew computing and traditional cloud computing

Aspect	Dew Computing	Cloud Computing
Location of Resources	Resources are distributed closer to data sources, often at the network edge.	Resources are centralised in remote data centers.
Latency	Low latency due to proximity to data sources, ideal for real-time applications.	Higher latency as data must travel to and from centralised data centers.
Scalability	Highly scalable, ideal for applications with variable resource needs.	Scalability may be limited by the capacity of data centers.
Data Processing	Supports real-time data processing and analytics at the edge.	Data processing typically occurs in centralised data centers.
Resource Efficiency	Optimises resource usage by distributing tasks across a network.	This may result in underutilised resources during periods of low demand.
Connectivity	Supports offline operation and communication in disconnected environments.	Requires a continuous internet connection for most services.
Use Cases	Ideal for IoT, real-time analytics, autonomous vehicles, and applications with low-latency requirements.	Well-suited for traditional web applications, software development, and data storage.

Table 58.2: Comparison of dew computing with existing work

S. No	Author	Technology Used		
		Cloud & Edge Computing	Cloud & Fog Computing	Cloud, Fog & Dew Computing
1	Singh, P et. al., 2020			
2	Ahammad, I et. al., 2021			
3	Ageed, Z. S et. al., 2021 [6]			
4	Gusev, M et .al., 2021 [7]			
5	Proposed Approach			

3. Impact of Dew Computing on Cloud Resource Management

Dew computing technologies profoundly impact cloud resource management, addressing many of the limitations and challenges associated with traditional cloud computing [8]. The reduced latency,

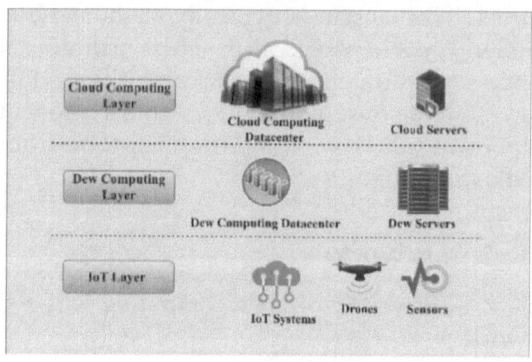

Figure 58.1: Dew computing in cloud computing

improved resource allocation, enhanced scalability, and edge analytics capabilities make Dew computing a crucial component in the evolving landscape of cloud resource management.

A. **Reduced Latency and Improved Responsiveness:** with its proximity to data sources at the network edge, Dew computing significantly reduces latency in data processing and retrieval. This results in quicker response times [9].

B. **Improved Scalability:** Dew computing enables the efficient allocation of resources to meet varying demands.

C. **Increased Reliability:** By dividing tasks over a network, Dew computing may provide reliability even in a failure. The availability of applications and services ensures that processing may continue on another node even if one node or server fails.

D. **Cost Savings:** The efficient use of resources, reduced data transfer, and optimised resource allocation contribute to cost savings in operational and infrastructure expenses.

4. Challenges in the Dew Computing

The integration of Dew computing into cloud resource management brings a set of critical challenges and considerations [10]. As resources are moved closer to the sources of data, issues regarding the security and confidentiality of the information grow. The seamless coordination of DEW and conventional cloud platforms creates issues with interoperability. To assure effective resource allocation and trustworthy service delivery in today's dynamic and spread computing environment, scalability and performance management issues must be solved.

5. Conclusion

Distributed Edge and Dew computing have emerged as a revolutionary technology, offering innovative solutions to the challenges faced by the conventional centralised cloud storage and processing paradigm. The novel methodology stands out by its ability to efficiently handle data at the edges of the network, carrying minimal latency, enhanced scalability, and optimal utilisation of existing resources. This research has demonstrated that Dew computing's systems, characterised by their low-latency, fault-tolerant, and cost-efficient nature, are particularly suitable for real-time applications and highly data-driven tasks. But still, the integration of Dew computing with cloud systems presents several challenges, particularly in the areas of security and interoperability. This research explored the remarkable impact of Dew computing on the operation of cloud resources, showcasing its ability to decrease latency, improve resource allocation, and facilitate efficient edge analytics. To fully utilise its potential, more developments are required in the future to combine the technology with artificial intelligence, maintain security regulations, and incorporate advanced machine learning algorithms. These actions are necessary to address the existing limitations and enhance operational effectiveness.

References

[1] Ahammad, I., Khan, A. R., and Salehin, Z. U. (2021). A review on cloud, fog, roof, and Dew computing: IoT perspective. International Journal of Cloud Applications and Computing (IJCAC), 11(4), 14-41.

[2] Singh, P., Kaur, A., Aujla, G. S., Batth, R. S., & Kanhere, S. (2020). Daas: Dew computing as a service for intelligent intrusion detection in edge-of-things ecosystem. IEEE Internet of Things Journal, 8(16), 12569-12577.

[3] Utomo, P. (2020, November). Dew computing: concept and its implementation strategy. In 2020 Fifth International Conference on Informatics and Computing (ICIC) (pp. 1-6). IEEE.

[4] Rajpoot, N. K., Singh, P., and Pant, B. (2023, April). Nature-inspired load balancing algorithms for resource allocation in cloud computing. In 2023 International Conference on Computational Intelligence and Sustainable Engineering Solutions (CISES) (pp. 827-832). IEEE.

[5] Sneha, S. P., and Tripathi, V. (2023, April). Green cloud computing: achieving sustainability through energy-efficient techniques, architectures, and addressing research challenges. In International Conference on Paradigms of Communication, Computing and Data Analytics (pp. 97-105). Singapore: Springer Nature Singapore.

[6] Ageed, Z. S., Zeebaree, S. R., Sadeeq, M. A., Ibrahim, R. K., Shukur, H. M., and Alkhayyat, A. (2021, September). Comprehensive study of moving from grid and cloud computing through fog and edge computing towards dew computing. In 2021 4th International Iraqi Conference on Engineering Technology and Their Applications (IICETA) (pp. 68-74). IEEE.

[7] Gusev, M. (2021, July). What makes Dew computing more than Edge computing for Internet of Things. In 2021 IEEE 45th Annual Computers, Software, and Applications Conference (COMPSAC) (pp. 1795-1800). IEEE.

[8] Prasad, G., Gujjar, P., Kumar, H. N., Kumar, M. A., and Chandrappa, S. (2023). Advances of cyber security in the healthcare domain for analyzing data. In Cyber trafficking, Threat Behavior, and Malicious Activity Monitoring for Healthcare Organizations (pp. 1-14). IGI Global.

[9] Prasad, M. G., Agarwal, J., Christa, S., Pai, H. A., Kumar, M. A., and Kukreti, A. (2023, January). An improved water body segmentation from satellite images using MSAA-Net. In 2023 International Conference on Machine Intelligence for GeoAnalytics and Remote Sensing (MIGARS) (Vol. 1, pp. 1-4). IEEE.

[10] Kumar, M. A., Pai, A. H., Agarwal, J., Christa, S., Prasad, G. M., and Saifi, S. (2023, January). Deep learning model to defend against covert channel attacks in the SDN networks. In 2023 Advanced Computing and Communication Technologies for High Performance Applications (ACCTHPA) (pp. 1-5). IEEE.

Innovative Framework for Domestic Waste Segregation to Alleviate Pressure on Waste Treatment Plants

Mrityunjay Jha[1], Prabh Deep Singh[2]

[1]Department of Computer Science & Engineering
[2]Graphic Era Deemed to be University Dehradun, India
E-mail: mrityunjayjha8055@gmail.com, ssingh.prabhdeep@gmail.com

Abstract

Waste is a material that is no longer needed or used. It covers all kinds of items that are thrown away after their first use or are considered worthless, useless and have no useful purpose. These non-usable materials when discarded by the people, go to waste treatment plants where they come under segregation process. But due to rapid consumption and waste generation at domestic level, these waste treatment plants are facing very high workload for waste processing leading to formation of huge landfills. In this paper a novel framework has been proposed which utilises the concept of automatic waste segregation at domestic level. Ultrasonic sensors, moisture sensor, electromagnet and plastic sensor have been used to segregate the waste into different categories that are wet waste, metal waste, plastic waste, and non-plastic i.e., biodegradable waste. This framework majorly benefits by automating the waste segregation at domestic level, decreasing workload at waste treatment plant.

Keywords: Waste, segregation, automation, sustainable waste management

1. Introduction

In contemporary society, effective waste management is imperative due to the substantial daily waste generation, posing challenges to the environment and public health. Conventional waste treatment facilities face mounting pressure, leading to extensive landfills and ecosystem consequences. To address this crisis, a pioneering framework is proposed, revolutionizing waste management through automated on-site sorting protocols. We propose to utilise sensors such as ultrasonic, humidity, electromagnets, and plastic into daily livelihoods, limiting dependence on manual collection and segregation of waste. Advantages include enhanced precision, reduced burden on treatment facilities, streamlined waste management, and fostering individual responsibility. This novel framework seeks to contribute to sustainable development. The research gap lies in the need for more efficient and automated domestic waste sorting. This framework's contribution lies in presenting a holistic solution, motivated by the urgency to mitigate environmental degradation caused by inefficient waste management practices.

2. Proposal Approach

2.1. Methodology and Model Specifications

The ultrasonic sensor placed at the dustbin detects the human standing in front of the dustbin and commands the servo motor to open the lid of the dustbin for the user to throw the garbage inside the bin.

The waste falls on the moisture sensor plate which reacts by rotating at 70-degree angle clockwise and anti-clockwise by the servo motor on moisture detection. If the moisture is detected in the waste, the moisture sensor sends the input to the servo motor to rotate 70-degree anti-clockwise and the waste gets collected in the wet waste compartment. If the moisture is not detected in the waste, the moisture sensor sends the input to the servo motor to rotate 70-degree clockwise, and the waste is further moved to the next level of segregation.

Now the waste comes on the electromagnetic plate and if the waste is metal, it gets stuck on the plate and the non-metal waste is sent to the next level segregation by rotating the plate 70 degree clockwise. The metal waste which is stuck

Chapter 59 DOI: 10.1201/9781003570349

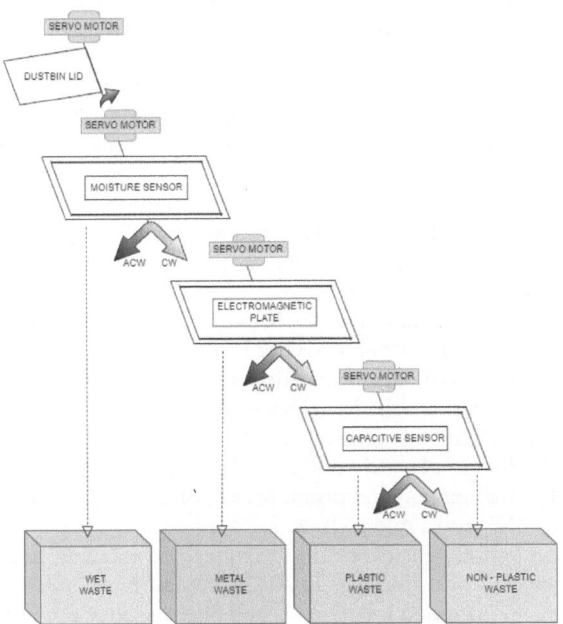

Figure 1: Waste flow diagram

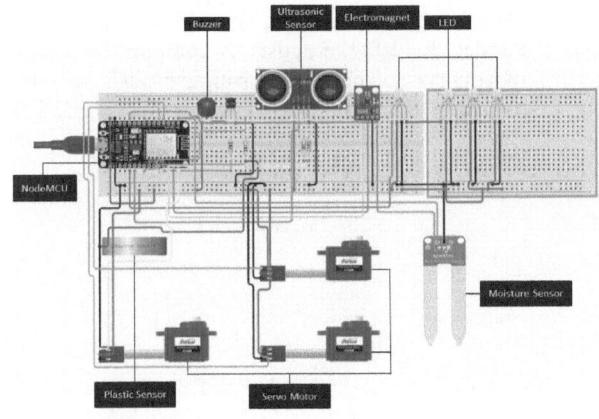

Figure 2: Circuit diagram

3. Comparison with Existing Work

S.No.	Author	Technology Used		
		Ultra-sonic Sensor	Electro-magnetic Plate	Plastic Sensor
1.	Sathyamoorthy, M. et. al., 2023 [1]	✓		✓
2.	Ismail, M. Z. et. al., 2023 [2]	✓	✓	
3.	Gautam, K. M. et. al., 2023 [3]	✓		
4.	Chitale, M. R. et. al., 2023 [4,5]	✓		
5.	Proposed Framework	✓	✓	✓

4. Result

The implemented framework, integrating ultrasonic sensors, humidity sensors, electromagnets, and plastic sensors for automated on-site waste sorting, has proven highly successful. Achieving precise waste categorisation, it substantially reduces the need for manual sorting, enhancing overall efficiency. Results indicate robust and dependable performance, diminishing the burden on waste treatment facilities and curbing landfill formation. This innovative approach fosters individual responsibility, encouraging the separation of recyclable from non-recyclable waste. The framework's success marks a significant stride towards sustainable waste management, mitigating environmental degradation and contributing to a conscientious and eco-friendly domestic waste management paradigm.

5. Conclusion

Waste production and disposal put great pressure on waste treatment facilities, leading to the creation of large landfills. Recognizing the urgent need for more sustainable waste management, this article outlines the first steps developed to address this issue. Combining ultrasonic sensors, humidity sensors, electromagnets and plastic sensors, the framework enables automatic separation of household waste. This new method effectively separates waste into different categories, including wet waste, metal waste, plastic waste, and biodegradable waste. One of the main benefits of this system is that it can reduce the burden on waste treatment plants by decentralizing the separation process. Therefore, it not only promotes better waste management, but also helps reduce the impact of environmental problems, ultimately promoting cleanliness and greater responsibility for the future environment.

on electromagnetic plate gets collected into the metal waste compartment by rotating the plate 70 degree anti-clockwise after the electromagnet is turned off.

On this level of segregation, the waste interacts with the capacitive sensor i.e., the plastic sensor. If the waste detected by the sensor is to be plastic, then the rotating plate rotates 70 degree anti-clockwise, and it gets collected into the plastic waste compartment and if the waste is sensed as non-plastic, then the plate rotates 70 degrees clockwise and gets collected into the biodegradable waste compartment i.e., non-plastic compartment.

References

[1] Sathyamoorthy, M., Vanitha, C. N., Raja, S. P., Sharma, A. K., Sharma, B., and Chowdhury, S. (2023, March). Smart city waste management system using IOT. In 2023 6th International Conference on Information Systems and Computer Networks (ISCON) (pp. 1-6). IEEE.

[2] Ismail, M. Z., Mohammad Zanawi, S. Z., and Yusof, M. R. (2023). Development of a smart trash can/dustbin using internet of things. In Advances in Technology Transfer Through IoT and IT Solutions (pp. 61-71). Cham: Springer Nature Switzerland.

[3] Gautam, K. M., Mahato, S., Sharma, H., and Soni, D. (2023). Automatic smart dustbin using IoT. Journal of Advances in Electrical Devices, 8(1), 17-23.

[4] Chitale, M. R., Chitpur, S. J., Chivate, A. B., Chopade, P. D., Deshmukh, S. M., and Marathe, A. A. (2023, September). Automated smart waste segregation system using IoT technology. In Journal of Physics: Conference Series (Vol. 2601, No. 1, p. 012015). IOP Publishing.

[5] Chitale, M. R., Chitpur, S. J., Chivate, A. B., Chopade, P. D., Deshmukh, S. M., and Marathe, A. A. (2023, September). Automated smart waste segregation system using IoT technology. In Journal of Physics: Conference Series (Vol. 2601, No. 1, p. 012015). IOP Publishing.

[6] Pendem, S., Suresh, K., Naqi, M., Sreekanth, D., and Chandana, D. (2023, May). Smart dustbin for efficient waste management using IOT. In AIP Conference Proceedings (Vol. 2477, No. 1). AIP Publishing.

[7] Prabhakaran, S., Yugeshkrishnan, M., Santhiya, M., and SM, D. K. (2023). Smart dustbin using IOT. The Scientific Temper, 14(02), 412-417.

[8] Rajpoot, N. K., Singh, P., and Pant, B. (2022, December). Cloud-IoT based smart healthcare framework: Application and trends. In Proceedings of the 4th International Conference on Information Management and Machine Intelligence (pp. 1-6).

[9] Rajpoot, N. K., Singh, P., and Pant, B. (2023). Load balancing strategies for cloud computing: A simulation-based study.

[10] Prasad, G., Gujjar, P., Kumar, H. N., Kumar, M. A., and Chandrappa, S. (2023). Advances of cyber security in the healthcare domain for analyzing data. In Cyber Trafficking, Threat Behavior, and Malicious Activity Monitoring for Healthcare Organizations (pp. 1-14). IGI Global.

[11] Bahuguna, A., Prasad, M. G., Bhagnal, Y., Yeole, A. N., and Prabhu, B. A. (2023, June). CNN-powered monument detection webapp for preserving India's Cultural Heritage. In 2023 International Conference on Applied Intelligence and Sustainable Computing (ICAISC) (pp. 1-6). IEEE.

[12] Singh, P., & Singh, D. P. (2023, March). A delay sensitive framework for effective healthcare using machine learning. In 2023 10th International Conference on Computing for Sustainable Global Development (INDIACom) (pp. 541-545). IEEE.

[13] Mittal, S., Mishra, A. K., Tripathi, V., Singh, P., and Pandey, P. (2023, August). A comparative analysis of supervised machine learning models for smart intrusion detection in IoT network. In 2023 3rd Asian Conference on Innovation in Technology (ASIANCON) (pp. 1-6). IEEE.

Unveiling Hate Speech

Harnessing ML to Address Online Discrimination

Abhiram Shukla, Prabh Deep Singh, Vaibhav Singh, Akash Chauhan

Computer Science and Engineering, Graphic Era University, Dehradun, India

E-mail: abhiramshukla19@gmail.com, ssingh.prabhdeep@gmail.com, vaibhavsingh99810@gmail.com, akash75457@gmail.com

Abstract

The prevalence of hate speech in online communication has led to the creation of algorithms based on machine learning to recognise and flag hateful content. This paper provides a unique approach for detecting hate speech that can reliably identify whether user input is hate speech or not. The model uses a three-step procedure that includes testing, training, and preprocessing. To maintain data integrity and reduce errors, preprocessing include cleaning the data and extracting features using methods like TF-IDF and count vectoriser. In the training phase, linguistic patterns and indications of hate speech are correlated using machine learning methods such as Support Vector Machines (SVM) and Naive Bayes. A confusion matrix is used to assess the model's performance and show how well it can discriminate between hate speech and non-hateful information. The suggested model is a useful instrument for preventing hate speech online since it can be used to various systems, generalises well, and can reliably identify whether a speech is hate speech.

Keywords: Hate speech detection, semantic nuances, computational fairness, cultural context

1. Introduction

The suggested approach tries to detect hate speech correctly while retaining the diversity of online discourse [1]. Hate speech, a damaging kind of prejudice, attacks individuals or groups based on their characteristics and can be found online in a variety of forms. While the internet is a great tool for sharing knowledge, it has also encouraged the spread of harmful content, such as hate speech [2]. To address this issue, technological technologies such as machine learning analyze linguistic patterns and semantic intricacies within large datasets to discriminate between hostile and acceptable interactions.

The creation of algorithms capable of automatically detecting and identifying hate speech in text data is critical to this strategy. This approach may extract significant components from text such as word frequencies, mood analyses, and contextual signals [3]. Then, to process these features and generate predictions about the presence of hate speech, machine learning algorithms are used, ranging from simple deep learning architectures like Convolutional Neural Networks and Recurrent Neural Networks to more sophisticated deep learning architectures like Support Vector Machines and Naive Bayes [4].

The creation of efficient hate speech detection models is not without difficulties, though. A continuous challenge is the constantly changing nature of online communication, which is characterised by shifting linguistic fashions and the creation of new forms of expression [5]. Models must be updated and retrained to adjust to changing linguistic environments to retain the effectiveness of hate speech detection.

2. Methodology

The entire procedure for the project in depth in this stage. This will make it easier for the reader to comprehend how the data flow works and what happens to it at each stage of the procedure [6]. The steps that our project involved are as follows: Data gathering, preprocessing, feature extraction, model choice, training, evaluation, application, and ongoing improvement as shown in Figure 1.

1. *Data Collection:* The first stage is to combine a varied and labeled dataset that includes instances of hate speech as well as samples of non-hateful and neutral language. The training data for the machine learning model comes from this dataset.

Chapter 60 DOI: 10.1201/9781003570349

2. ***Data Preprocessing:*** To make raw text data appropriate for the model's training, it must be cleaned and preprocessed. Tokenisation, lowercasing, stopword elimination, and stemming/lemmatisation are typical preprocessing operations.

3. ***Feature Extraction:*** For feature extraction, machine learning approaches require numerical input data. The text data must therefore be changed into a numerical form. Techniques for feature extraction in natural language processing often used include bag-of-words, TF-IDF, word embeddings, and more complex methods like BERT.

4. ***Model Selection:*** To identify hate speech, a variety of machine learning methods can be utilised, such as Support Vector Machines, Naive Bayes, Random Forest, and deep learning models like Convolutional Neural Networks or Recurrent Neural Networks. Deep learning models have shown promising results when understanding complex linguistic patterns.

5. ***Model Training:*** Training and Validation sets are created from the preprocessed data. The training set is used to develop the model, while the validation set is used to fine-tune it. The model gains the ability to recognise patterns and characteristics that separate hate speech from non-hateful discourse throughout training.

6. ***Model Evaluation:*** Following training, a separate test dataset is used to gauge the model's performance in terms of accuracy, precision, recall, F1-score, and other pertinent metrics. The

objective is to create a model that applies well to fresh, unexplored data.

7. ***Implementation:*** The model can be implemented in practical applications after it performs satisfactorily. It may be incorporated into social networking sites, online discussion boards, or any other venue where hate speech may be automatically detected and curbed.

8. ***Continuous Improvement:*** Because hate speech is dynamic and changes over time, it requires routine updating and retraining to be able to recognise new hate speech patterns.

3. Proposed Framework

Following will be the step-by-step data flow of the proposed framework Figure 2.

3.1. User Input

Here, for the sake of implementation, we made a Flask Web App that takes in user input and processes it to later classify it as hate speech or not [7]. When the framework is implemented on a larger scale, the user input can be any continuous text exchange that takes place online.

3.2. Model Deployment

Here, to let everyone use the model that we created, easily, we made an Flask Web App that takes in single user input, performs all the necessary preprocessing steps, extracts all the required features and feeds those features into the model [8].

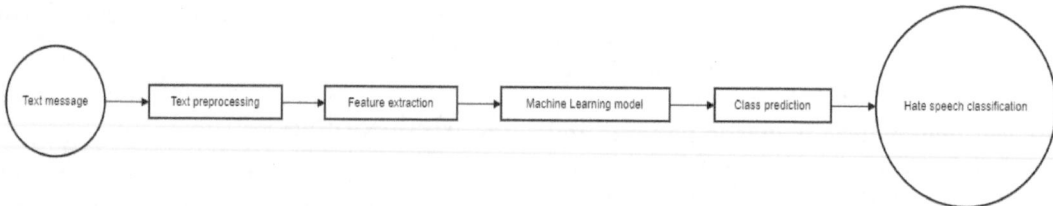

Figure 1: Methodology used in hate speech detection

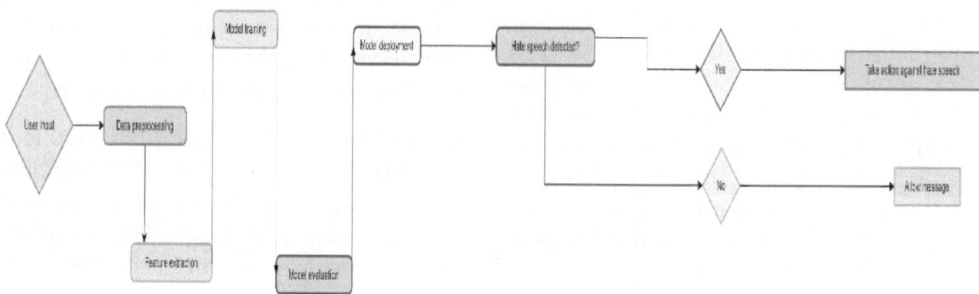

Figure 2: Proposed framework diagram

3.3. Hate Speech Detection

After feeding the extracted features into the model, our model will classify the user input into hate-speech or not. If the user input is classified as hate speech, action will be taken against it based on the platform that is using our model [9]. In case the user input is free of any hate speech, the data exchange will be allowed to happen normally.

4. Conclusion

The proposed strategy addresses the urgent problem of hate speech in the internet environment, to sum up [10]. The widespread use of hate speech requires a nuanced response due to its many different manifestations and changing semantics. With the use of this framework, we hope to develop a comprehensive method for classifying hate speech that can differentiate between it and legitimate speech while taking cultural context into account [11]. This framework uses machine learning methods to offer a safe and welcoming online environment while dealing with the difficulty of adjusting to evolving communication trends while minimizing computational bias. The technology works to accurately determine instances of hate speech using stringent preprocessing, feature extraction, and model training. By facilitating access to the model through an easy-to-use web interface, its utility is further enhanced. This approach helps to promote online harmony and combat the negative impacts of hate speech by addressing the ever-changing character of hate speech and enabling continual improvement.

References

[1] Boishakhi, F. T., Shill, P. C., and Alam, M. G. R. (2021, December). Multi-modal hate speech detection using machine learning. In 2021 IEEE International Conference on Big Data (Big Data) (pp. 4496-4499). IEEE.

[2] Rao, M. C., Yelavarti, K. C., and Kalyan, N. P. (2023, April). A Framework for Hate Speech Detection using Different ML Algorithms. In 2023 7th International Conference on Trends in Electronics and Informatics (ICOEI) (pp. 960-967). IEEE.

[3] Panchala, G. H., Sasank, V. V. S., Adidela, D. R. H., Yellamma, P., Ashesh, K., and Prasad, C. (2022, January). Hate Speech & Offensive Language Detection Using ML &NLP. In 2022 4th International Conference on Smart Systems and Inventive Technology (ICSSIT) (pp. 1262-1268). IEEE.

[4] Roy, S. S., Roy, A., Samui, P., Gandomi M. and Gandomi, A. H. "Hateful Sentiment Detection in Real-Time Tweets: An LSTM-Based Comparative Approach," in IEEE Transactions on Computational Social Systems, doi: 10.1109/TCSS.2023.3260217.

[5] Shukla, S., Nagpal, S., & Sabharwal, S. (2022, November). Hate Speech Detection in Hindi language using BERT and Convolution Neural Network. In 2022 International Conference on Computing, Communication, and Intelligent Systems (ICCCIS) (pp. 642-647). IEEE.

[6] Ahmed, U. and Lin, J. C. -W. "Deep Explainable Hate Speech Active Learning on Social-Media Data," in IEEE Transactions on Computational Social Systems, doi: 10.1109/TCSS.2022.3165136.

[7] Aoudni, Y., Donald, C., Farouk, A., Sahay, K. B., Babu, D. V., Tripathi, V., and Dhabliya, D. (2022). Cloud security based attack detection using transductive learning integrated with Hidden Markov Model. Pattern Recognition Letters, 157, 16-26.

[8] Wazid, M., Thapliyal, S., Singh, D. P., Das, A. K., and Shetty, S. (2022). Design and testbed experiments of user authentication and key establishment mechanism for smart healthcare cyber physical systems. IEEE Transactions on Network Science and Engineering.

[9] Rathod, R. G., Barve, Y., Saini, J. R., & Rathod, S. (2023, June). From data pre-processing to hate speech detection: An interdisciplinary study on women-targeted online abuse. In 2023 3rd International Conference on Intelligent Technologies (CONIT) (pp. 1-8). IEEE.

[10] Shawkat, N., Simpson, J., and Saquer, J. (2022, August). Evaluation of Different ML and Text Processing Techniques for Hate Speech Detection. In 2022 4th International Conference on Data Intelligence and Security (ICDIS) (pp. 213-219). IEEE.

[11] Huang, X.and Xu, M. (2021, December). An inter and intra transformer for hate speech detection. In 2021 3rd International Academic Exchange Conference on Science and Technology Innovation (IAECST) (pp. 346-349). IEEE.

IoT-Enabled Image Processing Approaches for Automated Plant Disease Detection

A Comparative Analysis

Gurleen Kaur Sandhu and Amardeep Singh

Computer Science and Engineering, Punjabi University, Patiala, India
E-mail: gurleensandhu08@gmail.com, amardeepdhiman@gmail.com

Abstract

This research paper delves into the intricate realm of IoT-enabled image-processing approaches for automated plant disease detection. In a world grappling with pressing issues surrounding food security and sustainable agricultural practices, the agricultural sector faces a significant challenge in managing and mitigating plant diseases. These diseases can wreak havoc on crop yields, posing threats to global food security and casting adverse economic and environmental impacts. In response to these challenges, the convergence of two disruptive technologies, the Internet of Things (IoT) and advanced image processing has opened up promising avenues for real-time disease monitoring, early detection, and data-driven decision-making. This paper comprehensively explores the methodologies that underlie these IoT-enabled image-processing techniques, shedding light on their critical steps, including image acquisition, preprocessing, feature extraction, and classification. It accentuates the pivotal role played by IoT in facilitating real-time data collection, enabling precise and efficient plant disease detection through the power of deep learning. This paper culminates in a comparative analysis of diverse IoT-enabled image processing methods, delineating their advantages and disadvantages. This analysis is a valuable resource for researchers, farmers, and stakeholders, aiding them in making informed decisions when selecting the most suitable approach for their specific needs in automated plant disease detection.

Keywords: IoT

1. Introduction

Agriculture is essential to meeting the nutritional needs of a constantly expanding global population. However, the agricultural sector faces various obstacles, the management and control of plant diseases being prominent among them. Crop yields can be drastically lowered due to these diseases, putting food security at risk and having negative economic and environmental effects [1]. There is a pressing need for novel strategies to address these issues and guarantee sustainable and efficient farming practices as the global population rises.

In recent years, promising remedies to plant diseases have emerged due to the convergence of two disruptive technologies: the Internet of Things (IoT) and image processing [2]. IoT is a network of interconnected physical objects with sensors and communication capabilities that are widely used across many industries. These technologies are potent tools for agriculture's real-time disease monitoring, early detection, and data-driven decision-making [3].

Also, it has allowed for the development of highly sophisticated and networked agricultural systems to keep tabs on and regulate all facets of the farming process. Furthermore, recent advances in image processing techniques have enabled automated picture analysis to detect and accurately diagnose plant diseases rapidly.

Detecting plant diseases in the past typically required human observation and time-consuming procedures, both of which can lead to missed diagnoses and delays in treatment. Large crop losses may occur between when a disease first appears and when it is discovered [4]. Technology's solution to this problem could improve speed and accuracy in disease diagnosis.

1.1. Image Processing in Agricultural

Scholars and researchers exploring various fields will find this collection of documents valuable for their academic pursuits and scholarly investigations [5]. The statistics derived from the Scopus database

Chapter 61 DOI: 10.1201/9781003570349

elucidate the distribution of document results within a specified period from 2013 to 2024, totalling 3,825 documents for analysis. These documents are categorised into various types: Articles, Conference Papers, Book Chapters, and Conference Reviews [6]. The largest category comprises Articles, constituting 1,919 documents, signifying a substantial focus on research published within scholarly journals during this period. Conference Papers follow closely behind, accounting for 1,760 documents, indicating a considerable emphasis on research findings and studies presented at academic conferences. Book Chapters are a smaller yet notable category, with 111 documents suggesting contributions to collective volumes or specialised publications offering in-depth insights. Finally, Conference Reviews account for 35 documents, likely providing critical evaluations and summaries of research presented at conferences. This breakdown of document types reveals a diverse array of sources and research outputs within the specified period, showcasing a range of scholarly endeavours and knowledge dissemination across academic platforms.

TITLE-ABS-KEY ("Image Processing" AND agricultural) AND PUBYEAR > 2012 AND PUBYEAR < 2025 AND (LIMIT-TO (DOCTYPE, "cp") OR LIMIT-TO (DOCTYPE, "ar") OR LIMIT-TO (DOCTYPE, "ch") OR LIMIT-TO (DOCTYPE, "cr"))

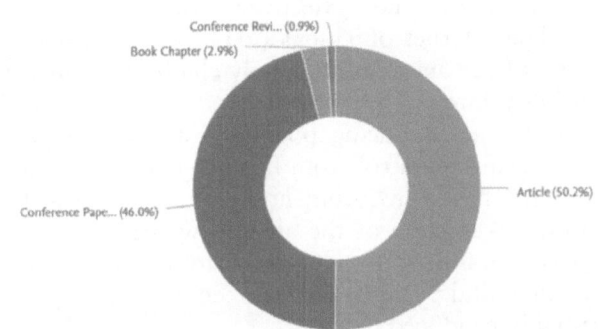

Documents by type

Conference Revi... (0.9%)
Book Chapter (2.9%)
Conference Pape... (46.0%)
Article (50.2%)

Figure 1: Available Documents by Type of Image Processing in Agricultural

The field of plant disease identification relies heavily on image processing. Acquiring, preparing, and analysing plant pictures for disease detection is a part of this process. Image capture methods, preprocessing procedures, feature extraction, and disease classification algorithms will all be discussed in this section, along with an overall overview of image processing in plant disease diagnosis.

Acquiring relevant images is the first step in image processing to detect plant diseases. High-resolution photographs of plants, leaves, or other botanical components are taken [7]. Several different methods of acquiring images are frequently employed here. High-resolution photographs of plants are easily accessible and inexpensive since they may be taken with standard digital cameras or cell phones. Drones with cameras can take pictures from above, giving farmers a bird's-eye view of their crops and possibly helping them catch diseases in their early stages [8]. Hyperspectral imaging takes pictures over a wide range of wavelengths to detect even the most minute changes in plant health. In addition, remote sensing based on satellites may capture expansive views of farmland, allowing for widespread crop health and disease monitoring.

Images are typically preprocessed and enhanced after acquisition to increase clarity and make them more amenable to analysis [9]. Noise reduction is frequently used to eliminate distracting artifacts; contrast enhancement is used to bring attention to important details; picture registration matches several images for consistency, and colour normalisation compensates for lighting differences.

The process of feature extraction is essential in the diagnosis of plant diseases. The process requires recognizing and quantifying features of photos that are diagnostic of the disease. Colour and texture analysis, leaf shape and structure measurement, pixel value statistics, and frequency domain analysis with tools like Fourier and wavelet transformations are all examples of such characteristics.

Classification is the final stage of feature extraction in plant disease detection. Classifying images as healthy or unhealthy is a typical use of machine learning and computer vision algorithms. Deep learning models like Convolutional Neural Networks (CNNs) for direct image analysis, ensemble learning by combining multiple classification models, and transfer learning by capitalizing on pre-trained models on large image datasets are just some algorithms used to analyse images.

In conclusion, image processing is a vital component in plant disease diagnosis. It helps farmers protect their crops from illness by quickly identifying the problem after taking, preprocessing, and analysing photos of affected plants. Recent advancements have greatly improved the efficacy of image processing in this crucial agricultural domain in computer vision and machine learning.

1.2. IoT in agriculture

Figure shows the statistics obtained from the Scopus database, offering insights into the distribution of document results during the specific year range spanning from 2013 to 2024.

TITLE-ABS-KEY (iot AND agriculture) AND PUBYEAR > 2013 AND PUBYEAR < 2024 AND (LIMIT-TO (DOCTYPE, "cp") OR LIMIT-TO

Documents by type

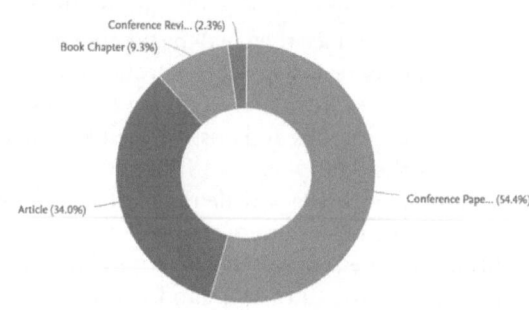

Figure 2: Available documents type on IoT in Agriculture

(DOCTYPE, "ar") OR LIMIT-TO (DOCTYPE, "ch") OR LIMIT-TO (DOCTYPE, "cr"))

The dataset comprises 6,160 documents, systematically categorised into distinct types: Conference Papers, Articles, Book Chapters, and Conference Reviews.

Predominantly, the category of Conference Papers stands out, encompassing 3,350 documents, suggesting a pronounced emphasis on research presented at academic conferences throughout this period. This prominence in conference-related research underscores a robust culture of active scholarly engagement and the dissemination of knowledge within the academic and scientific community. Concurrently, Articles represent the second most significant category, comprising 2,092 documents, underscoring the paramount importance of published research within reputable journals. The presence of 574 Book Chapters reflects meaningful contributions to collective volumes, often recognised for providing profound insights and specialised knowledge. Furthermore, 144 Conference Reviews indicate the presence of documents that likely encapsulate the summarisation and evaluation of research showcased at academic conferences. This comprehensive distribution of document types underscores the diverse sources of knowledge and research outputs within the delineated year range. It accentuates a dynamic and vibrant academic landscape that researchers and scholars across various fields will find invaluable for their scholarly pursuits and academic endeavours.

A revolutionary shift has occurred in farming practices due to IoT technology. Commonly referred to as "Smart Agriculture" or "Precision Agriculture," this technology revolution has allowed farmers and agricultural stakeholders to take advantage of data-driven insights and cutting-edge technologies to boost agricultural output, sustainability, and efficiency.

Connected physical devices and sensors with data gathering and communication capabilities are fundamental to IoT in agriculture. Soil moisture, weather, crop health, and livestock well-being are just some of the many factors these connected sensors track on farms [10]. These sensors capture data in real-time, send it for analysis, and then give the results back to farmers so they can make better decisions [11].

Internet of Things (IoT) uses in farming are numerous and varied. One major use case is precision farming, where information gathered by the Internet of Things (IoT) sensors informs the selection of irrigation, fertiliser, and pest control methods. This method helps to ensure the long-term viability of farming by reducing resource waste while increasing crop yields. In addition, the Internet of Things is useful in livestock monitoring since it lets farmers keep tabs on their animals' well-being, whereabouts, and behaviour, allowing for better breeding practices and more efficient feeding schedules.

Smart irrigation is another important use case that relies on IoT-enabled equipment to calculate the ideal amount of water to be applied based on soil moisture data and weather forecasts [3]. Over-irrigation, which can lead to soil damage, is avoided, energy costs are lowered, and water conservation is increased [12]. Drones, ground-based sensors with cameras, and spectrum imaging technologies can now be used for Things integration in agriculture . These devices can take pictures of crops and analyse them to spot nutrient deficiencies, plant stress, and other problems early on, allowing for more effective treatment.

The Internet of Things (IoT) has permeated all levels of the agricultural supply chain by improving visibility into the quality of goods while in transit and storage. Reducing post-harvest losses requires strict quality control from farm to market. In addition, IoT-based environmental monitoring is used to evaluate the state of the air and water, which aids farmers in following rules meant to protect the environment and promote environmentally friendly, sustainably produced food.

The Internet of Things is also being used to improve machinery servicing. Attached sensors on farm equipment capture data on how well it's functioning, allowing for predictive maintenance that keeps machines running smoothly and cuts down on unscheduled downtime. This saves money on repairs and ensures harvesting and planting go off without a hitch.

1.3. Existing IoT and Image Processing-Based Techniques

The current methods for detecting plant diseases using the Internet of Things and image processing are a huge step forward for farming. Researchers and farmers have created cutting-edge technologies for early and precise disease identification by harnessing the power of the Internet of Things (IoT) and image processing. These methods can significantly improve crop management and long-term food supply [13].

IoT-enabled sensor networks in farming are a leading technology in the current crop of IoT-based methods. Sensors in these networks track things like temperature, humidity, soil moisture, and air quality to keep tabs on the state of the world around us. These sensors send their data to a centralised system, which can be analysed and used to help farmers make real-time decisions [8]. For example, if a significant drop in soil moisture is detected, the system can prompt automated watering, reducing crop stress and potential disease outbreaks. Both resource efficiency and crop health benefit greatly from this constant monitoring and management.

Image processing methods have also developed in parallel with IoT software. Researchers have developed digital cameras, drones, and other imaging devices into image-based illness detection systems. Disease markers, including discoloration, lesion, or aberrant growth patterns, can be detected by processing these photos. These technologies use machine learning algorithms to precisely categorise photos as healthy or unhealthy, alerting farmers to impending epidemics.

Moreover, integrating IoT and image processing has led to sophisticated illness detection systems. The crop health and environmental factors can both be monitored by these systems. Image anomalies can be linked to specific environmental conditions, providing a more complete picture of disease dynamics. This all-encompassing method facilitates early intervention, accurate resource allocation, and environmentally sound farming methods.

Existing methods based on the Internet of Things and image processing have great potential but need some help. Protecting vital agricultural data necessitates resolving data security and privacy concerns. Large-scale sensor network implementation can be time-consuming and expensive; therefore, scalability remains an issue. Additionally, real-time monitoring systems require stable connectivity and infrastructure, which may need to be present in some agricultural regions.

The current methods for plant disease identification using the Internet of Things and image

Table 1: Plant disease identification Techniques

Authors	Preprocessing Technique	Description
[14, 15, 16]	Colour Space Conversion	Conversion of images to different colour spaces
[17, 12]	Noise Reduction (Denoising)	Utilisation of mean and median filters to reduce noise in images
[18, 19]	Cropping	Removal of unwanted areas in images
[7, 20, 21]	Enhancement Techniques	Application of filters to improve image quality
[22, 16]	Data Augmentation	Techniques to generate more data by applying transformations
[23, 24]	Edge-Based Segmentation	Utilises techniques like Canny edge detector, Sobel operator, and Prewitt operator for segmentation
[4, 25, 11, 15, 24]	Threshold-Based Segmentation	Involves techniques like Otsu thresholding, adaptive intensity-based thresholding, and entropy-based thresholding for segmentation. Combines region growing technique with local threshold for efficient lesion segmentation.
[26, 27, 28, 5]	K-Means Clustering	Utilises k-means clustering for diseased area segmentation, found efficient and suitable for segmentation.
[29, 30, 31]	Fuzzy C-Means Clustering	Preferred by some researchers for the segmentation of infected areas.
[32]	Super-Pixel Clustering (SLIC)	Combines SLIC with k-means clustering for disease spot segmentation.
[32, 33]	GrabCut Algorithm	Utilised for lesion segmentation
[34]	Weighted Lesion Segmentation Scheme. Fermi Energy-Based Segmentation	Suggested for efficient segmentation over Otsu, expectation maximisation (EM), active contour segmentation, and saliency segmentation approaches. Uses Fermi energy-based segmentation based on colour and grey-level intensity, performing better than Otsu and k-means approaches.
[35, 5]	Genetic Algorithm	Utilises genetic algorithm for plant disease detection through segmentation.

processing are a huge stride forward for contemporary agriculture. These innovations give farmers immediate feedback on the state of their crops and the surrounding environment, allowing for more accurate disease monitoring and better allocation of limited resources. Agricultural practices have the potential to be transformed towards higher efficiency, sustainability, and food security as the field develops and overcomes obstacles.

2. Biotic Infections in Plants with Their Categories in Various Crops

Understanding these different types of plant leaf diseases and their associated symptoms is essential for early detection and effective disease management in agriculture and horticulture [20, 14]. Early detection and appropriate control measures can help minimise crop damage and maintain healthy plant populations.

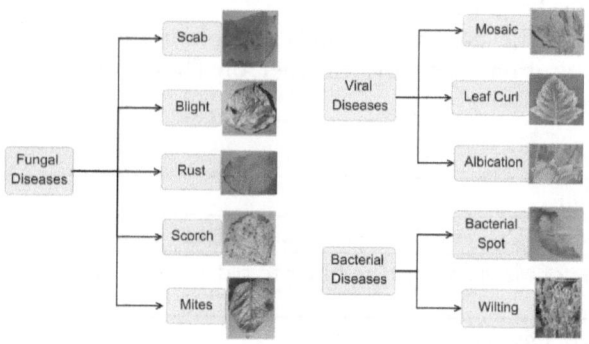

Figure 3: Plant leaf diseases and their associated symptoms

2.1. Bacterial Leaf Diseases

Bacterial leaf diseases are plant diseases caused by various species of bacteria. Tiny pale green spots on the affected leaves characterise the symptoms of bacterial leaf diseases [14]. These spots appear water-soaked and soon enlarge, eventually developing into dry, dead spots. Some bacterial leaf diseases exhibit brown or black water-soaked spots on the foliage, often surrounded by a yellow halo. Under dry conditions, these spots may take on a speckled appearance. Bacterial leaf diseases can significantly damage plants and reduce crop yields if not managed effectively. Management strategies often include using bactericides, crop rotation, and cultural practices.

Symptoms: Tiny pale green spots that become water-soaked and later enlarge into dry, dead spots, sometimes with a yellow halo. Under dry conditions, the spots may have a speckled appearance.

Example: Bacterial leaf spot with brown or black water-soaked spots on the foliage.

2.2. Viral Leaf Diseases

Plant-infecting viruses cause viral leaf diseases. These diseases are particularly challenging to diagnose because viruses typically do not produce easily observable signs. Instead, they may manifest as yellow or green stripes or spots on the affected plant's foliage [20]. Leaves infected with viruses might exhibit unusual characteristics, such as wrinkling, curling, and stunted growth. Insect vectors often spread viral diseases, including aphids, leafhoppers, whiteflies, and cucumber beetles. Effective management of viral diseases involves controlling the vectors and ensuring good sanitation practices to prevent disease spread.

Symptoms: Viruses do not produce readily observable signs but may manifest as yellow or green stripes or spots on foliage. Leaves can be wrinkled curled, and growth may be stunted.

Example: Mosaic Virus.

2.3. Fungal Leaf Diseases

Fungal leaf diseases are caused by various species of fungi and are characterised by distinct symptoms. Late blight, caused by the fungus Phytophthora infestans, results in water-soaked grey-green spots on lower, older leaves. As the disease matures, these spots darken, and white fungal growth forms on the undersides of the leaves [36]. Early blight, caused by the fungus Alternaria solani, initially presents as small brown spots with concentric rings, forming a bull's eye pattern on lower, older leaves. The disease spreads outward on the leaf surface with maturation, causing it to turn yellow. Downy mildew is another fungal disease with yellow to white patches on the upper surfaces of older leaves, which are covered with white to greyish growth on the undersides. Effective management of fungal leaf diseases often includes fungicide applications, proper plant spacing, and practices that promote good air circulation to reduce humidity, which is conducive to fungal growth.

Symptoms: Various symptoms associated with fungal diseases, including water-soaked grey-green spots that darken and develop white fungal growth on the undersides, small brown spots with concentric rings forming a bull's eye pattern, and yellow to white patches on the upper surfaces of older leaves.

Examples: Late blight caused by the fungus Phytophthora infestans, early blight caused by the fungus Alternaria solani, and downy mildew with yellow to white patches.

3. Methodology

The IoT-enabled image processing methodology leverages real-time data acquisition and deep learning techniques to automate the detection of plant

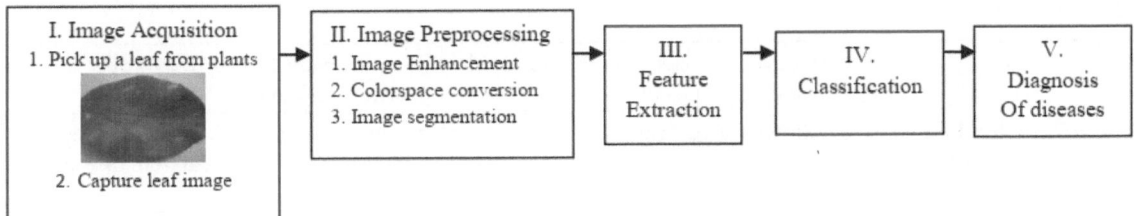

Figure 4: Basic Methodology of plant disease identification

diseases. The methodology outlines a multistep process shown in Figure for effectively identifying and classifying plant diseases using image analysis. First, the methodology begins with RGB image acquisition, which involves capturing high-quality images of plant leaves. IoT comes into play here by facilitating the remote and automated collection of these images through connected sensors and cameras deployed in agricultural fields. These sensors can transmit real-time data to a centralised system for analysis.

Next, the images are converted into a suitable colour space, followed by segmentation of the image components. IoT contributes by enabling real-time plant health monitoring by providing continuous image data, which can be processed at a central location or even on edge devices in the field. Integrating IoT and deep learning technologies plays a pivotal role in enhancing the efficiency and accuracy of this approach.

IoT is especially valuable in automated plant disease detection because it ensures that images are regularly and consistently obtained, allowing for continuous monitoring and early disease detection. The collected data can be transferred to a cloud-based or edge-based deep learning model configured for recognition, as described in the methodology [12].

Deep learning algorithms, typically implemented in neural networks, can learn intricate patterns and features within the images, making them highly suitable for plant disease classification. Integrating IoT and deep learning results in a system that can promptly and accurately detect plant diseases, thereby supporting precision agriculture, reducing the need for manual inspections, and improving crop yields through timely intervention [9].

Hence, IoT ensures the continuous supply of image data. At the same time, deep learning neural networks contribute to accurately identifying plant diseases, making this approach a powerful tool for enhancing agricultural practices and crop management.

Table 2: Comparative analysis of IoT-enabled image processing approaches

Sr. No	Approach	Advantages	Disadvantages
1	IoT + Convolutional Neural Networks (CNN) [20, 5]	Highly accurate in plant disease detection.	Requires significant computational resources for training and inference.
		Deep learning models can learn intricate patterns and features.	Extensive labelled data for training is necessary.
		Suitable for various plant species.	It may not perform well in cases of rare or novel diseases.
2	IoT + Transfer Learning [5]	Speeds up model development and training.	Performance heavily depends on the choice of a pre-trained model.
		Reduces the need for a large labelled dataset.	It may not be as accurate as finetuned models for specific diseases.
		Allows for quick adaptation to new disease types.	Limited flexibility compared to training from scratch.
3	IoT + Edge Computing [23, 24]	Realtime processing at the edge for immediate decisions.	Limited processing power and memory on edge devices.
		Reduces the need for continuous data transmission to the cloud.	It may not handle extremely large image datasets effectively.
		Suitable for remote or resource-constrained environments.	Limited flexibility in model updates and maintenance.

(Completed)

Table 2: (*Continued*)

4	IoT + Cloud Based Processing [36, 15, 23, 13, 12]	Access to extensive computational resources for analysis.	Data transfer latency may affect real-time decision-making.
		Allows for centralised model updates and maintenance.	Requires a reliable internet connection, which may only be available in some locations.
		Scalable for handling large datasets and multiple IoT devices.	Privacy and security concerns when handling sensitive agricultural data.

4. Comparative Analysis

Table presents a comparative analysis of IoT-enabled image processing approaches, highlighting their advantages and disadvantages. The approaches depend on specific application requirements, available resources, and the tradeoffs between accuracy, processing speed, and data handling capabilities.

5. Conclusion

In the dynamic landscape of agriculture, the fusion of IoT and image processing technologies has emerged as a transformative catalyst in the sphere of plant disease detection. The timely and accurate diagnosis of plant diseases carries profound implications for crop yields, food security, and the broader realm of environmental sustainability. This paper has underscored the central role played by IoT in facilitating real-time data acquisition and its seamless integration with cutting-edge deep learning techniques, particularly convolutional neural networks (CNNs) and transfer learning, in the automated detection of plant diseases. These technological advancements bear the potential to revolutionise the agricultural sector by expediting responses to disease outbreaks and optimizing the allocation of resources.

Our comparative analysis has provided valuable insights into the strengths and weaknesses of various IoT-enabled image processing approaches, underscoring the importance of tailoring the selection of methods to specific application requirements, available resources, and the complex interplay of accuracy, processing speed, and data handling capabilities. While these approaches hold great promise, they also confront notable challenges, including data security and privacy concerns, scalability of resources, and connectivity issues in remote agricultural regions. As technology advances, the agricultural sector stands to reap substantial benefits from these innovations. These advancements promise to enhance crop management, elevate agricultural efficiency, and secure the foundations for sustainable food production in the face of a burgeoning global population.

References

[1] Too, E. C., Yujian, L., Njuki, S., and Yingchun, L. (2019). A comparative study of fine-tuning deep learning models for plant disease identification. Computers and Electronics in Agriculture, 161, 272–279.

[2] Tripathy, P. K., Tripathy, A. K., Agarwal, A., and Mohanty, S. P. (2021). MyGreen: An IoT-enabled smart greenhouse for sustainable agriculture. IEEE Consumer Electronics Magazine, 10(4), 57–62.

[3] Ayaz, M., Ammad-Uddin, M., Sharif, Z., Mansour, A., and Aggoune, E. H. M. (2019). Internet-of-Things (IoT)-based smart agriculture: Toward making the fields talk. IEEE Access, 7, 129551–129583.

[4] Ashwinkumar, S., Rajagopal, S., Manimaran, V., and Jegajothi, B. (2021). Automated plant leaf disease detection and classification using optimal MobileNet based convolutional neural networks. Materials Today: Proceedings, 51, 480–487.

[5] Khamparia, A., Saini, G., Gupta, D., Khanna, A., Tiwari, S., and de Albuquerque, V. H. C. (2020). Seasonal crops disease prediction and classification using deep convolutional encoder network. Circuits, Systems, and Signal Processing, 39(2), 818–836.

[6] Dhingra, G., Kumar, V., and Joshi, H. D. (2018). Study of digital image processing techniques for leaf disease detection and classification. Multimedia Tools and Applications, 77(15), 19951–20000.

[7] Iqbal, Z., Khan, M. A., Sharif, M., Shah, J. H., ur Rehman, M. H., and Javed, K. (2018). An automated detection and classification of citrus plant diseases using image processing techniques: A review. Computers and Electronics in Agriculture, 153, 12–32.

[8] Manjula, G., Visu, P., and Chakaravarthi, S. (2018). IOT enabled weedicide control using image processing at agriculture field. EAI Endorsed Transactions on Smart Cities, 5(16), 170252.

[9] Li, L., Zhang, S., and Wang, B. (2021). Plant disease detection and classification by deep learning – A review. IEEE Access, 9, 56683–56698.

[10] Delnevo, G., Girau, R., Ceccarini, C., and Prandi, C. (2022). A deep learning and social IoT approach for plants disease prediction toward a sustainable agriculture. IEEE Internet of Things Journal, 9(10), 7243–7250.

[11] Ajaz, A., Taghvaeian, S., Khand, K., Gowda, P. H., and Moorhead, J. E. (2019). Development and evaluation of an agricultural drought index by

harnessing soil moisture and weather data. Water (Switzerland), 11(7), 1375.

[12] Sanjeevi, P., Prasanna, S., Siva Kumar, B., Gunasekaran, G., Alagiri, I., and Vijay Anand, R. (2020). Precision agriculture and farming using Internet of Things based on wireless sensor network. Transactions on Emerging Telecommunications Technologies, 31(12), e3978.

[13] Singh, A., Singh, K., Kaur, J., and Singh, M. L. (2023). Smart agriculture framework for automated detection of leaf blast disease in paddy crop using colour slicing and GLCM features based random forest approach. Wireless Personal Communications, 131(4), 2445–2462.

[14] Shrivastava, V. K., and Pradhan, M. K. (2021). Rice plant disease classification using color features: A machine learning paradigm. Journal of Plant Pathology, 103(1), 17–26.

[15] Wu, G., Li, B., Zhu, Q., Huang, M., and Guo, Y. (2020). Using color and 3D geometry features to segment fruit point cloud and improve fruit recognition accuracy. Computers and Electronics in Agriculture, 174, 105475.

[16] Shorten, C., and Khoshgoftaar, T. M. (2019). A survey on image data augmentation for deep learning. Journal of Big Data, 6(1), 1–48.

[17] Sindhu, P., and Indirani, G. (2022). Internet of Things enabled pomegranate leaf disease detection and classification using cuckoo search with sparse auto encoder. Mathematical Statistician and Engineering Applications, 71(3), 904–917.

[18] Yang, T., Siddique, K. H. M., and Liu, K. (2020). Cropping systems in agriculture and their impact on soil health – A review. Global Ecology and Conservation, 23, e01118.

[19] Waha, K., Dietrich, J. P., Portmann, F. T., Siebert, S., Thornton, P. K., Bondeau, A., et al. (2020). Multiple cropping systems of the world and the potential for increasing cropping intensity. Global Environmental Change, 64, 102131.

[20] Rahman, T., Khandakar, A., Qiblawey, Y., Tahir, A., Kiranyaz, S., Abul Kashem, S. B., et al. (2021). Exploring the effect of image enhancement techniques on COVID-19 detection using chest X-ray images. Computers in Biology and Medicine, 132, 104319.

[21] Qi, Y., Yang, Z., Sun, W., Lou, M., Lian, J., Zhao, W., et al. (2022). A comprehensive overview of image enhancement techniques. Archives of Computational Methods in Engineering, 29(1), 583–607.

[22] Chlap, P., Min, H., Vandenberg, N., Dowling, J., Holloway, L., and Haworth, A. (2021). A review of medical image data augmentation techniques for deep learning applications. Journal of Medical Imaging and Radiation Oncology, 65(5), 545–563.

[23] Guillén, M. A., Llanes, A., Imbernón, B., Martínez-España, R., Bueno-Crespo, A., Cano, J. C., et al. (2021). Performance evaluation of edge-computing platforms for the prediction of low temperatures in agriculture using deep learning. Journal of Supercomputing, 77(1), 818–840.

[24] Kc, K., Yin, Z., Li, D., and Wu, Z. (2021). Impacts of background removal on convolutional neural networks for plant disease classification in-situ. Agriculture, 11(9), 827.

[25] Mateen, A., and Zhu, Q. (2019). Weed detection in wheat crop using UAV for precision agriculture. Pakistan Journal of Agricultural, 56(3), 809–817.

[26] Javidan, S. M., Banakar, A., Vakilian, K. A., and Ampatzidis, Y. (2023). Diagnosis of grape leaf diseases using automatic K-means clustering and machine learning. Smart Agricultural Technology, 3, 100081.

[27] Amirul, M., Yusof, M., and Nazari, A. (2021). The disease detection for maize-plant using K-means clustering. Evolution in Electrical and Electronic Engineering, 2(2), 834–841.

[28] Ahmed, A. S., Obeas, Z. K., Alhade, B. A., and Jaleel, R. A. (2022). Improving prediction of plant disease using k-efficient clustering and classification algorithms. IAES International Journal of Artificial Intelligence, 11(3), 939–948.

[29] Sampathkumar, S., and Rajeswari, R. (2022). An automated crop and plant disease identification scheme using cognitive fuzzy C-means algorithm. IETE Journal of Research, 68(5), 3786–3797.

[30] Kohli, P., Kumar, I., and Vimal, V. (2022). Plant leaf disease identification using unsupervised fuzzy C-means clustering and supervised classifiers. In Studies in Computational Intelligence (pp. 281–293). Springer.

[31] Kumar Sahu, S., and Pandey, M. (2023). An optimal hybrid multiclass SVM for plant leaf disease detection using spatial Fuzzy C-means model. Expert Systems With Applications, 214, 118989.

[32] Zhang, S., You, Z., and Wu, X. (2019). Plant disease leaf image segmentation based on superpixel clustering and EM algorithm. Neural Computing & Applications, 31, 1225–1232.

[33] Lian, S., Guan, L., Pei, J., Zeng, G., and Li, M. (2023). Identification of apple leaf diseases using C-Grabcut algorithm and improved transfer learning base on low shot learning. Multimedia Tools and Applications, 1–23.

[34] Moussafir, M., Chaibi, H., Saadane, R., Chehri, A., El Rharras, A., and Jeon, G. (2022). Design of efficient techniques for tomato leaf disease detection using genetic algorithm-based and deep neural networks. Plant Soil, 479(1–2), 251–266.

[35] Alshammari, H., Gasmi, K., Krichen, M., Ammar, L. B., Abdelhadi, M. O., Boukrara, A., et al. (2022). Optimal deep learning model for olive disease diagnosis based on an adaptive genetic algorithm. Wireless Communications and Mobile Computing, 2022, 1–13.

[36] Foughali, K., Fathallah, K., and Frihida, A. (2018). Using cloud IOT for disease prevention in precision agriculture. Procedia Computer Science, 130, 575–582.